Springer Series in
MATERIALS SCIENCE 68

Springer
Berlin
Heidelberg
New York
Hong Kong
London
Milan
Paris
Tokyo

Springer Series in

MATERIALS SCIENCE

Editors: R. Hull R. M. Osgood, Jr. J. Parisi H. Warlimont

The Springer Series in Materials Science covers the complete spectrum of materials physics, including fundamental principles, physical properties, materials theory and design. Recognizing the increasing importance of materials science in future device technologies, the book titles in this series reflect the state-of-the-art in understanding and controlling the structure and properties of all important classes of materials.

61 **Fatigue in Ferroelectric Ceramics and Related Issues**
By D.C. Lupascu

62 **Epitaxy**
Physical Foundation and Technical Implementation
By M.A. Herman, W. Richter, and H. Sitter

63 **Fundamentals of Ion Irradiation of Polymers**
By D. Fink

64 **Morphology Control of Materials and Nanoparticles**
Advanced Materials Processing and Characterization
Editors: Y. Waseda and A. Muramatsu

65 **Transport Processes in Ion Irradiated Polymers**
By D. Fink

66 **Multiphased Ceramic Materials**
Processing and Potential
Editors: W.-H. Tuan and J.-K. Guo

67 **Nondestructive Materials Characterization**
With Applications to Aerospace Materials
Editors: N.G.H. Meyendorf, P.B. Nagy, and S.I. Rokhlin

68 **Diffraction Analysis of the Microstructure of Materials**
Editors: E.J. Mittemeijer and P. Scardi

69 **Chemical-Mechanical Planarization of Semiconductor Materials**
Editor: M.R. Oliver

70 **Isotope Effect Applications in Solids**
By G.V. Plekhanov

Series homepage – springer.de

Volumes 10–60 are listed at the end of the book.

E.J. Mittemeijer P. Scardi (Eds.)

Diffraction Analysis of the Microstructure of Materials

With 240 Figures and 39 Tables

Springer

Professor Dr. Ir. Eric J. Mittemeijer
Max Planck Institute for Metals Research
Heisenbergstrasse 3
70569 Stuttgart, Germany
E-mail: e.j.mittemeijer@mf.mpg.de

Professor Dr. Paolo Scardi
Università di Trento, Facoltà di Ingegneria
Dipartimento di Ingegneria
dei Materiali e Tecnologie Industriali
Via Mesiano 77, 38050 Mesiano, Italy
E-mail: Paolo.Scardi@unitn.it

Series Editors:

Professor Robert Hull
University of Virginia
Dept. of Materials Science and Engineering
Thornton Hall
Charlottesville, VA 22903-2442, USA

Professor Jürgen Parisi
Universität Oldenburg, Fachbereich Physik
Abt. Energie- und Halbleiterforschung
Carl-von-Ossietzky-Strasse 9–11
26129 Oldenburg, Germany

Professor R. M. Osgood, Jr.
Microelectronics Science Laboratory
Department of Electrical Engineering
Columbia University
Seeley W. Mudd Building
New York, NY 10027, USA

Professor Hans Warlimont
Institut für Festkörper-
und Werkstofforschung,
Helmholtzstrasse 20
01069 Dresden, Germany

ISSN 0933-033X

ISBN 3-540-40519-4 Springer-Verlag Berlin Heidelberg New York

Library of Congress Cataloging-in-Publication Data.

Diffraction analysis of the microstructure of materials / E.J. Mittemeijer, P. Scardi (eds.).
p. cm. – (Springer series in materials science, ISSN 0933-033X ; v. 68)
Includes bibliographical references and index.
ISBN 3-540-40519-4 (alk. paper)
1. Crystal optics. 2. Microstructure. 3. Diffraction. I. Mittemeijer, E.J. II. Scardi, P. (Paolo) III. Series.
TA418.9.C7D54 2003
620.1'1299–dc22 2003059237

Springer-Verlag is a part of Springer Science+Business Media.

springeronline.com

Printed in Germany

Typesetting by the editors
Data conversion, typesetting and production: PTP-Berlin Protago-TeX-Production GmbH, Berlin
Cover concept: eStudio Calamar Steinen
Cover production: *design & production* GmbH, Heidelberg

Printed on acid-free paper SPIN: 11383376 57/3111/YU 5 4 3 2 1

Preface

The Microstructure

Materials are substances that have now, or are expected to find in a not too distant future, practical use. Here we will confine ourselves to solid materials. The microstructure of materials then is a notion that comprises all aspects of the atomic arrangement in the material that should be known in order to understand its properties. Mostly we are concerned with crystalline materials. The conception microstructure then narrows to the description of the so-called *crystalline imperfection*; the idealized crystal structure, characterized by the filling of the unit cell, is considered to be known and thus its determination is not considered in this book.

Microstructure encompasses the compositional inhomogeneity, the amount and distribution of the phases in the material, the grain size and shape and the distribution functions of the grain-size parameters, the grain(crystal)-orientation distribution function (texture), the grain boundaries/interfaces and the surface of the material, the concentrations and distributions of crystal defects as vacancies, dislocations, stacking and twin faults, and, not least, lattice distortions due to strains/stresses, etc.

As may be anticipated from the above listing, the microstructure to a very large extent determines the properties of a material. The central issue of materials science may be described as: to develop models that provide the relation between the microstructure and the properties. To this end characterization of the microstructure of a material is a prerequisite.

Diffraction Analysis of the Microstructure

Diffraction analysis is perhaps the most powerful technique for investigating the microstructure by exploiting, especially, its sensitivity for the atomic arrangement and also the element specificity of the scattering power of an atom.

Each intensity maximum (called line profile or peak) in the diffraction pattern represents an average over the diffracting material; in the case of conventional X-ray diffractometry the diffracting volume is usually of the order 1 mm^3. This indicates the strength and, at the same time, the limitation of diffraction analysis: average values for structure parameters (microstructure parameters) are obtained (e.g. the dislocation density, the internal stress) which have a close bearing on the properties on mesoscopical and macroscopical scale, but the atomic configuration around an individual, isolated defect cannot be revealed in this way.

In this preface we do not intend to review and comment on the whole field of microstructural analysis by diffraction methods. However, with a view to the scope of this book, we wish to make the following remarks in order to set the (historical) record straight.

Already shortly after the discovery of diffraction of X-rays by crystals (1912), it was realized by Scherrer, at the time in Göttingen (1918), that the breadth of a reflection can be fruitfully used as a measure of the average finite size of the diffracting crystals. Also, as taught to undergraduates, by differentiating Bragg's law it seems obvious that information on lattice-parameter fluctuations is exhibited by diffraction-line broadening. Probably the first, seminal work in this area was performed by Dehlinger and Kochendörfer in Stuttgart (1927 and 1939). It is striking to observe that this work has been ignored largely in well known books reviewing the history of (X-ray) diffraction analysis (e.g. see "Early Papers on Diffraction of X-rays by Crystals" by Bijvoet, Burgers and Hägg (Vol. 1, 1969 and Vol. 2, 1972) and "Fifty Years of X-ray Diffraction" by Ewald (1962)); the name of Scherrer is recalled in these works, but in particular for his contribution in the development of the Debije–Scherrer technique (1916). This negligence may be understood from the underestimation, in particular by (classical) crystallographers in the first decades of the existence of the (X-ray) diffraction method, of the importance for mankind of knowledge on the real, imperfect structure and the crucial importance of diffraction methods to acquire such data. (Even Scherrer himself, in a personal reminiscence in the above mentioned book by Ewald, referred to his now famous formula only in passing).

In particular since, say, 1940, the full analysis of the width and shape of a diffraction line has become a topic of, until and beyond today, increasing importance. On the one hand, one observes the development of more and more advanced methods to extract microstructural parameters from the broad parameters of a number of diffraction lines using more or less realistic, general assumptions on the material imperfection/line shape: *line-profile analysis*. On the other hand, a recent, powerful approach of huge future potential appears to be *line-profile synthesis*, where the microstructural parameters are determined by fitting line profiles, calculated on the basis of a structure model specific to the material investigated (i.e. no line-shape assumptions are employed), to measured ones. Further, the emergence of the Rietveld method, where the whole diffraction pattern is used in a refinement of the filling of the unit cell (i.e. refinement of the idealized crystal structure), has made one aware that such an approach also allows the determination of microstructure parameters in a fitting procedure incorporating the *full diffraction pattern*.

The analysis of residual, internal (macro)stress is based on the measurement of the orientation dependence, with respect to the specimen frame of reference, of the lattice spacing in the (polycrystalline) material investigated. The *analysis of (macro)stress* has usually not been covered in textbooks on (X-ray) diffraction (there are a few exceptions where a marginal description is given). Reasons for this ignorance can be (i) that during decades the development of this method took place predominantly in Germany and corresponding publications were written in German and (ii), probably more important, that there was a strong, not rarely exclusively, engineering interest in this method and its results. Considering the earlier mentioned historical overviews on the development of the (X-ray) diffraction method, it will now be no surprise that only in a personal reminiscence by Glocker, in the earlier mentioned book by Ewald, the now so important diffraction method to determine stress is mentioned. Glocker, working in Stuttgart, as the above mentioned Dehlinger and Kochendörfer, can be considered as one of the pioneers of this method (see also his own textbook (in German, published in 1927)). The availability nowadays of two textbooks on this method in English and the participation of materials scientists worldwide, in further development of the method and its applications (see the proceedings of the (European and worldwide) conferences on "residual stress" (ECRS and ICRS)), as for example evidenced by the investigation of grain interaction and stresses in thin films, makes clear that stress analysis is an inseparable part of the field of microstructural diffraction analysis.

The Book

This book has been devised to offer an overview of currently "hot" topics in the field of the diffraction analysis of the microstructure of materials. We, as editors of this book, selected authors of prospective contributions on the basis of presentations at the International Conference on the "Analysis of microstructure and residual stress by diffraction methods" ("Size-Strain III", December 2–5, 2001, Trento, Italy), of which we together were the chairmen.

This book should not at all be considered as the proceedings of the conference mentioned:

- Firstly, only a fraction (corresponding to about 30%) of the presentations at the conference could be said to be represented in this book.
- Secondly, the prospective authors were asked, after the conference, to write a contribution for the book of monograph character. Thus, although the book requires a basic knowledge of materials science, or solid state physics or chemistry, it allows to get an impression of the current state of knowledge in the field. Moreover, the detail in the presentation of methods and techniques suffices for the readers of the book to apply them.
- Thirdly, all submitted manuscripts were subjected to a refereeing procedure involving at least two referees. A majority of the manuscripts were

revised on the basis of the comments of these referees. Thereafter the referees were asked for a second consideration of the now revised manuscript.

Thus, the book is neither a conference proceedings nor just a compilation of high quality research papers as could have been published in a first class, international journal. Instead, the book is a collection of twenty high quality contributions, which together provide *an overview of the frontline research on the "Diffraction Analysis of the Microstructure of Materials"*, where the detailedness of the information offered should allow application of the presented methods by Ph.D. students and professionals working in the field. To be complete is impossible. For example, we have no contribution dealing with only texture (preferred orientation) analysis. Yet, we hope to have achieved, by our choices and quality check and improvement, an important, timely book of monograph character.

Part I of the book is of special character. It deals with the history of line-profile analysis. In view of the deficiencies of the existing historical books (see above), such an overview is much needed to provide some background for those of the following chapters, which outline the current state of research in the field of line-broadening analysis.

Part II comprises six contributions, whose common denominator is the analysis of the whole diffraction pattern. This part of the book exposes the current debate on pattern *fitting* versus pattern *modelling*, the former approach is based on assumed (analytical) functions for peak-profile shape and the latter approach departs from suitable models (of the instrument and/or the sample microstructure), without imposing a priori a profile-shape function.

The first chapter provides a powerful approach, in principle, to the modelling of the instrumental peak profile on the basis of an analytical description of the spectral profile (wavelength distribution in the diffracted, recorded radiation) and of the various optical components in the beam path. The following chapter presents the fundamentals of Whole Powder Pattern Modelling, with applications to nanocrystalline materials and to highly defective metals. Full pattern simulation, based on physical models of line broadening due to line and planar defects, is the subject of the third chapter to study ceramic as well as metallic nanocrystalline materials. In some way as an alternative to the previous two contributions, Chap. 4 concerns the analysis of residual stress and crystallite size within the framework of a traditional Rietveld refinement.

The last two chapters in this part address the still controversial issue of the determination of an amorphous fraction in phase mixtures: the first one points out the success and the weakness of present-day techniques based on pattern fitting by the Rietveld method, including a discussion of the in the literature intensively discussed treatment of absorption effects, whereas the second one illustrates recent innovations based on the modelling of crystalline as well as amorphous phases.

The next part deals with a classical but still very important and developing topic of powder diffraction analysis: the determination of crystallite size and shape. Although diffraction analysis in principle allows determination of the size and shape of coherently scattering domains, the current trend involves adoption of appropriate assumptions, such as grain shape, to derive a grain size distribution from the measured data. Such an approach is especially appropriate for finely dispersed polycrystalline materials, which are the focal point of interest for the two chapters in this part. In the first chapter a recently introduced Bayesian/maximum entropy method is presented and the method is illustrated for the case of a nanocrystalline ceramic powder, whereas the second chapter concerns the use of small-angle diffuse scattering for the analysis of polydisperse samples of compact sub-microcrystalline metallic materials.

Line and planar defects are the subject of Part IV. The first contribution affords a detailed description of how to calculate the dislocation contrast factor for cubic materials, with examples of materials with face-centred, body-centred and primitive lattices. The next two contributions concern line broadening due to inhomogeneous dislocation distributions. The first one presents a generalization of Wilson's variance method for the case of dislocated single crystalline and polycrystalline materials, whereas the second one is based on the Krivoglaz/Wilkens' formulation and deals with the case of textured thin films. The fourth and final chapter in this part deals with the modelling of the diffraction pattern of faulted materials, discussing in detail the case of *fcc* materials exhibiting intrinsic, extrinsic and twin faulting.

The subject of Part V is grain interaction, in the case of elastic and plastic deformation. The first of the two contributions deals with the effect of grain interaction on the determination of residual stress, especially in thin films, showing how grain interaction leads to macroscopically elastically anisotropic behaviour, even in the absence of texture. In particular the consequences of surface anisotropy are indicated: an old problem that never has been dealt with conclusively; the approach presented may open a new route to settle the argument definitively. The second chapter concerns the role of grain interaction and dislocation structures in plastic deformation of polycrystalline materials, as described by the evolution of texture.

The following part, Part VI concerns the role of surfaces and interfaces in polycrystalline materials. The first contribution in this Part describes the effects of grain-surface relaxation, observed in nanocrystalline powders, on the diffraction-line broadening, within the Whole Powder Pattern Modelling approach (see Part II). The second chapter provides a determination of the grain-boundary stress from measured values for the grain size and the strain-free lattice spacing of a nanocrystalline material.

The last part, Part VII presents three contributions on thin film analysis. The first chapter provides a detailed overview of the existing techniques for measuring residual stress gradients in thin films by diffraction methods, also considering possible texture effects. The following two contributions concern thin film analysis by reciprocal space mapping using non-conventional X-ray sources, a subject of growing interest and important for many practical cases (in the microelectronic industry).

Acknowledgement

Finally, we would like to thank, for their assistance in preparing this book, our co-workers, Dr. Matteo Leoni and Dr. Udo Welzel, who themselves, in spite of their young age, are scientists of remarkable quality in the field covered by this book.

Stuttgart, Trento
August 2003

E.J. Mittemeijer
P. Scardi

Contents

List of Contributors

N. Armstrong
Department of Applied Physics
University of Technology Sydney
P.O.Box 123
Broadway, NSW 2007, Australia
Nicholas.Armstrong@uts.edu.au

D. Balzar
Department of Physics
and Astronomy
University of Denver
Denver, CO 80208, USA
and
Materials Science
and Engineering Laboratory
National Institute of Standards
and Technology
Boulder, CO 80305, USA
balzar@du.edu

F. Bernard
Laboratoire de Recherches
sur la Réactivité des Solides
UMR 5613 CNRS
Université de Bourgogne
BP 47870
21078 Dijon Cedex, France

M.A. Bertinetti
Instituto de Física Rosario
Facultad de Ciencias Exactas,
Ingeniería y Agrimensura
Conicet-Unr.
Bv. 27 de Febrero 210 bis. 2000
Rosario, Argentina

R. Birringer
Universität des Saarlandes
FR 7.3 Technische Physik
Postfach 151150, Geb. 43B
66041 Saarbrücken, Germany
r.birringer@nano.uni-saarland.de

R.E. Bolmaro
Instituto de Física Rosario
Facultad de Ciencias Exactas,
Ingeniería y Agrimensura
Conicet-Unr.
Bv. 27 de Febrero 210 bis. 2000
Rosario, Argentina
bolmaro@ifir.edu.ar

J. Bonevich
Metallurgy Division
National Institute
of Standards and Technology
Gaithersburg, MD 20899, USA

A. Borbély
Eötvös University Budapest
Department of General Physics
1518 Pázmány P. sétány 1/A
P.O.Box 32
Budapest, Hungary

A. Boulle
Science des Procédés Céramiques
et de Traitements de Surface
UMR CNRS 6638
ENSCI
47–73 Avenue Albert Thomas
87065 Limoges Cedex, France
a_boulle@ensci.fr

H.-G. Brokmeier
Institut für Werkstoffkunde
und -technik
Technische Universität Clausthal
GKSS-Forschungszentrum
Geesthacht
Max-Planck-Str., Geb. 03
21502 Geesthacht, Germany
Heinz-Guenter.Brokmeier@gkss.de

N.M. Budarina
Institute for Metal Physics
36, Vernadsky Str.
Kiev-142, 03142 Ukraine

M. Čerňanský
Institute of Physics
Academy of Science
of the Czech Republic
Na Slovance 2
Praha 8, Czech Republic

S.V. Cherepanova
Boreskov Institute of Catalysis
of SBRAS
pr. Lavrentieva 5
Novosibirsk 630090, Russia
svch@catalysis.usk.su

R.W. Cheary
Department of Applied Physics
University of Technology Sydney
P.O. Box 123
Broadway, NSW 2007, Australia

J.P. Cline
Ceramics Division
National Institute
of Standards and Technology
Gaithersburg, MD 20899, USA

A.A. Coelho
Bruker AXS GmbH
Östliche Rheinbrückenstr. 50
76187 Karlsruhe, Germany

A. Dauger
Science des Procédés Céramiques
et de Traitements de Surface
UMR CNRS 6638
ENSCI
47–73 Avenue Albert Thomas
87065 Limoges Cedex, France

R. Delhez
Department of Materials Science
and Technology
Delft University of Technology
Rotterdamseweg 137
2628 AL Delft, The Netherlands

P.F. Fewster
PANalytical Sussex
Inovation Centre
Science Park Square
University of Sussex
Falmer, Brighton
BN1 9SB, UK

A. Fourty
Instituto de Física Rosario
Facultad de Ciencias Exactas,
Ingeniería y Agrimensura
Conicet-Unr.
Bv. 27 de Febrero 210 bis. 2000
Rosario, Argentina

C. Genzel
Hahn-Meitner-Institute Berlin
Department of Structural Research
Glienicker Str. 100
14109 Berlin, Germany
genzel@hmi.de

M. Golshan
CLRC
Daresbury Laboratory
Warrington
Cheshire WA4 4AD, UK
m.golshan@dl.ac.uk

I. Groma
Eötvös University Budapest
Department of General Physics
1518 Pázmány P. sétány 1/A
P.O.Box 32
Budapest, Hungary
groma@metal.elte.hu

A.F. Gualtieri
Dipartimento di Scienze della Terra
Università di Modena
e Reggio Emilia
41100 Modena, Italy
alex@unimo.it

A. Guagliardi
C.N.R.-IC
c/o Dipartimento Geomineralogico
Campus Universitario
70125 Bari, Italy

R. Guinebretière
Science des Procédés Céramiques
et de Traitements de Surface
UMR CNRS 6638
ENSCI
47–73 Avenue Albert Thomas
87065 Limoges Cedex, France

M. Hoffmann
Universität des Saarlandes
FR 7.3 Technische Physik
Postfach 151150, Geb. 43B
66041 Saarbrücken, Germany

V. Holý
Masaryk University
Faculty of Science
Institute of Solid-State Physics
Kotlářská 2
61137 Brno, Czech Republic

A. Iseppi
Università die Modena
e Reggio Emilia
Dipartimento die Scienze della Terra
41100 Modena, Italy

W. Kalceff
Department of Applied Physics
University of Technology Sydney
P.O.Box 123
Broadway, NSW 2007, Australia

J.-D. Kamminga
Netherlands Institute
for Metals Research
Rotterdamseweg 137
2628 AL Delft, The Netherlands
j.d.kamminga@tnw.tudelft.nl

A. Kern
Bruker AXS GmbH
Östliche Rheinbrückenstr. 50
76187 Karlsruhe, Germany
Arnt.Kern@bruker-axs.de

G.N. Kryukova
Boreskov Institute of Catalysis
of SBRAS
pr. Lavrentieva 5
Novosibirsk 630090, Russia
kryu@catalysis.nsk.su

J. Kub
Academy of Sciences
of the Czech Republic
Institute of Physics
Na Slovance 2
Praha 8, Czech Republic

J. Kuběna
Masaryk University
Faculty of Science
Institute of Solid-State Physics
Kotlářská 2
61137 Brno, Czech Republic

R. Kužel
Charles University
Faculty of Mathematics
and Physics
Ke Karlovu 5
12116 Praha 2, Czech Republic
kuzel@karlov.mff.cuni.cz

J.I. Langford
University of Birmingham
School of Physics and Astronomy
Birmingham B15 2TT, UK
J.I.Langford@bham.ac.uc

D. Laundy
CLRC
Daresbury Laboratory
Warrington
Cheshire WA4 4AD, UK

M. Leoni
Università di Trento
Facoltà di Ingegneria
Dipartimento di Ingegneria
dei Materiali e Tecnologie Industriali
Via Mesiano 77
38050 Mesiano (TN), Italy
Matteo.Leoni@unitn.it

P. Lynch
Department of Applied Physics
University of Technology Sydney
P.O.Box 123
Broadway, NSW 2007, Australia

O. Masson
Science des Procédés Céramiques
et de Traitements de Surface
UMR CNRS 6638
Faculté des Sciences
123 Avenue Albert Thomas
87060 Limoges Cedex, France

E.J. Mittemeijer
Max Planck Institute
for Metals Research
Heisenbergstr. 3
70569 Stuttgart, Germany
e.j.mittemeijer@mf.mpg.de

M. Moore
Royal Holloway University
of London
Department of Physics
Egham, Surray TW20 0EX, UK

L.O. Olikhovska
Institute for Metal Physics
36, Vernadsky Str.
Kiev-142, 03142 Ukraine

N.C. Popa
National Institute
for Materials Physics
P.O.Box MG-7
Bucharest, Romania

P. Riello
Università di Venezia
Dipartimento di Chimica Fisica
Via Torino 155b
30170 Venezia-Mestre, Italy
Riellop@unive.it

P. Scardi
Università di Trento
Facoltà di Ingegneria
Dipartimento di Ingegneria
dei Materiali e Tecnologie Industriali
Via Mesiano 77
38050 Mesiano (TN), Italy
Paolo.Scardi@unitn.it

L.J. Seijbel
Bruker Nonius BV
Oostsingel 209
2612 HL Delft
The Netherlands

J.W. Signorelli
Instituto de Física Rosario
Facultad de Ciencias Exactas,
Ingeniería y Agrimensura
Conicet-Unr.
Bv. 27 de Febrero 210 bis. 2000
Rosario, Argentina

D. Šimek
Academy of Sciences
of the Czech Republic
Institute of Physics
Na Slovance 2
Praha 8, Czech Republic

S.V. Tsybulya
Boreskov Institute of Catalysis
of SBRAS
pr. Lavrentieva 5
Novosibirsk 630090, Russia
`tsybulya@catalysis.nsk.su`

A.I. Ustinov
Institute for Metal Physics
36, Vernadsky Str.
Kiev-142, 03142 Ukraine
`ustinov@imp.kiev.ua`

U. Welzel
Max Planck Institute
for Metals Research
Heisenbergstr. 3
70569 Stuttgart, Germany
`u.welzel@mf.mpg.de`

P. Zimmer
Universität des Saarlandes
FR 7.3 Technische Physik
Postfach 151150, Geb. 43B
66041 Saarbrücken, Germany

Part I

Retrospective on Line-Broadening Analysis

1 Line Profile Analysis: A Historical Overview

J.I. Langford

Polycrystalline materials invariably contain imperfections that modify the intensity of Bragg reflections. This departure from an ideal structure, known as *microstructure*, can profoundly influence the physical, mechanical and chemical properties of materials. Characterization of microstructure is thus of crucial importance in materials science and the study of structural imperfections by means of X-ray powder diffraction is known as Line Profile Analysis (LPA). Such a study is usually undertaken to investigate the nature of the microstructure arising from sample treatment, method of preparation, or some other influence.

Microstructure affects the breadth and shape of a diffraction line profile and may also introduce a displacement from its ideal position. The full distribution of intensity therefore needs to be taken into account, ideally for as many reflections as can reasonably be recorded, and LPA is thus one of the more demanding applications of powder diffraction. It is also a fascinating branch of materials science; as Bertram E. Warren said of imperfect structures, "As with humans, it is the deviations from regularity that are more interesting". Parameters defining line profiles provide information on the size and shape of diffracting domains ("crystallites"), together with the distribution of size, and on lattice deformation (microstrain), the nature of dislocations and the occurrence of faults or "mistakes" in the structure. This overview records the key developments in LPA and the life and work of some of the pioneers in the field.

1.1 Early Years

Although the term LPA appeared at a later date, the technique, albeit in a rudimentary form, is as old as X-ray powder diffraction itself. This was introduced independently in 1916, during the First World War, by Hull in the United States and Debye and Scherrer in Germany. Scherrer's main interest was in the structure and size of colloidal particles of silver and gold and his thesis on the subject included what came to be known as the Scherrer equation for obtaining "particle" (crystallite) size from diffraction data. This was published in 1918 [1]:

$$h = 2\sqrt{\frac{Ln2}{\pi}} \frac{\lambda}{\Lambda \cos\theta}, \tag{1.1}$$

where h is the full width at half maximum intensity (FWHM), Λ is a measure of crystallite size, λ is the wavelength and θ is the Bragg angle. The factor $2\sqrt{(Ln2/\pi)}$ converts h to the integral breadth, assuming that the diffraction line profile has the form of a Gaussian function. This factor was later replaced by the *Scherrer constant* that converted Λ to a "true" size, defined as the cube root of the crystallite volume, for a given crystallite shape.

Development of LPA during the next twenty years or so was mainly concerned with determining crystallite size and morphology. Two notable contributions were those of Laue in 1926 and Jones in 1938. Laue [2] considered the case of crystallites having the form of a parallelepiped, the basis of much subsequent work on size effects. He realised the practical and theoretical limitations of using the FWHM as a measure of line broadening and introduced the general form of the integral breadth[1]. He also used Lorentzian (Cauchy) and Gaussian functions to model line profiles. Jones [3] worked with data obtained by scanning film from a powder camera with a "microphotometer" to obtain "line profiles". He pointed out that an experimental line profile, $h(x)$, is the convolution of line profiles due to the sample, $f(x)$ and to instrumental effects, $g(x)$. Based on earlier work by Brill and Pelzer in the late 1920s, he devised an improved correction for obtaining the integral breadth β_f of $f(x)$ from the breadths β_g and β_h of $f(x)$ and $h(x)$, together with a method for removing the $K\alpha_2$ contribution. He also introduced the concept of an *integral-breadth apparent size* ε_β and showed that this is a volume-weighted average, proportional to the ratio of the fourth to the third moments of the size distribution.

Diffraction phenomena that could not be interpreted in terms of crystallite size were reported from time to time during the 1930s. For example, line broadening due to stacking faults in hexagonal cobalt was first mentioned in 1930 [4]. Indeed, there was considerable interest during the 1930s in imperfections that alter the regular periodicity of the crystal lattice (structural "mistakes"). For example, Jones and Sykes made a detailed study of the alloy Cu_3Au [5,6], a classic case of mistakes in three dimensions, but it was some years later before a rigorous theoretical treatment of such phenomena was published. These early workers laid the foundations of LPA; the period that followed could perhaps be regarded as its "golden years".

1.2 Bertram E. Warren (1902–1991)

Bertram E. Warren was a young researcher at the Massachusetts Institute of Technology between 1919 and 1929, except for a year with Ewald. A visit

[1] Defined as the width of a rectangle having the same height and area as the line profile

from W.L. Bragg in 1927 furthered his interest in crystallography and he then spent a year with Bragg at Manchester (1929–1930), working on the structures of metasilicates. He returned to MIT as Associate Professor in 1930 and for several years studied the structure of glasses and of non-periodic and nearly-periodic structures generally. This subsequently led to a study of other imperfect structures – order–disorder in binary alloys, cold-worked metals, etc., and to the widely used procedure for LPA that bears his name. Warren also founded one of the leading "schools" of X-ray diffraction. He was a great teacher and two graduate students who worked in Warren's group at MIT in the late 1930s, and who subsequently made significant contributions to powder diffraction, were William ("Bill") Parrish (1915–1991) and A.J.C. (Arthur) Wilson (1914–1995). Another alumnus of the MIT school was A.L. Patterson. It was Warren's reputation as a leading authority on Fourier analysis that drew Patterson to MIT. Although better known for his contribution to structure determination, he was the first to apply Fourier methods to the analysis of diffraction data from small crystallites [7].

1.3 Arthur J.C. Wilson (1914–1995)

The young Arthur Wilson went to MIT in 1936 as an experimental physicist and was awarded his first doctorate, for studies of the anomalous thermal behaviour of Rochelle salt, in 1938. In the same year he received an 1851 Exhibition Scholarship, which enabled him to join W.L. Bragg at the Cavendish Laboratory, Cambridge, where he remained until 1945. At the Cavendish Laboratory he used X-ray diffraction to make precise measurements of the thermal expansion coefficient of various metals and, in the course of analysing the data, he realised the importance of making allowance for systematic errors due to the camera geometry. This led to his seminal work in the years that followed on the effects of instrumental and other factors that influence the position, shape and intensity of X-ray reflections, of crucial importance in LPA. He was intrigued by the diffraction effects due to structural mistakes and successfully accounted for the diffraction phenomena observed for cobalt [8] and Cu_3Au [9]. This led to a profound interest in LPA theory and, together with Alec R. Stokes, he developed a general treatment, based on Fourier transforms, of diffraction effects due to small, distorted or otherwise imperfect crystals.

In the early days there was controversy regarding the origins of the line broadening observed for metallic samples and it was not until the 1940s that the concept of simultaneous crystallite-size and strain effects was generally accepted. It was then realized that broadening due to the former is independent of the order of a reflection, whereas for the latter it increases linearly with order (e.g. [10]). Wilson's theoretical treatment of diffraction effects due to structural imperfections was published in *X-ray Optics* in 1949 and the 1962 revision [11] is still a definitive work on the subject.

1.4 The Parrish Diffractometer

W.H. Bragg adapted an "ionisation spectrometer" to collect data from flat powder samples in 1921 [12] and this can perhaps be regarded as the first powder diffractometer. However, early LPA studies were made with powder cameras. These improved steadily over the years, culminating in the Guinier focusing camera [13] that employed a monochromator crystal to remove the $K\alpha_2$ component of the incident radiation. Data for LPA were obtained by scanning lines on photographic film with a photodensitometer. However, the breakthrough in obtaining good quality powder data routinely came when Bill Parrish, then of North American Philips, designed and constructed the parafocusing powder diffractometer after the Second World War. The first version was based on an ex-government spectrometer he had developed during the War for alignment of quartz oscillator crystals by unskilled operators. As a result, the yearly production rose from 50 000 plates to 30 million by 1944 and Parrish received a War Department Citation for his efforts. This rudimentary diffractometer was soon replaced by the familiar vertical-circle instrument, patented by Parrish in 1947. The introduction of this instrument was a milestone in powder diffraction; some forty years later there were an estimated 15 000 diffractometers of the Parrish design in laboratories throughout the world and it forms the basis of the majority of instruments in use today with conventional X-ray sources. Diffractometers based on other geometries, such as the Seemann–Bohlin method, are available, but they are not widely used for LPA. Details of the different types of powder diffractometer are given in [14].

For many years Parrish and Wilson, close friends since their days at MIT, worked together on powder diffractometry. Parrish continued to develop instrumental aspects – beam optics, procedures for alignment and so on. Wilson, on the other hand, who had moved to University College, Cardiff, in 1945, continued to work on the theory of geometrical and physical aberrations and their effect on diffraction data. Many of Parrish's publications on the subject are included in his *X-ray Analysis Papers* [15] and the contribution of Wilson and his co-workers was published in *The Mathematical Theory of X-ray Diffractometry* [16]. In the late 1950s and early 1960s an automatic system for controlling a powder diffractometer and collecting data was designed and built at Cardiff, the forerunner of the Automatic Powder Diffractometer (APD) of today.

1.5 Williamson–Hall Plot

In 1949 Hall [17] proposed a method, based on the work of Stokes and Wilson, for separating "size" and "strain" contributions to line breadths that was based on their order dependence. From a plot of the integral breadth β^* (in reciprocal units) as a function of d^* $(= 1/d)$, the reciprocal of the intercept gives an estimate of the apparent size ε_β of coherently-diffracting domains and the slope is a measure η of microstrain:

$$\beta^* \left(= \beta \cos \frac{\theta}{\lambda}\right) = \frac{1}{\varepsilon_\beta} + \left(\frac{\eta}{2}\right) d^* . \tag{1.2}$$

Hall studied various metals and deduced that the values of size roughly corresponded to the average distance between dislocations, a result that was later confirmed by electron microscopy. The method came to be known as the *Williamson–Hall Plot* [18]. Although it is based on the approximation that the line profiles due to size and strain are Lorentzian, which is unlikely to occur in practice, the plot is used to this day to give a qualitative indication of sample microstructure.

1.6 Fourier Methods

Allowance must be made for instrumental effects in all LPA methods and Stokes [19] introduced a correction based on the multiplicative property of the Fourier transforms of convoluted functions. In practice, Fourier transforms are necessarily replaced by the corresponding series and this method of representing line profiles, coupled with the introduction of the Parrish diffractometer, gave considerable impetus to the development of LPA. A key contribution was that of Bertaut in 1949 [20] for determining crystallite size by the Fourier method. Bertaut pointed out that the reciprocal of the initial slope of the Fourier "size" coefficients A_n^S with respect to n, the harmonic number, gives a measure of the thickness of crystallites, given as the length of columns of unit cells in the direction of the diffraction vector. It was noted later that this quantity is in fact an area-weighted average apparent size. Bertaut also showed that the second derivative of A_n^S *versus* n is proportional to the distribution of column lengths, a quantity that includes the distribution of crystallite size.

Bertaut took no account of lattice distortion in his analysis and, in the following year, Warren and Averbach [21], again using the formulation of Stokes and Wilson [10], introduced a procedure for separating the contributions from crystallite size and microstrain. Warren and Averbach assumed that the cosine coefficients A_n of the line profile due to imperfections are simply the product of A_n^S and A_n^D, the "distortion" (strain) coefficients, or

$$A_n = A_n^S A_n^D . \tag{1.3}$$

If l is the order of the reflection and $< e^2 >$ is the mean-square strain, the basic equation of the method devised by Warren and Averbach is

$$\ln A_n(l) = \ln A_n^S - 2\pi^2 n^2 l^2 < e^2 > . \tag{1.4}$$

The intercepts of plots of $\ln A_n(l)$ *versus* l^2 for different values of n thus gives the logarithm of A_n^S and the slopes give the variation of $< e^2 >$ with n, but rarely are more than two orders of a reflection available in practice. For convenience, n is usually replaced by L, a distance in the direction of the diffraction vector, where $L = n/2\hat{s}$, $\pm\hat{s}$ being the distance, in reciprocal space, over which the line profile is recorded. By taking into account peak displacement and asymmetry, Wagner [22] developed a procedure for obtaining the probabilities of the occurrence of deformation (stacking) faults and twin (growth) faults in cold-worked face-centred cubic metals. Warren modified Wagner's approach somewhat and extended the analysis to include body-centred-cubic and hexagonal close-packed structures. A detailed account of the Warren–Averbach method is given in *Warren's X-ray Diffraction* [23].

1.7 Variance Method

Experimental line profiles are necessarily truncated at a finite range and this introduces systematic errors in A_n. These errors, which are particularly severe if there is measurable overlap of the tails of reflections, and other factors that affect the reliability of A_n were considered by Young, Gerdes and Wilson [24]. Wilson had long been conscious of the limitations of methods based on Fourier coefficients and in 1962 published an alternative method, based on a suggestion by Tournarie [25], that used the variance as a measure of line broadening [26]. This is based on the linear variation of the second central moment of line profiles with range of truncation, a consequence of an inverse-square variation in intensity. The slope and intercept of such a plot provide information on crystallite size, microstrain and structural mistakes [27]. There are sound theoretical reasons for this approach, but a number of practical difficulties emerged and it was not widely used.

1.8 Further Development of Fourier Methods

By the end of the 1960s the basic theory underlying diffraction effects due to sample microstructure had largely been established, though it has since been refined by numerous authors, notably in the study of dislocations. A definitive handbook on powder diffraction, *X-ray Diffraction Procedures for Polycrystalline and Amorphous Materials* by Klug and Alexander was published in 1954 and the substantial revision of 1974 contains a useful account of LPA theory and basic methods [28]. Alexander also derived analytical expressions for line profiles arising from diffractometer geometry [29].

The Warren–Averbach method soon became a standard procedure in LPA, notably in metallurgical applications and other cases of simple structures with minimal line-profile overlap. However, when size effects are accompanied by strain broadening, the latter is simply interpreted in terms of diffraction effects and in general its physical interpretation is unclear. Furthermore, Wilkens [30] pointed out that Warren and Averbach's approximation used to describe strain broadening, namely the truncation of (1.4) at the second term, is incompatible with the nature of the microstrain in plastically deformed materials. The theory of line broadening due to dislocations had been considered by Wilson [31], Krivoglaz and Ryaboshapka [32] and others and, by means of a modification of the Warren-Averbach method, Wilkens devised a procedure for studying them. This was developed further by Groma, Ungár and Wilkens [33], who showed that the dislocation structure can be described by five parameters, namely the dislocation density, the mean spatial fluctuation of the density, the effective outer cut-of radius, the dipole polarization and the spatial fluctuation of the polarization. The use of Fourier methods to study dislocations is now one of the principal metallurgical applications of LPA.

A different approach to characterising lattice distortion was proposed by Vogel, Haase and Hosemann [34]. By using the logarithm of the ratio of A_n for two orders of a reflection, the size term vanishes and a hypothesis regarding the nature of the disorder present can then be based on the behaviour of $\ln\{A_n(l)/A_n(l+1)\}$ as a function of n. A parabolic trend, equivalent to a constant mean-square strain $< e^2 >$, is interpreted as a fluctuation in lattice parameter in different crystallites, whereas a linear trend ($< e^2 > \propto 1/n$, approximately) is attributed to the presence of dislocations or to paracrystallinity [35]. Indeed, there was considerable interest at this time in obtaining information on lattice deformation given by the variation of microstrain with n, and hence L.

A simplification of the basic Warren–Averbach method was introduced by Houska and his co-workers [36, 37], whereby Fourier series for two orders of a reflection were obtained by fitting only five parameters. This is possible because of simple relationships between crystallite-size and strain Fourier coefficients and the authors claimed that such a simplification was of value when overlapping line profiles occur. The method was used in a detailed study of the dependence of $< e^2 >$ on n for various metals and alloys. Various other improvements have been made to the Warren–Averbach method over the years, particularly by Delhez, de Keijser and Mittemeijer of the Technical University, Delft. Equation (1.4) is exact if the variation of $e(n)$ is Gaussian, but there is no particular reason a priori that the strain distribution should be of this form. In order to avoid this limitation, these authors used only the cosine expansion to separate size and strain contributions [38]. This entailed plotting A_n versus l^2 (i.e. neglecting the third and higher terms in (1.4). These and related procedures require data for at least two orders of a reflection, which is not always possible or practicable to achieve, and various authors adapted the Warren–Averbach method to analyse data from a single peak. For example, Mignot and Rondot [39, 40] fitted a quadratic to the A_n versus n curve to obtain A_n^S and A_n^D. Single-line methods have been listed and reviewed by Delhez, de Keijser and Mittemeijer [41]. Another modification by these authors uses the first derivatives of the A_n versus n curve. By assuming that $g(x)$ and $h(x)$ can be approximated by some analytical function, this approach can be applied to calculate initial derivatives, from which size and strain parameters can be extracted [42].

1.9 Powder Pattern Fitting

Much of the progress in LPA during the last thirty years or so has been concerned with data analysis, made possible by a dramatic improvement in computing power that was denied the pioneers in the field. Indeed, this has been a period of renaissance for powder diffraction methods generally. An important development during the 1970s was the introduction of alternative analytical functions to model line profiles. These include the Voigt [43], pseudo-Voigt [44] and Pearson VII [45] functions that superseded the Lorentzian and Gaussian that had been used earlier. (Definitions and properties of these functions are given in Appendix 1 of [46].) An important consequence was the development of programs using analytical functions to model all reflections in a powder pattern, e.g. [47], to obtain their position, maximum intensity, area, breadth and some shape parameter for use in LPA and other applications. This procedure is known as *Pattern Decomposition or Whole Powder Pattern Fitting.* Unlike earlier LPA methods, the approach is applicable to powder patterns with overlapping reflections and can thus be used to study materials of any crystal system. Indeed, truncation errors are largely avoided. The use of these more flexible functions takes into account the fact that the shape of line profiles changes with angle and often with hkl or lattice direction. If a "split" function is used to allow for asymmetry, e.g. [48], an adequate fit throughout a powder pattern can usually be obtained. Reliability of the method depends on the precision with which the function models experimental line profiles, but it is at best only an approximation. Theoretical line profiles due to structural imperfections do not have a simple analytical form.

Correction of observed breadths for instrumental effects in pattern decomposition is straightforward if a Voigt function is used [49], since the convolution of two such functions is also a Voigtian. However, the convolution of pseudo-Voigt or Pearson VII functions has no simple analytical form. One way of making an instrumental correction when using these functions to model line profiles is to obtain the equivalent Voigt parameters [41]. A different approach was devised by Enzo, Fagherazzi, Benedetti and Polizzi [50]. This is based on the convolution of pseudo-Voigt functions, to give parameters defining $f(x)$ for subsequent analysis based on Fourier coefficients or the integral breadth. Similar methods have since been described by other authors, e.g. [51], and if an adequate fit is achieved, reliable microstructural parameters based on the integral breadth can usually be obtained. However, if Fourier coefficients are used, these relate to analytical functions or to their convolution product, and do not necessarily define $f(x)$ precisely. This can lead to significant systematic errors in the coefficients An and in any microstructural information derived from them.

The introduction of Pattern Decomposition was something of a milestone in LPA. Since line-profile parameters can be obtained for most, if not all, reflections, a self-consistent three-dimensional representation of microstructure can often be deduced, even for low symmetry materials. Physically meaningful quantities can be obtained and, unlike methods based on one or two reflections, the precision of results can be evaluated.

1.10 Whole Powder Pattern Modelling

With some notable exceptions, application of LPA over the years has often been phenomenological, in that the results obtained do not necessarily define actual physical properties. However, an important recent development is the comparison of a complete experimental pattern, by some refinement procedure, with a calculated powder pattern based on postulated physical quantities. For example, van Berkum et al. [52] simulated powder patterns for Al–Si alloys with finely dispersed misfitting Si inclusions and Langford, Louër and Scardi [53] obtained the crystallite-size distribution in CeO_2 by comparing theoretical and experimental patterns. Ungár et al. [54] adopted a different approach to characterize dislocations; Fourier transforms of experimental line profiles were compared with theoretical functions of size and strain transforms. The most general procedure to date is that of Scardi, Leoni and Dong [55], who included deformation and twin faults with dislocations and a crystallite-size distribution in the physical model. Such procedures for obtaining realistic quantities to define sample microstructure are known as *Whole Powder Pattern Modelling* and eventually this approach may well become standard practice. However, it has hitherto only been applied to the relatively simple cases of high-symmetry materials containing approximately equiaxial crystallites. Despite its limitations, Whole Powder Pattern Fitting is likely to remain the most powerful general method for undertaking LPA for the time being.

References

1. P. Scherrer: Göttinger Nachrichten **2**, 98 (1918)
2. M. von Laue: Z. Krystallogr. **64**, 115 (1926)
3. F.W. Jones: Proc. Roy. Soc. Lond. A **166**, 16 (1938)
4. S.B. Hendricks, M.E. Jefferson, and J.F. Schultz: Z. Kristallogr. **73**, 376 (1930)
5. C. Sykes and F.W. Jones: Proc. Roy. Soc. Lond. A **157**, 213 (1936)
6. F.W. Jones and C. Sykes: Proc. Roy. Soc. Lond. A **166**, 376 (1938)
7. A.L. Patterson: Phys. Rev. **56**, 972 (1939)
8. A.J.C. Wilson: Proc. Roy. Soc. Lond. A **180**, 277 (1942)

9. A.J.C. Wilson: Proc. Roy. Soc. Lond. A **181**, 360 (1943)
10. A.R. Stokes and A.J.C. Wilson: Proc. Camb. Phil. Soc. **40**, 174 (1944)
11. A.J.C. Wilson: *X-Ray Optics* (Methuen, London 1962)
12. W.H. Bragg: Proc. Phys. Soc. Lond. **33**, 222 (1921)
13. A. Guinier: C.R. Acad. Sci. Paris **204**, 1115 (1937)
14. W. Parrish and J.I. Langford, in *International Tables for Crystallography*, Vol. C. Ed. by A.J.C. Wilson and E. Prince (Kluwer Academic Publishers, Dordrecht 1999) pp. 42–79
15. W. Parrish: *X-Ray Analysis Papers* (Philips Technical Library, Eindhoven 1965)
16. A.J.C. Wilson: *Mathematical Theory of X-Ray Diffractometry* (Philips Tech. Library, Eindhoven 1962)
17. W.H.Hall: Proc. Phys. Soc. Lond. A **62**, 741 (1949)
18. G.K. Williamson and W.H. Hall: Acta Metall. **1**, 22 (1953)
19. A.R. Stokes: Proc. Phys. Soc. Lond. **61**, 382 (1948)
20. E.F. Bertaut: C.R. Acad. Sci. Paris **228**, 187–189, 492–494 (1949)
21. B.L. Averbach and B.E. Warren: J. Appl. Phys. **20**, 885 (1949)
22. C.N.J. Wagner: Acta Metall. **5**, 427–434, 477–482 (1957)
23. B.E. Warren: *X-Ray Diffraction* (Addison-Wesley, Reading MA 1969) pp. 251–314
24. R.A. Young, R.G. Gerdes, and A.J.C. Wilson: Acta Cryst. **22**, 155 (1967)
25. M. Tournarie: C.R. Acad. Sci. Paris **242**, 2161 (1956)
26. A.J.C. Wilson: Nature **193**, 568 (1962)
27. A.J.C. Wilson: Proc. Phys. Soc. Lond. **80**, 286 (1962); ibid. **81**, 41 (1963)
28. H.P Klug and L.E. Alexander: *X-ray Diffraction Procedures for Polycrystalline and Amorphous Materials* (Wiley, New York 1954) Revised 1974 pp. 618–708
29. L.E. Alexander: J. Appl. Phys. **19**, 1068 (1948); ibid. **21**, 126 (1950)
30. M. Wilkens: Phys. Stat. Sol. A **2**, 359 (1970)
31. A.J.C. Wilson: Acta Cryst. **5**, 318 (1952)
32. M.A. Krivoglaz and K.P. Ryaboshapka: Fiz. Metall. Metalloved. **15**, 18 (1963)
33. I. Groma, T. Ungár and M. Wilkens: J. Appl. Cryst. **21**, 47 (1988); ibid. **22**, 26 (1989)
34. W. Vogel, J. Haase and R. Hosemann: Z. Naturforsch. A **29**, 1152 (1974)
35. B. Crist and J.B. Cohen: J. Polym. Sci., Polym. Phys. Ed. **17**, 1001 (1979)
36. T. Adler and C.R. Houska: J. Appl. Phys. **50**, 3282 (1979)
37. C.R. Houska and T.M.Smith: J. Appl. Phys. **52**, 748 (1981)
38. R. Delhez, T.H. de Keijser and E.J. Mittemeijer, in *Accuracy in Powder Diffraction*. Ed. by S. Block and C.R. Hubbard. NBS Spec. Pub. No. 567 (US Dept of Commerce, Gaithersburg 1980) pp. 231–253
39. J. Mignot and D. Rondot: Acta Metall. **23**, 1321 (1975)
40. J. Mignot and D. Rondot: Acta Cryst. A **33**, 327 (1977)
41. R. Delhez, T.H. de Keijser and E.J. Mittemeijer: Fresenius Z. Anal. Chem. **312**, 1 (1982)
42. T.H. de Keijser, E.J. Mittemeijer and H.J.C. Rozendaal: J. Appl. Cryst. **16**, 309 (1983)
43. J.I. Langford: J. Appl. Cryst. **11**, 10 (1978)
44. G.K. Wertheim, M.A. Butler, K.W. West and D.N.E. Buchanan: Rev. Sci. Instrum. **11**, 1369 (1974)
45. M.M. Hall, V.G. Veeraraghavan, H. Rubin and P.G. Winchell: J. Appl. Cryst. **10**, 66 (1977)

46. J.I. Langford and D. Louër: Rep. Prog. Phys. **59**, 131 (1996)
47. J.I. Langford, D. Louër, E.J. Sonneveld and J.W. Visser: Powder Diffr. **1**, 211 (1986)
48. S.A. Howard and R.L. Snyder, in *Advances in Materials Characterization II*. Ed. by R.L. Snyder, R.A.Condrate and P.F. Johnson (Plenum, New York 1985) pp. 43–56
49. J.I. Langford, in *Accuracy in Powder Diffraction*. Ed. by S. Block and C.R. Hubbard. NBS Spec. Pub. No. 567 (US Dept. of Commerce, Gaithersberg 1980) pp. 255–269
50. S. Enzo, G. Fagherazzi, A. Benedetti and S. Polizzi: J. Appl. Cryst. **21**, 536 (1988)
51. D. Balzar: J. Appl. Cryst. **25**, 559 (1992)
52. J.G.M. van Berkum, R. Delhez, T.H. de Keijser and E.J. Mittemeijer: Phys. Stat. Sol. A **134**, 335 (1992)
53. J.I. Langford, D. Louër and P. Scardi: J. Appl. Cryst. **33**, 964 (2000)
54. T. Ungár, J. Gubicza, G. Ribárik and A. Borbély: J. Appl. Cryst. **34**, 298 (2001)
55. P. Scardi, M. Leoni and Y.H. Dong: IUCr Commission on Powder Diffraction Newsletter **24**, 23 (2000)

Part II

Analysis of the Full Diffraction Pattern

2 Convolution Based Profile Fitting

A. Kern, A.A. Coelho, and R.W. Cheary

In convolution based profile fitting profiles are generated by convoluting functions together to form the observed profile shape. In this chapter a general convolution approach to X-ray and neutron powder data is described, which can accurately convolute and refine a wide variety of profile shapes numerically, including user defined profiles, without the need to convolute analytically. With functions representing both the aberration functions of the diffractometer as well as the various specimen contributions a fundamental parameters based synthesis of line profile shapes is achieved, but more generally any combination of appropriate functions may be used in this context for modelling profiles empirically.

In the literature, convolution based profile fitting is normally associated with microstructure analysis where the sample contribution needs to be separated from the instrument contribution in an observed profile. This is no longer the case. The present convolution approach can be also performed on a fully empirical basis to provide better fits to data and a greater variety of profile shapes.

2.1 Introduction

Since the introduction of the Rietveld method [1,2], one of the most challenging problems in profile fitting has been the accurate characterization of the peak shape and its variation with 2θ. Commonly used profile fitting methods can be divided into three broad categories:

1. *Analytical Profile Fitting.* This method is generally characterized by peak shape functions (PSF) with an explicit and relatively simple mathematical form which can be differentiated analytically with respect to each of the refined parameters within the least squares refinement procedure. The most common PSFs in analytical profile fitting are the Gaussian, Lorentzian, sums of Gaussians or Lorentzians, Voigt, pseudo-Voigt and Pearson VII functions where the angular dependence of the peak shape is described by smoothly varying functions with 2θ (e.g. [3,4]). Although these functions are convenient to use they are often unable to fit X-ray profiles well over the whole 2θ range without using a large number of pa-

rameters. This frequently leads to correlation problems, loss of uniqueness and instability of the refinement procedure.

2. *Learned Peak Shape Fitting.* In this approach the PSF is generated from the actual shapes of the peaks in a pattern. The procedure requires a number of well-resolved profiles so that PSFs at any 2θ can be generated by interpolation (e.g. [5]). A learned PSF therefore gives an optimal fit to almost any diffraction profile as it is not based on an analytical expression. One of the main drawbacks of this method is that it requires non-overlapped peaks well distributed over the whole 2θ range and this is often not possible in practice. In addition learned PSFs needs to be re-determined for each pattern owing to different specimen properties or instrument set-up.
3. *Direct Convolution Approach.* In this method profiles are generated by a process where various functions are convoluted to form the observed profile shape. For a convolution of n functions this process can be written as

$$Y(2\theta) = W \otimes F_1(2\theta) \otimes F_2(2\theta) \otimes \ldots \otimes F_i(2\theta) \otimes \ldots \otimes F_n(2\theta)\,, \qquad (2.1)$$

   where
   - $Y(2\theta)$ is the observed line profile shape,
   - W is the source emission profile, and
   - $\otimes$ denotes the convolution process.

In powder diffractometry the functions $F_i(2\theta)$ can be interpreted as both the aberration functions of the diffractometer [6] as well as the various specimen contributions leading to the so-called Fundamental Parameters Approach (FPA). More generally any combination of appropriate functions $F_i(2\theta)$ may be used in this context. Most direct convolution fitting methods are restricted to combinations of $F_i(2\theta)$ that can be convoluted analytically (e.g. GSAS [7]); typically these comprise Lorentzians, Gaussians, the hat (impulse) function and the exponential function. However, software such as TOPAS [8] can accurately convolute and refine a wide variety of profile shapes both numerically and analytically, including user defined profiles, without the need for explicit analytical convolution.

Convolution based profile fitting is well known since the landmark publication of Klug and Alexander [9], its use however has been delayed due mainly to computing demands. Two of the first studies using direct convolution were reported by Taupin [10] and Parrish et al. [11]. Since then convolution has been used as a fundamental parameters approach to profile fitting notably to TOF neutron data [12] and X-ray data [13, 14]. Nevertheless, convolution based profile fitting has only been implemented as a standard fitting procedure during the last ten years since freeware and commercial software packages such as XFit–KoalaRiet [15] and TOPAS have become available.

The present chapter deals with the convolution approach as implemented in TOPAS [13,16–19]. Of significance is its ability to construct profile shapes with any shape dependence on angle and *hkl* direction out of a pool of functions (including user-defined functions) using a mixture of analytical and numerical convolutions and to fit the PSFs to the observed line profiles with all parameters refineable.

The use of physically based instrument aberration functions as opposed to empirically based functions leads directly to the special case of the fundamental parameters approach with all instrument parameters refineable. This approach is widely used within all currently employed profile fit methods including (i) single line fitting, (ii) whole powder pattern fitting, (iii) whole powder pattern decomposition according to Pawley [20] and Le Bail et al. [21], (iv) Rietveld refinement [1,2] and (v) ab-initio structure determination from step intensity data [22].

2.2 Convolution Based Profile Fitting

2.2.1 General Considerations

In the literature, convolution based profile fitting is typically associated with microstructure analysis where the specimen contribution is separated from the instrument contribution (e.g. [3,4]). It can however also be used in an empirical manner to provide a greater variety of profile shapes resulting in better fits. The two general areas of application for convolution based profile fitting are:

1. The purely empirical parameterization of line profile shapes by convolution of appropriate functions $F_i(2\theta)$ according to (2.1); a schematic of which is shown in Fig. 2.1.
2. The explicit discrimination of instrument and specimen contributions [23]:

$$Y(2\theta) = (W \otimes G) \otimes S\,, \tag{2.2}$$

   where G and S are instrument and sample contributions respectively which are modeled separately using appropriate $F_i(2\theta)$ functions according to (2.1).
   In general, a particular target system or instrumental setup $(W \otimes G)$ is seen as an entity called the instrument function $I(2\theta)$, which can be either measured (conventional approach) or calculated from first principles based on the known instrument geometry (fundamental parameters approach, which is discussed in detail later).

In the conventional convolution approach an instrument function $I(2\theta)$ is determined experimentally from a standard material, which ideally should (but does not) meet the following requirements: (i) large crystallite size to minimize size broadening, (ii) small crystallite size to give sufficient particle statistics, (iii) no strain and (iv) possess the same linear absorption coef-

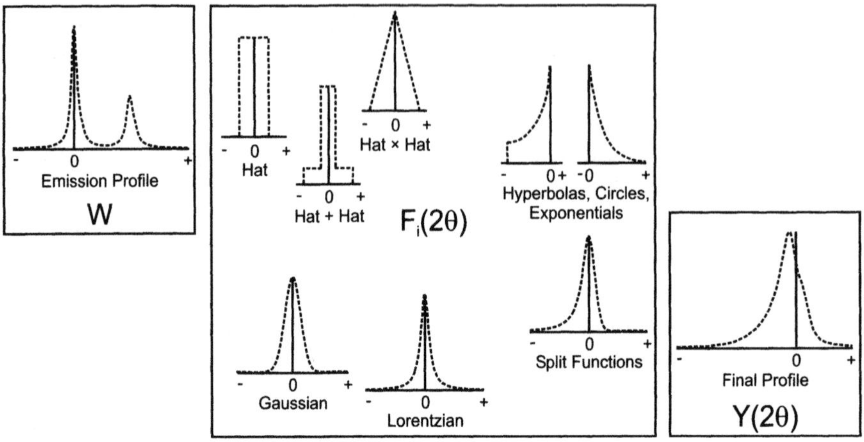

$$Y(2\theta) = W \otimes F_1(2\theta) \otimes F_2(2\theta) \otimes \ldots \otimes F_i(2\theta) \otimes \ldots \otimes F_n(2\theta)$$

Fig. 2.1. Schematic representation of the convolution approach as given in (2.1). The final profile $Y(2\theta)$ is described by the selection of appropriate functions $F_i(2\theta)$ and convoluting them on top of the emission profile W. Note the capabilities of this approach to add functions before convoluting them as shown for the sum of two hat functions as well as to convolute split-type functions

ficient as the specimen. The instrument function is then convoluted with specimen related functions to represent the "real" specimen properties. In simplified approaches Lorentzian and Gaussian functions are employed to describe both crystallite size and strain effects. Obviously the requirement to have both large and small crystallites simultaneously is a contradiction and limits the accuracy of microstructure analysis at small levels of specimen broadening.

In principle any instrument function can be described using the conventional convolution approach. However, the quality of the instrument function is directly related to the quality of the standard material and thus additional broadening in the instrument function is unavoidable for reasons described above. As a result specimen broadening is underestimated resulting in erroneous specimen characteristics such as over-estimated crystallite size estimates.

A further drawback is that the instrument function needs re-determination when relevant instrument settings, such as slit settings, change. In addition the effects of sample penetration by the incident beam is often overlooked resulting in increased broadening in the instrument function. This is an intrinsic error to the conventional convolution approach whereby the instrument function is linked to sample preparation (packing density).

2.2.2 Numerical Procedures of Convolution Based Profile Fitting

Convolution based profile fitting requires accurate numerical procedures for carrying out multiple convolution integrals. Typically a calculated profile is reduced to a histogram and the convolution integral is reduced to a summation. To avoid systematic errors the angular step size needs to be very small but in so doing the operation becomes very time consuming [13]. In the TOPAS implementations of the FPA, a semi-analytical procedure has been developed for convoluting the aberration functions. In this procedure two aberration functions are folded together by first evaluating them at equal 2θ intervals and then forming a continuous function comprising straight line segments connecting adjacent points. As the calculated functions are a series of linear sections it is possible to calculate the convolution integral analytically [13]. Some difficulties are experienced with aberration functions that possess singularities but in these cases the functions are integrable and singularities are overcome [17]. In cases where one of the aberration functions can be analytically integrated with straight line segments, then the integration is performed in a semi-analytical manner where the aberration function in question is analytically convoluted with the series of straight-line segments of the other. For example, Gaussian, Lorentzian and exponential aberration functions are convoluted analytically with straight-line segments.

One of the most important achievements of the convolution technique for practical purposes is speed of calculation. Accurate multiple convolution calculations over large 2θ ranges can be very time intensive and it is of central importance that the calculation time is minimized to enable Rietveld refinement to be completed in a reasonable amount of time without loss of accuracy. Various procedures have been implemented some of which are described by Cheary and Coelho [13, 16–18].

2.2.3 The Wavelength Distribution in Laboratory Diffractometers

The knowledge of the exact shape of the source emission profile is of central importance to the development of an accurate powder profile description. In the absence of instrumental effects the spectral shape represents the highest possible resolution of a diffractometer and above 90° 2θ the emission profile is the dominant contribution to the line profile shape. For sealed tubes and rotating anodes the source emission profile width for $2\theta \leq 40^\circ$ is $\leq 0.01^\circ$ 2θ. This is considerably smaller than the FWHM values of actual diffraction lines from commercial diffractometers which depending on the choice of slits are typically in the range 0.04° to 0.10° 2θ. In this region the contribution of the emission profile is swamped by the geometrical aberrations. At $2\theta > 60^\circ$, the emission profile tends to dominate over the geometrical aberrations and at $2\theta > 100^\circ$ the total breadth of the geometrical aberrations are relatively small and the profile shape conforms closely to the emission profile.

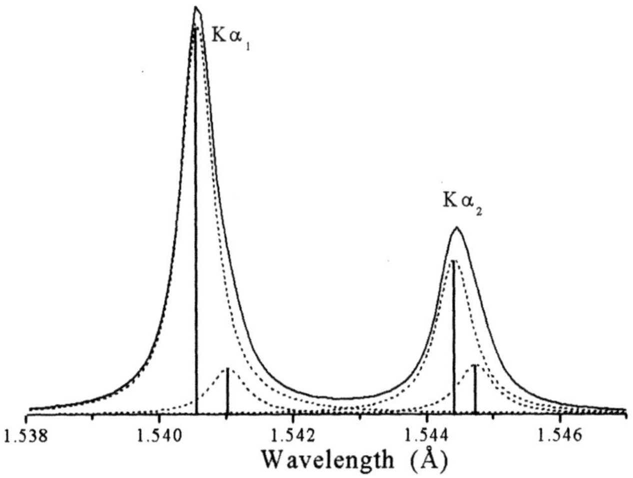

Fig. 2.2. Phenomenological representation of the Cu K_α emission profile based on four Lorentzians (from [19])

Unlike synchrotron and neutron diffraction, the source emission profiles of the transition element anodes used in laboratory X-ray diffraction do not have simple descriptions. The form of the $K\alpha_{1,2}$ emission profile from Cu is shown in Fig. 2.2. For all of the transition element anodes used in X-ray diffraction neither the $K\alpha_2$ nor the $K\alpha_2$ lines are Lorentzian in shape and both lines are asymmetric with extended high angle tails. Moreover the asymmetries and FWHM values of the $K\alpha_1$ and $K\alpha_2$ peaks are different.

The natural asymmetry of the emission lines arises from the multiplet structure of the transitions. In addition to the main transitions involving the change in vacancy state $\underline{1s} \rightarrow \underline{2p}$, it has been recognised that $\underline{3d}$ spectator transitions also contribute up to 30% of the $K\alpha_1\alpha_2$ emissions [24]. In these transitions the atom is doubly ionised and the actual vacancy transition is still $\underline{1s} \rightarrow \underline{2p}$, but the second vacancy in the $3d$ level is not directly involved in the transition. The notation for this transition is $\underline{1s3d} \rightarrow \underline{2p3d}$. A phenomenological representation for accurately describing the asymmetric Cu emission profile was first used by Berger [25]. In this model the $K\alpha_1$ and $K\alpha_2$ lines are each represented as the sum of two Lorentzian profiles as shown in Fig. 2.2.

For Cr, Mn, Fe, Co, Ni and Cu the phenomenological representations are used to accurately represent the emission profiles from these elements with up to seven Lorentzians being used in some cases.

Another aspect of $K\alpha$ emission lines required for accurate line profile analysis is the satellite multiplet structure in the high energy tail as shown in Figs. 2.3 and 2.4. For Cu these collectively have an intensity of $\approx 0.6\%$ of the $K\alpha_1$ emission line. In X-ray diffraction studies it is sufficient to represent

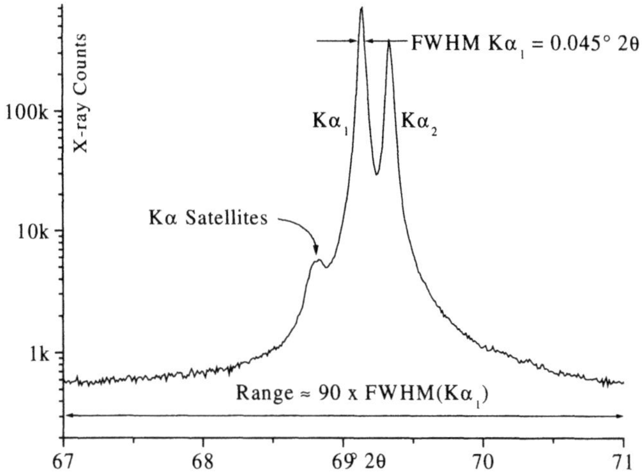

Fig. 2.3. Cu$K\alpha$ emission profile showing the satellite group of lines and the extent of the tails from the $K\alpha1$ and $K\alpha2$ emission lines. This profile was recorded using the 400 line from a silicon single crystal wafer (from [19])

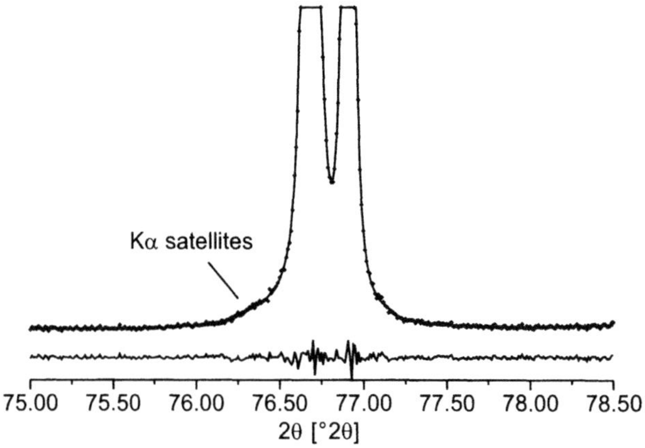

Fig. 2.4. Single line fundamental parameters fit to the 331 line of CeO_2 [27] at about 76.7° 2θ. Divergent beam data have been taken with a D8 ADVANCE (Bruker AXS). The $K\alpha$ satellites contribute significant intensity to the total profile, which is typically neglegted in analytical profile fitting approaches

the $K\alpha$ satellite group as a single broad Lorentzian; thus, for the case of Cu $K\alpha$ a total of five Lorentzians are used to accurately describe the spectrum.

As the natural shape of an emission profile is Lorentzian, the tails can extend a considerable distance from the central peak as shown in Fig. 2.3. Most diffractometers, however, operate with $K\beta$ filtering or some form of monochromatisation and as such the tails are attenuated to varying degrees as illustrated in Fig. 2.5. When a Ni $K\beta$ is inserted in the beam of Cu$K\alpha$

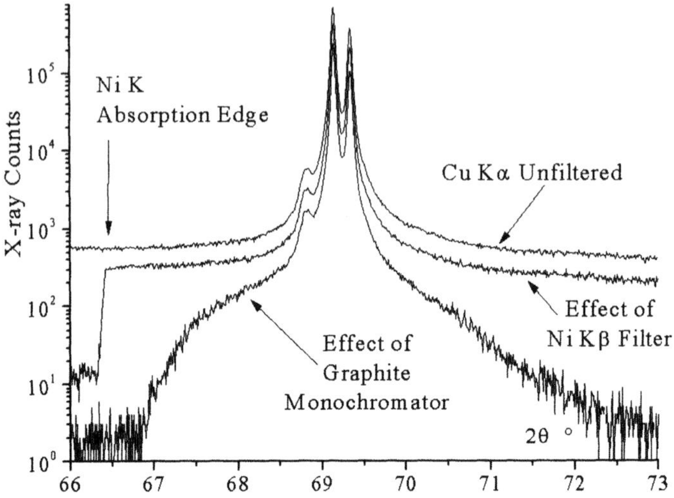

Fig. 2.5. Cu$K\alpha$ emission profile obtained using the 400 line from a silicon single crystal wafer. Each pattern was recorded sequentially using the same sample, first with no filter or monochromator in the beamline, then with a Ni $K\beta$ filter, and finally with a standard curved graphite diffracted beam monochromator (from [19])

the attenuation appears to be more or less uniform across the profile except for below the Ni K absorption edge. The inclusion of a curved graphite monochromator in the diffracted beam greatly reduces the range of wavelengths entering the detector resulting in profile tails that diminish more rapidly than the natural emission profile. These monochromators can also affect the relative intensity $I_{K\alpha 2}/I_{K\alpha 1}$ by up to $\pm 10\%$ depending on the alignment and setting of the crystal. For example, in the monochromated spectrum shown in Fig. 2.5 the relative intensity of the Cu $K\alpha_1/K\alpha_2$ is approximately 0.46 rather than about 0.5 as in the unfiltered spectrum.

"Ground and bent" Johansson monochromators achieve perfect focusing when correctly aligned and are able to select a very narrow band of wavelengths. Figure 2.6 shows an example spectrum for Cu$K\alpha$ from an asymmetrically cut ground and bent Ge crystal used as an incident beam monochromator. The wavelength band-pass is narrow enough to remove 99.98% of the $K\alpha_2$ component and 100% of the $K\alpha$ satellite group. Also seen is that the Lorentzian tails are effectively eradicated.

The inclusion of a parabolic multilayer mirror in the incident beam of a diffractometer can also distort the wavelength spectrum [26]. This happens because the $K\alpha_1$ and $K\alpha_2$ components reflect off the mirror in slightly different directions as shown in Fig. 2.7. The separation of the $K\alpha_1$ and $K\alpha_2$ peak maxima, $\Delta 2\theta_{K\alpha_{21}} = 2\theta(K\alpha_2) - 2\theta(K\alpha_1)$, in a line profile is then either too large or too small than the same profile acquired from a diffractometer with

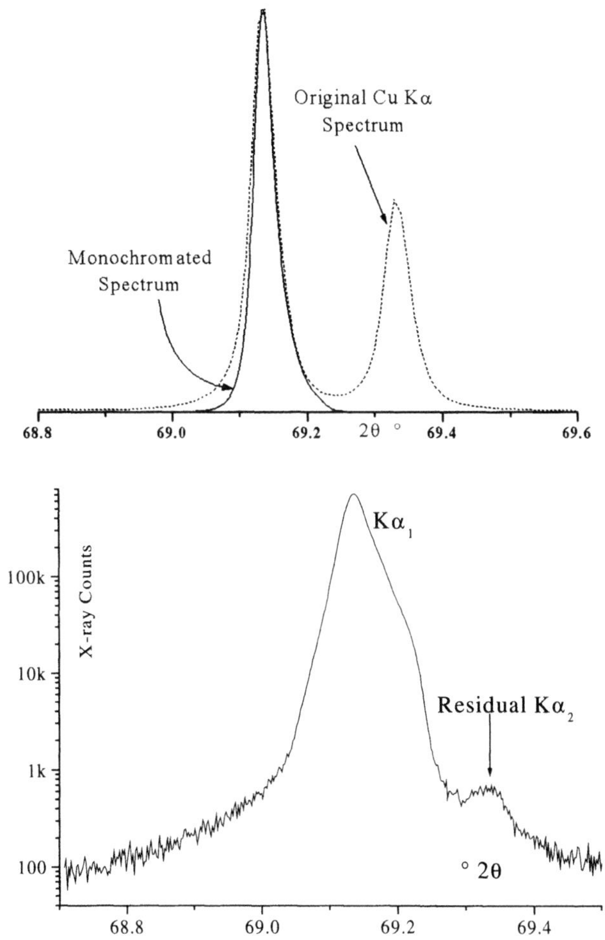

Fig. 2.6. Wavelength spectrum emerging from an asymmetrically cut Ge111 ground and bent incident beam monochromator presented linearly and logarithmically. The $K\alpha$ satellites are completely removed, but the $K\alpha_2$ is still present at $\approx 0.02\%$ of $K\alpha_1$ even in a well aligned system (from [19])

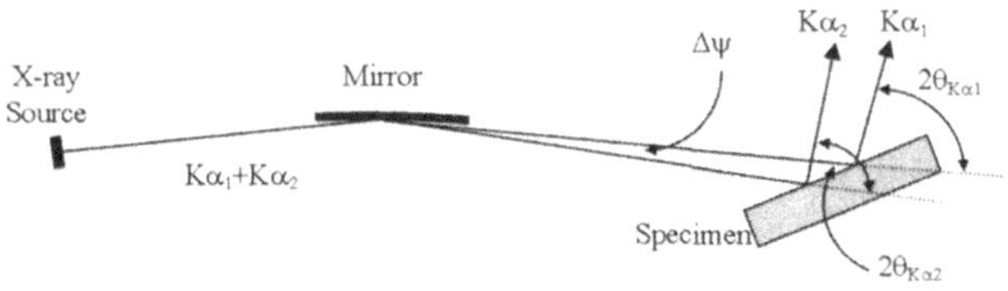

Fig. 2.7. Reflection of $K\alpha_1$ and $K\alpha_2$ wavelengths from a parabolic multilayer mirror diffracting off a powder specimen (from [19])

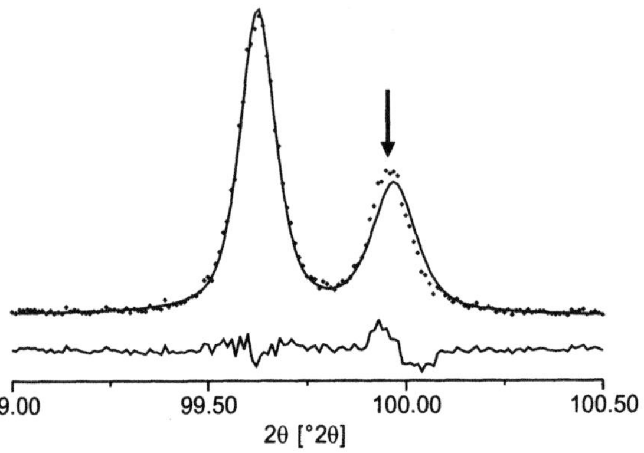

Fig. 2.8. Single line fit to the (041)/(322) line of LaB_6 (NIST SRM 660a) at about 99.6° 2θ. Parallel beam data have been taken with a D8 ADVANCE (Bruker AXS) equipped with a single Göbel mirror in the primary beam and Parrish–Hart analyser slits in the diffracted beam. The $K\alpha_2$ peak maximum (*arrow*) is significantly shifted towards the $K\alpha_1$ reflection resulting in a significant misfit, if the source emission profile is not modified accordingly as shown here

no mirror. When a diffractometer is set up with its mirror in the orientation shown in Fig. 2.7 the separation $\Delta 2\theta_{K\alpha_{21}}$ is smaller than the value expected from the known $K\alpha_1$ and $K\alpha_2$ wavelengths by an amount corresponding to the difference $\Delta\psi$ in the directions of the two incident beams on to the specimen. $\Delta\psi$ for a conventional long fine focus tube with a projected target width of 0.04 mm can rise to 0.0017° and leads to significant aberrations at high 2θ angles as shown in Fig. 2.8.

2.2.4 Examples

As an illustration of the power of the empirical convolution approach, it will be shown that all of the CeO_2 data made available as part of the size-strain round robin conducted by the IUCr CPD[1] [27] can be fitted accurately. The data for the following illustration include

- laboratory X-ray data (D8 ADVANCE, Bruker AXS),
- synchrotron X-ray data (NSLS X3B1, ESRF BM16),
- CW neutron data (ILL D1A, NCNR BT1),
- TOF neutron data (ISIS HRPD).

Two datasets were collected from each instrument; one well crystallized specimen to determine the instrument functions and one specimen exhibiting strong specimen broadening.

[1] International Union of Crystallography, Commission on Powder Diffraction

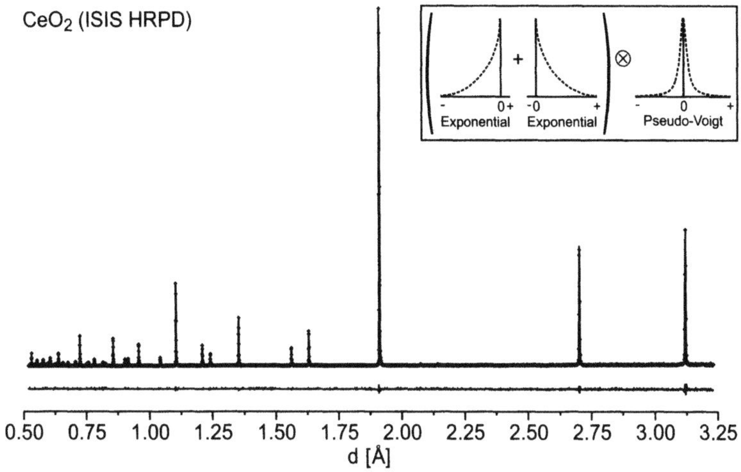

Fig. 2.9. Empirical full pattern Pawley fit to the CeO_2 TOF neutron data (ISIS HRPD), measured instrument function [27]

The functions used to fit the data were chosen from the "function pool" shown in Fig. 2.1. An attempt was made to minimize the number of functions and the number of parameters used.

Figure 2.9 shows an example Pawley fit to the ISIS TOF neutron data. The upper right corner shows a schematic of the sum of two exponential functions (two refineable parameters each), used to describe *d*-spacing dependent asymmetry towards small and large *d*-values, convoluted with a single Pseudo-Voigt function (one refineable parameter). Of note is that only 5 refineable parameters are sufficient to model the whole pattern providing an excellent fit.

In Figs. 2.10–2.13 the refinement results are shown for the most asymmetric peaks for the lab X-ray data (D8 ADVANCE, Fig. 2.10), synchrotron X-ray data (NSLS X3B1, Fig. 2.11), CW neutron data (ILL D1A, Fig. 2.12), and TOF neutron data (ISIS HRPD, Fig. 2.13). The data representing the measured instrument functions are shown in Figs. 2.10a–2.13a, respectively (in the following referenced as "sharp data"). The broadened data are shown in Figs. 2.10b–2.13b, respectively, together with the instrument functions determined from the measured standard data scaled to the maximum peak intensity.

A maximum of 5 refineable parameters were required to fit the sharp data including its dependence on angle/d. To describe the broadened data across the whole angular range and to obtain microstructure information two Voigt functions comprising 4 refinable parameters were convoluted on top of the instrument function (Double–Voigt Approach, e.g. [28]). The microstructure results together with the round robin equivalents [27] are shown in Table 2.1. The excellent agreement seen for all datasets quantitatively demonstrate the

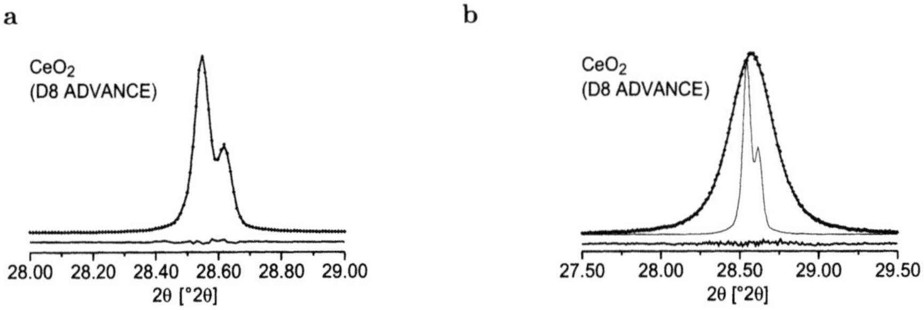

Fig. 2.10. (**a**) Laboratory X-ray data, measured instrument function (D8 ADVANCE). (**b**) Laboratory X-ray data, broadened data (D8 ADVANCE)

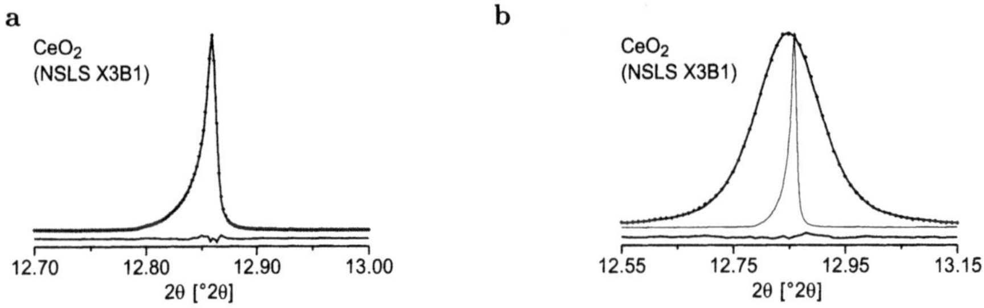

Fig. 2.11. (**a**) Synchrotron data, measured instrument function (NSLS X3B1) (**b**) Synchrotron data, broadened data (NSLS X3B1)

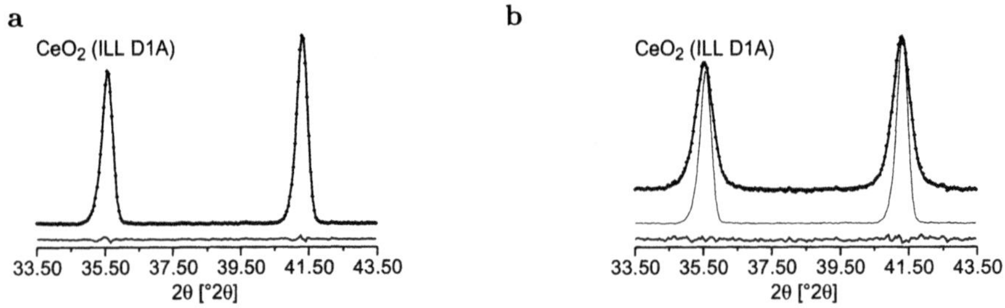

Fig. 2.12. (**a**) CW neutron data, measured instrument function (ILL D1A) (**b**) CW neutron data, broadened data (ILL D1A)

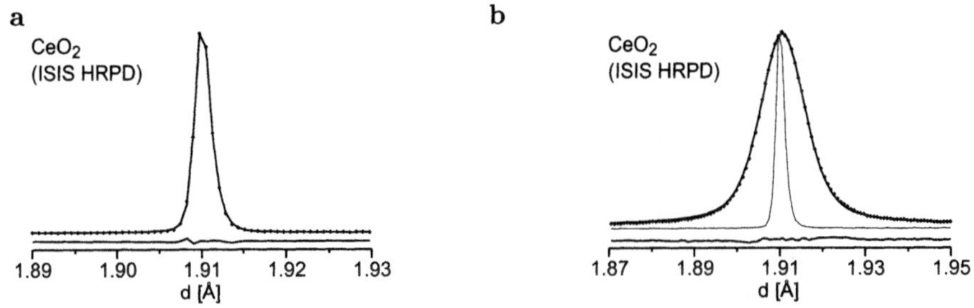

Fig. 2.13. (**a**) TOF neutron data, measured instrument function (ISIS HRPD). Same data as in Fig. 2.9 (**b**) TOF neutron data, broadened data (ISIS HRPD)

Table 2.1. Microstructure results for the size-strain round robin data [27]. Results are also given for the ESRF BM16 and NCNR BT1 data. The round robin results are provided in the last row. L_{Vol} represents the volume weighted column length

	L_{Vol}		Microstrain	
D8 ADVANCE	23.43	(0.08)	0.0149	(0.0014)
NSLS X3B1	23.72	(0.08)	0.0307	(0.0014)
ESRF BM16	22.59	(0.05)	0.0143	(0.0010)
ILL D1A	23.29	(0.18)	0.0273	(0.0034)
NCNR BT1	23.88	(0.34)	0.0259	(0.0052)
ISIS HRPD	22.93	(0.06)	0.0193	(0.0021)
Balzar (2001)	22.60	(0.90)	"nearly strain free"	

superior performance of convolution based profile fitting on a wide variety of profile shapes and diffraction instruments. The absolute microstructure values obtained by the Double-Voigt Approach can certainly be disputed; this however is by no means a limitation of convolution based profile fitting but rather a limitation of the microstructural model being used.

2.3 Fundamental Parameters Approach

2.3.1 General Considerations to FPA

FPA line profile shapes are characterized in terms of the physical parameters describing the diffractometer or the diffraction process [13, 16–19]. Similar to (2.2) observed line profile shapes $Y(2\theta)$ are described by convoluting the source emission profile with the instrument aberrations; sample broadening is also incorporated by convolution. In other words FPA represents $Y(2\theta)$ in terms of the dimensions of the diffractometer and the physical properties of the sample according to

$$Y(2\theta) = (W \otimes G) \otimes P \otimes S\,, \tag{2.3}$$

where P represents sample contribution function. The most important instrument, sample and microstructure contributions to the line profile shape are provided in Table 2.2.

The instrument function $I(2\theta) = (W \times G)$ in terms of the individual aberrations $G_i(2\theta)$ can be written as

$$I(2\theta) = W \otimes G_1(2\theta) \otimes G_2(2\theta) \otimes \ldots \otimes G_i(2\theta) \otimes \ldots \otimes G_n(2\theta)\,. \tag{2.4}$$

Thus it is seen from (2.1) and (2.4) that FPA is a special case of convolution based profile fitting. The important difference is that $I(2\theta)$ for FPA is determined from prior knowledge of the emission profile and diffractometer

Table 2.2. Most important instrument, sample, and micro structure effects contributing to powder diffraction line profiles. Note, that absorption is classified as a sample contribution, and does therefore not contribute to the instrument function unlike in conventional approaches

Instrument	Sample	Micro structure
Equatorial plane	Equatorial plane	
- Target width	- Absorption	- Crystallite Size
- Divergence slit angle	- Sample thickness	- Microstrain
- Receiving slit angle	- Tilt	- Strain
		- Stacking faults
		...
Axial plane	Axial plane	
- Soller slit(s)	- Sample length	
- Target length		
- Receiving slit length		

geometry resulting in far fewer refinable profile parameters. This results in decreased correlations between the profile, background and in particular microstructural parameters leading to an overall improvement of the refinement process within all profile fitting methods.

It is clear from (2.3) that FPA explicitly distinguishes between sample transparency and related microstructure broadening. While sample contributions are unavoidable, FPA allows from profile shape considerations the independent treatment of aberrations such as the refinement of the linear absorption coefficient or the thickness of a non-infinitely thick sample (e.g. [29]).

Diffraction broadening analysis is an intrinsic property of FPA. Even in well crystallized powders with crystallite sizes up to 2 µm, crystallite size broadening is detectable in high angle lines and it is necessary to include a physical broadening function in the refinement.

When more than one type of sample broadening is present, FPA affords a multiplicity of broadening functions that can be convoluted together to form $S(2\theta)$. For example when crystallite size (xtal size) and stacking fault (s. fault) broadening are present then

$$S(2\theta) = S_{\mathrm{xtal\,size}}(2\theta) \otimes S_{s.\,\mathrm{fault}}(2\theta) \tag{2.5}$$

The TOPAS implementation of FPA makes available many different types of component broadening functions. These include the standard Lorentzian and Gaussian functions with a full width at half maximum that can be constrained to fit a particular 2θ or hkl dependence. It is also possible to incorporate any number of user-defined functions to accommodate other types of line broadening.

2.3.2 Applicability of the FPA

Most powder diffractometers are either divergent beam instruments operating in Bragg–Brentano geometry or parallel beam instruments equipped with Göbel mirrors operating in Debye–Scherrer geometry. Over the past ten years the number of diffractometer options available from manufacturers have increased dramatically in particular for parallel beam instruments and in general users operate with a wider range of X-ray optical designs.

The basic objectives of the FPA technique is to be able to fit any powder diffraction profile using physically based models to describe both the instrument profile and diffraction broadening generated by the specimen. In principle the technique can be applied to any powder diffractometer and profiles of widely differing shapes should be accommodated by simply modifying the physical parameters of the diffractometer used to describe the profiles.

Although most applications of FPA have focused on the conventional divergent beam diffractometer, it has also been utilized for analyzing neutron diffraction data [30] and synchrotron data [31, 32].

In general instrument parameters are not refined as they are measurable and known. Second order effects however often lead to slight variations in some of the instrument parameters and refinement of one or two parameters is sometimes necessary. For example, strongly oriented thin films and powders generally produce narrower than expected line profiles simply because the diffraction cone is incomplete. During refinement, this type of problem can be accommodated by refining on one of the axial divergence parameters, such as the diffracted-beam Soller angle, which typically refines to a value that is slightly different from the measured value.

Because the instrument contribution to line profile shapes can be predicted, instrument optimization is possible where an appropriate balance between intensity and resolution can be deduced. It is also possible to assess instrument alignment by comparing expected and observed profile shapes.

2.3.3 Diffractometer Configurations and Their Geometrical Aberrations

Divergent Beam Laboratory Diffractometers

In the conventional divergent beam laboratory diffractometer, diffraction is symmetric and the principal geometric aberrations contributing to a profile alongside the wavelength distribution are,

- the finite width of the X-ray source,
- flat specimen error,
- finite width of the receiving slit,
- specimen transparency,
- axial divergence.

As discussed above, both primary and secondary monochromators modify the source emission profile significantly and also reduce axial divergence [17–19].

Position sensitive detectors (PSDs) lead to additional contributions to the instrument function. PSDs used in divergent beam laboratory diffractometers are of the linear, step-scanned type with a small acceptance angle of up to 10° 2θ, which are mounted tangentially to the detector circle. For such a detector the only position on the detector that is normally in focus is its centre. All beams diffracted from planes angled relative to the specimen surface will enter the detector at off-centre positions (asymmetric diffraction). Additional aberrations are

- defocusing due to asymmetric diffraction,
- discharge resolution of the detector,
- parallax error.

Parallel Beam Diffractometers

In parallel beam diffractometers with analyser foils and/or analyser crystals and/or channel cut monochromators in the primary or diffracted beam there are only two active geometric aberrations namely,

- axial divergence,
- finite angular acceptance of the receiving optics.

PSDs used in parallel beam diffractometers are either linear, step scanned detectors as described above, or large curved stationary detectors with acceptance angles of up to 150° 2θ. In case of a PSD there are additional aberrations dependent on its type; these are

- discharge resolution for both detector types,
- parallax error for linear detectors.

In laboratory diffractometers, the parallel beam is produced by a parabolic graded multilayer mirror (so-called "Göbel mirror" [33]); with the line X-ray source positioned at the focus of the mirror. Although the beam may be parallel in the equatorial plane, it will not be parallel in the axial plane and axial divergence is present in both the incident and diffracted beams. Low angle profiles will therefore be asymmetric although not to the same extent as diverging beam instruments. Göbel mirrors as well as analyzer crystals and/or channel cut monochromators will in addition significantly change the source emission profile as discussed above.

2.3.4 Geometric Instrument Aberrations

Geometric aberrations tend to determine the shape of line profiles at low to medium 2θ angles (ie. $2\theta < 60°$). At high 2θ angles ($2\theta > 100°$), the profile conforms primarily to the shape of the emission profile. With the exception of the aberrations associated with the receiving optics and the finite X-ray source width (laboratory diffractometers), all geometric instrument aberration profiles vary with 2θ. In the following sections the shapes of the major instrument aberrations used in FPA to describe the various diffractometer configurations are briefly discussed for conditions that are typical of those encountered in practice.

Finite X-ray Source Width

The finite X-ray source width aberration is angle independent and its effects are present when sealed tubes or rotating anodes are employed (laboratory diffractometers). For long fine focus tubes the source width aberration profile is generally expressed as an impulse function of width $\Delta 2\theta_x$ as shown in Fig. 2.14. Although this function may not be strictly valid the exact shape is not critical due to its small broadening contribution. For a target width ≈ 0.4 mm at a take-off angle of 6° the projected width $W_x \approx 0.04$ mm and the aberration profile width $\Delta 2\theta_x \approx 0.01°$.

In broad focus tubes the target width ≈ 2 mm and $W_x \approx 0.2$ mm at 6° take-off so that the aberration profile width $\Delta 2\theta_x \approx 0.056°$. At this level the source width contribution dominates and a more accurate form for the aberration profile shape is necessary. A good approximation under these circumstances is a Gaussian shape rather than an impulse function. In diffractometers with a curved crystal incident beam monochromator the source width can also have a greater contribution because of the magnification effect introduced by the monochromator. This occurs with asymmetrically-cut Johansson incident beam monochromators where the source-crystal and crystal-focal point distances are typically ≈ 120 mm and ≈ 230 mm, respectively. A fine focus tube with a projected width of 0.04 mm is then effectively magnified to ≈ 0.08 mm. Under these conditions the effective source width can be trimmed by reducing the width of the focal line slit.

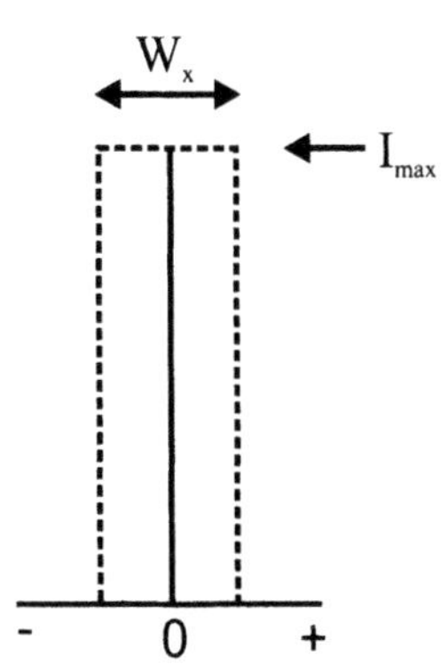

Fig. 2.14. Simple aberation model for a source of projected width W_x

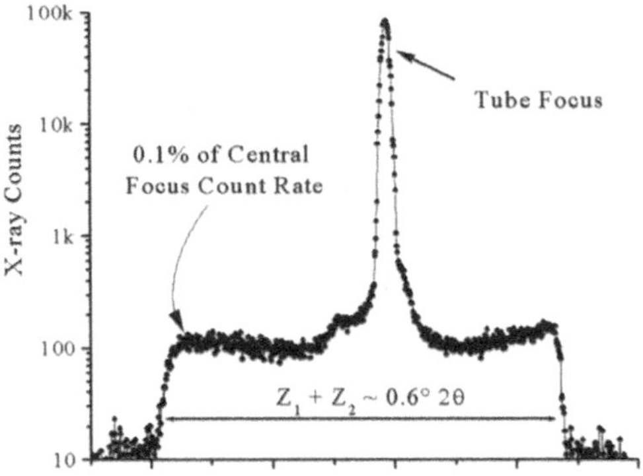

Fig. 2.15. Intensity scan with 50 µm wide slit of an image formed through a 10 µm pinhole in platinum of the 0.4 mm wide long fine focus in a Cu anode X-ray tube set at 40 kV, 40 mA (from [19])

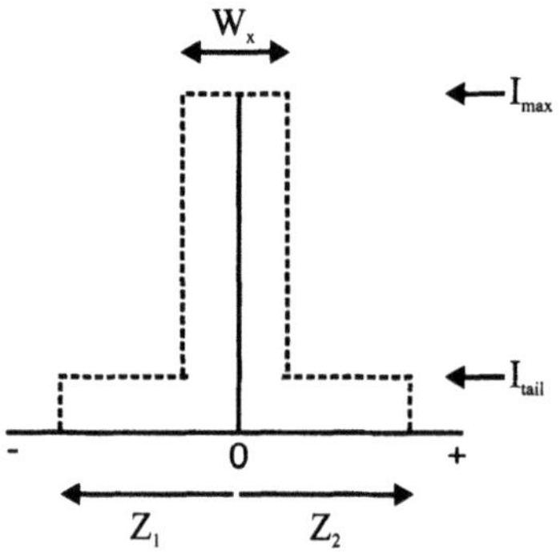

Fig. 2.16. Aberration model with "tube tails"containing additional parameters $f = I_{\text{tail}}/I_{\text{max}}$ and angular widths Z_1 and Z_2 from the central maximum

For accurate line profile analysis it is also necessary to modify the simple impulse model even with long fine focus tubes [34]. This is illustrated in Fig. 2.15 which shows the intensity recorded by scanning with a 50 µm slit across the image of a X-ray source formed through a 10 µm pinhole in platinum. A better approximation therefore to the source width aberration function is a sharp impulse function superimposed on a broad impulse function to represent the so called "tube tails" as shown in Fig. 2.16. The parameters introduced to describe "tube tails" are the extents of the high and low angle tails, Z_1 and Z_2, and the intensity of the tail f relative to the intensity at the tube focus. In most instances the intensity of the tails is $\approx 0.1\%$ of the peak intensity and is only significant when analyzing intense lines. The tails themselves are not necessarily symmetric with respect to the tube focus and can extend over a 2θ range up to 0.6° (2θ).

Receiving Optics

In divergent beam diffractometers a receiving slit is placed at the focus of the diffracted beam. The count rate incident on the detector increases with increasing slit width but at the expense of resolution (for further discussion see e.g. [35]). Most commercial diffractometers have a selection of receiving slits ranging in width from 0.05 mm up to 0.6 mm. When the slit size is larger than 0.15 mm, the receiving slit aberration is often the dominant aberration over the angular range $2\theta = 15°$ to $60°$. The aberration function for a perfectly aligned receiving slit is an impulse function that is angle independent.

In parallel beam diffractometers the receiving slit system is based on either Parrish–Hart analyzer slits or analyzer crystals. Analyzer slits act as an angular filter in the diffracted beam. The aberration function, or transmission function, for these is a triangle function (Fig. 2.17), in which the base width W_r is given by the angular aperture Δ of the slits. A problem often encountered with analyzer slits is specular X-ray reflection from the analyzer foils [36]. Weak satellite peaks appear on both the high angle and low angle profile tails but not necessarily of the same intensity as shown in Fig. 2.18. This effect can be incorporated into the aberration profile by adding two Voigt functions of unequal intensity, one on each side of the triangular aberration function, to represent the satellite reflections [19]. The parameters of the satellite peaks can then be determined by fitting profiles from a reference material such as the NIST reference standard LaB_6, SRM 660a. An alternative approach to determining the aberration profile of an analyzer slit system, without the effects of the wavelength profile, is to simply carry out a 2θ scan across the incident beam. Provided the axial divergence of the incident beam is kept small, by including axial Soller slits, and the equatorial divergence is negligible then the incident beam scan will have the same shape as the aberration profile of the analyzer slits.

The inclusion of an analyzer crystal in the diffracted beam of a parallel beam diffractometer gives high-resolution diffraction patterns with low background; the intensity however is invariably less than the Parrish–Hart configuration. The aberration profile introduced by the analyzer crystal is generally very narrow and can be determined by measuring the rocking curve

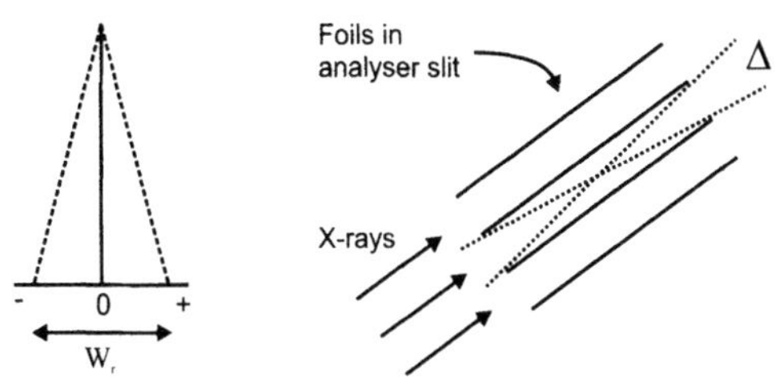

Fig. 2.17. Triangle shaped aberration function for a set of analyser slits with an angular aperture Δ, where $\Delta/2$ = spacing between the foils/length of the foils

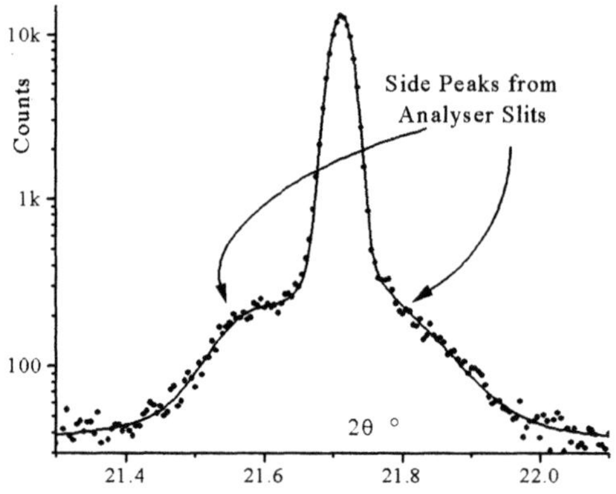

Fig. 2.18. Reflection satellite peaks from analyser slit recorded using the 310 line from NIST standard material LaB_6 SRM 660a using the diffractometer on Station 2.3 at Daresbury synchrotron (from [19])

of the crystal. For a perfect analyzer crystal the aberration profile will be determined by the Darwin profile of the analyzer crystal. In practice however the aberration profile is broadened by the mosaic structure of the crystal, stresses in the crystal and deviations in the eveness or curvature of the crystal surface. As a consequence, the aberration profile can be dependent on the size of the beam incident on the crystal. A first approximation to the shape of the aberration profile of an in-situ analyzer can be obtained from a 2θ scan of the analyzer/detector using a very fine incident beam as shown in Fig. 2.19. Although the recorded profile includes the wavelength distribution, the result does at least give an indication of the shape and upper limit of the FWHM of the aberration profile.

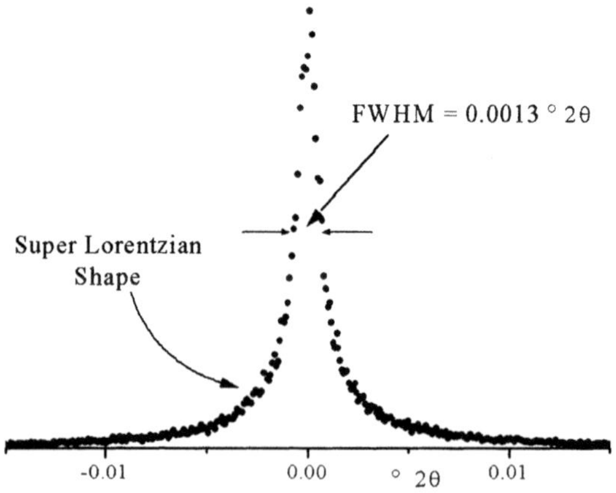

Fig. 2.19. 2θ scan across a 0.1 mm × 0.1 mm incident beam using a Ge111 analyser crystal on beamline BM16 at the ESRF, Grenoble (from [19])

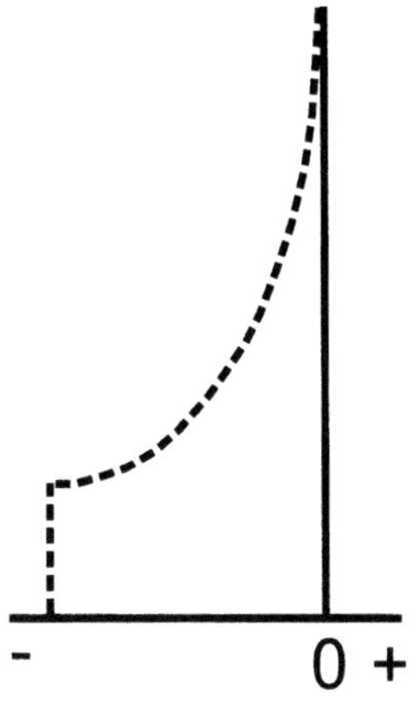

Fig. 2.20. Flat specimen aberration function

Flat Specimen Error

The flat specimen error occurs because the surface of the sample is tangential to the goniometer focusing circle rather than on the focusing circle leading to defocusing. The effects of the flat specimen error are discernable as an increase in asymmetry and breadth towards low 2θ angles and can be represented by the aberration function shown in Fig. 2.20.

Divergence slits that vary as a function of 2θ (variable divergence slits) and used to illuminate a fixed sample length results in a flat specimen aberration profile that is generally small at low 2θ values and increases with increasing 2θ. With an illuminated length of 20 mm on the specimen, flat specimen error is clearly discernable at $2\theta > 20°$. It is not always possible to collect data with a fixed illuminated length over a large 2θ range as the required beam divergence at large 2θ may be larger than can be accommodated by the diffractometer. For example, to maintain a fixed beam length of 20 mm on a specimen over the range $2\theta = 0°$ to $90°$, the required angle of divergence α increases from $0°$ up to $\approx 4°$. A divergence angle $\alpha = 4°$ is close to the maximum value at which most diffractometers can operate when a pyrolytic graphite monochromator is installed in the diffracted beam. At angles of α greater than 4° the diffracted beam may extend beyond the graphite crystal and thus not diffracted into the detector (see also e.g. [37]).

Specimen Transparency

In divergent beam diffractometers, specimen transparency produces asymmetry and broadening of the instrumental profile function. For perfect focusing all diffraction should occur on the focusing circle, but when the beam penetrates the surface, diffraction will occur over a range of depths within the specimen.

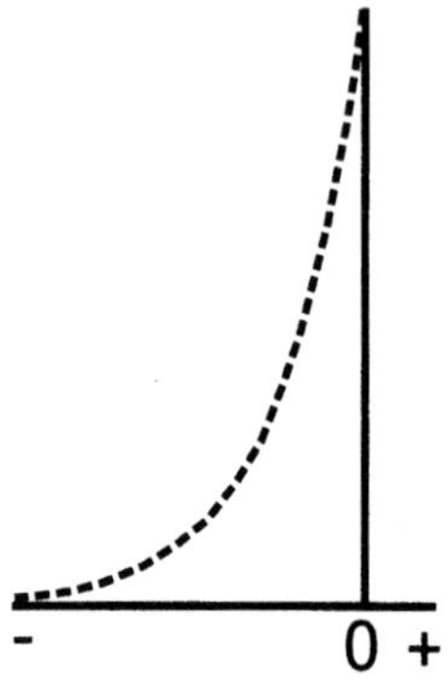

Fig. 2.21. Specimen transparency aberration function

Asymmetry and broadening from specimen transparency is greatest for low absorption materials and is clearly evident for linear attenuation coefficients $\mu < 50\,\mathrm{cm}^{-1}$. The contribution of specimen transparency is greatest at $2\theta \approx 90°$ and at this angle the aberration profile has a FWHM $\approx 0.03°$ 2θ when $\mu \approx 50\mathrm{cm}^{-1}$; this drops to $\approx 0.005°$ 2θ when $\mu \approx 200\,\mathrm{cm}^{-1}$. The aberration profile for an infinitely thick specimen is shown in Fig. 2.21. Specimen transparency effects are quite strong in organic materials with very low attenuation coefficients ($\mu \approx 30\,\mathrm{cm}^{-1}$ or less). They can also be observed in loosely bound powders of low atomic number materials where, depending on preparation, the packing density can be as low as 50%. This is why powdered standard reference materials such Al_2O_3 (e.g. SRM674a) or Silicon (e.g. SRM 660a) are not ideal materials for the measurement of the instrumental function or for checking instrument alignment. In this case, the effective linear absorption coefficient is not known, but required to properly correct the sample transparency effect. Preparation of a thin film on a zero background holder minimizes this effect; however such preparation may lead to uncertainties in sample height and surface flatness resulting in sample displacement errors, making lattice parameter constrained refinements more difficult. Materials better suited for the determination of the instrumental function are well annealed and highly absorbing powders such as CaF_2, Y_2O_3, CeO_2 or LaB_6 (ideally the NIST SRM660a which can be used as received), prepared in conventional sample holders with cavity.

Diffractometer Defocusing

Defocusing in divergent beam diffractometers results in broadened diffraction lines and occurs when the receiving slit is not positioned at the focus of the diffracted beam. The most common causes of defocusing are,

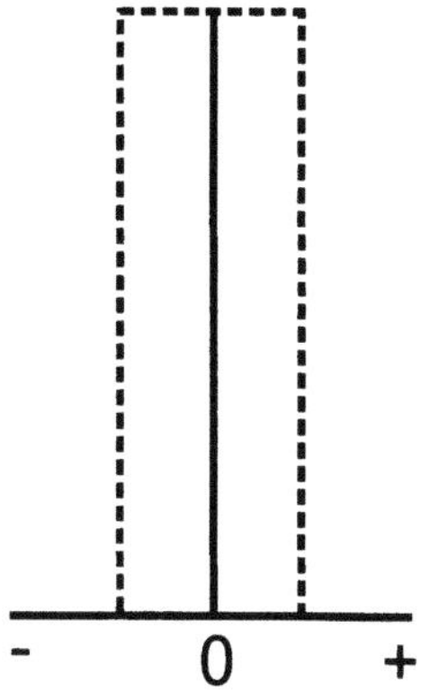

Fig. 2.22. Defocussing aberation function

- mis-setting of the incident beam angle,
- wrong position of the receiving slit where the distance of the slit to the sample is larger or smaller than the nominal radius of the diffractometer.

For both of these conditions the focus of the diffracted beam will be either in front of or behind the receiving slit and in both cases the aberration profile is an impulse function as shown in Fig. 2.22 with the same width as the defocused beam at the receiving slit.

In asymmetric diffraction, the breadth of diffraction lines increases as the deviation from the symmetric condition increases. Defocusing is larger and varies more rapidly at low 2θ values. Conversely, at high 2θ values, diffractometers will tolerate large defocusing errors with no detectable increase in the breath of the line profiles. Also, by reducing the angle of divergence, defocusing can be reduced but at the expense of diffracted intensity.

Axial Divergence

The axial divergence aberration is due to divergence of the X-ray beam along the diffractometer axis in the plane of the specimen and is one of the most dominant instrument contributions. In practice, axial divergence is most readily recognized by the asymmetry it introduces into low angle diffraction lines where the low angle tails extend further than the high angle tails.

In general a diffractometer will record X-ray counts over a range of measured angles 2ϕ rather than the true diffraction angle 2θ. The only rays for which $2\phi = 2\theta$ will be those propagating parallel to the equatorial plane and incident on the diffractometer axis. For a particular ray path the measured diffraction angle 2ϕ for a true diffraction angle 2θ depends on the axial divergence β and γ in the incident and diffracted rays (see Fig. 2.23). In the absence of Soller slits the maximum axial divergence in the incident and diffracted beams is determined by the source length L_x, sample length L_s and receiving

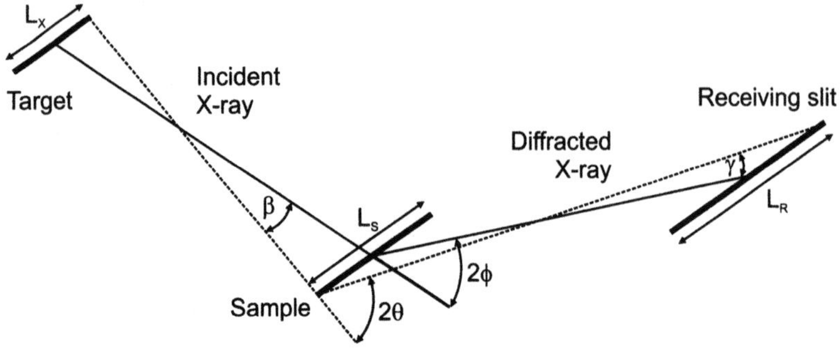

Fig. 2.23. An illustration of axial divergence in which the incident ray and diffracted ray are at angles of β and γ relative to the equatorial plane. Although the diffraction angle at the sample is 2θ, the diffracted beam is recorded at an angle 2ϕ. Adapted from [19]

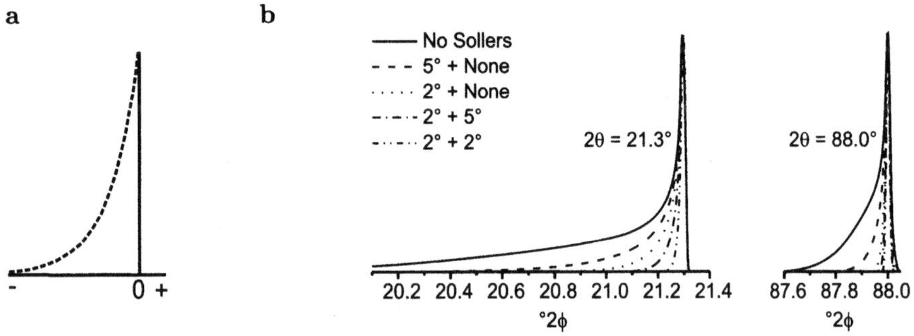

Fig. 2.24. (**a**) Axial divergence aberration function. (**b**) An illustration of axial divergence profiles calculated for a laboratory diffractometer with various combinations of incident-beam and diffracted beam Soller slits (or no Soller slits) for two 2θ values (corresponding to the 001 and 321 reflections of LaB_6). Commercial Soller slits range from 2 to 5° [17]

slit length L_r. In the latter case the axial divergence can be as large as 10° in the incident beam and 10° in the diffracted beam for a diffractometer with a 12 mm long X-ray source, a 12 mm long receiving slit and a 20 mm long specimen (all lengths in the axial direction). Most commercial diffractometers are supplied with Soller slits having angular apertures between 2° and 5° and in such cases the breadth of the aberration profile (Fig. 2.24a) is considerably reduced (Fig. 2.24b). When narrow Soller slits are included in the beam path, the dimensions L_x, L_r and L_s tend to be redundant parameters in the calculation of the aberration profile as the Soller slits control the maximum axial divergence of the incident and diffracted beams.

The major change to profiles brought on by axial divergence is the extension of the low angle tails for $2\theta < 50°$. At very high angles, $2\theta \geq 150°$, the asymmetry reverses and the extension of the high angle tails increases with increasing 2θ. Broadening from axial divergence is evident at all 2θ values but passes through a minimum in the region $2\theta \approx 110°$ and not at $90°$ as frequently approximated by ignoring axial divergence in the incident beam (for details refer to [17,18]; see also Fig. 2.24b). However, the shift of the peak maximum is small at all 2θ values unlike the shift in centre of gravity, which varies considerably from large negative values, near $2\theta \approx 0°$, to large positive values as 2θ approaches $180°$.

In practice it is not always possible to calculate the exact form of the axial divergence aberration function for a particular specimen/diffractometer configuration. The two main reasons for this are,

- in specimens with strong preferred orientation, such as thin films and rolled or extruded metals, the diffraction cones are no longer of uniform intensity along the arcs of the diffraction cones,
- the inclusion of a monochromator in the beam path reduces the axial divergence. When diffracted or incident beam monochromators are present the optical path length of the beam is extended. With a diffracted beam monochromator the optical path length of the diffracted beam is extended by the optical path between the receiving slit and the detector slit (for graphite monochromators tuned to Cu Ka radiation the optical path length in the diffracted beam is typically increased by ≈ 100 mm). Incident beam monochromators also increase the optical path length of the incident beam. For the most common type of monochromator in use, the asymmetrically ground and bent Ge monochromator, the path is increased by over 300 mm. Monochromators also act as angular intensity filters and their effect on profile shape is similar to the addition of Soller slits in the beam path. The effect of a monochromator can therefore be represented as a Soller slit in a profile fitting model.

Linear Position Sensitive Detector (LPSD) Aberrations

LPSDs with angular windows up to $10°$ (at 200 mm instrument radius) are used on commercial diffractometers to increase the data collection rate e.g. for kinetic studies. LPSDs can be used in a stationary mode where only a fixed angular region of a pattern is recorded, or in a scanning mode where the detector is scanned in θ–2θ mode with its centre maintained at the focusing condition. As the detector is scanned the total diffraction pattern is formed by adding and averaging the patterns recorded at each step [38].

In LPSD systems, the receiving slit aberration is no longer relevant and the flat specimen aberration is replaced by an aberration function that embraces three effects that are folded together:

- flat specimen error including defocusing,
- parallax error,
- thermal noise.

A full treatment of these aberrations and how they are modified by a scanning LPSD system is given by Cheary and Coelho [16].

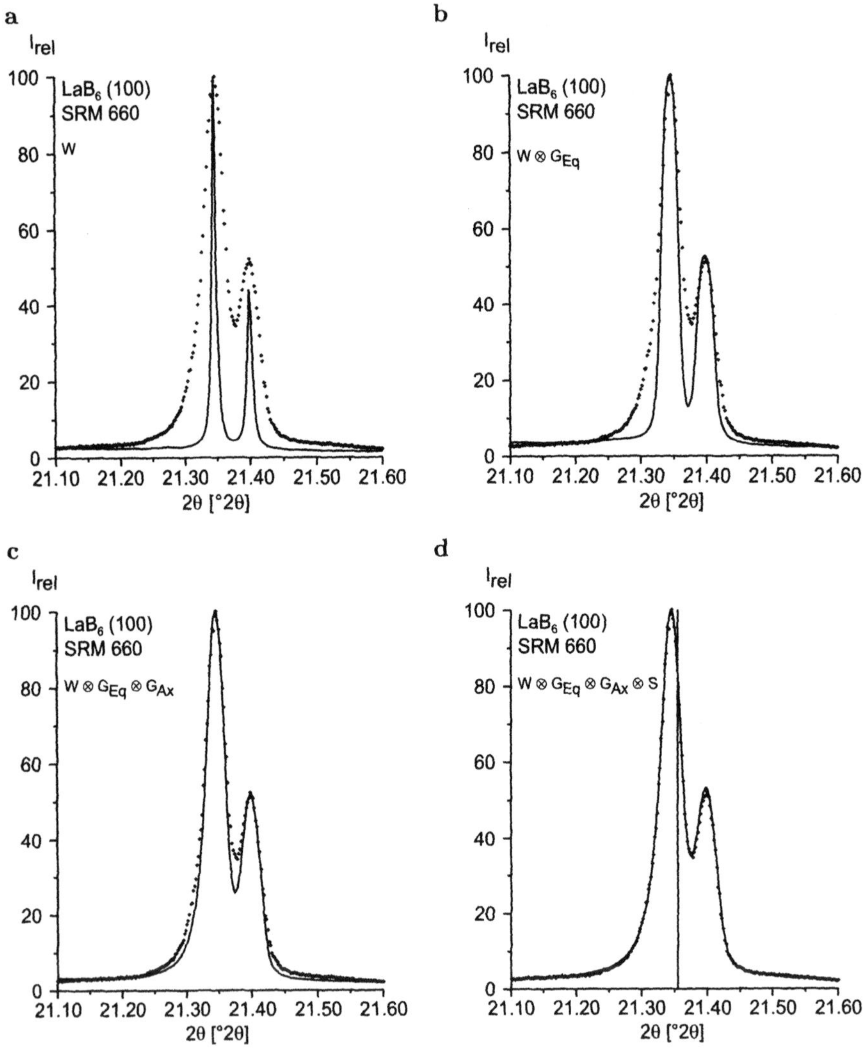

Fig. 2.25. Step-by-step synthesis of an observed line profile (LaB_6, 100, SRM 660) by convoluting the source emission profile W (**a**) with equatorial G_{Eq} (**b**) as well as axial instrument contributions G_{Ax} (**c**) and finally crystallite size and micro-strain effects S (**d**). Relative intensities, all calculated intensities in (**a**–**c**) scaled to the observed peak maximum intensity. From [29]

2.3.5 Examples

Example FPA Single Line Fitting

Figure 2.25 demonstrates the FPA based convolution process for the (100) reflection of LaB_6 at 21.35° 2θ. The CuKα emission profile scaled to the peak intensity maximum is shown in Fig. 2.25a. The emission profile accounts for the asymmetric line shapes of $K\alpha_1$ and $K\alpha_2$, and the "$K\alpha_3$" satellite as well. Fig. 2.25b convolutes equatorial instrumental contributions with the emission profile, Fig. 2.25c further convolutes the axial instrumental contributions. Further convolutions for microstructure contributions, represented by crystallite size and microstrain, result in the excellent fit seen in Fig. 2.25d. Note that the calculated peak maximum position is offset from the Bragg 2θ position; this is expected and there are two reasons for it. First the centroids of some of the aberration profiles are different to the Bragg 2θ position and secondly the manner in which Bragg 2θ is calculated is based on the wavelength of $K\alpha_1$ and not the peak maximum of the emission profile. This intrinsically corrects for errors in 2θ due to instrument and sample aberrations. Thus provided the aberration models for the aberration functions are correct, FPA gives a Bragg 2θ position matching that determined as if the data were collected with a "perfect sample" on a "perfect instrument" with monochromatic radiation. Therefore, especially at low angles, the calculated 2θ positions do not generally coincide with the peak maximum.

With FPA the excellent profile shape description as well as the improved determination of peak positions lead to more accurate lattice parameters in the Pawley, Le Bail and Rietveld procedures. The instrument function constraint, which improves the description of profile asymmetry, eliminates the need to refine an asymmetry parameter, which is mandatory in the "analytical functions approach", and also leads to the refinement of physically meaningful zero point or sample height errors. While the zero point of modern diffractometers can be easily aligned to better than 0.005° 2θ, the "analytical functions approach" typically refines zero point errors in the order of several hundreds of a degree 2θ which is due to poor asymmetry modelling. With FPA this is no longer the case.

Example FPA Full Pattern Fitting

Figure 2.26 shows an FPA based Pawley fit to the well crystallized CeO_2 size-strain round robin data (laboratory X-ray data, D8 ADVANCE). The convolution of the instrument and specimen contributions are shown schematically in the upper right corner of Fig. 2.26.

Notable but expected is the fact that the pure instrument function as calculated by FPA is not sufficient to describe these "standard data" as shown in Fig. 2.27. Indeed, because of inevitable specimen contributions such as size broadening, an additional Lorentzian function is required on top of the instrument function to achieve a satisfactory fit; this yields a volume weighted column length L_{Vol} of 253.75 ± 2.45 nm, but no microstrain. Nevertheless, the whole pattern in Fig. 2.26 has been described almost perfectly with a single refineable parameter for profile shape.

Fitting to the broadened CeO_2 data using two Voigt functions with 4 refineable parameters gave 22.59 ± 0.08 nm for L_{Vol} and 0.0149 ± 0.0014 for the microstrain coefficient. This is very much in agreement with the results given

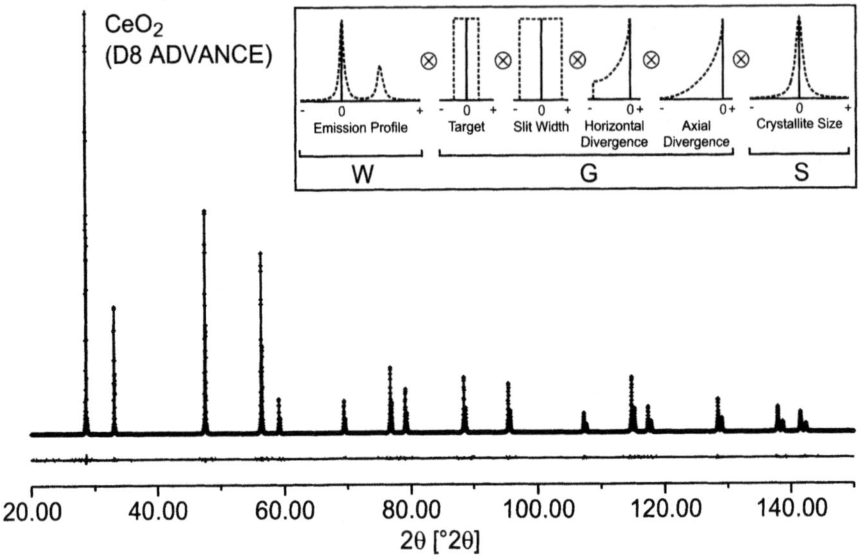

Fig. 2.26. Fundamental parameters based full pattern Pawley fit to the CeO_2 laboratory X-ray data (D8 Advance, Bruker AXS), measured instrument function [27]

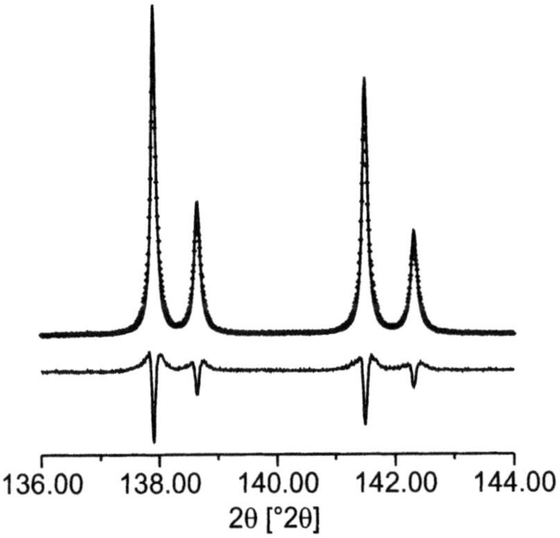

Fig. 2.27. Magnified region of Fig. 2.25 between 136° and 144° 2θ. The *fitted line* represents the calculated instrument function and demonstrates significant broadening of the "standard data"

in Table 2.1 for the conventional convolution approach using the same microstructure modelling. However, the same agreement will not be found when analyzing specimen broadening at smaller levels. The excellent agreement between the conventional convolution approach and FPA is due to the fact that the size broadening of the standard material is insignificant compared to the sample of interest. At smaller levels of specimen broadening the FPA will always provide better results compared to approaches using measured instrument functions.

Treatment of Satellite Peaks and Tube Tails

An important advantage of FPA in practice is the ability to easily account for additional scattering contributions coming from $K\alpha$ satellite peaks and tube tails; these contributions are typically ignored in present "analytical function approaches".

In Fig. 2.28 an example is shown for the 211 line of LaB_6, NIST SRM660a. Divergent beam data have been taken with a D5000 (Bruker AXS) equipped with an AEG long fine focus tube. The broad impulse function of tube tails has a width of about 1° 2θ and an intensity that is approximately 2% of the total line profile intensity. The R_{WP} of the fit shown in Fig. 2.28 is 3.6%, the Durbin-Watson parameter DW [39–41], which is a measure of serial correlation and therefore the quality of the fit model, is 1.57, which is close to the ideal value of 2. Ignoring tube tails yields an R_{WP} of 8.9% while DW drops to 0.27; when also ignoring the $K\alpha$ satellites R_{WP} becomes 12.0% with a DW of 0.16.

Although tube tail effects as large as that shown in Fig. 2.28 are rare, the $K\alpha$ satellites are always present in diffraction data taken with sealed tubes and rotating anodes and, for accurate work, should not be ignored particularly at low levels of specimen broadening.

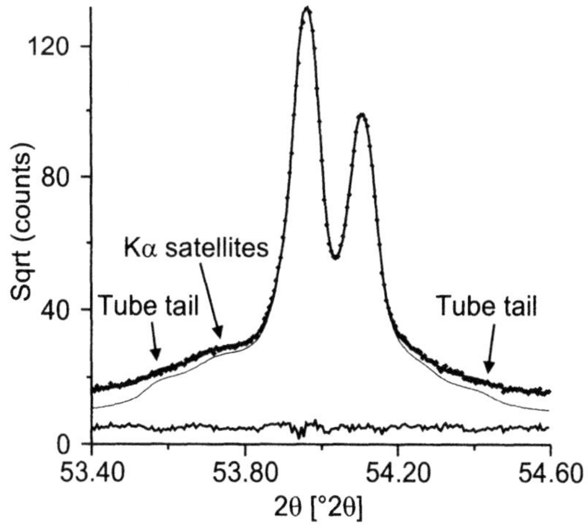

Fig. 2.28. Single line fundamental parameters fit to the 211 line of LaB_6 (NIST SRM 660a) at about 54° 2θ with tube tails. The *thin line* represents the calculated data without background. Courtesy by Cline [42]

2.4 Convolution Based Profile Fitting versus Analytical Profile Fitting

2.4.1 Quality of Fit

One of the outstanding features of convolution based profile fitting is the quality of fit, which is directly visible from the difference plot, but can also be quantified using agreement indices such as R-values or the DW parameter. As an example for a single line fit, Fig. 2.29 compares a FPA fit to a split-Pearson VII (SPVII) fit. The improvement of the various agreement indices provided in Table 2.3 is obvious.

Table 2.3. Agreement indices obtained by the split-Pearson VII (SPVII) and fundamental parameters approach (FPA) in Fig. 2.29

	R_{P}	R_{WP}	R_{Exp}	GOF	DW
SPVII	4.8	6.1	1.9	3.2	0.44
FPA	2.2	3.1	1.9	1.6	1.35

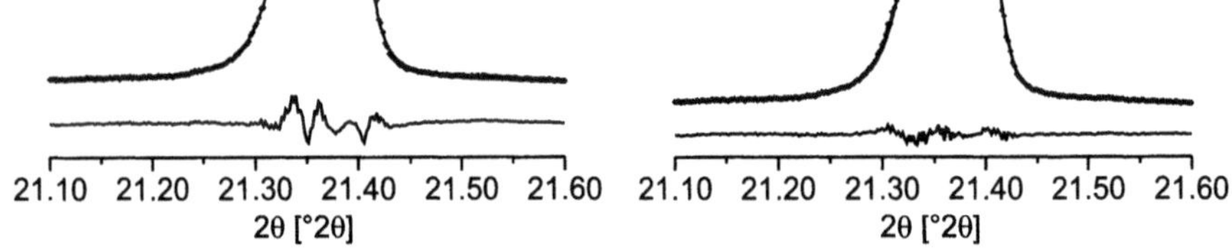

Fig. 2.29. Comparison of a single line fit to the 100 line of LaB_6 (SRM 660a) with a SPVII function (**a**) versus a FPA fit of the same data in (**b**)

2.4.2 Number of Fit Parameters Required

Convolution based profile fitting generally reduces the number of refinable profile parameters and thus many problems related to over-parametrization such as refinement of redundant parameters and parameter correlations are minimized. This is illustrated in Fig. 2.30 for a single line fit of the well known "5 fingers" of quartz using 3 unconstrained Pearson VII functions in Fig. 2.30a and fundamental parameters in Fig. 2.30b. The Pearson VII functions exhibit an almost perfect fit as indicated by the R_{WP} of 2.2%; this however is at the expense of largely divergent half-width and shape parameters particularly for the second and third peaks. Using fundamental parameters to introduce the instrument function as a constraint (thin line in Fig. 2.30b) and fitting independent crystallite size contributions to each of the peaks results in a slightly worse R_{WP} of 2.8%; however, the half-width and shape divergence problems are removed thus realizing a more stable and meaningful determination of integrated intensities and analysis of anisotropic peak broadening.

A comparison of the number of profile parameters required in (i) single line fitting and (ii) whole powder pattern fitting techniques (where profile parameters are constrained by smoothly varying functions with 2θ) such as the Pawley, Le Bail or Rietveld method, is shown in Table 2.4. As can be seen the use of the instrument function concept within convolution based profile fitting, preferentially with the FPA, drastically reduces the number of refineable profile parameters, resulting in

- improved refinement stability,
- increased refinement speed due to faster convergence, and
- reduced parameter correlations and therefore more physically meaningful parameters.

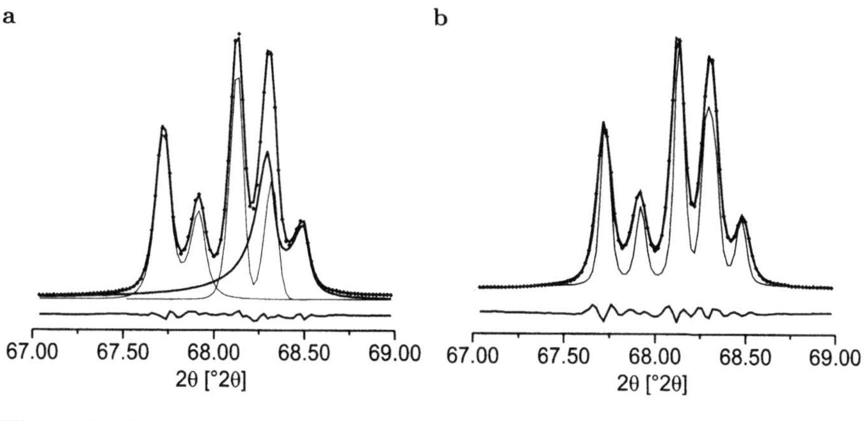

Fig. 2.30. Comparison of an unconstrained single line fit to the Quartz "5 finger peak" with SPVII functions (**a**) versus a FPA fit of the same data in (**b**)

Table 2.4. Minimum number of refineable profile parameters required in traditional Analytical Profile Fitting (APF) and Direct Convolution Approaches (DCA). $2\theta_k$: line position; I: line intensity; the split-Pearson VII function is a composite function consisting of two Pearson VII functions and uses two different full widths at half maximum $fwhm1$, $fwhm2$ and PearsonVII exponents $m1$, $m2$; U, V, W: parameters describing the angle dependent full widths at half maximum [43]; na, nb, nc: parameters describing the angle dependent peak shape (e.g. [44]); A: asymmetry parameter; L, G: Lorentzian and Gaussian type crystallite size contributions

Single Line Fitting		Rietveld, LeBail or Pawley method	
APF (SPVII)	DCA	APF (PVII)	DCA
$2\theta_k$	$2\theta_k$	U, V, W	L, G
I	I	na, nb, nc	
$fwhm1$, $fwhm2$	L, G	A	
$m1$, $m2$			
6	4	7	2

2.5 Conclusions

The benefits that follow from synthesizing and fitting X-ray line profiles by means of convolution are many and varied and are clearly evident from the accurate description of observed line profile shapes across the whole 2θ range for a wide range of instrument and sample conditions. Some of the most important advantages of convolution based profile fitting are:

- Peak shape and angle dependence can be described using minimal profile parameters in laboratory and synchrotron X-ray data as well as CW and TOF neutron data. This is possible because numerical convolution and numerical differentiation is used within the refinement procedure so that a wide range of functions can be easily incorporated into the convolution equation.
- The number of parameters required to describe a pattern is typically smaller than the "analytical function approach" and thus parameter correlations are significantly reduced. In addition, increasing profile complexity does not necessarily lead to an increase in the number of profile parameters.
- The quality of fit by convolution is normally better than using other methods. The uncertainty in derived parameters is therefore reduced.

Using physically based diffractometer models for the instrument aberration functions rather than functions empirically selected directly leads to the Fundamental Parameters Approach. FPA has already been implemented for most laboratory based X-ray powder diffractometer configurations including conventional divergent beam instruments, parallel beam instruments and diffractometers used for asymmetric diffraction. It can also accommodate various optical elements (e.g. multilayers and monochromators) and detector systems (e.g. point and position sensitive detectors) and has already been applied to neutron powder diffraction systems (e.g. Australian Nuclear Science and Technology Organisation, ANSTO) as well as synchrotron based diffractometers (e.g. BM16 at the European Synchrotron Radiation Facility, ESRF, and Station 2.3 at the Synchrotron Radiation Source, SRS, Daresbury).

- The FPA is a universal approach to profile fitting in which parameters fitted are physically identifiable and measurable. The validity of the fitted terms is therefore self-evident.
- The excellent profile shape description with FPA significantly improves the accuracy of derived lattice parameters. As a result, refined zero point and sample height errors no longer serve as garbage can for misfits.
- A small number of profile parameters are required to describe a pattern and therefore parameter correlation is typically less than with purely empirical convolution based profile fitting.
- Knowledge of the instrument contribution improves the discrimination between profile and background intensity in complex peak overlaps, unless sample broadening becomes dominant.
- It is possible to identify whether or not a diffractometer is operating at its optimum resolution and determine unequivocally whether or not an observed line profile posses specimen broadening. The FPA is the ideal tool for instrument verification particularly in regulated laboratories.
- Synthesised profiles rather than profiles from a reference material can be used as the basis for deconvolution of instrumental effects from observed profiles.

References

1. H.M. Rietveld: Acta Cryst. **22**, 151 (1967)
2. H.M. Rietveld: J. Appl. Cryst. **2**, 65 (1969)
3. S.A. Howard and K.D. Preston: *Reviews in Mineralogy* **20**, 217 (1989)
4. R.L. Snyder: *The Rietveld Method.* (Oxford University Press, 1993) pp. 111–131
5. A. Hepp and C. Bärlocher: Aust. J. Phys. **41**, 229 (1988)
6. A.J.C. Wilson: *Mathematical Theory of X-ray Powder Diffractometry.* (Gordon and Breach, New York 1963)
7. A.C. Larson and R.B. von Dreele: Los Alamos National Laboratory Report No. LA-UR-86-748 (1987)
8. A.X.S. Bruker: TOPAS V2.0: *General profile and structure analysis software for powder diffraction data.* (Bruker AXS GmbH, Karlsruhe 2000)
9. H.P. Klug and L.E. Alexander: *X-ray Diffraction Procedures.* (John Wiley and Sons, New York 1954)
10. D. Taupin: J. Appl. Cryst. **6**, 266 (1973)
11. W. Parrish, T.C. Huang, G.L. and Ayers: IBM Research Rpt. RJ1761 (1976)
12. W.I.F. David and J.D. Jorgensen: *The Rietveld Method.* (Oxford University Press, 1993) pp. 197–226
13. R.W. Cheary and A.A. Coelho: J. Appl. Cryst. **25**, 109 (1992)

14. P. Suortti: *The Rietveld Method.* (Oxford University Press, 1993) pp. 167–185
15. R.W. Cheary and A.A. Coelho: *Software: Xfit-Koalariet CCP14 Library.* (http://www.ccp14.ac.uk, 1996)
16. R.W. Cheary and A.A. Coelho: J. Appl. Cryst. **27**, 673 (1994)
17. R.W. Cheary and A.A. Coelho: J. Appl. Cryst. **31**, 851 (1998)
18. R.W. Cheary and A.A. Coelho: J. Appl. Cryst. **31**, 862 (1998)
19. R.W. Cheary, A.A. Coelho, and J.P. Cline: *Accuracy in Powder Diffraction III.* NIST, Gaithersburg, USA. April 22-25 2001, proceedings, (2002) in press
20. G.S. Pawley: J. Appl. Cryst. **14**, 357 (1981)
21. A. Le Bail, H. Duroy, and J.L. Fourquet: Mat. Res. Bull. **23**, 447 (1988)
22. A.A. Coelho: J. Appl. Cryst. **33**, 899 (2000)
23. F.W. Jones: Proc. R. Soc. Lond. Ser. A **166**, 16 (1938)
24. G. Hölzer, M. Fritsch, M. Deutsch, J. Härtwig, and E. Förster: Phys. Rev. A **56**, 4554 (1997)
25. H. Berger: X-ray Spectrometry **15**, 241 (1986)
26. H. Toraya and H. Hibino: J. Appl. Cryst. **33**, 1317 (2000)
27. D. Balzar: *Size-Strain Round Robin.* http://www.boulder.nist.gov/div853/balzar/s-s_rr.htm (2001)
28. D. Balzar and H. Ledbetter: J. Appl. Cryst. **26**, 97 (1993)
29. A. Kern and A.A. Coelho: *Powder Diffraction.* (Allied Publishers, New Delhi 1998) pp. 144–151
30. A.A. Coelho, R.W. Cheary, and K.L. Smith: J. Solid State Chem. **129**, 346 (1997)
31. R.W. Cheary, E. Dooryhee, P. Lynch, N. Armstrong, and S. Dligatch: J. Appl. Cryst. **33**, 1271 (2000)
32. R.W. Cheary, C.C. Tang, M.A. Roberts, and S.M. Clark: Mat. Science Forum **378–381** , 254–259 (2001)
33. M. Schuster and H. Göbel: J. Phys. D. Appl. Phys. **28**, A270 (1995)
34. J. Bergmann, R. Kleeberg, A. Haase, and B. Breidenstein: Mat. Sci. Forum **347–349** , 303–308 (2000)
35. R. Jenkins and R.L. Snyder: *Introduction to X-ray Powder Diffractometry.* (Wiley-Interscience, New York 1996) pp. 403
36. M.A. Roberts and C.C. Tang: J. Synchrotron Rad. **5**, 1270 (1998)
37. M.E. Bowden and M.J. Ryan: Powder Diffraction **6(2)**, 78 (1991)
38. H. Göbel: Adv. in X-ray Anal. **22**, 255 (1979)
39. J. Durbin and G.S. Watson: Biometrika **37**, 409 (1950)
40. H.D. Flack and M.G. Vincent: Acta Cryst. A **36**, 495 (1980)
41. R.J. Hill and I.C. Madsen: J. Appl. Cryst. **19**, 10 (1986)
42. J.P. Cline, (2001) Personal communication
43. G. Cagliotti, A. Paoletti and F.P. Ricci: Nucl. Inst. **3**, 223 (1958)
44. R.A. Young: *Introduction to the Rietveld method.* In: The Rietveld Method, ed. by R.A. Young. IUCr Book Series (Oxford University Press, Oxford 1993) pp. 1–39

3 Whole Powder Pattern Modelling: Theory and Applications

P. Scardi and M. Leoni

In close analogy to the structural refinement provided by the Rietveld method, Whole Powder Pattern Modelling (WPPM) is proposed as a general technique for microstructure refinement. WPPM consists in a simultaneous modelling of all peak profiles in the diffraction pattern from (single or multi-phase) polycrystalline materials, without using arbitrary profile functions. The entire diffraction pattern is modelled directly in terms of physical parameters describing the main microstructural features influencing peak profile width, shape and position, also considering instrumental effects and background.

Main parameters are lattice parameters and defect content, expressed in terms of dislocation density, effective outer cut-off radius, dislocation character (edge/screw), contrast factor, twin and deformation fault probabilities, together with mean and variance of a grain size distribution. Different models for lattice distortions and grain size and shape can be tested, to simulate or to model diffraction data from several systems of practical interest, including highly deformed metals and finely dispersed crystalline powders.

A further development is envisaged in the direction of the Rietveld method, whose basic algorithm can be easily incorporated in a WPPM approach. In fact, besides the per se utility as an analytical technique for materials science studies, the WPPM approach can improve the quality of structural refinements since it provides a better control of the peak profiles in terms of physically measurable quantities. Besides discussing the theoretical basis of WPPM, the present work illustrates the application to a typical case of study, a series of ball milled Ni powder samples. Results are successfully compared with information obtained from TEM pictures and literature data.

3.1 Introduction

Most current methods for powder diffraction data analysis include a pattern modelling stage, necessary because of the intrinsic nature of diffraction, which provides information in a strongly integrated form. The diffracted signal, in fact, is the result of a convolution of several (sample-related) physical and instrumental effects, further complicated by the overlapping between reflections and by the presence of a background, and in many cases by the use of multiple spectral components (e.g., $K\alpha_1/K\alpha_2$) in X-ray beams.

As a consequence, discriminating different contributions to the experimental pattern by simply delimiting an angular range totally attributed to a given reflection (single-peak analysis) can be a rather difficult and subjective procedure. When studying the effects related to width and shape of diffraction profiles, this is a serious limitation to the applicability of classical methods based on integral breadths (e.g. the Scherrer formula (SF) or the Williamson–Hall (WH) plot [1–4]) or numerical Fourier analysis (e.g., Warren–Averbach (WA) method [5, 6]). Analogous limitations are encountered in structural studies, where exact peaks positions and integrated intensity of single diffraction peaks are required.

All these reasons led in the last decades to the development and ever-increasing use of profile fitting techniques. Most methods and algorithms proposed so far follow one of the two basic, and opposite in some respect, strategies: (a) pattern decomposition (PD) (separation in single diffraction profiles) and (b) pattern modelling (PM) (modelling of the entire observable pattern, or of a large part of it).

PD is carried out by fitting analytical functions to the various identified reflections and representing the background by a suitable polynomial, without conditions on fitting parameters resulting from structural or microstructural constraints (e.g., integrated peak intensity is not related to structure factors and peak width and shape parameters do not directly depend on domain size and lattice defects) [7]. PD can be quickly performed by most software packages of academic or commercial production, and it provides direct information on peak position, intensity, width and shape to be used in subsequent data analysis [8]. As a drawback, PD can be rather arbitrary in the definition of the actual number of reflections (certainly an issue in low symmetry and multiple-phase specimens) and in the definition of the starting parameters to be used in the fitting routine (typically based on a non-linear least-squares numerical algorithm). Uniqueness of solution can be difficult to assess and error estimates tend to be unreliable, given the preponderance of systematic errors and unknown parameter correlation. As a consequence, despite the unquestioned flexibility and frequent application, PD can easily fall in the category of ill-conditioned problems.

A possible variant to the "blind" PD procedure described so far involves the use of a structural constraint to peak positions. According to this method (sometimes referred to as Pawley method [9, 10]), lattice parameters are refined instead of peak positions, leading to a generally more robust analysis and better convergence [8]. To be applicable, though, this method requires all observed reflections be known and indexed.

This is in the direction of PM, which aims at a simultaneous modelling of all reflections in a powder pattern, based on the parameters describing the physical properties of the specimen. So far, the most renowned and absolutely successful example of this philosophy is the Rietveld method [11, 12]. Structural parameters describing the crystalline structure are directly refined by adapting model profiles to the experimental pattern by means of a suitable minimization algorithm [12]. Like in PD, diffraction profiles are described by analytical functions, but with the important difference of being strongly constrained in position and intensity by a structural model. Peak width and shape are also constrained by suitable parametric expressions accounting for their change with the diffraction angle (e.g., by the well-known Caglioti et al. formula [13]). The strong point of the Rietveld method, however, is the direct comparison of structural parameters and experimental data, without intermediate steps (e.g. preliminary extraction of integrated intensity). This provides a direct assessment of the quality of the structural model and of the refinement, and also allows a correct use of counting statistics for estimating standard deviations.[1]

PD and PM share the basic hypothesis, which is also a critical limit, that experimental profiles can be represented by a-priori chosen analytical functions. Profile parameters are adjusted by a minimisation routine, but the functional form, whatever flexible, is fixed and usually selected independently of the physical properties of the sample under study. This is a strong and arbitrary constraint that can bias the results in a somewhat unpredictable way.

Pattern decomposition and Rietveld refinement algorithms frequently assume profiles to be Voigtian, which essentially means that the shape is intermediate between a Lorentzian and a Gaussian (several bell-shaped curves of common use, like Voigt, pseudo-Voigt and Pearson VII functions belong to this category) [12, 14]. Actually, there is no physical limit to the shape of diffraction profiles, which is largely determined by the physical properties of the sample and by the instrumental effects of the measurement geometry. Profiles need not be Voigtian: this is the case, for instance, of line profiles produced by nanocrystalline powders dispersed according to a lognormal size distribution [15, 16], but it is also true of profiles related to the presence of dislocations [17].

These discrepancies, when overlooked, can lead to considerable errors in line profile analysis, and negative effects can also be envisaged on structural refinements, since systematic (model) errors are introduced [16].

Whole Powder Pattern Fitting (WPPF) [18, 19] and Whole Powder Pattern Modelling (WPPM) [17, 19] are possible solutions to overcome some limitations of conventional PD and PM methods. Following the Rietveld method philosophy, WPPF and WPPM provide information on specimen microstructure by directly comparing model-related peak profiles with the experimental pattern. Main features of interest are size and shape of diffraction domains and nature and concentration of lattice defects. The difference between WPPF and WPPM is that in the former experimental profiles

[1] It should be recognized, however, that estimated standard deviations (e.s.d.)s tend to be underestimated owing to the parameter correlation and to the generally unknown systematic (model) errors.

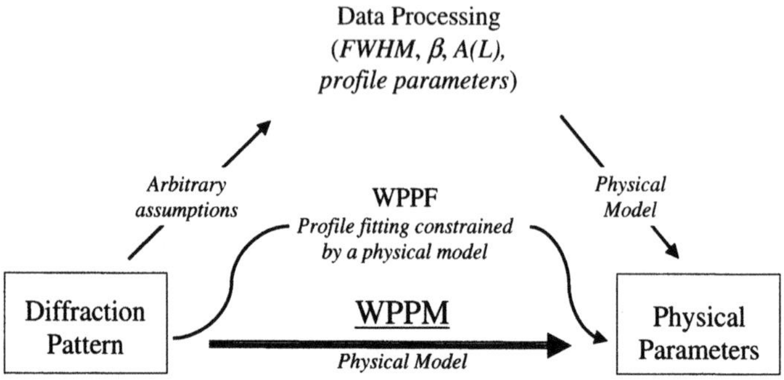

Fig. 3.1. Assumption levels for traditional methods versus WPPF and WPPM

are modelled by analytical functions, even if profile parameters are directly linked to microstructure parameters which are the actual fitting parameters; WPPM, instead, makes no use of arbitrary analytical profile functions, and directly models the experimental data by means of a suitable microstructural model.

While the Rietveld method mostly depends on the relation between structural parameters and peak intensity and position, WPPF and WPPM directly compare microstructure with peak profile width, shape and position. As a consequence, WPPF/WPPM can be regarded as *microstructural refinement* methods in the same way as the Rietveld is a structural refinement method.

Trying to summarise concepts relative to line profile analysis, traditional methods and WPPF/WPPM can be classified according to the assumptions level diagram shown in Fig. 3.1. To bridge experimental data (the diffraction pattern) and model parameters, traditional methods require some substantially arbitrary procedure to extract profile parameters (typically in the form of Full Width at Half maximum (FWHM), integral breadths (β) or Fourier coefficients ($\mathbb{C}(L)$)) from the data; these parameters are then used for the analysis of lattice defects and domain size and shape in a separate procedure. Therefore, results of well-known methods like the SF, WH plot or WA Fourier analysis, are not obtained by a direct analysis of the experimental data.

WPPF relies on profile fitting with analytical functions of all profiles encompassing the studied angular range (usually a rather wide interval); the important difference with respect to classical PD or PM is that profile parameters are related to (and constrained by) physically meaningful parameters like mean and variance of a distribution of crystalline domain size (also accounting for grain shape), concentration and character (edge/screw) of dislocations, deformation and twin faults, lattice parameters. Constraints provided by the microstructure reduce the arbitrariness of using analytical profile function. Finally, WPPM is straightforward as no a-priori fixed analytical profile function is used, and an analytical model of the microstructure is directly compared with (and refined against) the experimental data.

The present work reviews the theoretical basis of WPPF and WPPM techniques together with practical examples of applications and simulations of diffraction patterns by nanocrystalline and highly defected metal powders. To assess the validity of WPPM, results are compared with Transmission Electron Microscopy (TEM) observations.

3.2 Theoretical Basis of WPPF and WPPM

3.2.1 Basic Expressions for the Diffracted Intensity from a Defected Material

The diffraction profile from a polycrystalline material is the result of a convolution of sample-related (physical) effects and instrumental profile [4]. For this reason, diffraction line profiles can be conveniently written in terms of Fourier transform (FT) as:

$$I(d^*) = k(d^*) \int \mathbb{C}(L)^{2\pi i L \cdot d^*_{hkl}} dL\,, \tag{3.1}$$

where d^* is the diffraction vector in reciprocal space (d^*_{hkl} in Bragg condition) and $k(d^*)$ includes geometrical and structural terms which are constant or known functions of d^* (e.g., Lorentz-polarization (LP), $|F|^2$, etc.)[2]. Microstructure-related effects, with the exception of absorption, are included in $\mathbb{C}(L)$.

The advantage of using the FT formalism is in the properties of convolution integrals: $\mathbb{C}(L)$ can be obtained by simply multiplying contributions from the various physical effects (mainly: domain size, lattice distortion, faulting) and from the instrumental component. Since all necessary terms can be described by known relations, the general expression of the FT can be written in analytical form for several cases of practical interest [16–19].

Even if this approach is valid independent of crystal symmetry, so far the WPPF/WPPM algorithm was worked out for *fcc* materials[3]. Studied materials are assumed to be free of preferred orientations and of applied, thermal or residual macrostrain, in the hypothesis (largely verified for the considered class of finely dispersed and/or defected materials) that the coherent diffraction domains (crystallites) are sufficiently small and numerous to produce appropriate counting statistics [4].

Equation (3.1) is the basis of the WPPF/WPPM algorithm. However, in the specific application to defected *fcc* materials it is necessary to use a more

[2] Strictly speaking $k(d^*)$ partly relates to the microstructure, as it includes absorption and may also account for other effects, like extinction, which depend on sample density.

[3] An extension to other crystal structures does not involve major conceptual problems, and is the object of current developments.

general expression for the diffraction line profile, considering the diffraction signal from a set of crystallographic planes $\{hkl\}$ in terms of sub-components relative to each hkl combination. In fact, in the most general case, the diffraction profile of each sub-component can have a different width and shape and can be centred in a (slightly) different position of the reciprocal lattice.

For this reason we distinguish between the diffraction vector in Bragg condition for the $\{hkl\}$ family of planes in absence of lattice defects and strain, $d^*_{\{hkl\}} = 2\sin\theta_{\{hkl\}}/\lambda$,(where θ is the diffraction angle and λ the wavelength), and the diffraction vector for a hkl combination, d^*_{hkl}, related to a specific profile sub-component. According to these definitions, the line profile from a set of $\{hkl\}$ planes, $I_{\{hkl\}}$, is given by the sum of contributions from the hkl profile sub-components, I_{hkl}:

$$I_{\{hkl\}}(d^*, d^*_{\{hkl\}}) = k(d^*) \cdot \sum_{hkl} \omega_{hkl} I_{hkl}(s_{hkl}) \,, \tag{3.2}$$

where ω_{hkl} is a weight function, $s_{hkl} = d^* - (d^*_{\{hkl\}} + \delta_{hkl})$ is the distance, in reciprocal space, from the peak centroid for the hkl component and δ_{hkl} is the shift from the reciprocal-space point corresponding to the Bragg condition in absence of defects. Both ω_{hkl} and δ_{hkl} depend on the specific defects present in the material: explicit expressions are given in the following for the effect of faulting. The sum in (3.2) is over the independent profile sub-components (up to 4 for a *fcc*, depending on the specific combination of Miller indices).

According to (3.1), the line profile for a sub-component is given by:

$$I_{hkl}(s_{hkl}) = \int_{-\infty}^{\infty} \mathbb{C}_{hkl}(L) e^{2\pi i L \cdot s_{hkl}} dL \,, \tag{3.3}$$

where the Fourier length L is the conjugate variable of s_{hkl}.

The key-point to model experimentally observed profiles by means of (3.2) and (3.3) is introducing suitable expressions for the complex function $\mathbb{C}_{hkl}(L)$. Here we consider a rather general model with four profile components related respectively to instrumental effects, domain size and shape, lattice distortions and (twin and deformation) faulting. Additional or different terms, however, can be considered [17,20]. The FT can then be written as:

$$\mathbb{C}_{hkl}(L) = T^{IP}(L)^I A^S_{hkl}(L) < e^{2\pi i \cdot \psi_{hkl}(L)} >< e^{2\pi i \cdot \phi_{hkl}(L)} > \,, \tag{3.4}$$

where $T^{IP}(L)$ and $A^S_{hkl}(L)$ are the FT of the instrumental profile (IP) and domain size components, respectively, and the two terms in brackets ($<>$) are average phase factors related to lattice distortions (ψ) and faulting (ϕ). Equation (3.4) implicitly assumes that no interactions between components are present (separate averages for planar defects and lattice distortions). While the phase factors are generally complex quantities, the first two components are real by definition [21], unless an asymmetric IP is considered.

3.2.2 Line Profile Components

Crystallite Size and Shape

Line broadening due to domain size was the first microstructural effect studied by powder diffraction. In fact, not only the Scherrer formula [1] dates back to 1918 (just six years after the discovery of X-ray diffraction), but the entire subject was already brought to a considerable level of understanding before 1950, mostly owing to the work of Wilson and co-workers [21–23] and of Bertaut [24]. Despite this early understanding, dealing with size effects in line profile analysis is not a trivial issue. In fact, no physical principle can be invoked a priori to assess grain morphology, simply because crystallite size is not a tensor property [25, 26]. In other words, grain shape can be any, in principle, irrespective of crystal symmetry. In addition, grain size in real polycrystalline systems is hardly uniform: most likely a size distribution exists (if not a size and a shape distribution).

Bertaut [24] proposed an elegant interpretation to consider size broadening without making assumptions on grain shape and size distribution. This can be done by introducing the concept of column length, ideally considering grains in a polycrystalline system as made of columns of cells along the scattering direction. Line broadening due to particle size can then be treated in terms of column length distribution, from which mean column length can be related to the profile integral breadth. Even if this approach is correct, some external hypothesis is always required if one wishes to express the results in terms of size and shape of physically observable grains.

All of this is to support the need for models of grain shape and size distribution to interpret line broadening due to size effect. The approach shown in the following can be used for any grain shape and size distribution, but we confine our attention to a few convex shapes (sphere and simple polyhedra), and to a few commonly used distributions. In this way we can deal with a reasonable number of parameters to be refined by WPPF/WPPM, and we can profit of some interesting properties of polyhedra: details of this approach can be found in the cited literature [26].

We consider crystals with cubic symmetry and simple convex shapes like sphere, cube, octahedron and tetrahedron, for which the volume can be written in terms of a single length parameter, D (diameter for sphere and edge for the other solids). The intensity diffracted by a small crystal, in absence of other broadening sources, is given by

$$I_c(d^*, D) = k(d^*) \int_0^{D'(hkl)} A_c^S(L, D) e^{2\pi i L d^*_{hkl}} dL \,, \tag{3.5}$$

where $A_c^S(L, D)$ is the FT of the diffraction profile for a crystallite c and $D'(hkl)$ is the maximum possible Fourier length, that for the crystal shapes considered can conveniently be defined as [26]

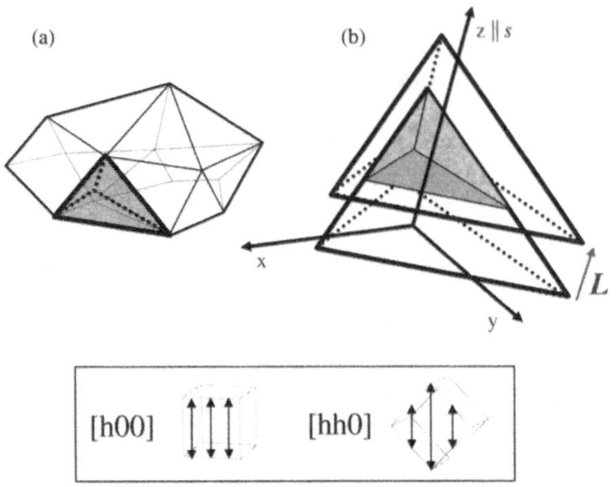

Fig. 3.2. Schematic representation of a polyhedral crystal and decomposition in tetrahedra (**a**). "Ghost" concept (shaded intersection volume) applied to a generic tetrahedral crystal displaced of L along the scattering direction $\boldsymbol{s}$ (**b**). In the insets: schematic picture of a cubic crystal and its projection along [h00] and [hh0], respectively

$$D'(hkl) = D/K_c(hkl)\,. \tag{3.6}$$

$K_c(hkl)$ depends on crystal shape and lattice direction (vide infra).

Expressions for $A_c^S(L, D)$ can be calculated for any crystal shape by means of the "ghost" concept proposed by Stokes and Wilson [22]. The FT, as shown in Fig. 3.2, is proportional to the intersection volume $V(L, D)$ between a crystal and its "ghost", an equivalent crystal translated a distance L along the scattering direction $\boldsymbol{s}$. The normalisation constant is the crystal volume, $V_c(D)$.

By simple geometrical reasoning it can be proved that the FT of a tetrahedron is a cubic polynomial of L; since polyhedra can always be subdivided in a finite number of tetrahedra, this important FT property extends to any polyhedral crystal [26]. For the crystal shapes considered here (sphere and single length-parameter polyhedra) the FT is a cubic function of L/D:

$$A_c^S(L, D) = \frac{V(L, D)}{V_c(D)} = \sum_{n=0}^{3} H_n^c (L/D)^n\,. \tag{3.7}$$

Coefficients H_n^c for the studied crystal shapes are reported in Table 3.1, together with corresponding $K_c(hkl)$ [26]. Following the ghost concept illustrated in Fig. 3.2, it can be easily understood why the integration limit in (3.5) is finite and how to calculate $K_c(hkl)$ for the various shapes and crystallographic directions (see the inset in Fig. 3.2).

The polynomial coefficients also enter known expressions relating grain size to Bertaut's mean column lengths (area-weighted $\langle L\rangle_s$ and volume

Table 3.1. H_n and $K_c(hkl)$ coefficients for different crystal shapes: sphere, cube, tetrahedron and octahedron ($A = h/\sqrt{h^2+k^2+l^2}$, $B = k/\sqrt{h^2+k^2+l^2}$, $C = l/\sqrt{h^2+k^2+l^2}$) [26]

	Sphere	Cube	Tetrahedron		Octahedron	
			$A \geq B+C^{(*)}$	$A \leq B+C$	$A \geq B+C$	$A \leq B+C$
H_0	1	1	1	1	1	1
H_1	$-\frac{3}{2}$	$-(A+B+C)$	$-3\sqrt{2}A$	$-\frac{3(A+B+C)}{\sqrt{2}}$	$-\frac{3A}{\sqrt{2}}$	$-\frac{2A+B+C}{\sqrt{2}}$
H_2	0	$AB+BC+CA$	$6A^2$	$\frac{3(A+B+C)^2}{2}$	$\frac{3(A^2-B^2-C^2)}{2}$	$\frac{A^2-3(B-C)^2+2A(B+C)}{4}$
H_3	$\frac{1}{2}$	$-ABC$	$-2\sqrt{2}A^3$	$-\frac{(A+B+C)^3}{2\sqrt{2}}$	$\frac{-A^3+3A(B^2+C^2)+2(B^3+C^3)}{2\sqrt{2}}$	$\frac{(A+B+C)(A^2+3(B-C)^2)}{4\sqrt{2}}$
$K_c(hkl)$	1	A	$\sqrt{2}A$	$\frac{A+B+C}{\sqrt{2}}$	$\frac{A+B+C}{\sqrt{2}}$	$\frac{A+B+C}{\sqrt{2}}$

$^{(*)}$ Conditions: $h \geq k+l$, $h \leq k+l$ are equivalent, respectively, to $A \geq B+C$, $A \leq B+C$.

weighted $\langle L\rangle_v$):

$$\langle L\rangle_s = -\left[\left.\frac{dA_c^S(L,D)}{dL}\right|_{L=0}\right]^{-1} = \frac{D}{K_k} = -\frac{D}{H_1} \tag{3.8}$$

$$\begin{aligned}\langle L\rangle_v = [\beta(s)]^{-1} = 2\int_0^{D/K} A_c^S(L,D)dL = \frac{D}{K_\beta} \\ = \frac{D\{3H_3 + 2K_c[2H_2 + 3K_c(H_1 + 2K_cH_0)]\}}{6K_c^4},\end{aligned} \tag{3.9}$$

where $\beta(s)$ is the integral breadth and K_k, K_β are the integral breadth and initial-slope Scherrer constants, respectively [26,27].

To account for the grain size dispersion, suitable size distributions need be introduced. In principle, any distribution function can be equivalently used, but to limit the number of refinement parameters, we consider some common and experimentally observed two-parameter distributions: Lognormal (g_l), Gamma (g_p) and an alternative distribution recently proposed by York [28] for normal growth phenomena (g_y). Explicit expressions for the three distributions, with the corresponding expression for the moments of nth order, are[4]

$$g_l(D) = \frac{1}{D\sigma_l\sqrt{2\pi}} \mathrm{e}^{-\frac{(\ln D-\mu_l)^2}{2\sigma_l^2}}, \quad M_{l,n} = \mathrm{e}^{n\mu_l + \frac{n^2}{2}\sigma_l^2} \tag{3.10}$$

$$g_p(D) = \frac{\left(\frac{\sigma_p D}{\mu_p}\right)^{\sigma_p - 1} \sigma_p \mathrm{e}^{-\frac{\sigma_p D}{\mu_p}}}{\mu_p \cdot \Gamma(\sigma_p)}, \quad M_{p,n} = \left(\frac{\mu_p}{\sigma_p}\right)^n \frac{\Gamma(n+\sigma_p)}{\Gamma(\sigma_p)} \tag{3.11}$$

$$g_y(D) = \frac{\left(\frac{\sigma_y D}{\mu_y}\right)^{\sigma_y} \mathrm{e}^{-\frac{\sigma_y D}{\mu_y}}}{\mu_y \cdot \Gamma(\sigma_y)}, \quad M_{y,n} = \left(\frac{\mu_y}{\sigma_y}\right)^n \frac{\Gamma(1+n+\sigma_y)}{\Gamma(1+\sigma_y)}. \tag{3.12}$$

If the sample is a polydisperse system of grains with shape c (e.g., sphere, cube, etc.) described by a distribution of length parameter $g_i(D)$, the scattered intensity is

$$I(d^*) = k(d^*)\frac{\int_0^\infty I_c(d^*,D)g_i(D)V_c(D)dD}{\int_0^\infty g_i(D)V_c(D)dD}. \tag{3.13}$$

By using (3.5) in (3.13), simplifying constants and swapping the integration order we can write

[4] The *incomplete gamma function* is defined as $\Gamma(x,a) = \int_a^\infty y^{x-1}\mathrm{e}^{-y}dy$, whereas $\Gamma(x) = \Gamma(x,0)$.

$$I(d^*) = k(d^*)[M_{i,3}]^{-1} \int\limits_{L=0}^{\infty} \int\limits_{D=LK}^{\infty} A_c^S(L,D) g_i(D) D^3 dD e^{2\pi i L d^*_{hkl}} dL$$

$$= k(d^*) \int\limits_0^{\infty} A^S(L) e^{2\pi i L d^*_{hkl}} dL \,. \tag{3.14}$$

In conclusion we obtain an expression equivalent to (3.5), where $A^S(L)$, the FT for the system of crystals with shape c and size distributed according to $g_i(D)$, is given by

$$A^S(L) = [M_3]^{-1} \int\limits_{LK}^{\infty} A_c^S(L,D) g_i(D) D^3 dD \,. \tag{3.15}$$

This general expression can be used for the different combinations of crystal shapes and distributions. Solving the integral using (3.7) and introducing (3.10), (3.11) and (3.12) we obtain the FT for Lognormal, Gamma and York distributions, respectively:

$$A_l^S(L) = \sum_{n=0}^{3} Erfc\left[\frac{\ln(L \cdot K_c) - \mu_l - (3-n)\sigma_l^2}{\sigma_l \sqrt{2}}\right] \frac{M_{l,3-n}}{2M_{l,3}} \cdot H_n^c L^n \tag{3.16}$$

$$A_p^S(L) = \sum_{n=0}^{3} \left(\frac{\sigma_p}{\mu_p}\right)^n \frac{\Gamma(\sigma_p + (3-n), (K_c L \sigma_p/\mu_p))}{\Gamma(\sigma_p + 3)} \cdot H_n^c L^n \tag{3.17}$$

$$A_y^S(L) = \sum_{n=0}^{3} \left(\frac{\sigma_y}{\mu_y}\right)^n \frac{\Gamma(\sigma_y + (4-n), (K_c L \sigma_y/\mu_y))}{\Gamma(\sigma_y + 4)} \cdot H_n^c L^n \,. \tag{3.18}$$

These expressions find a practical application in the WPPF/WPPM scheme of (3.4), and the only free refinement parameters are μ_i and σ_i for the selected size distribution.

It can be useful to relate the parameters of a polydisperse system described by a distribution $g_i(D)$ to mean column lengths. To this purpose it is not necessary to calculate derivatives and integrals in (3.8) and (3.9), since it can be proved that the presence of a distribution of sizes does not alter the relation between crystal shape and diffraction line profile [27], which is given in terms of Scherrer constants and distribution moments as [26]

$$\langle L \rangle_S = \frac{1}{K_k} \frac{M_{i,e}}{M_{i,2}} \tag{3.19}$$

$$\langle L \rangle_S = \frac{1}{K_\beta} \frac{M_{i,4}}{M_{i,3}} \,. \tag{3.20}$$

From the FT expressions in (3.16), (3.17) and (3.18) it is also possible to calculate an analytical expression for the corresponding column length distributions, simply obtained by deriving the FT two times with respect to L. After normalisation:

$$P_l(L) = \sum_{n=0}^{3} \frac{H_n^c}{H_1^c} \cdot \left\{ n(1-n)Erfc\left[\frac{\ln(L\cdot K_c) - \mu_l - (3-n)\sigma_l^2}{\sigma_l\sqrt{2}}\right] \right.$$
$$+\frac{1}{\sigma_l^3}\sqrt{\frac{2}{\pi}}\left[-\ln(L\cdot K_c) + \mu_l + (2+n)\sigma_l^2\right] \tag{3.21}$$
$$\left. \times\mathrm{Exp}\left[-\left(\frac{\ln(L\cdot K_c) - \mu_l - (3-n)\sigma_l^2}{\sigma_l\sqrt{2}}\right)^2\right]\right\} \frac{M_{l,3-n}}{2M_{l,2}} \cdot L^{n-2}$$

$$P_p(L) = \sum_{n=0}^{3} \frac{H_n^c}{H_1^c} \cdot \left\{ n(1-n)\cdot\Gamma\left(\sigma_p + (3-n), \frac{K_c L \sigma_p}{\mu_p}\right) \right.$$
$$+\left[\mathrm{Exp}\left(-\frac{K_c L \sigma_p}{\mu_p}\right)\left(\frac{K_c L \sigma_p}{\mu_p}\right)^{\sigma_p+(3-n)} \right. \tag{3.22}$$
$$\left.\left. \times\left(2+\sigma_p+n-\frac{K_c L \sigma_p}{\mu_p}\right)\right]\right\} \frac{(2+\sigma_p)}{\Gamma(\sigma_p+3)}\left(\frac{\sigma_p}{\mu_p}\right)^{n-1} \cdot L^{n-2}$$

$$P_y(L) = \sum_{n=0}^{3} \frac{H_n^c}{H_1^c} \cdot \left\{ n(1-n)\cdot\Gamma\left(\sigma_y + (4-n), \frac{K_c L \sigma_y}{\mu_y}\right) \right.$$
$$+\left[\mathrm{Exp}\left(-\frac{K_c L \sigma_y}{\mu_y}\right)\left(\frac{K_c L \sigma_y}{\mu_y}\right)^{\sigma_y+(4-n)} \right. \tag{3.23}$$
$$\left.\left. \times\left(3+\sigma_y+n-\frac{K_c L \sigma_y}{\mu_y}\right)\right]\right\} \frac{(2+\sigma_y)}{\Gamma(\sigma_y+4)}\left(\frac{\sigma_y}{\mu_y}\right)^{n-1} \cdot L^{n-2}.$$

Even if these formulae can be used for comparing results of WPPM/WPPF with other ones obtained by traditional methods (e.g., WA), they are not strictly necessary for data analysis. In the following, we directly refer to the grain size distribution which has a more direct physical meaning and can be compared with TEM observations in a simple and intuitive way.

Lattice Distortions

The average phase term due to lattice distortions can be written in a quite general form as the sum of real and imaginary terms ($A_{\{hkl\}}^D(L)$ and $B_{\{hkl\}}^D(L)$):

$$\begin{aligned}\left\langle e^{2\pi i\psi_{hkl}(L)}\right\rangle &\equiv \left\langle e^{2\pi i\psi(Ld^*_{\{hkl\}},\varepsilon_{\{hkl\}}(L))}\right\rangle \\ &= \left\langle \cos(2\pi Ld^*_{\{hkl\}}\varepsilon_{\{hkl\}}(L))\right\rangle + i\left\langle \sin(2\pi Ld^*_{\{hkl\}}\varepsilon_{\{hkl\}}(L))\right\rangle \\ &= A^D_{\{hkl\}}(L) + iB^D_{\{hkl\}}(L)\,, \end{aligned} \tag{3.24}$$

where $\varepsilon_{\{hkl\}}(L)$ is the projection along the scattering direction $\boldsymbol{s}$ of the relative difference between the distortions of two cells separated by a distance L (projection of the cell distance along $\boldsymbol{s}$). The way the average is calculated depends on the specific model for the distortion source, and usually involves the calculation of a mean-square strain $\langle\varepsilon^2_{hkl}(L)\rangle$.[5] The two FT terms refer to the family of planes because, according to our basic hypothesis of absence of texture and macrostrain, lattice distortions are assumed to have the same effect on each profile sub-component in a given family:

$$\varepsilon_{hkl}(L) \equiv \varepsilon_{\{hkl\}}(L)\,. \tag{3.25}$$

The literature reports few cases where this condition was not used: more than polycrystalline materials, of interest in the present work, examples concern single crystals or highly oriented thin films [29,30].

Several mean-square strain expressions have been proposed for different sources of lattice distortions, including dislocations, [31–33], inclusions in a crystalline matrix [20], or grain-surface relaxation in nanocrystalline powders [34,35]. Moreover, lattice parameter (a_0) variations from grain to grain, typically due to chemical composition fluctuations, can produce an apparent strain $\varepsilon = (a - a_0)/a_0$ with the peculiarity of being isotropic and space-independent (independent of L) within each crystalline domain.

A classical approach to strain broadening is that proposed by Warren and Averbach [5,6]. If one takes the first terms in the power expansion of (3.24), then for small L it holds

$$A^D_{\{hkl\}}(L) \cong 1 - 2\pi^2L^2d^{*2}_{\{hkl\}}\langle\varepsilon^2_{\{hkl\}}(L)\rangle \tag{3.26}$$

$$B^D_{\{hkl\}}(L) \cong -\frac{4}{3}\pi^3L^3d^{*3}_{\{hkl\}}\langle\varepsilon^3_{\{hkl\}}(L)\rangle\,. \tag{3.27}$$

Additional terms involve higher order moments ($\langle\varepsilon^n_{\{hkl\}}(L)\rangle$) of the strain distribution [6,20]. Since the imaginary term involves odd-order moments only, it is rigorously zero if the strain distribution $p(\varepsilon_{\{hkl\}}(L))$ is an even function of L. Otherwise, since $B^D_{\{hkl\}}(L)$ depends on $(Ld^*_{\{hkl\}})^3$, it is frequently neglected, thus restricting the analysis to determining the mean-square strain (second order moment) only.

The WA has been very successful and is still a reference method in line profile analysis. Beyond the limits in the power expansion approximation,

[5] The root-mean-square strain $\langle\varepsilon^2_{hkl}(L)\rangle^{1/2}$ or *microstrain* is also frequently considered.

however, the WA theory does not account for a rather basic physical property of crystalline systems: being related to the elastic behavour of the crystalline medium, lattice distortions can produce anisotropic effects, i.e., line broadening does not increase monotonically with d^*.

This is the case of dislocations, whose contribution to line broadening depends on the relative orientations of dislocation line, Burgers vector and diffraction vector, as well as on the elastic anisotropy of the studied material [32,33].

For the distortion Fourier transform of the peak profile we follow the approach proposed by Wilkens [32,33], which still provides, despite the limiting hypothesis of a restrictedly random distribution of screw dislocations, the best available expression, whose validity has been verified for different arrangements and different types of dislocations [32,33,36–39]. The real term of the peak profile FT can be written as

$$A^D_{\{hkl\}}(L) = \exp\left[-\frac{1}{2}\pi b^2 \bar{C}_{\{hkl\}} \rho d^{*2}_{\{hkl\}} L^2 f\left(\frac{L}{R'_e}\right)\right] , \tag{3.28}$$

where $\boldsymbol{b}$ is the Burgers vector, whose modulus is $b = a_0/\sqrt{2}$ for the rather common $\{111\} < 110 >$ slip system in *fcc* materials, $\bar{C}_{\{hkl\}}$ is the average contrast factor, ρ is the dislocation density and R'_e is a length related to the outer cut-off radius of the dislocations (R'_e is an *effective* outer cut-off radius). The function $f(L/R'_e)$ is given by Wilkens in terms of $f^*(\eta)$, with [33]:

$$\eta = \frac{1}{2} e^{-1/4} \frac{L}{R'_e} \cong 0.3894 \frac{L}{R'_e} . \tag{3.29}$$

For $f^*(\eta)$ we use an equivalent but simpler expression obtained by van Berkum as power expansion of the original Wilkens' formula [20]:

$$f\left(\frac{L}{R'_e}\right) = f^*(\eta) \cong \begin{cases} -\ln\eta + \frac{7}{4} - \ln 2 + \frac{\eta^2}{6} - \frac{32\eta^3}{225\pi} & \text{for } \eta \le 1 \\ \frac{256}{45\pi\eta} - \left(\frac{11}{24} + \frac{\ln 2\eta}{4}\right)\eta^{-2} & \text{for } \eta > 1 . \end{cases} \tag{3.30}$$

For small L values, the leading term in the exponential of (3.28) is of the type $-L^2 \ln(R'_e/L)$. It is worth noting that this functional dependence on the Fourier length is common among the models proposed by several authors. The difference between the various formulations is in the meaning of R'_e with respect to the assumed dislocation distribution and interaction mechanism. For instance, Krivoglaz et al. [36] introduced different definitions of R'_e : they proved it is proportional to the crystal size for a totally random distribution of straight dislocations, but need be replaced by the radius R_0 for dislocation loops. These authors came to the conclusion that the functional dependence $-L^2 \ln(R'_e/L)$ is a general result in models with dislocation screening and single-parameter correlation function; an effective cut-off radius can be introduced for more general models than the uniform dislocation distributions [36].

Groma et al. [38], arrived at a similar conclusion for the case of arbitrary one- and two-particle distribution functions: the modulus of the (complex) FT, also contains the same $-L^2 \ln(R_{\text{eff}}/L)$ dependence [20].

A relation between R_{eff} and R'_e can be obtained by using the approximation [33]:

$$f\left(\frac{L}{R'_e}\right) \cong \ln\left(\frac{R'_e}{L}\right) + 2\,. \tag{3.31}$$

In the hypothesis of a restrictedly random dislocation distribution, equivalence between Wilkens' and Groma et al. models implies that

$$-L^2 f\left(\frac{L}{R'_e}\right) \cong -L^2\left[\ln\left(\frac{R'_e}{L}\right) + 2\right] \equiv -L^2 \ln\left(\frac{R_{\text{eff}}}{L}\right), \tag{3.32}$$

from which one obtains $R_{\text{eff}} = e^2 R'_e \cong 7.389 R'_e$.

Despite the analogy in the functional dependence of the FT on the Fourier length and on the appropriate length parameter (e.g., R'_e, R_{eff}, R_0) according to the models of Krivoglaz et al., Wilkens and Groma et al., the exact meaning of the effective outer cut-off radius depends on the specific model. As a consequence, great care should be used in the physical interpretation of this parameter and derived ones (e.g., Wilkens' parameter $M' = R'_e\sqrt{\bar{\rho}}$), especially if treated as absolute quantities.

From the practical point of view of the WPPF/WPPM approach, an advantage of using (3.28) and (3.30) is that the FT is finite also for large L (whereas $\exp[-L^2 \ln(R_{\text{eff}}/L)]$ diverges), and can be easily computed analytically. We can also consider higher order terms in the exponential: the leading term is $(Ld^*_{\{hkl\}})^4 g(L/R) \cdot \rho^2 \bar{C}^2_{\{hkl\}}$, where the function $g(L/R)$ is given by Wilkens [33]. As an alternative, the model proposed by van Berkum or the one by Groma et al. can be adopted [20,38]. However, the weight of this term in the exponential is much lower than that of $L^2 f^*(L/R'_e)$ and it is significant only for high diffraction angles and around the peak top, where large d^* and large L, respectively, are important. Peak overlapping, typical of powder patterns, and presence of other significant contributions to line broadening (e.g., domain size and faulting) make it difficult to refine parameters of higher order terms, whose correlation with other physical parameters can be rather strong, leading to poor convergence.

The anisotropic nature of the distortion line broadening due to dislocations is taken into account by the contrast factor (also known as orientation factor [39,40]), C_{hkl} [32,33,41,42]. Also the outer cut-off radius includes a (hkl) dependence, but this is weaker than that of the contrast factor [39,40], so we consider R'_e as a modelling parameter independent of the scattering direction.

Under the assumption of (3.25), implying all dislocation slip systems be equally populated, anisotropy can be described by an average contrast factor,

$\bar{C}_{\{hkl\}}$, obtained by single contrast factors C_{hkl}, as a weighted average over the dislocation population and independent permutations of Miller indices.

For cubic materials a general relation exists between contrast factor and Miller indices, equivalent to the expression originally proposed by Stokes and Wilson in 1944 to account for the anisotropy in the "apparent strain" contribution to line broadening [23]. In particular, the anisotropy can be expressed by a linear dependence on the orientational parameter H[6]

$$\bar{C}_{\{hkl\}} = (A + B \cdot H) = A + B\frac{h^2k^2 + h^2l^2 + k^2l^2}{(h^2 + k^2 + l^2)^2} . \tag{3.33}$$

$A(= \bar{C}_{\{h00\}})$ and B can be calculated from the elastic constants (c_{ij} or s_{ij}) of the studied material, for each scattering direction and given dislocation system (dislocation line and Burgers vector) [39–42]. As a consequence, the only refinement parameters in the distortion term of (3.28) are ρ and R'_e.

Additionally, the dislocation character can be refined. As recently shown by Ungar et al. [43], A and B for edge and screw dislocations (respectively, A_E, B_E and A_S, B_S) can be expressed in a parametric form of the elastic constants. If the latter are known, we can write the average contrast factor as a linear combination of the average contrast factors of the two dislocation types:

$$\begin{aligned} \bar{C}_{\{hkl\}} &= \left[\varphi\bar{C}_{E,\{hkl\}} + (1-\varphi)\bar{C}_{S,\{hkl\}}\right] \\ &= \left[\varphi A_E + (1-\varphi)A_S\right] + \left[\varphi B_E + (1-\varphi)B_S\right] H , \end{aligned} \tag{3.34}$$

where φ, the edge dislocation fraction, is a refinement parameter to be adjusted for the specific sample under study. The effect of the average contrast factor on the anisotropic profile broadening can be appreciated in Fig. 3.3, where simulated patterns of several *fcc* materials are compared by fixing all line broadening parameters but the contrast factor. Different B are used for three representative materials, considering spherical grains with $\langle D\rangle = 25$ nm, $\rho = 10^{16}\,\mathrm{m}^{-2}$, $\varphi = 0.5$ for the three patterns. Same value of A (= 0.3) was used to highlight differences, while keeping same profile width and shape for the (400) reflections:

- Cu: $A_i > 1$, strong anisotropy ($A_i = 2c_{44}/(c_{11} - c_{12}) = 3.21$).
- Al: $A_i \approx 1$, low anisotropy ($A_i = 1.23$. For isotropic materials: $A_i = 1$).
- YSZ (Yttria Stabilised-Zirconia): $A_1 < 1$, moderate anisotropy ($A_i = 0.53$).

Even if this approach (refining B or φ) is quite flexible and led to successful results [17], one should always be aware of the simplification in the

[6] In a previous work [17] we used a different expression for the contrast factor: $\bar{C}_{\{hkl\}} = \bar{C}_{\{h00\}}(1 + q \cdot H)$, which is equivalent to (3.33) with $\bar{C}_{\{h00\}} = A$ and $q = B/A$. However this formulation, similar to that introduced by Ungar et al. (for instance, see [43]), requires $A \neq 0$ which is an unnecessary restriction (cases are reported where $A = 0$ [42]). For this reason (3.33) is preferred.

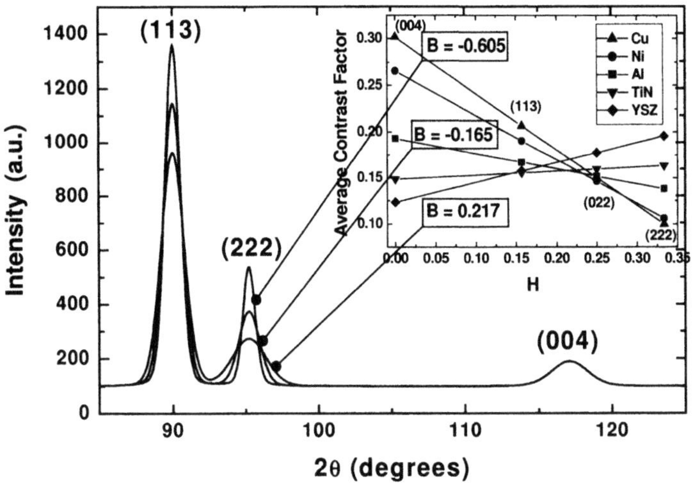

Fig. 3.3. WPPM simulated patterns for some representative cases of average contrast factors (Cu, Al and Yttria Stabilised-Zirconia (YSZ))

assumption that the systems contains only a fraction of edge and of screw dislocations, with no interaction terms [32,33] and without considering other, more complex configurations.

As a final remark, it is worth mentioning that expressions equivalent to (3.33) can be written for different cell symmetries. Besides cubic, the case of hexagonal materials has been treated in detail by Klimanek and Kuzel [39,40]. More generally, the contrast factor dependence on the Miller indices can be expressed by the Laue-group invariants whose expressions are available from the literature [44,45].

Twin and Deformation Faults

For the line broadening due to stacking faults we follow the approach proposed by Warren [6], suitably corrected and extended to higher faulting probabilities by Velterop et al. [46]. For the *fcc* case, the basic hypothesis is that faulting planes belong to the {111} family. However, since we consider powder samples free of texture, without loosing in generality we can assume that in each crystalline grain faulted planes are on a given {111} plane only, say (111).

Under these assumptions, the average phase term due to faulting in (3.4), $\langle e^{2\pi i\phi \cdot {hkl}(L)}\rangle$, can be written as a complex quantity:

$$\left\langle \mathrm{e}^{2\pi i\phi(L;d^*_{\{hkl\}},L_0/h_0^2)} \right\rangle = A^F_{hkl}(L)\left[1 + iB^F_{hkl}(L)\right] , \qquad (3.35)$$

where the dependence on the Miller indices has been explicitly indicated ($L_0 = h + k + l$, $h_0^2 = h^2 + k^2 + l^2$). $A^F_{hkl}(L)$ and $A^F_{hkl}(L) \cdot B^F_{hkl}(L)$ are, respectively, cosine and sine terms due to faulting.

In general, a peculiar (hkl) dependence of line broadening is observed when faulting is present. Deformation and twin faults, expressed by the respective probabilities (α and β), lead to an anisotropic line broadening, but also to a characteristic peak shift and asymmetry. Unlike the case of the distortion line broadening considered before, the sine term is not zero (unless twin faults are absent):

$$A^F_{hkl}(L) = (1 - 3\alpha + 2\beta + 3\alpha^2)^{\left|\frac{1}{2} L d^*_{\{hkl\}} \frac{L_0}{h_0^2} \sigma_{L_0}\right|}, \quad (3.36)$$

$$B^F_{hkl}(L) = -\sigma_{L_0} \frac{L}{|L|} \frac{L_0}{|L_0|} \frac{\beta}{(3 - 12\alpha - 6\beta + 12\alpha^2 - \beta^2)^{1/2}} . \quad (3.37)$$

The sign function, σ_{L_0}, is defined as

$$\sigma_{L_0} = \begin{cases} +1 \text{ for } L_0 = 3N + 1 \\ \;\;0 \text{ for } L_0 = 3N \\ -1 \text{ for } L_0 = 3N - 1 \end{cases} \qquad N = 0, \pm 1, \pm 2, \ldots . \quad (3.38)$$

Peak shift due to faulting, expressed in reciprocal space, is also a (hkl)-dependent quantity[7]:

$$\delta_{hkl} = \left\{ \frac{1}{2\pi} \arctan \left[\frac{(3 - 12\alpha - 6\beta + 12\alpha^2 - \beta^2)^{1/2}}{(1 - \beta)} \right] - \frac{1}{6} \right\} d^*_{\{hkl\}} \frac{L_0}{h_0^2} \sigma_{L_0} . \quad (3.39)$$

It is worth noting that refinement parameters related to faulting are the two faulting probabilities only. An important feature of faulting is that different effects are expected for the various profile sub-components, in terms of different broadening, peak shift and asymmetry. Independent profile sub-components in cubic systems can be one to four, depending on the specific combination of Miller indices, with different weight ω_{hkl} to be used in the summation of (3.2). In particular, one can distinguish between fault unaffected ($L_0 = 3N$) and affected ($L_0 = 3N \pm 1$) sub-components, with different weight ω_{hkl} to be used in the summation of (3.2), which in absence of texture can be directly related to the multiplicity of the reflection and to the rule defined by the sign function σ_{L_0} (a list is presented in Tables 1 and 2 of Velterop et al. [46]).

As a consequence, a rather complex shape can be observed for the diffraction peaks from faulted materials, which cannot be correctly reproduced by means of a simple $\{hkl\}$ profile fitting with arbitrary analytical functions. This is quite evident from the simulations shown in Fig. 3.4. As a further remark, it is expected that the present approach, based on different line profiles and positions for the various sub-components, leads to results different from those given by the conventional WA method and Warren's theory.

[7] In the limit of low probabilities, (3.39) is equal to the expression given by Warren ($\delta_{hkl} = (\sqrt{3}/4\pi)\alpha \cdot d^*_{\{hkl\}}(L_0/h_0^2)\sigma_{L_0}$) [6]. Likewise, Warren's expressions for the FT terms can be obtained from (3.36) and (3.37), by dropping quadratic probability terms.

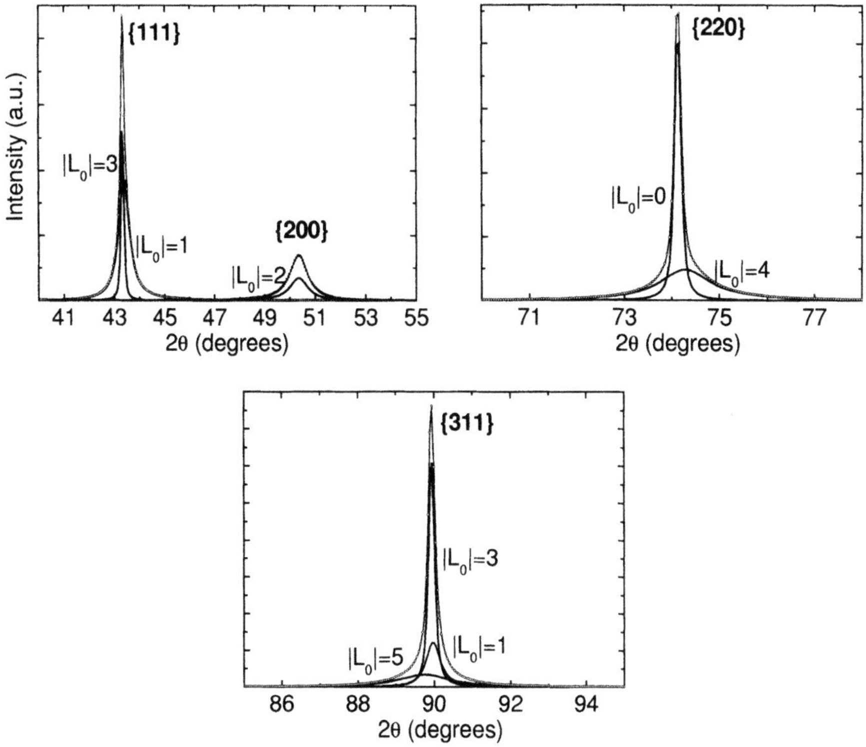

Fig. 3.4. WPPM simulated pattern for a generic polycrystalline Cu phase with $\langle D \rangle = 59\,\text{nm}$, $\rho = 10^{13}\text{m}^{-2}$ ($\varphi = 0.5$), $\alpha = 2\%$, $\beta = 5\%$. For each $\{hkl\}$ reflection, independent hkl profile sub-components are reported with the corresponding $|L_0|$ value

Instrumental Profile and Calibration Curve

It is generally assumed in line profile analysis that instrument contribution is confined to a line broadening effect, mainly due to spectral dispersion of the radiation, optical components and diffraction geometry (including sample transparency and other aberrations). As the instrumental and sample-related physical effects are convoluted [4], the instrumental profile is important when it is comparable with the line broadening from the sample: broad profiles, like those observed for nanocrystalline materials or heavily deformed materials, are usually not much affected by the instrument.

Actually this is not completely true for the WPPF/WPPM approach, which is based on a simultaneous refinement of all profile parameters, including lattice parameters. Peak positions are clearly related to lattice parameters but also to faulting, whose determination can be biased by experimental errors, for instance associated to the 2θ zero of the goniometer or to the position and transparency of the sample under study. For this reason an internal standard, like NIST SRM 640c (Silicon) [47], should be used. As an alternative, instrument resolution should be tested prior to measurements, by using a suitable standard, in order to build an external calibration curve.

A convenient choice as a powder standard is NIST SRM 660a (LaB_6) [48], an instrumental line broadening standard with certified lattice parameter. Figure 3.5a shows the external calibration curve (difference between observed and expected peak maximum position as a function of 2θ) which is modelled by a polynomial, subsequently embodied in the WPPF/WPPM algorithm. Experimental peak positions were obtained by profile fitting using symmetrical pseudo-Voigt (pV) functions, a choice justified by the negligible asymmetry of the profiles (see Sect. 3.4). It is worth noting that part of the aberrations leading to the curve of Fig. 3.5a is intrinsic in the sample and instrumental geometry features. Main contributions in the low 2θ region are given by sample transparency and axial divergence [4, 21], which in the present case are very low. However, even if our experimental set-up limits such aberrations, a residual effect on peak position is always present and need be considered.

The use of a standard is also useful to obtain a parametric description of the Instrumental Profile (IP). Since the relative procedure has been already described several times and can be considered rather standard, details can be found elsewhere [49, 50]. Figure 3.5b shows the trend, as a function of 2θ, of the Full Width at Half Maximum (FWHM $= 2\omega$) and of the Gaussian fraction (complement of the pV mixing parameter (η)). Both trends in Fig. 3.5c, β^2 and $(2\omega)^2$ as a function of 2θ, are properly modelled by a $U \tan^2\theta + W$ function (β^2: $U = 0.00231(3)$, $W = 0.0333(10)$, $N = 24$, $r^2 = 0.9970$; $(2\omega)^2$: $U = 0.00211(2)$, $W = 0.00169(6)$, $N = 24$, $r^2 = 0.9989$), an indication that, with the employed optics, wavelength dispersion is the dominant 2θ-dependent effect [51], responsible for the instrumental profile broadening toward high diffraction angles. Further line broadening terms include finite size of focal line and receiving slit, which are independent of 2θ. Errors associated to the sample, like those related to transparency, which is very low given the high absorption coefficient of LaB_6, are negligible, as well as axial divergence (no $\sin^2(2\theta)$ and no $\cot^2\theta$ dependence, respectively) [21, 51].

The parametric expressions shown in Fig. 3.5a–c are embodied in the WPPF/WPPM algorithm so that the FT component due to the IP can be calculated and used in (3.4) for any required peak position:

$$T_{pV}^{IP}(L) = (1-k)\cdot\exp\left(\frac{-\pi^2\sigma^2L^2}{\ln 2}\right) + k\exp(-2\pi\cdot\sigma L)\,. \qquad (3.40)$$

$\sigma = \omega \cdot \cos\theta/\lambda$ is the half width at half maximum in the reciprocal space, whereas k is related to the pV mixing parameter (η) by [50]

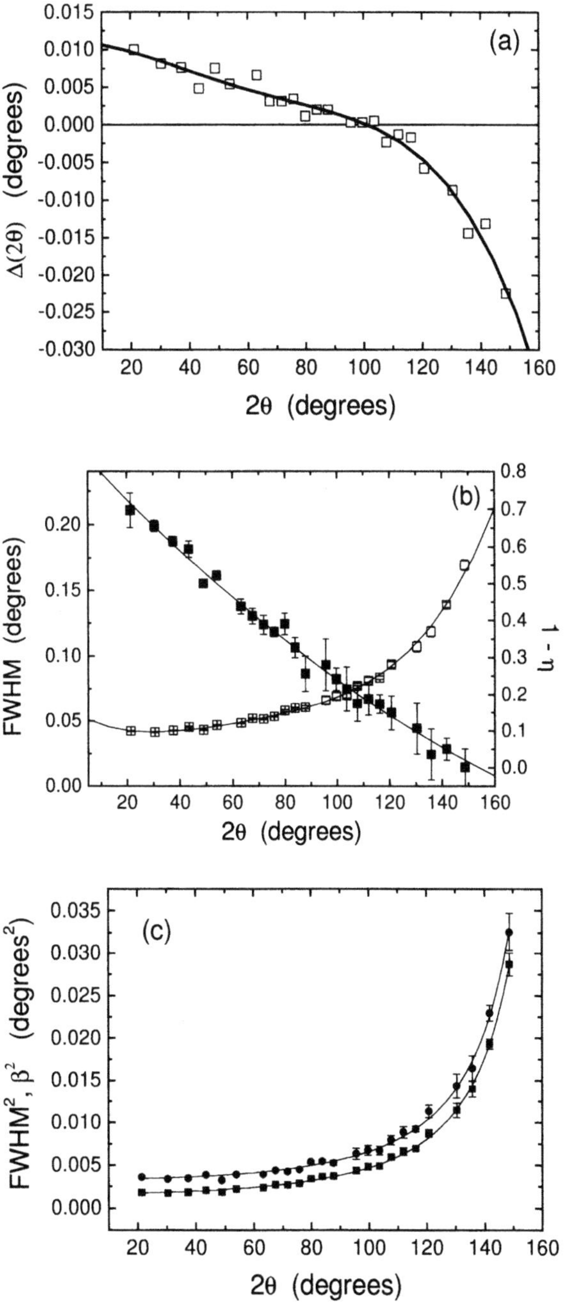

Fig. 3.5. Difference between certified and measured (as obtained by fitting pV profile functions) peak position for a SRM 660a (LaB_6) (**a**). FWHM (■) and $(1-\eta)$(□) (**b**) and $FWHM^2$ (■) and β^2 (●) as a function of 2θ (**c**)

$$k = \frac{1}{1+\frac{\phi_C}{\phi_G}\frac{(1-\eta)}{\eta}} = \frac{1}{1+\frac{1}{\sqrt{\pi \ln 2}}\frac{(1-\eta)}{\eta}}, \tag{3.41}$$

where the shape factors ($\phi = 2\omega/\beta$) for the Gaussian (G) and Lorentzian (C) components have been introduced, respectively [3, 7]:

$$\phi_G = 2\sqrt{\frac{\ln 2}{\pi}} = 0.93949\,, \quad \phi_C = \frac{2}{\pi} = 0.63662\,.$$

As an alternative to the experimental determination and parameterisation of the IP, a fundamental parameters approach [52,53] can be used to synthesise the IP from physical and geometrical parameters of the instrument/sample system. In this case the IP can be convoluted with the physical broadening profile component, or suitable expressions can be worked out and used in (3.4) in place of $T_{pV}^{IP}(L)$. Finally, asymmetric instrumental profiles can also be considered by using a complex $T_{pV}^{IP}(L)$, therefore introducing an additional sine term. If required, the extension to this case (not considered here in detail), is straightforward.

3.2.3 WPPM and WPPF

The basic expression of WPPF/WPPM (3.3) can be written in explicit form by considering real and imaginary terms of the FT related to the various sources of line broadening. If we consider (3.14), (3.24), (3.35) and (3.41), a general expression for the intensity of a profile sub-component hkl can be written as[8]

$$\begin{aligned} I_{hkl}(s_{hkl}) = & \int_{-\infty}^{\infty} T_{pV}^{IP} A^S A_{hkl}^F \\ & \times \left\{ \left[A_{\{hkl\}}^D - B_{\{hkl\}}^D B_{hkl}^F \right] \cos(2\pi L s_{hkl}) \right. \\ & \left. - \left[A_{\{hkl\}}^D B_{hkl}^F + B_{\{hkl\}}^D \right] \sin(2\pi L s_{hkl}) \right\} dL\,. \end{aligned} \tag{3.42}$$

In the assumption that $B_{\{hkl\}}^D \cong 0$ (implicitly adopted here for the dislocation distortion model), the expression above reduces to

$$I_{hkl}(s_{hkl}) = \int_{-\infty}^{\infty} T_{pV}^{IP} A^S A_{hkl}^F A_{\{hkl\}}^D \left[\cos(2\pi L s_{hkl}) - B_{hkl}^F \sin(2\pi L s_{hkl}) \right] dL\,. \tag{3.43}$$

Explicit expressions for the various terms are: (3.16), (3.17) and (3.18) for the size broadening of grains (of different shapes) distributed according to a

[8] A typing error is present in the original formulation reported in [17]

Lognormal, Gamma or York distribution, respectively; (3.28) for distortion line broadening due to dislocations; (3.36) and (3.37) for the contribution of twin and deformation faults; (3.40) for the instrumental contribution.

The use of additional or alternative line broadening models is possible by adding or substituting terms in (3.42) or (3.43). The WPPM/WPPF approach can then be used to test and compare different models of the real microstructure of materials.

WPPF and WPPM can be distinguished in the way (3.3) (or (3.43)) is used. WPPF involves the use of an analytical function (a Voigtian in the present formulation) to fit the experimental profiles. The fitting procedure, carried out on the entire pattern, is done by using a suitable minimisation condition on the FT, which makes the Voigtian profile a particularly convenient choice. In fact, the FT of a Voigt curve is a simple exponential function of L:

$$T_V(L) = e^{-\pi\beta_{VG}^2 L^2 - 2\beta_{VC} L}\,, \tag{3.44}$$

where β_{VG} and β_{VC} are, respectively, the integral breadths of the Gaussian and Lorentzian components of the Voigtian in the reciprocal space. The values of β_{VG} and β_{VC} for each observed profile can be refined by means of the condition [18]:

$$\sum_L w_L \left[T_{V,hkl}(L) - A_{hkl}(L)\right]^2 \stackrel{!}{=} \min\,. \tag{3.45}$$

The model FT ($A_{hkl}(L)$) is given by the product of the real terms due to size, distortion, faulting and IP:

$$A_{hkl}(L) = T_{pV}^{IP}(L) \cdot A_{\{hkl\}}^S(L) \cdot A_{\{hkl\}}^D(L) \cdot A_{hkl}^F(L)\,. \tag{3.46}$$

Equations (3.45) and (3.46) are used for each profile sub-component, thus allowing a proper evaluation of faulting effects. Profile asymmetry due to faulting is introduced in a simplified way, by using split-Voigt functions (profile functions with different left and right integral breadth components) with a condition related to faulting probabilities, obtained following an approach similar to that proposed by Warren [6]. The relative algorithm is described in [19].

The FT can be used as weight function in the minimisation: $\omega_L \equiv A(L)$. As the FT is a monotonically decreasing function of L, this choice gives more weight to low L values, which essentially means to overweight the information in the peak tails with respect to peak top (which is also consistent with the counting statistics).

Since the minimisation in WPPF can be performed analytically, calculation can be rather fast thus suited to deal with large data sets and a large number of reflections.

WPPM makes a direct use of (3.2) and (3.43) to model the experimental pattern, considering the required *hkl* profile sub-components. This is the most straightforward and unbiased solution, even if it needs to be treated numerically, including lengthy calculations of a convolution integral. Modern PCs however, can provide good results in reasonable computation time (figures are given below).

Even if WPPM provides a direct connection between experimental pattern and physical parameters (see Fig. 3.1), whereas WPPF retains a somewhat arbitrary element of using analytical profile functions, the results obtained by the two approaches are frequently in reasonable agreement [19]. The reason is in the strong constraint to profile fitting due to the physical model, through the condition used in the WPPF minimisation (3.45) and (3.46) [16, 18, 19].

In any case, in both WPPF and WPPM the entire pattern is refined on the basis of few physical parameters:

- μ_i and σ_i – parameters of the grain size distribution (with a choice between grain shapes and possible distribution functions)
- ρ, R_e' and φ – dislocation density, effective outer cut-off radius and fraction of edge dislocations (effective dislocation character). As an alternative, Wilkens parameter $M' = R_e'\sqrt{\rho}$ can also be considered as a refinement parameter if one of the two – ρ or R_e' – is fixed. If the elastic constants of the studied material are not known, B in (3.33) can be refined instead of φ.
- α and β – probabilities of twin and deformation faulting, respectively.
- a_0 – lattice parameter.

In addition, WPPF/WPPM allow the refinement of an analytical background (power series or Chebyshev polynomial of the first kind, up to a reasonable maximum of nine terms), relative intensity of spectral components of the X-ray radiation (up to five components), 2θ zero error of the goniometer, sample displacement from the goniometer axis and normalised peak intensities (refined values are integrated intensities). In addition, an external calibration curve can be adopted together with the parametric description of the IP (cf. Sect. 3.2.2), and a Lorentz-polarization factor can be applied to correct intensities.

Some practical considerations can be made concerning the use of (3.3) and (3.43). Since the integrand in (3.43) is an even function of L, then

$$\int_{-\infty}^{\infty} dL = 2\int_{0}^{\infty} dL\,.$$

The upper integration limit is formally infinity, but can actually be replaced by a sufficiently high value of L, which can be established from a condition on the grain size distribution. In particular, the Fourier length can never exceed the diameter of the largest grain, $D_{\max}$. This value can be defined by means of the following relations:

$$\int_{D_{\max}}^{\infty} g(D)dD \le \tau\,, \tag{3.47}$$

where τ is a suitably small threshold value (e.g., 10^{-3} or 10^{-4}).

As to the software, the core of the minimisation algorithm, both for WPPM and for WPPF, is a non-linear least squares (NLSQ) routine written in Fortran (Compaq Visual Fortran, Ver. 6.6), based on the MINPACK library [54] implementation of the Levenberg-Marquardt method. Convergence requires a few iterations (usually less than 10).

A Fast Fourier Transform algorithm was used in WPPM to reduce calculation time, down to values comparable to those of profile fitting. With a 2.0GHz Pentium-IV PC, a complete refinement for a typical *fcc* phase (7–8 reflections) takes less than 10 minutes (CPU time).

3.3 Experimental

XRD data were collected on a Rigaku PMG-VH powder diffractometer, using Cu radiation produced at 40 kV/45 mA and the following geometrical setup: 185 mm goniometer radius, 1/2° divergence slit, 2° incident beam and diffracted beam Soller slits, 0.15 mm receiving slit, 1/2° Anti-Scattering slit, graphite curved-crystal analyser (in the diffracted beam-path). This arrangement of optical components provided narrow and symmetrical instrumental profiles across the 2θ range of interest ($\approx 30 \div 150°$), so that the IP could be easily parameterised [49,50]. Each powder pattern was collected with 60 s counting time per 0.05° step.

Ball milled Ni samples were prepared in a Fritsch Pulverisette 7 planetary micromill, using an agate grinding bowl and 5 zirconia balls. Starting powder material was 99.9% pure Ni (Johnson–Matthey). Milling was conducted in static air with the same amount of powder (2.5 g) for different time (6 h, 12 h, 24 h, 48 h and 96 h), keeping a fixed powder-to-ball ratio of 7.5. Further details on the experimental procedure can be found in [17].

3.4 Application of WPPM

3.4.1 WPPM Results for Ball-Milled Ni Powders

To illustrate the application of WPPM we consider a recently proposed case of study concerning severely deformed Ni powder obtained by ball milling (b.m.) [17]. The example is particularly appropriate, as ball milling leads to increasingly defected and finely dispersed materials [55–57], ideal to test the various features of WPPM and to verify the presence and relative weight of the different lattice defects.

TEM and XRD are the main techniques to study ball milled materials. Concerning XRD, most of the literature reports results obtained by simplified methods like the SF or the WH plot [1–4]. A Fourier analysis of the line profiles is less frequent, and is usually carried out in the frame of the conventional WA method [5,6]. The above mentioned traditional methods are very popular, probably because relatively straightforward and known since many years, but they are not specific for the case of nanocrystalline and/or severely deformed materials. Basic hypotheses are frequently in contrast with the real features of the studied materials. For instance, the SF considers crystalline domain size effects only, and provides a volume weighted apparent size more than a true grain size. Specific hypotheses on the domain shape can be introduced [21, 26, 27] but are rarely used, and in any case the presence of a size distribution is not accounted for. The WH plot is conceived to evaluate size effects and lattice distortions, but it includes the same limitations as the SF, and it is based on the further limiting assumption that the domain size broadening is related to a Lorentzian peak-profile component, which is rarely true in real systems. For instance, this is in direct conflict with the frequent case of a lognormally-distributed nanocrystalline grain size [15–17]. Finally, the WA method, even if more rigorous in the line profile analysis and separation of domain size and lattice distortion effects, is not suitable to describe the line broadening due to dislocations (see [20] and references therein), and need to be supported by a fitting stage to separate profiles in the frequent case of peak overlapping.

WPPM can provide detailed information on the nature and density of lattice defects and on the evolution of the microstructure. A clear advantage over traditional methods is that all relevant line broadening components can be considered simultaneously, by means of a model based on physical parameters describing the real microstructure.

Figure 3.6a shows the XRD patterns of some of the ball-milled Ni powder samples (as-received, 24 h, 48 h and 96 h). It is clearly visible the effect of the mechanical treatment on line profiles, as well as the anisotropic nature of the line broadening, with $(h00)$ lines markedly broader than (hhh) lines (Fig. 3.6b), corresponding to the soft and stiff directions in Ni crystals (cf. Fig. 3.3). Even if the observable pattern includes few reflections, on a sufficiently resolved scale (e.g., in a log scale plot) peak overlapping is always present, even for short b.m. times and in the as-received sample. As a consequence, single-peak analysis is inappropriate, in this as well as in many other real cases of study where peak profile overlapping is present as a rule more than as an exception. An a priori background subtraction and peak profile separation can be a risky procedure, therefore conventional methods always involve a certain degree of uncertainty and unreliability of the results.

A minor but important feature in Fig. 3.6a is the presence of some weak reflections in the tail regions of the main Ni phase. These can be attributed to a small amount of cubic NiO, resulting from a partial oxidation of the ball milled metal (main NiO reflections are marked by asterisks in Fig. 3.6a). Since this minority phase may influence the modelling, we included NiO as a second phase, profiting of the circumstance that it is *fcc* like Ni. Due to the low intensity, lattice parameter and intensities only were refined for NiO.

Crystalline domains are assumed to be spherical, and consequently the size effect broadening is isotropic, i.e., independent of (hkl) ($K^{c} = 1$, $H_0^{c} = 1$,

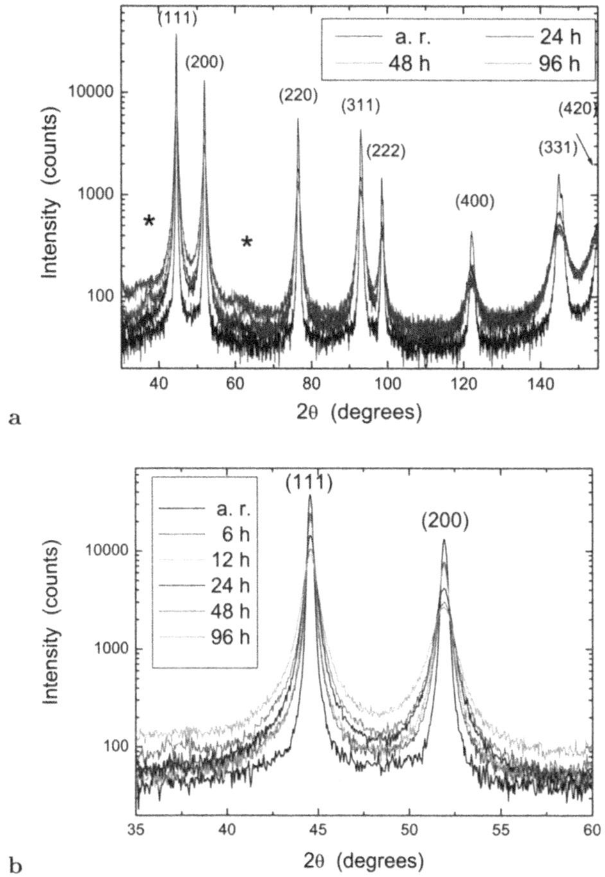

Fig. 3.6. XRD patterns of Ni powder in the as-received condition and after 24, 48 and 96 h ball milling. Asterisks indicate the position of the most intense lines of the NiO traces (**a**). Detail of (111) and (200) peaks for as-received Ni and after 6, 12, 24, 48 and 96 h ball milling (**b**)

$H_1^c = -3/2$, $H_2^c = 0$, $H_3^c = 1/2$). As a distribution of the polydisperse sphere system we used a lognormal function, which seems appropriate to several real cases including highly deformed metals [56, 58, 59] or finely dispersed powders [15, 60–62]. In this case, the FT expression of the size broadening component (3.16) is rather simple. Elastic constants for Ni, necessary to calculate A_E, B_E, A_S and B_S, are: $c_{11} = 247(5)$ $c_{12} = 153(5)$ $c_{44} = 122(2)$ (GPa) (Elastic anisotropy factor: $A_i = 2c_{44}/(c_{11} - c_{12}) = 2.54$) [63]. In this way, the edge/screw fraction can be refined.

The refinement strategy included a few preliminary iterations by WPPF (generally less than 10, depending on the choice of starting parameters), before switching to the WPPM mode for the last 2–3 iterations required to reach

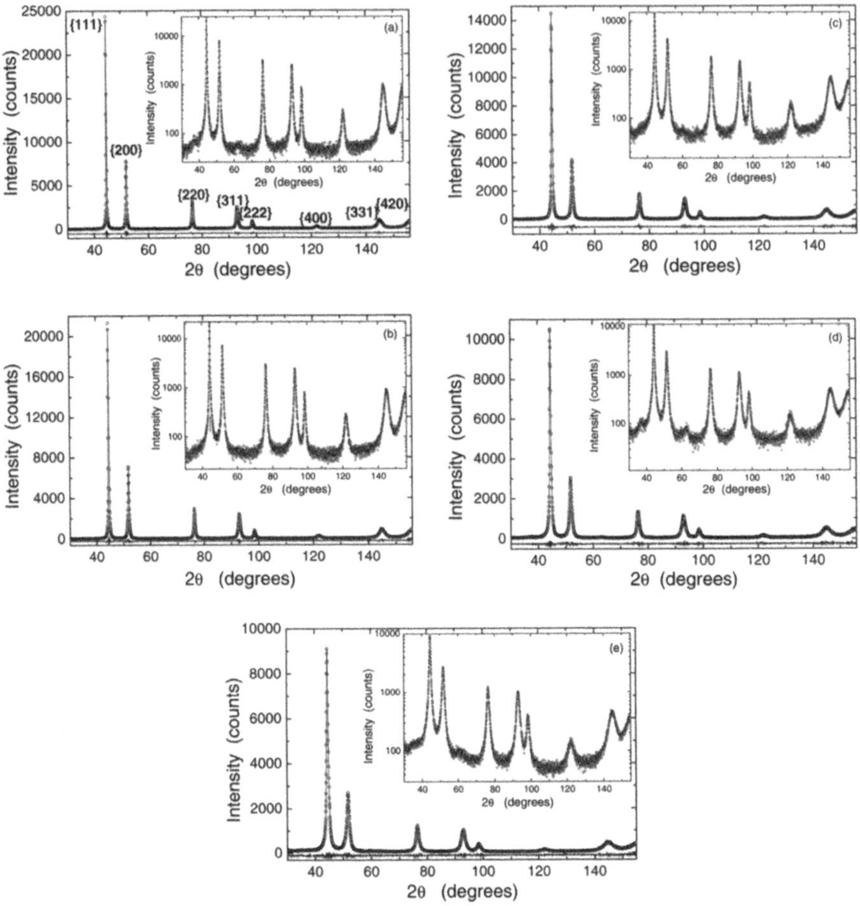

Fig. 3.7. WPPM results for different ball milling times: 6 h (**a**), 12 h (**b**), 24 h (**c**), 48 h (**d**) and 96 h (**e**). Miller indices of Ni peaks are reported in (**a**)

a convergence. In this way computation time was reduced considerably. The graphical result of WPPM is shown in Fig. 3.7 for some of the samples ball milled for different time. Physical parameters obtained by the refinement are shown in Fig. 3.8 as a function of the milling time. Results can be summarised in the following points:

- Grain diameter decreases with time (Fig. 3.8a), tending to saturate to 8–10 nm beyond 24 h ball milling.
- Dislocation density follows an opposite trend (Fig. 3.8a), as it increases rapidly in the early stages of ball milling, but the rate decreases with the milling time.
- The percentage of deformation faults (Fig. 3.8b) is rather low (always below 1%).

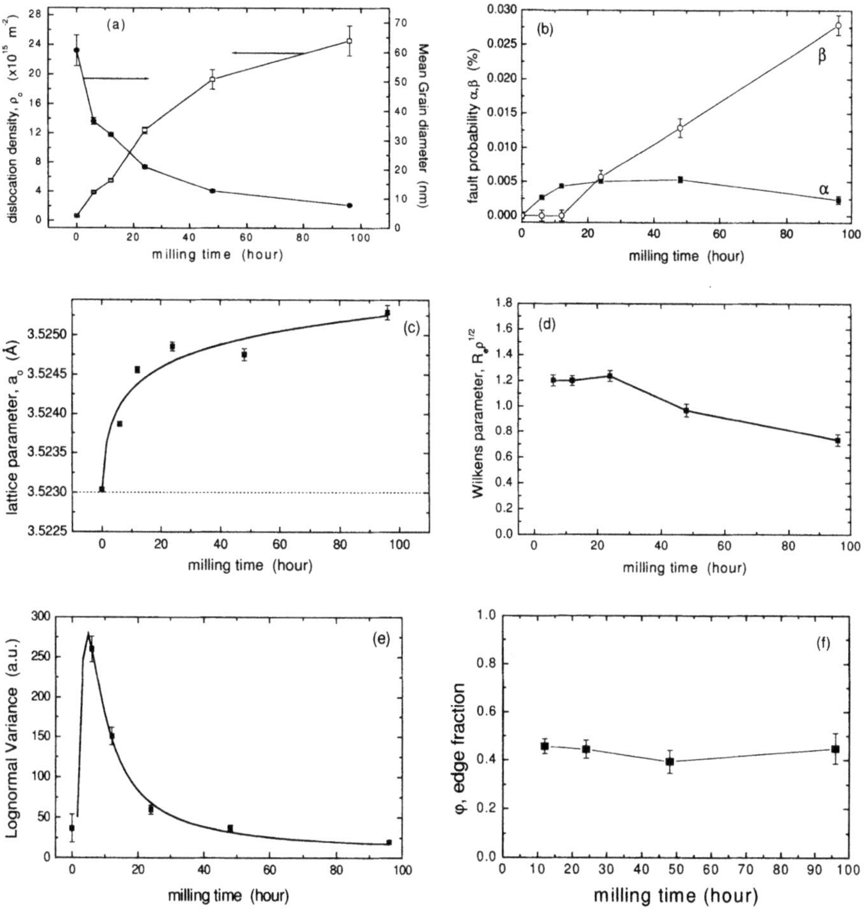

Fig. 3.8. WPPM results: dislocation density, mean grain diameter (**a**), deformation (α) and twin (β) faulting probabilities (**b**), lattice parameter (dotted line corresponds to the value for bulk Ni [73]) (**c**), Wilkens parameter (**d**), lognormal distribution variance (**e**) and dislocation character (edge fraction) (**f**) as a function of the ball milling time

- In the early stages of ball milling no twin faults are produced (Fig. 3.8b); twins appear after 12 h, and their probability increases steadily for longer milling time, reaching $\approx 3\%$ after 96 h.
- The trend of the Ni lattice parameter (Fig. 3.8c) is similar to that of the dislocation density, with a rapid increase during the early stages of b.m., and then tending to saturate after ≈ 24 h.

Ball milling produces a high density of dislocations. Severe plastic deformation occurs in the initial stage, leading to a sharp increase in dislocation density and grain size refinement. Longer milling time does not add plastic deformation to the same rate and the dislocation density tends to saturate.

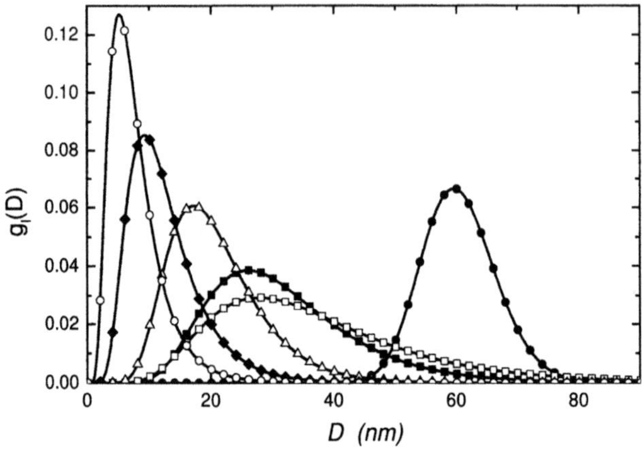

Fig. 3.9. Lognormal grain diameter distributions for Ni powders: as-received (•), after 6 h (□), 12 h (■), 24 h (△), 48 h (◊) and 96 h (○) of ball milling [17]

The Wilkens parameter [32] $M' = R'_e\sqrt{\rho}$, can provide an insight into the dislocation arrangement. Values around or below unity indicate strong interaction between dislocations, observed for instance in dislocation-cell walls [64]. In the present case, the decreasing value of the Wilkens parameter down to values below unity with increasing milling time (Fig. 3.8d), supports the hypothesis that dislocations are not randomly distributed in nanocrystals but are densely contained in GBs. Low values (< 1) of the Wilkens parameter have already been reported for ball milled metal powders [20]. However, it is worth underlining that the interpretation of the data and the general validity of Wilkens hypotheses is questionable for $M' < 1$ [32]. Actually this threshold should refer to $M = R_e\sqrt{\rho}$, involving the 'true' outer cut-off radius R_e, whereas in our expressions we consider an effective value, R'_e, which is just proportional to it.

This observation indicates that even if the trend of the dislocation density and grain size are relatively smooth and continuous, microstructure evolution can be a more complex process, where dislocation arrangement and interaction change with the milling time [55,65]. We can hypothesise a progressive evolution from a dislocation-cell structure to a fully nanocrystalline state, with formation of non-equilibrium grain boundaries made of high density of extrinsic dislocations.

This picture is supported by the grain size evolution. In fact, grain reduction due to ball milling tends to stop after 24–48 h, saturating to a value of 8–10 nm, and analogous results have been observed for most *fcc* metals [55,56,66]. In addition to the mean grain diameter we can also consider the effect of ball milling on the shape of the grain size distribution, shown in Fig. 3.9 for the various milling times. Useful features of the size distributions can also be appreciated in Fig. 3.8e, where the variance of the distribution is reported as a function of the b.m. time.

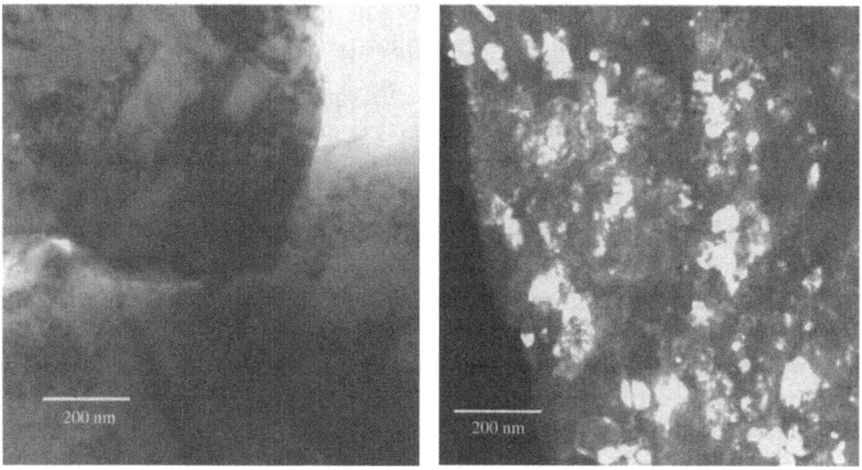

Fig. 3.10. TEM bright field picture of the as-received sample (**a**) and dark field picture of sample ball milled for 12 h (**b**)

The as-received sample is apparently made of evenly dispersed domains, with an average value about 60 nm, and a nearly symmetrical, relatively narrow distribution. Dislocation density is low ($\approx 6 \times 10^{14}\,\mathrm{m}^{-2}$) and faulting probabilities below detectable limit ($< 0.1\%$). TEM pictures of the as-received sample (Fig. 3.10a) suggests that the average domain size given by WPPM in this case refers to sub-grain structures in a polycrystalline sample with large grains and relatively low (at least from the point of view of XRD) defect density.

In the early stages of ball milling (0–12 h) the grain size distribution tends to widen and to shift to lower diameter values. In this regime b.m. exhibits the highest efficiency in the comminution process and in increasing defect density. Large plastic deformation is introduced, and grain size is reduced to a mean value of ≈ 30 nm, with a large dispersion about this value (cf. Fig. 3.10b and Fig. 3.9). For longer milling time, grain size reduction is less pronounced and the size distribution tends to narrow again. For this reason we can observe a peak in the variance trend of Fig. 3.8e. This means that extensive b.m. tends to eliminate large grains (above 40 nm), and correspondingly the size distribution slowly shifts to smaller diameters with a reduced dispersion about the mean value. A nanocrystalline state is then reached when grain size is around 8–10 nm, and the system exhibits relatively low dispersion about this value. After long milling time, a large fraction of the material is therefore part of the GB, which tends to store most of the excess energy [57]. An estimate of this effect can be provided by considering a simple GB model like that proposed by Nazarov et al. [67], based on disordered arrays of extrinsic dislocations and disclinations quadrupoles.

The excess elastic energy of GBs per unit area associated to disclinations and to random GB dislocations are, respectively [67]:

$$\gamma_{\mathrm{Discl}} = \frac{G\langle \Omega^2 \rangle D \ln 2}{16\pi(1-\nu)} , \tag{3.48}$$

$$\gamma_{\mathrm{Disl}} = \frac{G b^2 \rho D}{12\pi(1-\nu)} \ln\left(\frac{D}{b}\right) , \tag{3.49}$$

where G is the shear modulus, ν is the Poisson ratio and Ω is the power of the disclinations. We can calculate γ_{Disl} by using the WPPM results in Fig. 3.8 for ρ and D, and taking $G = 76$ GPa and $\nu = 0.3$ for Nickel. To estimate γ_{Discl} we follow the indications of Nazarov et al. [67], that in heavily deformed metals $\langle \Omega^2 \rangle^{1/2} \approx 0.03$. This leads to the result shown in Fig. 3.11a, where the total excess GB energy ($\gamma_{\mathrm{EX}} = \gamma_{\mathrm{Disl}} + \gamma_{\mathrm{Disc}}$) is shown together with the two components of (3.48) and (3.49). Figure 3.11a also shows the trend of the energy calculated for the same WPPM data, but considering randomly (and non-interacting) dislocations [57]:

$$\gamma_{RD} = AGb^2 \rho \ln\left(\frac{R_e}{b}\right) , \tag{3.50}$$

where $A = A_S = 1/4\pi$ for screw dislocations and $A = A_E = 1/[4\pi(1-\nu)]$ for edge dislocations.

From the WPPM results, line broadening anisotropy is such to give an edge dislocation fraction constantly around 0.5 (Fig. 3.8f), which corresponds to an uniform distribution of edge and screw dislocations. Therefore, the trend shown in Fig. 3.11a was calculated assuming screw and edge dislocations in the same proportion: $A = (A_S + A_E)/2$.

The most striking features in Fig. 3.11a are the maximum around 24 h and the much higher energy that can be attributed to non-equilibrium GBs (from (3.48) and (3.49)) with respect to the "bulk" value (from (3.50)), for random non-interacting dislocations. TEM evidence and literature data clearly point out that dislocations in ball milled nanocrystalline materials are not found inside the crystalline domains, but tend to form disordered arrays at GB, a conclusion also supported by the present study (TEM pictures and Wilkens parameter). As to the trend of γ_{EX}, this is in a very good agreement with the results of Zhao et al. [57] for nanocrystalline Fe obtained by ball milling. This also supports the above formulated hypothesis that prolonged ball milling (> 24 h in our case) drives the system from a dislocation-cell structure to a fully nanocrystalline state where a progressive GB relaxation occurs. Computer simulations support the hypothesis of a GB softening in the nanocrystalline state [68,69].

A low deformation fault probability (well below 1%) is observed in all b.m. samples (Fig. 3.8b). The reason for the observed low values of probably reasonably lies in the high Stacking Fault Energy (SFE) of pure Ni ($200\,\mathrm{mJm^{-2}}$) [55].

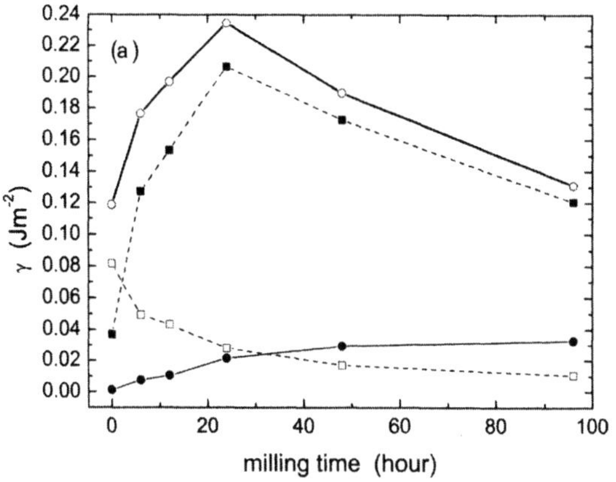

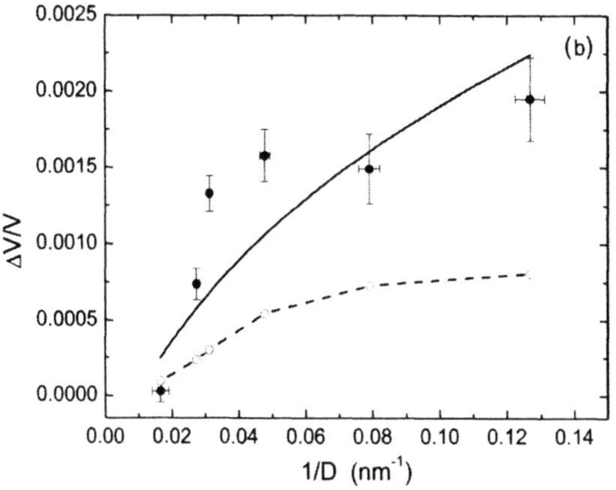

Fig. 3.11. Excess elastic energy of grain boundary (GB) from (3.48) and (3.49): total (○), dislocation term (■) and disclination term (□); corresponding energy term in the case of random (non-interacting) dislocations (3.50) (•) (**a**). Volume change calculated from the lattice parameter of Fig. 3.8 (•) and as volume expansion for dislocation/disclination GB (*dash*) and best fit (*line*) using (3.51) and (3.52) (see text) (**b**)

The SFE, however, can be considerably lowered by impurities. This can be the case, also considering that oxygen incorporation is very likely, given the observed presence of nickel oxide. Stacking (deformation) faults are typically produced in plastic deformation regime, as a consequence of dislocation motion in *fcc* materials. Dislocations slip is progressively made more difficult by increasing ball milling, and becomes virtually absent both when a dislocation-cell structure forms and in nanocrystals [70]; as a consequence α is not expected to increase with b.m. time.

Twins faults present a different evolution. Their formation is observed after the early stage of b.m. (> 12 h), probably as a consequence of a tendency to reduce surface energy. Other studies on b.m. metals show a similar evidence [70, 71].

Also the trend of the lattice parameter has been already observed in analogous studies on ball milled metals [55–57] and can be attributed to several mechanisms related to or induced by plastic deformation.

Liu et al. [72] reported a supersaturation of vacancies in evaporated thin films, resulting from a general increase in solubility of finely dispersed grains. In fact, according to the Gibbs–Thomson equation, vacancy and defect solubility ($C(T, D)$) increase as D^{-1} [72]:

$$C(T, D) = \frac{4\omega\gamma C_0}{kTD} = \frac{K}{D}, \tag{3.51}$$

where ω is the atomic volume, γ is an interfacial energy and C_0 is the equilibrium solute stability in bulk. From the above (3.51), we can thus expect a volume change $(\Delta V/V)_{\mathrm{SOL}} \propto 1/D$.

The presence of a small fraction of NiO in our ball milled samples supports the hypothesis of oxygen incorporation in the Ni matrix, to the point that excess oxygen forms oxide [55]. If we consider the grain size evolution (Figs. 3.8 and 3.9), however, we cannot expect a simple hyperbolic trend of the lattice parameter as a function of the average grain size. A further contribution to lattice expansion is related to the excess energy. In fact, a high density of extrinsic dislocations at grain boundaries leads to a lattice expansion, that according to Nazarov et al. [67] can be written as

$$\left(\frac{\Delta V}{V}\right)_{\mathrm{Disl}} = 0.14 b^2 \rho \ln\left(\frac{D}{b}\right). \tag{3.52}$$

A corresponding volume increase due to disclinations can be calculated by assuming that the ratio $\gamma_{\mathrm{Disl}}/\gamma_{\mathrm{Disc}}$ is equal to the ratio of volume changes due to dislocations and disclinations [56].

Figure 3.11b shows the volume change calculated from the lattice parameter change as

$$\left(\frac{\Delta V}{V}\right)_{\mathrm{Exp}} = 3\frac{a - a_0}{a_0},$$

where $a_0 = 0.3523\,\mathrm{nm}$ (for bulk Ni [73]). Together with the experimental data we have plotted the volume expansion for dislocation/disclination GB, and a best fit of a trend given by (3.51) and (3.52). The proportionality constant K was refined. As can be seen, the modelled trend very roughly reproduces the experimental values.

It is therefore reasonable that high elastic strains produced by non-equilibrium grain boundaries and increased oxygen solubility both contribute to the increase of lattice parameter with the b.m. time, but the relative weight of the two effects is difficult to estimate.

In conclusion, ball milling increases lattice defects as expected, but the trend is rather peculiar. After $\approx$ 24 h b.m. tends to lose efficiency in introducing lattice damage, and the trends of mean grain diameter, dislocation density, lattice parameter and deformation fault probability tend to flatten out. Correspondingly, twin faults appear and their probability increases steadily with milling time. The observed formation of low-energy boundaries can provide an energy release mechanism to the highly deformed microstructure. These observations are therefore in agreement with the conclusion of Fecht et al. [65] that prolonged ball milling promotes a transition from a dislocation cell structure to a properly nanocrystalline state.

3.5 Comparison between TEM and WPPM Data

TEM perfectly integrates XRD in the analysis of ball milled materials [56,62]. In fact, TEM is ideal to measure low-medium defect densities (dislocation densities up to $\approx 10^{14}\,\mathrm{m}^{-2}$), whereas XRD line profile analysis in polycrystalline materials is usually sensible above this threshold value. Grain size data may be different, since XRD is sensitive to the size of coherently scattering crystalline domains, whereas TEM pictures do not always provide a clear distinctions between grains and crystalline domains in highly defected metals, so the observation may not be interpreted in terms of true domain structure. Actual domain sizes can be difficult to quantify by TEM when grain size is comparable to the size and thickness of the specimen, but also when a fine dislocation-cell structure is present [74] and grains are subdivided in several coherent domains (cf. Fig. 3.10). This is the microstructure expected for the early stages of ball milling, and in this case a domain size distribution can hardly be obtained by TEM in an objective way.

Quite differently, dark field TEM pictures are very effective in showing crystalline grain distributions on the nanometer scale. After long milling time (96 h in the present study), a properly nanocrystalline state is obtained, where crystalline grains are dispersed and loosely packed together. Under these conditions, as shown in Fig. 3.12, grains are mostly made of single domains, clearly visible and sufficiently delineated to allow a quantitative analysis. Grains are reasonably equiaxed and much smaller than the thickness of the TEM specimen, so stereological corrections are not needed. Grain size values were obtained as equivalent-circle diameters [62] ($D_e = 2\sqrt{A/\pi}$, where A is the projected area) using several dark-field pictures like that shown in Fig. 3.12, up to approximately 850 grains for sample 96 h.

The TEM distribution of grain diameters is shown in Fig. 3.13 (histogram), together with the lognormal distribution whose parameters (mean and variance) were refined by WPPM (line). The agreement is remarkably good, especially if we consider that several and different experimental errors are introduced by the two techniques.

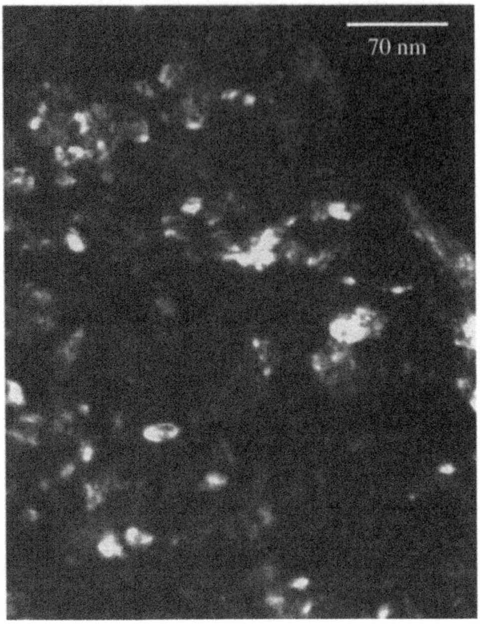

Fig. 3.12. TEM dark field picture for the Ni sample ball milled for 96 h

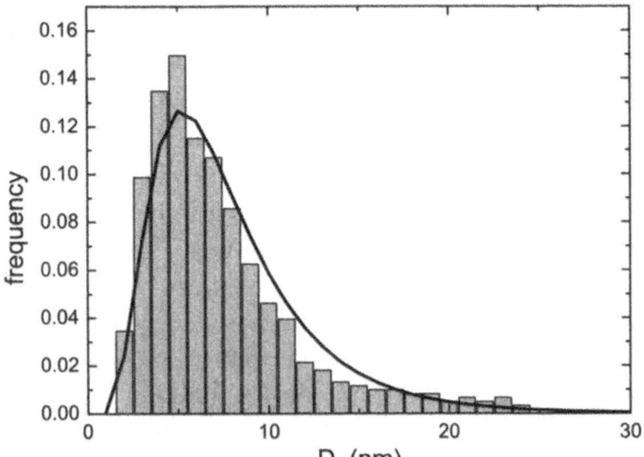

Fig. 3.13. Grain size distribution for sample 96 h: TEM (histogram) and lognormal distribution refined by WPPM (*line*) [17]

The TEM distribution is not exactly a lognormal, with a larger disagreement with respect to the WPPM result in the 10–15 nm size range. Beyond experimental uncertainty, it is worth noting that faulting may have an influence in this discrepancy. While WPPM accounts for the presence of faulting, so that the refined diameter corresponds to the overall grain size (including defects which do not interrupt coherence), TEM can show the presence of

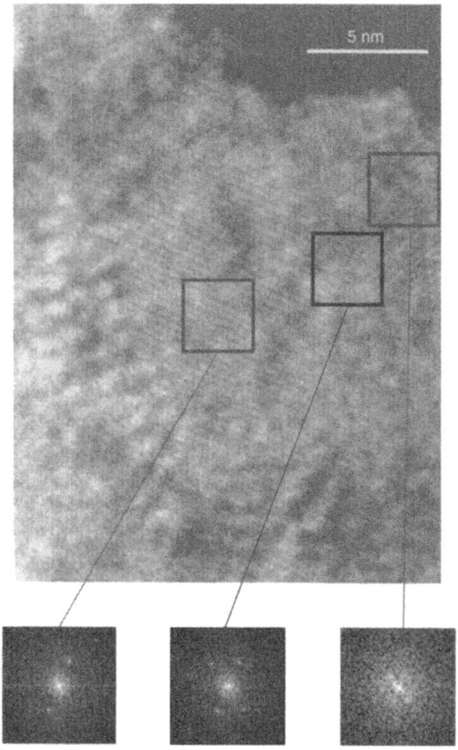

Fig. 3.14. High resolution TEM (HREM) picture. HREM picture for the Ni sample ball milled for 96 h, with indication of the selected area diffraction (SAD) from regions corresponding to different domains separated by a low angle grain boundary (see insets)

twins, but resolution in our dark-field pictures is probably too low to disclose deformation faults [62]. Twinned grains, which are accounted for in the WPPM distribution as medium-large grains, do not probably appear simultaneously in the same dark-field picture because of their different orientation. Given the relatively high percentage of twin faults ($\approx 3\%$) we may assume several medium-large grains being crossed by the planar defect, so that they appear smaller in dark field pictures.

The presence of twins is reported in other studies on nanocrystalline metals obtained by severe plastic deformation (e.g., see [70, 71]). Their formation is justified in terms of energy reduction: extensive deformation, like after prolonged ball milling, promotes the formation of low-energy grain boundaries to release part of the excess energy stored in the grain boundaries. Direct evidence of a low energy boundary is found in the high resolution TEM picture of Fig. 3.14. Selected Area Diffraction (SAD) for three different regions, highlighted by corresponding squares, points out the different orientations across the boundary.

3.6 Conclusions

WPPF and WPPM are conceived as general procedures for simultaneous refinement of microstructural parameters of polycrystalline materials by an analysis of the entire diffraction pattern. In particular, lattice defects and size and shape of crystalline grains are considered in terms of their influence on line profile position, shape and width. WPPF/WPPM can then be considered as microstructural refinement techniques, in analogy to the Rietveld method that provides a refinement of structural parameters (actually the Rietveld method, in modern sense, also considers non-structural information like, e.g., texture [11,75,76], residual stress [77–79] and phase composition [80,81]). The two methods (WPPF (or WPPM) and Rietveld), however, can be joined in a single, powerful procedure for the refinement of structure and microstructure of pure or multi-phase materials.

In the present formulation, physical parameters like dislocation density, effective outer cut-off radius, dislocation character, mean and variance of a suitable distribution of grain size, deformation and twin faulting probabilities and lattice parameters are refined by a direct comparison with the experimental data, by means of a minimisation algorithm weighted on the appropriate counting statistics. Profiting of the convolutive approach, including additional or alternative models is relatively simple (e.g., see [35]). As a consequence, besides the direct application to the modelling of XRD patterns of finely dispersed powders, nanocrystalline and severely plastically deformed materials, WPPF/WPPM can also be used as a powerful simulation tool, and to test the merit of different, alternative or concurrent, models of the material microstructure.

Basic difference between WPPM and WPPF is the use in the latter of analytical profile functions. From this point of view, WPPM can be seen as a virtually unbiased and fully general approach to the modelling of powder data, since it does not employ arbitrary analytical profile functions: the entire diffraction pattern (with the exception of a background polynomial function) is directly modelled in terms of physical parameters related to the microstructure. However, the difference in the results obtained by the two approaches is usually not large [19, 35], because the constraints provided by the microstructural model and the simultaneous refinement of several physical parameters, drastically reduce the arbitrariness of using analytical profile functions.

WPPM application was illustrated by means of a typical case of study, consisting in severely plastically deformed metal powders. WPPM provided a detailed picture of the effect of extensive ball-milling of *fcc* Ni powder samples. Results were discussed in the light of existing literature and also compared with TEM observations on the same samples. The agreement between grain size distributions provided by WPPM and by TEM for a nanocrystalline Ni sample obtained by prolonged ball milling was remarkable, and certainly contributes to validate the procedure. Analogous agreement was observed in completely different systems (nanocrystalline ceria powders) [35], further extending the survey of WPPF/WPPM successful applications.

So far the model is fully developed for *fcc* materials only. Extension to *bcc* cases and other cubic structures is currently in progress, as well as inclusion of the case of hexagonal materials [82]. A complete generalisation to any possible crystallographic phase is certainly possible, on the very same basis shown in this work, but will require further efforts that are the object of current and future research activity. Further developments can also be envisaged in the direction of including the model based on a recently proposed maximum entropy approach [83], in order to remove the need for an a-priori choice of a size distribution function.

References

1. P. Scherrer: Nachr. Ges. Wiss. Göttingen **26**, 98 (1918)
2. G.K. Williamson and W.H. Hall: Acta Metall. **1**, 22 (1953)
3. J.I. Langford: *Accuracy in Powder Diffraction II.* In: NIST Spec. Pub. No. 846. Ed. by E. Prince and J.K. Stalick. (US Dept of Commerce, Gaithersburg MA 1992) pp. 110–126
4. H.P. Klug and L.E. Alexander: *X-ray Diffraction Procedures for Polycrystalline and Amorphous Materials.* 2nd edn. (Wiley, New York 1974)
5. B.E. Warren and B.L. Averbach: J. Appl. Phys. **21**, 595 (1950); B.E. Warren and B.L. Averbach: J. Appl. Phys. **23**, 497 (1952)
6. B.E. Warren: *X-ray Diffraction* (Addison-Wesley, Reading, MS 1969)
7. J.I. Langford and D. Louër: Rep. Prog. Phys. **59**, 131 (1996)
8. Y.H. Dong and P. Scardi: J. Appl. Cryst. **33**, 184 (2000)
9. G.S. Pawley: J. Appl. Cryst. **14**, 357 (1981)
10. H. Toraya, in *The Rietveld Method.* Ed. by R.A. Young (Oxford University Press, Oxford 1993) pp. 254–275
11. H.M. Rietveld: J. Appl. Cryst. **2**, 65 (1969)
12. R.A. Young (Ed:): *The Rietveld Method.* (Oxford University Press, Oxford 1993)
13. G. Caglioti, A. Paoletti and F.P. Ricci: Nucl. Instrum. Methods **35**, 223 (1958)
14. J.I. Langford: J. Appl. Cryst. **11**, 10 (1978)
15. J.I. Langford, D. Louër, and P. Scardi: J. Appl. Cryst. **33**, 964 (2000)
16. P. Scardi, M. Leoni and Y.H. Dong: CPD Newsletter **24**, 23 (2001)
17. P. Scardi and M. Leoni: Acta Cryst. A **58**, 190 (2002)
18. P. Scardi, M. Leoni and Y.H. Dong: Eur. Phys. J. B **18**, 23 (2000)
19. P. Scardi, M. Leoni and Y.H. Dong; Mat. Sci. Forum. **378–381**, 132 (2001)
20. J.G.M. van Berkum: *Strain Fields in Crystalline Materials.* (Technische Universiteit Delft, Delft, 1994) Ph.D. Thesis
21. A.J.C. Wilson: *X-ray Optics*, 2nd edn. (Methuen, London 1962)
22. A.R. Stokes and A.J.C. Wilson: Proc. Cambridge Phys. Soc. **38**, 313 (1942)
23. A.R. Stokes and A.J.C. Wilson: Proc. Cambridge Phys. Soc. **41**, 174 (1944)
24. E.F. Bertaut: C.R. Acad. Sci. Paris **228**, 187, 492 (1949); E.F. Bertaut: Acta Cryst. **3**, 14 (1950)

25. J.F. Nye: *Physical Properties of Crystals: Their Representation by Tensors and Matrices.* Reprint edn. (Oxford Univ. Press, Oxford 1987)
26. P. Scardi and M. Leoni: Acta Cryst. A **57**, 604 (2001)
27. J.I. Langford and A.J.C. Wilson: J. Appl. Cryst. **11**, 102 (1978)
28. B.R. York: Advances in X-ray Analysis **41** 544 (1999)
29. A.C. Vermeulen, R. Delhez, T.H. de Keijser, E.J. Mittemeijer: J. Appl. Phys. **77**, 5026 (1995)
30. J.-D. Kamminga, L.J. Seijbel, R. Delhez: This volume, Chapter 12.
31. M.A. Krivoglaz and K.P. Ryaboshapka: Fizika Metall. **15**, 18 (1963)
32. M. Wilkens: Phys. Stat. Sol. (a) **2**, 359 (1970)
33. M. Wilkens: *Fundamental Aspects of Dislocation Theory.* Ed. by J.A. Simmons, R. de Wit, R. Bullough. Vol. II. Nat. Bur. Stand., (US) Spec. Publ. No. 317 (Washington, DC 1970) pp. 1195–1221
34. A.C. Nunes and D. Lin: J. Appl. Cryst. **28**, 274 (1995)
35. M. Leoni and P. Scardi: This volume, Chapter 16.
36. M.A. Krivoglaz, O.V. Martynenko and K.P. Ryaboshapka: Phys. Met. Metall. **55**, 1 (1983)
37. J.-D. Kamminga and R. Delhez: J. Appl. Cryst. **33**, 1122 (2000)
38. I. Groma, T. Ungar and M. Wilkens: J. Appl. Cryst. **21**, 47 (1988)
39. P. Klimanek and R. Kuzel Jr.: J. Appl. Cryst. **21**, 59 (1988)
40. R. Kuzel Jr. and P. Klimanek: J. Appl. Cryst. **22**, 299 (1989)
41. M. Wilkens: Phys. stat. sol. (a) **104**, K1 (1987)
42. N. Armstrong and P. Lynch. This volume, Chapter 10.
43. T. Ungar, I. Dragomir, A. Revesz and A. Borbely: J. Appl. Cryst. **32**, 992 (1999)
44. N.C. Popa: J. Appl. Cryst. **31**, 176 (1998)
45. P.W. Stephens: J. Appl. Cryst. **32**, 281 (1999)
46. L. Velterop, R. Delhez, T.H. de Keijser, E.J. Mittemeijer and D. Reefman: J. Appl. Cryst. **33**, 296 (2000)
47. J.P. Cline, R.D. Deslattes, J.-L. Staudenmann, E.G. Kessler, L.T. Hudson, A. Henins and R.W. Cheary: *Certificate SRM 640c.*(NIST Gaithersburg, MD 2000)
48. J.P. Cline, R.D. Deslattes, J.-L. Staudenmann, E.G. Kessler, L.T. Hudson, A. Henins and R.W. Cheary: *Certificate SRM 660a.* (NIST Gaithersburg, MD 2000)
49. M. Leoni, P. Scardi, J.I. Langford: Powder Diffr. **13**, 210 (1998)
50. P. Scardi and M. Leoni: J. Appl. Cryst. **32**, 671 (1999)
51. J.I. Langford: Prog. Cryst. Growth and Charact. **14**, 185 (1987)
52. R.W. Cheary and A.A. Coelho: J. Appl. Cryst. **25**, 109 (1992)
53. A.A. Kern and A.A. Coelho: *A New Fundamental Parameters Approach in Profile Analysis of Powder Data.* In: Proceedings of the international school on powder diffraction (ISPD'98), edited by S.P. SenGupta, Allied Publishers Ltd., New Delhi, 1998. pp. 144–151.
54. B.S. Garbow, K.E. Hillstrom and J.J. More: *MINPACK Project.* (Argonne National Laboratory, November 1996) `http://www.netlib.org/minpack`
55. A.I. Salimon, A.M. Korsunsky, A.N. Ivanov: Mat. Sci. Eng. **A271**, 196 (1999)
56. R.Z. Valiev, R.K. Islamgaliev, I.V. Alexandrov: Prog. Mat. Sci. **45**, 103 (2000)
57. Y.H. Zhao, H.W. Sheng and K. Lu: Acta Mater. **49**, 365 (2001)
58. R.Z. Valiev, E.V. Kozlov, Y.F. Ivanov, J. Lian, A.A. Nazarov and B. Baudelet: Acta Metall. Mater. **42**, 2467 (1994)

59. R.K. Islamgaliev, F. Chmelik, R. Kuzel Jr.: Mat. Sci. Eng. A **237**, 43 (1997)
60. C.G. Granqvist and R.A. Buhrman: J. Appl. Phys. **47**, 2200 (1976)
61. L.B. Kiss, J. Söderlund, G.A. Niklasson and C.G. Granqvist: Nanostr. Materials **12**, 327 (1999)
62. C.E. Krill and R. Birringer: Phil. Mag. **77**, 621 (1998)
63. A.G. Every and A.K. McCurdy (Eds.): *Landolt-Börnstein*. Vol. 29 (Springer, Berlin Heidelberg New York 1992) p. 14
64. T. Ungar, H. Mughrabi, D. Rönnpagel and M. Wilkens: Acta Metall. **32**, 333 (1984)
65. H.J. Fecht, E. Hellstern, Z. Fu and W.L. Johnson: Metall. Trans. A **21**, 2333 (1990)
66. H.H. Tian and M. Atzmon: Acta Mater. **47**, 1255 (1999)
67. A.A. Nazarov, A.E. Romanov, R.Z. Valiev: Scripta Materialia **34**, 729 (1996)
68. J. Schiøtz, T. Rasmussen, K.W. Jacobsen and O.H. Nielsen: Phil. Mag. Lett. **74**, 339 (1996)
69. J. Schiøtz, F.D. Di Tolla, K.W. Jacobsen: Nature **391**, 561 (1998)
70. P.G. Sanders, A.B. Witney, J.R. Weertman, R.Z. Valiev, R.W. Siegel: Mat. Sci. Eng. A **204**, 7 (1995)
71. P. Chatterjee and S.P. Sen Gupta: Phil. Mag. A **81**, 49 (2001)
72. X.D. Liu, H.Y. Zhang, K. Lu and Z.Q. Hu: J. Phys. Condens. Matter **6**, L497 (1994)
73. International Center for Diffraction data: PDF card #4–850. (ICDD, Newtown Square, PA)
74. T. Ungar, J. Gubicza, G. Ribarik, and A. Borbely: J. Appl. Cryst. **34**, 298 (2001)
75. N.C. Popa: J. Appl. Cryst. **25**, 611 (1992)
76. M. Järvinen, in *Defect and Microstructure Analysis by Diffraction*. Ed. by R.L. Snyder, J. Fiala, H.J. Bunge. IUCr series (Oxford University Press, New York 1999) pp. 556–559
77. C.J. Howard and E.H. Kisi: J. Appl. Cryst. **32**, 624 (1999)
78. N.C. Popa: J. Appl. Cryst. **33**, 103 (2000)
79. D. Balzar and N.C. Popa. This volume, Chapter 5.
80. R. Hill, in *The Rietveld Method*. Ed. by R.A. Young (Oxford University Press, Oxford 1993) pp. 61–101
81. I.C. Madsen, N.V.Y. Scarlett, L.M.D. Cranswick and T. Lwin: J. Appl. Cryst. **34**, 409 (2001)
82. P. Scardi and M. Leoni: (2004). In preparation
83. N. Amstrong, J.P. Cline, and W. Kelceff. This volume, Chapter 8; N. Armstrong, J.P. Cline, W. Kalceff, and J. Bonevich, in *Proceedings of the 3rd Conference on Accuracy in Powder Diffraction*. Ed. by J.P. Cline, J. Post and P. Scardi (NIST, Gaitherburg, MD 22.–25.4.2001)

4 Full Profile Analysis of X-ray Diffraction Patterns for Investigation of Nanocrystalline Systems

S.V. Tsybulya, S.V. Cherepanova, and G.N. Kryukova

The peculiarities of the structural arrangement of nanocrystalline materials, which are the specific objects of structural analysis, are discussed. For their structural investigation, two approaches for the full profile analysis of X-ray powder diffraction patterns are proposed. Use of the first one (the modified Rietveld algorithm) allows us to refine the average atomic structure parameters, including the atomic coordinates, the Debye–Waller factors, and the occupation of atomic positions for materials for which the X-ray diffraction patterns demonstrate the effects of anisotropic peak broadening, regardless of the nature of these effects. The second method involving full profile analysis based on the statistical structural model of the nanoparticle, taking account of particle size and shape, the probability of planar defects and the correlation in their distribution, the strain field parameters, enables us to obtain quantitative information on the real structure of the nanocrystalline systems. The results of the real structure study of some practically important nanomaterials such as oxides, carbon materials, metals and alloys are also described.

Nowadays, interest in nanocrystalline materials keeps growing enormously. Catalysts, nanoceramics, and nanocrystalline heat-resisting alloys represent only the beginning of the list of materials that are produced on an industrial scale and widely used in the different fields of science and technology. The growing interest in ultradispersed materials has stimulated a distinct burst of activity in their structural investigation.

Conventional methods of X-ray diffraction (XRD) analysis, including the Rietveld refinement procedure [1], are based on the principle of three-dimensional structural ordering of the crystalline materials. From the viewpoint of such an approach, different kinds of the crystal structure disorder ("defects") are the factors preventing the determination of the average atomic structure. In the case of nanocrystalline materials, local imperfections in the regular structure can be considered as elements of the nanostructure, being integral or specially imposed parts of the latter. Diffraction effects caused by these imperfections are of great interest, since these effects may be considered as an information source on the arrangement and concentration of such "defects".

The principal feature of nanocrystalline materials is the extremely high density of their defects. This means that the nanocrystalline state of the matter may demonstrate collective properties of particle (or defect) ensembles with respect to the diffraction effects, as well as other physical phenomena that lead to the appearance of specific diffusion scattering that is not characteristic of crystals with perfect structure or a low defect density. In the case of a high concentration of structural defects in the materials under study, some diffraction peaks can become very broad and even disappear in the spectra, which makes it impossible to use line profile analysis (LPA) methods based on the analysis of the shape, shift and integral breadth (or full width at half maximum intensity, FWHM) of the individual XRD peaks [2–9]. At the same time, along with the disappearance of some X-ray diffraction reflections, new diffusion peaks irresponsible for the strict Bragg scattering are observed in the X-ray diffraction spectra of compounds with a high defect density, and some characteristic examples of this situation will be given below. Owing to these facts, it is not always possible to use LPA methods for the determination of different kinds of interacting defects over the wide range of their concentrations. Obviously, one can gain more detailed structural information using the full profile of the XRD pattern, and therefore the development of methods based on the analysis of the full profile becomes vital in the field of XRD structural analysis of partially disordered and nanocrystalline materials.

4.1 The peculiarities of Structural Analysis of Nanocrystalline Materials

Materials with nanoscale-level sizes (below 100 nm) are specific objects of XRD structural analysis. Siegel [10] proposed the following classification, which is depicted in Fig. 4.1.

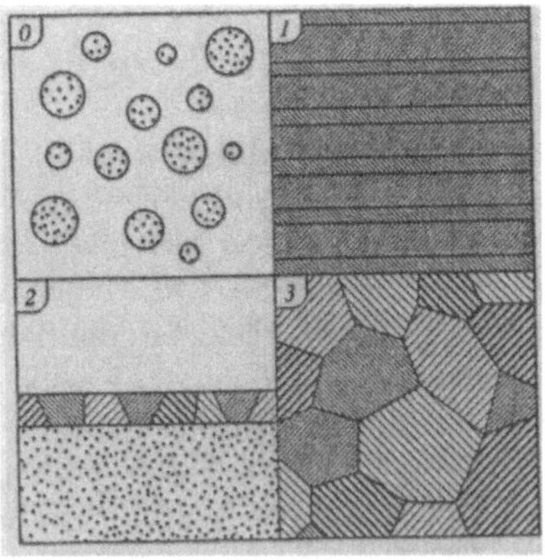

Fig. 4.1. The classification of nanomaterials by R.W. Siegel [10]: *0* – zero-dimensional systems; *1* – one-dimensional nanocrystals; *2* – two-dimensional systems; *3* – three-dimensional nanocrystals

The first category is zero-dimensional (0D) systems, such as isolated nanoparticles supported on carriers or embedded into matrices. Objects composed of alternating coherent domains of nanometer-sized thickness belong to the category of one-dimensional (1D) materials. The best examples of two-dimensional (2D) systems are nanocrystalline films supported on the surface of massive carriers. And, finally, there are three-dimensional (3D) systems represented by compacted materials with coherently scattering domains (CSD) of a few nanometers in size, or highly porous aggregates composed of nanoparticles.

A researcher is inevitably faced with the problem of determinating the mean particle size (or the CSD size) as a primary task in the XRD analysis of nanocrystalline materials. This value could sometimes be obtained with sufficient accuracy by using the width of the X-ray diffraction peaks. Problems arise due to pronounced overlapping of strongly broadened reflections, especially in structural studies of crystals that have low crystallographic symmetries. The analysis of 0D systems is complicated by the fact that diffraction by the support or the matrix may contribute significantly to the XRD patterns of these materials.

Along with the mean particle size, the particle size distribution function is another characteristic of nanocrystalline materials, being principally important in some particular cases, e.g. for some nanocrystalline alloys with a particle size lying in the range from 8 to 20 nm, which is the reason for their unique magnetic properties [11, 12]. The procedure of determination of the particle size distribution requires a precise definition of the shape of the XRD peaks.

Nanoparticles supported on carriers often have anisotropic forms. Detailed analysis of the several broadened diffraction reflections is necessary for the determination of the particle shape parameters, but very often it is not possible.

Usually, nanoparticles possess a crystal structure that is different from that of the massive phase, and this may be caused by size effects or matrix influence. This results in a decrease or increase of the lattice parameters of the nanoparticles [13]. For nanocrystals, the Debye–Waller factors are usually larger than those of monocrystalline analogs due to the static deviations of atoms from their average positions in the nanoparticle lattice.

1D nanostructures themselves represent unique objects for XRD analysis, since their structure is arranged in a specific way: there are numerous planar defects, such as stacking faults, twin boundaries, and interfaces between different phases in heterogeneous systems. These defects lie perpendicular to one crystallographic direction, and the distances between them are not equal but lie within the nanoscale range. In 1D nanomaterials, the lattice ordering in other two crystallographic directions is retained, and individual domains are coherently bounded with each other. By their structural arrangement, 0D, 2D, and 3Dtypes of nanocrystalline systems go beyond the above situation – their structural blocks, nanoparticles or CSDs possess nanoscale-level sizes in all three crystallographic directions and, in the general case, these blocks may be located at random with respect to each other. The specific structural packing of 1D nanostructures causes various rather complicated diffraction effects, which will be considered further in detail. The structural study of 1D nanomaterials is aimed at the determination of the domain crystal structure, the type of their interfaces, and the average domain thickness, as well as the thickness distribution.

Particular attention should be focused on the nature of microstrains in nanocrystalline systems. The main sources of lattice deformation in monocrystals and powder materials are dislocations that appear randomly in the crystal structure and preserve the ability to diffuse over the crystal lattice. Dislocations are nonequilibrium defects, but they are retained in the crystal structure owing to different mechanisms of their multiplication, in particular the Frank–Read mechanism [14]. As a rule, the size of nanocrystalline particles appears to be smaller than the characteristic size of the Frank–Read loop ($l = Gb/\tau$, where G is the shear modulus, b is the modulus of the Burgers vector, and τ is the critical shearing strain). Theoretical calculations have revealed that the value of l lies between 20 and 2000 nm [15], which constrains the possibility of dislocation multiplication by the Frank–Read mechanism in the nanoparticles. The authors [16] have performed a theoretical analysis of the stability of dislocations in nanocrystals, and they found that the probability of the presence of moving dislocations inside the nanoparticle declines with a decrease in the particle size. Analysis of the literature [15–19] and our experience in the TEM study of nanocrystalline metal and oxide systems [20–23] confirms these conclusions: single dislocations have rarely been observed in nanoparticles with sizes below 40 nm, although a few examples of the visualization of isolated dislocations in the nanocrystals can be found elsewhere [15, 19]. These references provide evidence that single dislocations cannot be considered as the prime source of the microstrains in the crystal lattice of nanomaterials. At the same time, the role of the dislocations concentrated in the vicinity of the inter-grain boundaries is of great importance. Valiev and others [24–27] have claimed that inter-grain boundaries could be considered as sources of long-range strain fields that extend through the crystal area, the dimension of which depend on the structure of the grain boundary. Thus, inter-grain boundaries indirectly affect the shape and width of the XRD peaks, owing to the effect of distortions caused by strains due to these defects.

Finally, let us specify the main tasks of the XRD structural analysis of nanocrystalline materials. Firstly, with the help of this analysis important crystallographic data, such as lattice parameters, atomic coordinates, interatomic distances, occupation of the atomic positions, and Debye–Waller factors, should be refined. Secondly, it should be possible to determine important nanostructural parameters (the mean particle or CSD size, the particle size distribution, the type and density of planar defects, and the microstrain values) within sufficient accuracy on the basis of XRD experiments.

To our knowledge, there is no algorithm that can take into account the specific nature of the structural arrangement of nanocrystalline materials in a general way, but we are perfectly confident that any solution should be based on analysis of the full profile of the XRD pattern. Also, we do not pretend that both of the methods of full profile analysis that will be described in this chapter are universal, but they may be useful for the structural investigation of a large set of materials that possess complicated structural arrangement. Thus, the modified Rietveld method, which will be outlined in Sect. 4.2, is aimed at the refinement of the main structural parameters (atomic coordinates, occupation of the structural positions, Debye–Waller factors, etc.) for objects for which the diffraction patterns show anisotropic broadening of the X-ray diffraction peaks, no matter what the nature of the peak broadening. In contrast, the Kakinoki–Komura method (see Sect. 4.3) allows us to analyze the reasons for peak broadening by using a model of the structural arrangement of nanocrystalline materials that involves the finite size and shape of the nanoparticle, planar defects, coherently intergrown domains, and microstrains of differing natures.

4.2 Crystal Structure Refinement Using the Modified Rietveld Method

It is well known that the Rietveld method [1] is the best structural refinement procedure for 3D-ordered crystalline materials using diffraction data. The main idea of this method is to represent the profile of the XRD pattern as the sum of some empirically chosen approximation functions with specified or refined parameters. In this case, the XRD profile can be expressed as

$$y_i = \sum_{j=1}^{N} I_j \Omega_{ij} + y_i^b \,, \tag{4.1}$$

where y_i is the calculated intensity at point i, y_i^b is the background intensity, I_j is the calculated intensity of reflection i, Ω_{ij} is the peak shape function, and N is the number of Bragg reflections.

The shape of the XRD peak is well described by Voigt, pseudo-Voigt, and Pearson VII functions [28–30]. In the case of polycrystalline materials with relatively large individual particles that possess perfect structure, the width and shape of the XRD peaks depends totally on the instrumental factors.

Usually, the FWHM scattering angle dependence is given by a monotonic function [31]:

$$H_j^2 = U \mathrm{tg}^2\theta_j + V \mathrm{tg}\theta_j + W \,, \tag{4.2}$$

where U, V and W are parameters responsible for the instrumental broadening of X-ray diffraction peaks.

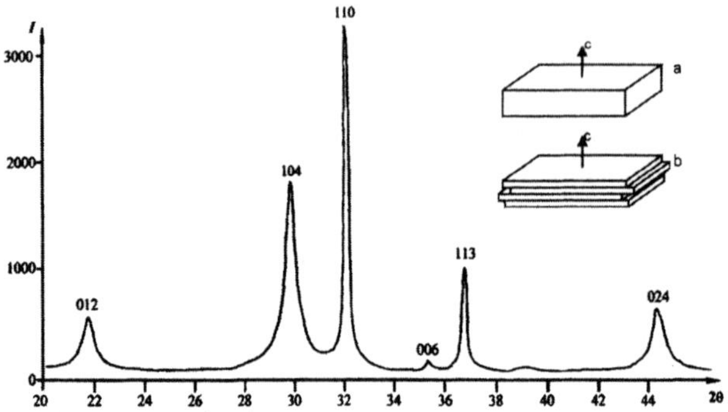

Fig. 4.2. The X-ray powder diffraction pattern of α-Fe_2O_3 obtained by thermal decomposition of hematite, and a microstructural model of the oxide nanoparticle (*insert*) [40]

The particle size and different lattice disorders in nanocrystalline materials appear to be the main reasons for the broadening of their diffraction maxima and the change of shape of the XRD peaks. In the simple case of an isotropic form of the particle with isotropic microstrains, the FWHM relationship remains a monotonic function. Evidently, (4.2) can be used as written but with new U', V', and W' values, or with some modifications (for example, when the particle size is taken into account as an explicit value [32]). However, anisotropic broadening of peaks such as that shown in Fig. 4.2 is often observed in the XRD patterns of powder materials and may be caused by different reasons – (i) anisotropic particle shape, (ii) anisotropic strains, and (iii) the presence of stacking faults in the sample structure.

Some authors have proposed the introduction of an anisotropic particle shape and an anisotropic microstrain distribution using second-rank symmetric tensors [33–37] or other direction-dependent models [38, 39] for the description of the anisotropic broadening of the XRD peaks. In the light of these approaches, the parameters of models (e.g. components of tensors) are related to each other by the symmetry of the crystal class. This is valid for many practical situations and useful in that sense that one could obtain important structural characteristics such as the atomic coordinates, the occupation, and the Debye–Waller factors, along with the microstructural characteristics of the materials under study.

Nevertheless, even the modified Rietveld method fails to describe some cases of anisotropic broadening of XRD peaks. Let us consider the results of the structural analysis of α-Fe_2O_3 prepared by the dehydration of FeOOH as an example of such a situation [40]. Figure 4.2 shows the XRD pattern of the oxide sample. As is evident from TEM results, the particles of the parent hydroxide, FeOOH, are crystallized in the shape of rectangular platelets in accordance with the orthorhombic crystallographic symmetry of this structure. After dehydration, particle shape looks almost the same as shown in the insert of Fig. 4.2 and marked "a", but the symmetry of the formed oxide phase, α-Fe_2O_3, is hexagonal. As a result, particle dimensions in the crystallographically equivalent directions are different; therefore, it is impossible to use the formalism given in [33–38], where the authors proposed describing the particle shape according to crystal symmetry rules. The second reason for the very complicated character of the peak broadening in this XRD pattern is the presence of coherent interfaces between the nanocrystalline domains lying in the (001) plane of the α-Fe_2O_3 structure. Schematically, this kind of domain stacking is depicted in the insert of Fig. 4.2 and marked "b". The FWHM of the XRD reflections of $00l$ type of a thin plate-like crystal possessing the smallest size in the [001] direction is expected to be larger than that of the hkl reflections. But when nanocrystalline domains are coherently stacking in the crystal structure, the FWHM of the $00l$ reflections should be smaller than that of the hkl reflections (and this is indeed observed in the experimental XRD patterns of our sample [40] for an 006 reflection), since the peak width depends on the whole thickness of the particle instead of the thickness of one isolated domain.

Thus, we encounter the problem of reliable description of the broadening of XRD peaks:

1. If the shape of the nanoparticle under study does not coincide with the symmetry of its crystal lattice (the classic case of pseudomorphism).
2. If some planar defects, including stacking faults, exist in the crystal structure, and their presence results in anisotropic broadening of the XRD reflections.

We have developed a computer program [41] based on the modified Rietveld algorithm that allows us to use the individual profile parameters of every diffraction peak for the simulation of XRD patterns. They include the positions of the diffraction peaks ($2\theta_j$), the peak halfwidth (H_j), and parameters describing the peak shape, such as the m_j power index for the Pearson VII function. These and other structural parameters can be refined during the process of the fitting of theoretical data to the experimental ones. In our opinion, such a formalized approach is a logical extension of the Rietveld method based on the use of empiric functions.

Even without knowledge of the nature of peak broadening in XRD patterns, we may successfully use the proposed algorithm for the refinement of the main structural parameters of some powder materials, including atomic coordinates, the occupation of structural positions, etc. We would like to show here the results of the crystal structure refinement of nonstoichiometric magnesia–alumina spinel, as an example of such a study [42,43]. The sample, of $Mg_{0.36}Al_{2.44}O_4$ composition, was prepared by the low-temperature calcination of a mixture of magnesium nitrate with aluminum hydroxide, followed by thermal treatment at 1000°C. Closed vacancy loops, which are the specific defects of the crystal structure, are clearly seen in the TEM micrograph of

Table 4.1. Refinement of the cation distribution in nonstoichiometric magnesia–alumina spinel (chemical composition $Mg_{0.36}Al_{2.43}O_4$)

Atom	Position	Coordinates	Occupation	Cation distribution
A	8a	1/8, 1/8, 1/8	0.95	0.35 Mg + 0.6 Al
B	16d	1/2, 1/2, 1/2	0.87	1.74 Al
A^*	8b	3/8, 3/8, 3/8	0.05	0.05 (Al,Mg)
B^*	16c	0, 0, 0	0.025	0.05 (Al,Mg)
O	32e	0.258, 0.258, 0.258	1	–

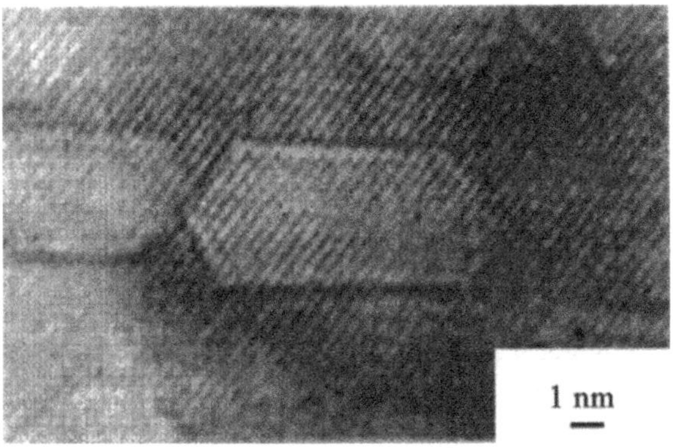

Fig. 4.3. TEM micrograph of dislocation loops in nonstoichiometric magnesia–alumina spinel [42]

the spinel sample, as is evident from Fig. 4.3. Stacking faults form the loop walls, and cation vacancies are concentrated mainly in the vicinity of these defects. Due to their presence, anisotropically broadened peaks have been observed in the XRD pattern of this sample, and therefore the profile parameters of the individual peaks were used for the theoretical calculation of the XRD patterns. Table 4.1 and Fig. 4.4 show some results of the refinement of the cation distribution over the structure of nonstoichiometric spinel. From these data it is evident that a portion of the cations occupies the A^* and B^* positions, which are empty in the regular spinel structure. Note that the number of A^*cations is equal to that of the B^*cations, as well as to the number of vacant tetrahedrons in A positions. We have suggested that the empty positions and interstitial atoms are produced by translating one part of the crystal with respect to the other along the (111) plane, i.e. according to the classic mechanism of stacking fault formation. The interstitial atoms are not located randomly, as point defects, in the spinel structure, but are aligned in a row, thus becoming part of the loop structure. Indeed, analysis of the anisotropic broadening of the diffraction peaks using the method described in [44] indicated that stacking faults bounded by half-dislocations where the

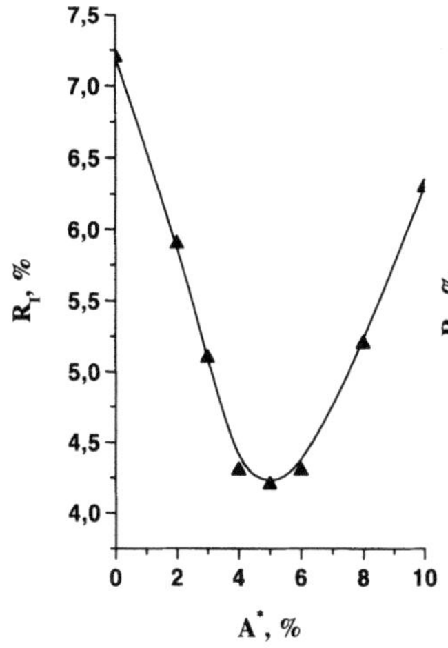

Fig. 4.4. The change in the R_I-factor with the cation occupation of additional tetrahedral (A^*) and octahedral (B^*) positions in the structure of magnesia–alumina spinel [42]

Fig. 4.5. Projection of the spinel structure onto the (100) plane (**a**) and a model of the stacking fault defect (**b**) [42]

dominant structural defects in this sample. A model of this defect, simulated with the help of XRD structural refinement and TEM analysis, is shown in Fig. 4.5.

Thus, our modified version of the Rietveld method allows us to make structural refinement of materials for which the XRD patterns are characterized by anisotropic broadening of the diffraction peaks. Sometimes, analysis of the results of refinement of the structure (occupation) and profile (peak halfwidth) parameters succeeds in providing reliable models of planar defects. However, to check the reliability of these models, one needs to use other simulation methods of XRD patterns. These methods should take into consideration not only the general principles of periodic crystal structure but also different kinds of structural imperfection. We will discuss one of these approaches in detail in the next section.

4.3 Microstructure Modelling and Full Powder Pattern Simulation Using the Kakinoki–Komura Method

4.3.1 The Choice of the Simulation Algorithm

The above-listed examples of the structural study of some disordered materials clearly show that simulation of models of nanocrystals with stacking faults as well as 1D nanostructures is the main problem in the XRD analysis of nanocrystalline materials. Before going on to the description of the algorithm realized in our program package, let us briefly consider some approaches to the calculation of X-ray diffraction effects induced by the presence of faulting in the structure of the materials under investigation. In principle, scattering intensity functions for any structure types, as well as for any kind of planar defects, may be obtained in explicit forms. This approach has been realized for f.c.c., h.c.p. and b.c.c. crystals [45,46], some materials possessing corundum- and spinel-like arrangements [44]. One can also find in the literature examples of the study of some compounds with other structural types; however, one can hardly consider that above-mentioned approach forms the basis of a generalized algorithm for the calculation of diffraction effects from imperfect crystals.

Some calculation programs provide the possibility of correcting diffraction effects induced by planar defects. Thompson and co-authors [33] made some modifications to the Rietveld method that enable the incorporation line broadening from microtwins. Solovyov [37] also proposed some alterations to the Rietveld procedure, but only for the special case in which crystals are composed of two types of unit cells that differ in atomic arrangement and/or content and have the same shape and size. Moreover, just broadening effects, rather than the effects of peak shifts and asymmetry resulting from planar defects, can be calculated. Scardi, Leoni and Dong [47] introduced a different approach, Whole Powder Pattern Modelling (WPPM, see Chapter 3), for the calculation of the diffraction profile without using arbitrary analytic profile functions. In the framework of this method, simulation of the XRD pattern is performed directly in terms of physical and instrumental effects influencing the line profile. But in the framework of this approach, analytical expressions of Fourier coefficients of the profile components should again be specified in accordance with the type of planar defect. Thus, it is necessary to change the calculation procedure for every new material with another kind of structural defect, i.e. in every particular case.

In the field of study of nanostructural materials, principal limitation of the Rietveld (including our procedure described in Sect. 4.2) and WPPM methods lies in their main concept, which implies a structural model based on the notion of 3D regular crystal ordering. Consequently, it is imperative to perform a calculation of the structural amplitudes only for Bragg reflections with integer *hkl* indices. Then calculated amplitudes are blurred in the vicinity of these Bragg reflections by the use of different functions – empirically chosen analytical functions (Rietveld procedure) or numerical functions obtained by the convolution of different broadening effects (WPPM). But it should be noted that at a high degree of structural defects and given a complicated character in their correlation in the sample structure, diffusion scattering effects lying beyond the positions of the Bragg peaks arise in the X-ray diffraction patterns. For the analysis of such diffraction effects, methods that do not possess the limitations of the 3D model and that make possible the calculation of the scattering intensity at any point of the reciprocal lattice would be highly satisfying.

Screening the literature and choosing the algorithm, we came to the conclusion that approaches for the analysis of stacking faults in close-packed structures [45] as well as the theory of diffraction by one-dimensionally disordered (lamellar) materials [48,49] could be used for simulation of the XRD patterns of 1D nanosystems. In the framework under consideration, the 1D nanostructure is described as an imperfect crystal, the statistical sequence of the biperiodic layers being disordered with respect to each other in a single direction. The particle size and crystallite shape are taken into account in an explicit form. Different kinds of disorders are specified with the use of parameters responsible for the variations in the layer positions.

Nowadays, there are two basic approaches to the calculation of the scattering intensity due to a 1D disordered structure. The first approach involves a direct computer simulation of the ensemble of particles with stacking faults, using the Monte Carlo method [50–53], see also Chapter 13. The alternative approach is based on simulation of the 1D structure by introducing order–disorder parameters and some probability coefficients that allow us to take into account the statistical character of the defect distribution. In principle, both approaches should provide similar (or close) results. However, while it is relatively simple to realize, the Monte Carlo method often requires a long running time for the generation of all possible configurations.

4.3.2 The Theory of Scattering by a 1D Disordered Crystal

The theory of scattering by 1D disordered lamellar structures based on the probabilistic approach is outlined in reference [48]. Crystals are represented as a stack of biperiodic layers that alternate along the layer normal (usually the $\mathbf{c}$-axis of the layer) in a way specified by some probability coefficients. With such a kind of crystal stacking, the lack of crystal periodicity in one direction leads to a specific type of scattering intensity distribution: the intensity of the diffracted X-ray beams in the reciprocal space differs from zero not only at points with integer h, k, and l indices, but also along lattice rods parallel to the $\mathbf{c}^*$-axis and coming through the hk nodes of the 2D reciprocal lattice. For one isolated crystal composed of N layers, the total amplitude of the diffracted X-ray beams is given by the following expression:

$$A(\boldsymbol{s}) = \sum_{m=1}^{N} \Phi_m(\boldsymbol{s}) \exp(-2\pi i \boldsymbol{s} \boldsymbol{r}_m)\,, \tag{4.3}$$

where $\boldsymbol{s}$ is a diffraction vector that defines the position of any point with (x^*, y^*, z^*) coordinates in the reciprocal space, and $\Phi_m(\boldsymbol{s})$ is the scattering amplitude from layer m, which is shifted from the coordinate origin by the vector $\boldsymbol{r}_m$.

For convenience, the scattering intensity normalized to the crystal unit cell can be defined as the sum of two terms:

$$i(\boldsymbol{s}) = \frac{\Omega}{N\sigma} A(\boldsymbol{s}) A^*(\boldsymbol{s}) = \frac{\Omega}{N\sigma} \left\{ \sum_{m=1}^{N} \Phi_m(\boldsymbol{s}) \Phi_m^*(\boldsymbol{s}) \right.$$
$$\left. + \sum_{m=1}^{N} \sum_{m'=1}^{N} \Phi_m(\boldsymbol{s}) \Phi_{m'}^*(\boldsymbol{s}) \exp\left[-2\pi i \boldsymbol{s}(\boldsymbol{r}_m - \boldsymbol{r}_{m'})\right] \right\}, \tag{4.4}$$

where $\Omega = [\boldsymbol{ab}]$ is the area of the 2D unit cell; and σ is the area of the layer, $m \neq m'$.

The first term in (4.4) is responsible for the intensity contribution from independent scattering by all layers; the value of the second term depends on the relative displacement between the m and m' layers. If $\boldsymbol{r}_n = \boldsymbol{r}_m - \boldsymbol{r}_{m'}$ and $n = |m - m'|$, the scattering intensity is

$$i(\boldsymbol{s}) = \frac{\Omega}{N\sigma} \left\{ \sum_{m=1}^{N} |\Phi_m(\boldsymbol{s})|^2 \right.$$
$$\left. + 2\mathrm{Re} \left[\sum_{n=1}^{N-1} \sum_{m=1}^{N-n} \Phi_m(\boldsymbol{s}) \Phi_{m+n}^*(\boldsymbol{s}) \exp(-2\pi i \boldsymbol{s} \boldsymbol{r}_n) \right] \right\}, \tag{4.5}$$

Equation (4.5) is suitable for calculations in the case of a definite layer sequence, i.e. when it is known which layer is located in the n-position. For average values,

$$\left\langle |\Phi(\boldsymbol{s})|^2 \right\rangle = \frac{1}{N} \sum_{m=1}^{N} |\Phi_m(\boldsymbol{s})|^2 \tag{4.6a}$$

$$\langle \Phi(\boldsymbol{s}) \Phi_n^*(\boldsymbol{s}) \exp(-2\pi i \boldsymbol{s} \boldsymbol{r}_n) \rangle = \frac{1}{N-n} \sum_{m=1}^{N-n} \Phi_m(\boldsymbol{s}) \Phi_{m+n}^* \exp(-2\pi i \boldsymbol{s} \boldsymbol{r}_n) \tag{4.6b}$$

Equation (4.5) is transformed into

$$i(\boldsymbol{s}) = \frac{\Omega}{\sigma}\left\{\left\langle |\Phi(\boldsymbol{s})|^2 + 2\mathrm{Re}\left[\sum_{n=1}^{N-1}\frac{N-n}{N}\langle\Phi(\boldsymbol{s})\Phi_n^*(\boldsymbol{s})\exp(-2\pi i\boldsymbol{s}\boldsymbol{r}_n)\rangle\right]\right\rangle\right\}. \tag{4.7}$$

Let us now consider peculiarities of X-ray diffraction from a single biperiodic layer. In this case, the scattering geometry may be depicted as reciprocal rods coming through the nodes of the 2D reciprocal lattice. Their cross-section is indefinitely small when the layer size is indefinite. If the layer has a defined dimension, the size and shape of the rod cross-section become dependent on those of the layer.

The scattering amplitude for a layer, normalized to the unit cell area, can be written as

$$\Phi(\boldsymbol{s}) = \frac{F_{hk}(l)D(\varepsilon_h, \varepsilon_k)}{\Omega}, \tag{4.8}$$

where $F_{hk}(l)$ is scattering amplitude along hk rod for a infinite layer, $D(\varepsilon_h, \varepsilon_k)$ is a function centered on the hk rods and dependent on the size and shape of the layers in their own plane, $(\varepsilon_h, \varepsilon_k-)$ are coordinates along the a^*- and b^*-axes, and $\boldsymbol{s} = (h+\varepsilon_h)\boldsymbol{a}^* + (k+\varepsilon_k)\boldsymbol{b}^* + l\boldsymbol{c}^*$.

To calculate the scattering intensity using (4.7), it is necessary to prescribe the probability parameters for appearance of any layer combination in the stack composed of two, three or more layers. The types of the first and last layers in each subsequence, as well as position of the last layer with respect to the first one in common coordinates, are also to be specified. These requirements may be fulfilled if a probability rule for the generation of the layer succession is known. For this purpose, Kakinoki and Komura [54, 55] proposed to use the statistical Markov chain.

The probability process comprises a Markov chain of order m ($m \geq 1$) if the probability of a new state depends only on the number m of former states preceding the new one [56]. Use of the Markov chain allows us to generate any statistical sequences of the layers as well as to take into account short-order parameters upon alternation the layer. Crystals are suggested to have layers of differing compositions and structural arrangements. Their combination along the layer normal and the character of their superposition in the basal plane may be specified using order-disorder parameters.

Jagodzinski [57] was the first person to consider short-range order in the layer sequence as an important characteristic. For this purpose, he introduced a new parameter, defined as the number of preceding layers upon which the probability of finding a layer of a given type at any place in the stack depends. This parameter, denoted by S, designates *a short-order factor in the alternation of the layers.* If $S = 0$, the layer sequence is completely disordered and defined only by the relative content of a definite layer type, W_i. If $S = 1$, one should prescribe W_i and add a new parameter, P_{ij}, which is responsible for the conditional probability of the appearance of layer j after layer i (or pair ij). In the case of $S = 2$, an additional probability parameter P_{ijk}, which defines the conditional probability of the appearance of layer k after the pair ij, is used for the description of the layer sequence, along with the W_i and P_{ij} values.

On the other hand, the manner of the superposition in any pair of neighboring layers in the stack may depend on that of the preceding layer pairs. Sakharov and others [58, 59] proposed a parameter equal to the number of preceding layer pairs which affects the manner of the layer superposition in definite pair. This parameter, designated by G, is called the *short-order factor of the layer superposition* in the basal plane. If $G = 0$, the manner of the superposition, m, in every pair of layers i and j is described only by the W_m^{ij} probability parameter. If $G = 1$, an additional P_{mn}^{ijk} parameter, along with W_m^{ij} value, should be used for simulation of the layer sequence. This parameter specifies the conditional probability of the appearance of the nth manner of layer superposition in the jk pair after the mth manner of layer superposition in the preceding ij pair.

Thus, a model of an disordered structure can be represented in two ways; namely, as an irregular layer sequence composed of the different layers, or as a layered stack in which the layers are displaced along their own plane. We would like to emphasize here that in the general case one should simultaneously take into account different types of layers and various manners of their superposition. Kakinoki and Komura [54, 55] provided an elegant and suitable matrix form of (4.7) for the description of the diffraction effects from a finite-sized crystal composed of alternating layers at $S \geq 0$, $G = 0$. Sakharov and others [58, 59] confirmed that (4.7) is also valid for the structures with $G \geq 1$. The only difference is associated with the rules for matrix filling.

Combining $\Phi(\boldsymbol{s})$ from (4.8) and the matrix form of (4.8), one can write the intensity distribution for:

1. A sequence of infinite layers

$$i_{hk}(l) = \frac{1}{\Omega\sigma}\left\{\mathrm{Trace}\boldsymbol{FW} + 2\mathrm{Re}\sum_n^{N-1}\frac{N-n}{N}\mathrm{Trace}\boldsymbol{FWQ}^n\right\}. \quad (4.9a)$$

2. A crystal composed of layers of a definite size

$$i_{hk}(\boldsymbol{s}) = i_{hk}(l)G(\varepsilon_h, \varepsilon_k)\,, \quad (4.9b)$$

where $G(\varepsilon_h, \varepsilon_k) = D(\varepsilon_h, \varepsilon_k)D^*(\varepsilon_h, \varepsilon_k)$.

$\boldsymbol{W}$, $\boldsymbol{F}$, and $\boldsymbol{Q}$ are square matrices. Matrix $\boldsymbol{W}$ is diagonal. Each nonzero element represents the product of the full probability for layer subsequences of length S and the full probability of different manners of superposition in the layer subsequence of length $(G+1)$. Matrix $\boldsymbol{F}$ consists of products of the layer structure factors. Each element of the matrix $\boldsymbol{Q}$ constitutes a product of the conditional probability for a layer subsequence composed of $S + 1$ layers of different types, the conditional probability for different manners of superposition in the sequence composed of $G+2$ layers, and the corresponding phase shift factor.

One can find elsewhere $\boldsymbol{F}$, $\boldsymbol{W}$, and $\boldsymbol{Q}$ matrices for different values of short-order factors [48]. Equations (4.9a) and (4.9b) provide a principal possibility for the calculation of the X-ray scattering intensity for a statistical layer sequence composed of an arbitrary number of biperiodic layers with different degrees of their short-order at the alternation of layers of variable composition and manners of superposition.

If different orientations of crystallites in the powder sample are taken into account, calculation of the intensity at each point $s = |\boldsymbol{s}| = 2\sin\theta/\lambda$ reduces to calculation of the intensity distributed at the intersection of a Ewald sphere of radius s and each hk rod:

$$I(s) = \frac{1}{4\pi s^2}\sum_{hk}\int_A i_{hk}(\boldsymbol{s})dA\,, \tag{4.10}$$

where $i_{hk}(\boldsymbol{s})$ is the intensity diffracted by a single crystallite at the point defined by vector $\boldsymbol{s}$ on the hk rod, dA is the element of the surface common to the sphere of radius s and the hk rod.

4.3.3 Disorders of the First and Second Types: A Generalized Algorithm for Microstrain Account

Besides the disorder in the alternation of layers of different types and the disorder in their superposition, random deviations of layers from their average positions in the normal and tangential directions may occur. According to Guinier's terminology [60], the fluctuations in the layer positions correspond to disorders of the first and second types. Disorder of the first type is characterized by the average ideal lattice, with interplanar distances that are very close to those of the real lattice. Thus this disorder type implies the presence of a long-range order, which is no more disturbed than a short-range order. Disorder of the second type is characterized by the fact that distances between neighboring atoms are subjected to weak fluctuations, but these fluctuations increase in amplitude depending on a number of intermediate atoms. In other words, atomic deviations from their equilibrium positions are accumulated.

X-ray powder diffraction patterns are distinctly different for these two types of lattice distortions [61]. Disorder of the first type results in an intensity decrease of Bragg reflections and an increase in the diffuse scattering. Disorder of the second type gives rise to broadening of the diffraction peaks.

In 1D disordered crystals disorders of the first and second types are caused by fluctuations in the positions of the layers. In the framework of the model, this means that the vectors of the inter-layer translations, which define the positions of the layers, vary around their average values. Taking account of disorders of the first and second types produces changes to the elements of the matrix $\boldsymbol{Q}^n$ in 4.9a) for the intensity distribution. The elements of matrix $\boldsymbol{Q}$ include the factor of the phase shift between the nearest layers as one of the multipliers. This factor is equal to $\exp(-2\pi i \boldsymbol{s} \boldsymbol{t}_m^{ij})$, where $\boldsymbol{t}_m^{ij}$ is the vector of translation of j-type layers with respect to i-type layers at the mth manner of their superposition. Assuming that the deviations of vectors of inter-layer translations are distributed over a 3D Gauss function, the phase shift factor can be written as

$$\int_{\Delta t} \exp\left[-2\pi i \boldsymbol{s}\left(\langle \boldsymbol{t}_m^{ij}\rangle \boldsymbol{\Delta t}\right)\right] \exp\left(\frac{\boldsymbol{\Delta t}^2}{2\boldsymbol{\sigma}_1^2}\right)$$
$$= \exp\left(-2\pi i \boldsymbol{s}\langle \boldsymbol{t}_m^{ij}\rangle\right) \exp\left[-2\pi^2(\boldsymbol{s}\boldsymbol{\sigma}_1)^2\right] , \quad (4.11)$$

where $\boldsymbol{\Delta t}$ is the deviation of the vector of inter-layer translations, $\boldsymbol{\sigma}_1^2$ is the variance of the distribution. Thus, in the case of fluctuations of vectors of inter-layer translations, the matrix $\boldsymbol{Q}$ should be multiplied by $\exp[-2\pi^2(\boldsymbol{s}\boldsymbol{\sigma}_1)^2]$. $\boldsymbol{Q}^n$ is introduced to account for the phase shift between two layers divided by n layers. By analogy with matrix $\boldsymbol{Q}$, the matrix $\boldsymbol{Q}^n$ should be multiplied by $\exp[-2\pi^2(\boldsymbol{s}\boldsymbol{\sigma}_n)^2]$, where $\boldsymbol{\sigma}_n^2$ is the variance of the distribution for this case.

Consider how the $\boldsymbol{\sigma}_n^2$ value depends on n. Suppose that we know the γ_1 law of the distribution of deviations of vectors of inter-layer translations for the m- and $(m+1)$-layers. To account for fluctuations of vectors of inter-layer translations in the calculation of the intensity distribution, it is necessary to find the γ_n law of distribution of deviations of vectors of inter-layer translations for the m- and $(m+n)$-layers. This is a simple procedure if all γ_1 distributions are independent or not correlated. In this case, the γ_n distribution is described as an n-fold convolution of γ_1 with itself:

$$\gamma_n = \underbrace{\gamma_1 \times \gamma_1 \times \ldots \times \gamma_1}_{n} \quad \text{for any } n .$$

Correspondingly, the additivity law is valid for variances:

$$\sigma_n^2 = n\sigma_1^2 . \quad (4.12)$$

Such independent fluctuations in the layer positions are thought to be disorders of the second type [48] and represent a model of a paracrystalline lattice proposed by Hosemann [62]. But it is only one particular case of disorders of the second type. It is possible to extend this approach to general disorder analysis. In the general case, the following expression for $\boldsymbol{\sigma}_n^2$ is valid:

$$\sigma_n^2 = \sum_{i=1}^{n}\sum_{j=1}^{n} \rho_{ij}\sigma_1^2 , \quad (4.13)$$

where ρ_{ij} is a correlation coefficient between deviations of layers of types i and j. The correlation coefficients ρ_{ij} are equal to 1 if $i = j$, by definition. One can prove the following:

1. In the absence of correlation between the γ_1 distributions of the fluctuations of the vectors of the inter-layer translations (when $\rho_{ij} = 0$ for all $i \neq j$), (4.12) is valid,
2. In the case of a maximal positive correlation of γ_1 the distributions of the fluctuations of the vectors of the inter-layer translations (when $\rho_{ij} = 1$ for all $i \neq j$) we deduce the next equation:

$$\sigma_n^2 = n^2 \sigma_1^2 . \tag{4.14}$$

Such fluctuations also correspond to disorders of the second type, namely microstrains which give a tangential dependence of the integral breadth of the XRD peak on the diffraction angle. It is evident that different dependencies of $\boldsymbol{\sigma}_n^2$ on n can be deduced for different values of the correlation coefficients.

Such fluctuations when $\gamma_n = \gamma_1$ and, correspondingly,

$$\sigma_n^2 = \sigma_1^2 = \text{const} , \tag{4.15}$$

are assigned to disorders of the first type. In this case, the multiplier

$$\exp[-2\pi^2 (\boldsymbol{s}\boldsymbol{\sigma}_1)^2]$$

is similar to the Debye–Waller factor and leads to a decrease in the calculated intensities of the Bragg peaks and an increase in diffuse scattering.

We suggest that in the general case the following equation will be true:

$$\sigma_n^2 = n^k \sigma_1^2 . \tag{4.16}$$

One can use k as an additional parameter that may change in the range from 0 to 2. This makes it possible to account for both disorders of the first type and all cases of disorder of the second type.

4.4 Program Description

The method for simulation of the full profile of XRD patterns described in the previous section has been realized as a software package with a user-friendly interface for IBM-compatible personal computers [63]. The program involves the following calculation steps:

1. Calculation of the distribution of the scattering amplitudes along rods in the reciprocal space for a layer of any type, taking account of the layer's crystallographic symmetry.
2. Calculation of the distribution of the scattering intensity diffracted by one stack along rods in the reciprocal space, taking account of: (a) the short-range order in the alternation and superposition of the layers; (b) disorders of the first and second types; and (c) the normal or lognormal distributions of stacks over the layer number.

3. Calculation of the distribution of the scattering intensity taking account of the size and shape of the layer. The shape of the layer may be specified as a circle or a rectangle.
4. Calculation of the XRD pattern from the particle ensemble, taking account of (a) the instrumental function, (b) the absorption, and (c) the background corrections.
5. A comparison between the theoretical and experimental XRD patterns, with the calculation of their quantitative correspondence parameters.
6. Model refinement.

There is also the possibility of calculations of the theoretical XRD patterns for biphase systems. This procedure may be useful for the structural study of nanoparticles dispersed over supports.

To take account of the instrumental factors, convolution of the theoretical XRD pattern with the instrumental function is performed. The form of the instrumental function, $G(s)$, depends on the scattering angle, but in practice it does not change over the intervals from 5 to 10° on 2θ. It enables us to divide the full profile of the XRD pattern into small sections and to calculate the instrumental functions for every isolated part of the spectrum. The file for the instrumental functions may be compiled in any convenient way; for example, the instrumental functions may be calculated on the basis of the geometry of the XRD apparatus [64], or one can obtain them experimentally with the help of standards. The background function is calculated by using experimental values via cubic spline approximation.

For the comparison of the theoretical and experimental XRD patterns, numerical criteria are used as follows:

$$R = \frac{\sum\limits_n |I^e(n) - KI^c(n) - I^b(n)|}{\sum\limits_n I^e(n)}, \tag{4.17a}$$

$$F = \sum_n \left[I^e(n) - KI^c(n) - I^b(n)\right]^2, \tag{4.17b}$$

$$D = \max\left|I^e(n) - KI^c(n) - I^b(n)\right|, \tag{4.17c}$$

where $I^e(n)$ is the experimental X-ray diffraction intensity at point n, $I^c(n)$ is the calculated X-ray diffraction intensity, $I^b(n)$ is the background intensity, and K is a normalizing factor determined by the least-squares method.

4.5 Simulation of the XRD Patterns of Nanocrystalline Materials: Some Examples

4.5.1 Filamentary Carbons

There are huge amounts of different nanostructural carbons that could be described within framework of disordered graphite-like materials. Figure 4.6

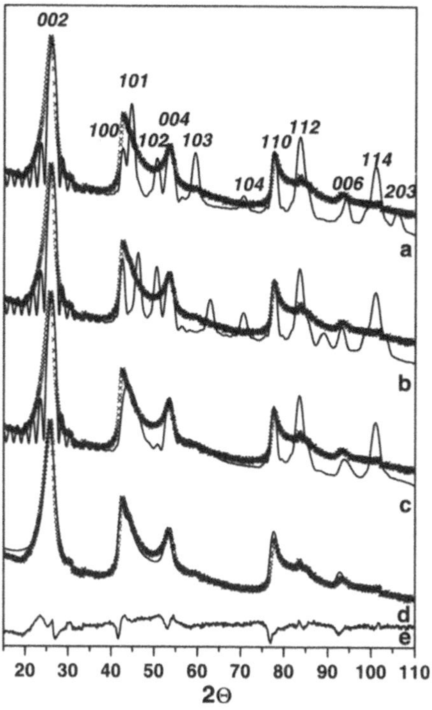

Fig. 4.6. Experimental (*cross*) and theoretical (*solid line*) XRD patterns (*logarithmic scale*) of filamentary carbon calculated with different models of carbon structure: an ordered 2H-polytype (**a**); an ordered 3R-polytype (**b**); a statistical sequence of layers A, B, and C ($W_A = W_B = W_C = 1/3$, $P_{AB} = P_{AC} = 1/2$) without fluctuation in the layer positions (**c**); a random sequence of layers A, B, and C with fluctuations in the layer positions (**d**); a difference curve for case d (**e**)

shows a typical XRD pattern (on a logarithm scale) of nanocarbons obtained via catalytic decomposition of hydrocarbons. These materials, which are widely used as adsorbents and catalyst supports, are highly porous systems with particle sizes lying between 3 and 10 nm. The carbon nanoparticle structure may be represented as a sequence of graphite nets, but the degree of disordering upon their stacking may vary from one sample to the other. Our method allows us to evaluate the character of the structural disordering in such samples.

The XRD pattern shown in Fig. 4.6 cannot be described using models of the regular hexagonal ($ABAB\ldots$) or rhombohedral ($ABCABC\ldots$) graphite 2H- and 3R-polytypes, respectively (Fig. 4.6a,b). Here A designates a layer with a structure similar to that of the single graphite net; layers B and C also have a graphite-like structure, but they are displaced from layer A by vectors $\boldsymbol{t}_1 = (1/3, 2/3)$ and $\boldsymbol{t}_2 = (2/3, 1/3)$, respectively. It is more likely that the XRD pattern given in Fig. 4.6 corresponds to the situation in which layers A, B, and C are statistically alternated with each other. This layer sequence may be specified by the following parameters: $S = 1$, $G = 0$, W_i, P_{ij} $(i, j = 1, 2, 3)$. In the framework of this model, the B or C layer may appear after the A layer with equal probability. But the theoretical XRD pattern calculated with this model does not fit well with the experimental XRD spectrum (Fig. 4.6c). Good correspondence between the experimental and theoretical XRD patterns is observed for the model, including the statistical deviations of graphite layers from their regular positions (Fig. 4.6d). Such a type of disorder results in the disappearance of the diffraction lines with hkl Miller indexes (besides $00l$ and $hk0$) and their transformation into diffusion scattering in a simulated XRD pattern that is similar to experimental one. Actually, the proposed model looks the same as the well-known turbostratic model of graphite-like materials [65].

4.5.2 Metallic Cobalt

Different kinds of structural disorders in metals have long been subjects of numerous studies. Metallic Co appears to be the first sample for which stacking faults randomly distributed over the volume of metallic particles have been experimentally observed [66]. However, the formation of more complicated defects such as coherent states, which are alternating microdomains with close-packed structures (microtwins and microinclusion of one phase into another), is of no surprise, as is evident from the references [67, 68]. An example of such a system is shown in Fig. 4.7. These coherent systems are

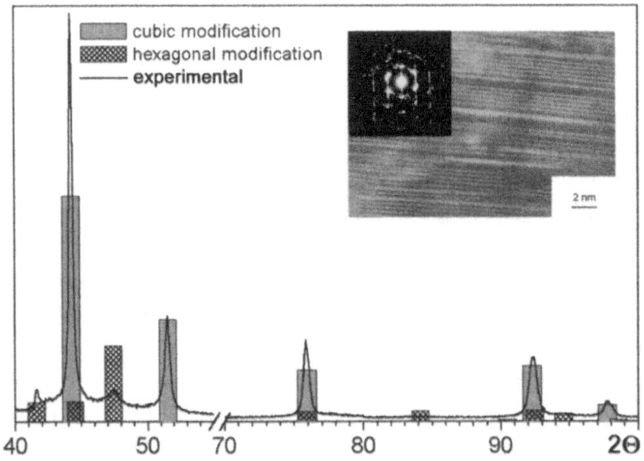

Fig. 4.7. XRD pattern and TEM structural image (*insert*) of nanocrystalline Co obtained after treatment in the catalytic reaction [68]

typical 1D nanostructures, and their diffraction effects were studied earlier by the authors of this chapter [68].

The diffraction peaks observed in the XRD patterns of these specimens were assigned to the reflections of cubic and hexagonal modifications of the metallic Co phase (Fig. 4.7). Formally, this diffraction pattern could be interpreted as being obtained from a physical mixture of Co particles with different structural types. But according to TEM analysis, isolated Co particles having alternatively pure hexagonal or cubic structures have not been identified. Instead, the microstructure of every crystallite is rather complicated: thin (about 2–3 nm) microdomains are clearly visible in the TEM micrograph of this sample, as shown in the insert of Fig. 4.7. It seems likely that these domains are either microtwins or alternating crystal regions with cubic or hexagonal packing. The appearance of ACB fragments in the regular layer sequence of $ABCABC$ type describes the former case well, whereas the latter situation can be represented as the formation of structural fragments with $ABAB$ hexagonal packing inside the cubic $ABCABC\ldots$ stack. Theoretical XRD patterns were calculated using both models.

Theoretical XRD Patterns for Cubic Co with Different Types of Microtwin Inclusions

In this case "layer 1" of ABC type and "layer 2" of ACB type (twin configuration of ABC) were specified as the building blocks of the models used for the calculations. Theoretical XRD patterns for models of cubic Co modification with the inclusion of a single twinning fragment are shown in Fig. 4.8. These models may be described by using the following values of different parameters: $S = 1$, $G = 0$, and $P_{22} = 0$ (see the parameter notations given in Sect. 4.3). When P_{22} is equal to zero, this means that the ACB layer (layer 2) cannot appear after a layer of similar structural composition. Let us see how the XRD pattern will change along with the varying of the W_2 parameter responsible for the relative content of the microtwins. At $W_2 = 0$, the model describes a regular cubic packing (see Fig. 4.8a). When the W_2 value increases from 0 to 0.2, additional strongly broadened peaks appear in the calculated XRD patterns (Fig. 4.8b). Upon further increase of the same parameter, the diffusion peaks became narrower (Figs. 4.8c,d). At $W_2 = 0.5$, the structural model represents an ordered 6H-polytype with regular $ABCACB\ldots$ packing.

If we let W_2 be constant, e.g. $W_2 = 0.3$, we begin to increase the probability of the appearance of layers ACB after each other, P_{22}, thus varying the thickness of microtwining in the cubic matrix. Upon an increase of the domain thickness, the diffraction peaks gradually disappear and are transformed into the diffusion background. At $W_2 = 0.3$ and $P_{22} = 0$, the theoretical pattern is almost similar to the experimental one; good fitting is observed especially for "cubic" reflections and diffusion scattering with 2θ equal to 47.4^0. But at the same time, the first diffusion peak in the theoretical XRD pattern is

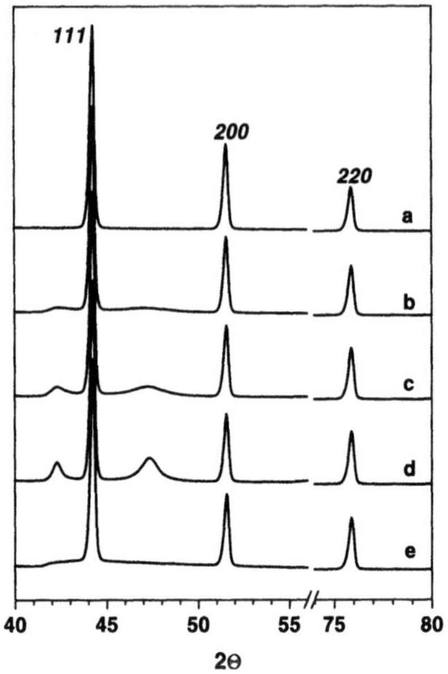

Fig. 4.8. Theoretical XRD patterns of Co cubic modification with different types of microtwin inclusions: regular cubic packing (**a**); randomly distributed ACB fragments (**b**); model (**b**) with $W_2 = 0.3$ (**c**); model (**b**) with $W_2 = 0.4$ (**d**); microtwining with a relative probability for the appearance of an ACB fragment after a ACB fragment, $P_{22} = 0.3$ ($W_2 = 0.3$) (**e**)

shifted into the high-angle range in comparison with the position of the same peak in the experimental spectrum (Fig. 4.10a).

Theoretical XRD Patterns for Cubic Co with Different Types of Inclusions of Hexagonal Microdomains

The AB type of layer packing was chosen as the building block. Then the cubic (AB–CA–BC) and hexagonal (AB–AB) structural modifications may be represented by sequential shifts of the AB layers by vectors $\boldsymbol{t}_1 = (1/3, 2/3)$ and $\boldsymbol{t}_2 = (0, 0)$, respectively. This means that both modifications can be simulated using one layer of the same structure (layer 1) and two different manners (1 and 2) of the superposition of this layer onto itself. How will the XRD patterns change if the cubic structure includes single hexagonal $ABAB$ fragments or, in other words, if intrinsic stacking faults appear at random in this structure? This model is specified by the following parameters: $S = 0$, $G = 1$, $W_2^{11} \neq 0$, and $P_{22}^{111} = 0$, where W_2^{11} is a parameter responsible for the probability of the superposition 2 of layer 1 onto itself. P_{22}^{111} is the conditional probability of the appearance of a pair comprised of 1-layers and superimposed by 2-way after the same layer pair. At $P_{22}^{111} = 0$, only the formation of $ABAB$ fragments (instead of $ABABAB$ and longer) is possible.

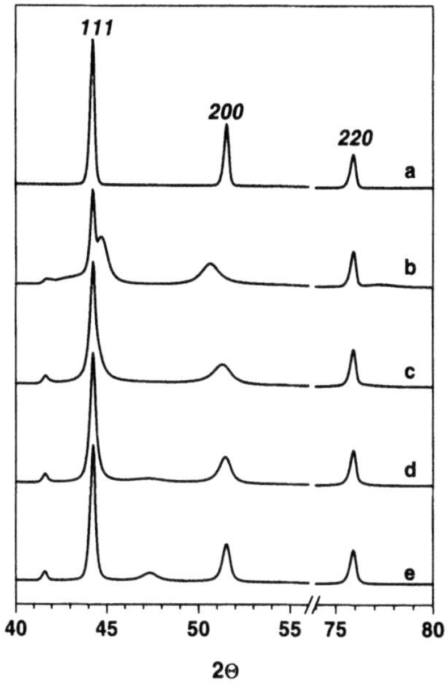

Fig. 4.9. Theoretical XRD patterns of Co cubic modifications with different types of inclusions of hexagonal microdomains: regular cubic packing ($W_2^{11} = 0$) (**a**); statistical distribution of t_2 vectors, i.e. randomly distributed stacking faults, $W_2^{11} = 0.2$, $W_{22}^{111} = 0$ (**b**); and inclusions of the long fragments of hexagonal close packing at $W_2^{11} = 0.2$, $P_{22}^{111} = 0.4$ (**c**), $P_{22}^{111} = 0.6$ (**d**), and $P_{22}^{111} = 0.8$ (**e**)

Figure 4.9a shows the XRD pattern calculated with the use of the model with $W_2^{11} = 0$. At $W_2^{11} = 0.2$ one can see a splitting of the 111 reflection in the diffraction pattern (Fig. 4.9b). The peak with indices 200 shifts into the small-angle region and becomes wider. Additionally, a weak diffusion peak appears at a 2θ angle of 41.7°. The observed XRD pattern is similar to that obtained from material characterized by a random stacking faults distribution analogous to the case described in [5, 60]. Very close results were obtained by Ustinov [69], who used the Monte Carlo method for the calculation of the theoretical XRD patterns.

Let us increase the microdomain thickness while W_2^{11} is fixed (e.g. equal to 0.2) and P_{22}^{111} varies from 0. When the latter parameter increases, the degree of splitting of the 111 reflection decreases. The peak with indices 200 becomes narrower, and its position is close to that which is characteristic for a regular cubic structure (Fig. 4.9c). At $P_{22}^{111} = 0.6$, a diffusion halo appears at a 2θ value of 47° (Fig. 4.9d). With a further increase in the probability parameter, this diffusion peak becomes narrower (Fig. 4.9e).

Our results obtained for the model that implies cubic inclusions into a hexagonal matrix are in line with data provided by Jagodzinski [57, 70], who claimed that in this case both "cubic" and "hexagonal" reflections should appear in the XRD pattern.

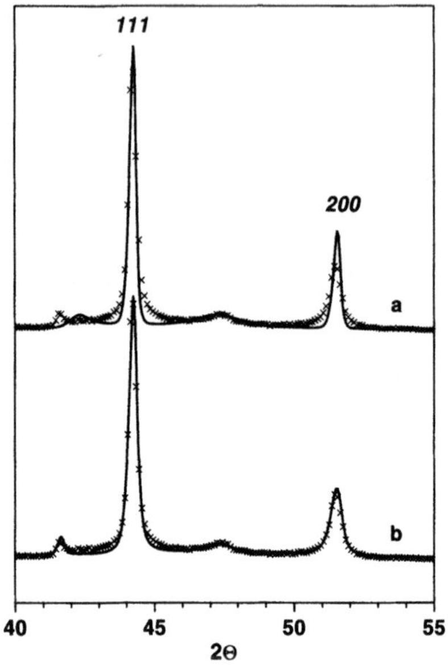

Fig. 4.10. Experimental (*cross*) and theoretical (*solid line*) XRD patterns calculated using different models of nanocrystalline Co: thin microtwins, $W_2 = 0.3$, $P_{22} = 0$ (**a**); alternating microdomains with cubic and hexagonal close packings, $W_2^{11} = 0.2$, $P_{22}^{111} = 0.8$ (**b**)

The best correspondence between the calculated and experimental XRD patterns is achieved at a P_{22}^{111} value equal to 0.8 (Fig. 4.10b). For the model involving alternating microdomains of different Co modifications, the positions of all calculated peaks in the XRD pattern are found to be the same as the experimental ones, which was not observed for the model based on the inclusion of microtwin fragments into the cubic Co structure. Therefore, the structure of metallic Co particles is thought to be composed of coherent microdomains possessing different structural arrangements that are characteristic of cubic and hexagonal modifications of metallic Co.

All of the theoretical XRD patterns have been calculated for an average particle size equal to 60 nm. The mean thickness of the hexagonal domains, $\langle l \rangle$, can be evaluated with the help of the following equation:

$$\langle l \rangle = \frac{\sum_m \varpi_m l_m}{\sum_m \varpi_m},$$

where ω_m is the probability of the appearance of a domain the thickness of which is specified by l_m ($m = 1, 2, \ldots$). l_1 is the thickness of the *AB–AB* fragment, l_2 is the thickness of *AB–AB–AB* fragment, and so on. In our case $l = 2$ nm, which is in good accordance with the value obtained directly from the TEM micrograph for this sample [68].

4.5.3 Metastable In–Ni Alloys

Nonequilibrium substitutional solid solutions obtained by mechanochemical alloying are systems that possess excess free energy due to nonstoichiometric defects as well as defects caused by plastic deformation. This stage in the formation of intermetallic compounds comes before the formation of solid solutions of metals with negative enthalpy of mixing [71]. The concentrations of In and Ni in the alloys were 11.3 and 88.7% at., respectively. The sequential formation of $Ni_{10}In_{27}$, Ni_2In_3, and NiIn is observed in In–Ni systems upon their mechanochemical treatment for 1–4 min. Further activation of the intermetallic compounds with a metal solvent leads to formation of the solid solution, with highly disordered structures [72].

TEM analysis has revealed that size of the particles formed is over 100 nm and that their structure is highly distorted. A high concentration of microstrains, distributed rather nonuniformly over the particle volume, was found to be the main peculiarity of the alloy microstructure. There are crystal regions with condensed strains along with areas of the regular lattice without any distortions, as shown in Fig. 4.11. It seems likely that the distortions appear due to the static deviations of atoms from their average positions in the crystal lattice, since Ni and In atoms differ strongly in their metallic radii. The second reason for microstrain formation is defects caused by plastic deformation (dislocations and stacking faults). But single dislocations in the TEM micrographs were practically invisible, probably due to the strongly distorted sample structure.

We would like to know whether our method will be useful for the study of such systems as the above-described alloys, which are structures in which a break in the long-range ordering is caused by the accumulated short-order distortions. Figure 4.12 shows the theoretical and experimental XRD patterns for In–Ni alloys. Simulating the crystal model, we have suggested that coherent scattering domains (CSD) with effective sizes similar to the average lattice distances at which long-range order is not disturbed form the XRD pattern. In other words, there is an assumption that for any combination of

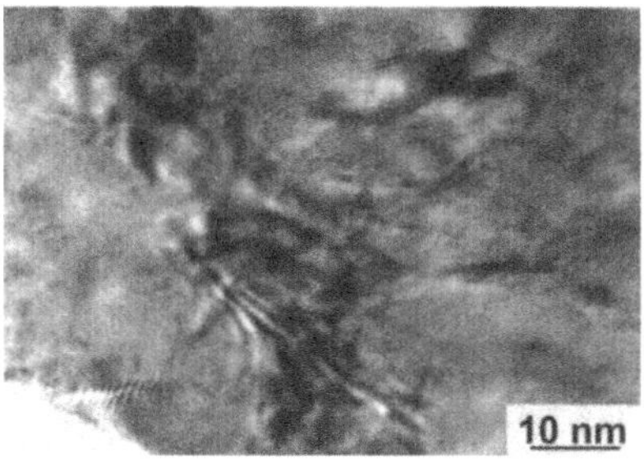

Fig. 4.11. A TEM micrograph of the In–Ni alloy obtained by mechanochemical synthesis [72]

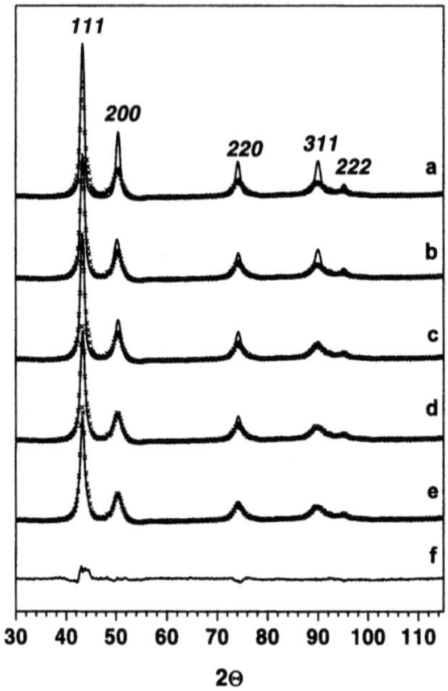

Fig. 4.12. The experimental (*cross*) and theoretical (*solid line*) XRD patterns for In–Ni alloy: a particle of isotropic shape (cylinder with height equal to its diameter, 10 nm) (**a**); the same as (**a**) with the inclusion of stacking faults (**b**); the same as (**a**) with the inclusion of microstrains (disorders of the second order) (**c**); a model combined from (**b**) and (**c**) (**d**); a particle of anisotropic shape (cylinder height and diameter equal to 12 nm and 6 nm, respectively) with the inclusion of stacking faults and microstrains (**e**); the difference curve for model (**e**) (**f**) [72]

the layers following one after another, the first (or initial) layer and layer N are scattering coherently, whereas the scattering due to the layer $N+1$ is incoherent with respect to the first layer. Close-packed atomic layers alternating along the directions of type [111] were chosen as the building blocks for the model simulation. Model 1 (Fig. 4.12a) describes the scattering by CSD of a cylindrical shape, with the cylinder height equal to its diameter. The particle size obtained from the halfwidth of the 111 reflection in the experimental XRD pattern was used as an approximate value of the CSD parameter for the calculations. As a result, good correspondence between the theoretical and experimental XRD patterns is only observed for 111 reflection. Let us introduce planar defects such as stacking faults. As is evident from Fig. 4.12b, they affect the shape and FWHM of all of the diffraction peaks, especially the 200 reflection, which becomes wider and shifts into the small-angle region of the theoretical XRD pattern. The same effect is observed in the experimental XRD spectrum. At the same time, variations in the layer positions lead, as expected, to the broadening of all of the reflections, especially those with indices 311 and 222 (Fig. 4.12c). Taking into account stacking faults and strains simultaneously, we obtained a reasonable fit between the calculated and experimental XRD patterns, except for the 220 peak (see Fig. 4.12d).

It should be noted that simulation of the X-ray diffraction patterns of these systems with the help of our software package showed some correlation between size parameters and microstrain values. At the same time, it was not possible to get a good fit between the experimental and calculated X-ray diffraction patterns of the samples by introducing only microstrain parameters. That is why we also varied CSD during simulation. It seems likely that this effect arises as a result of the nonuniform distribution of microstrains; namely, the existence of a crystal region with condensed strains along with crystal areas of regular structure without distortions. Such nonuniformity in the microstrain distribution may lead to the effect of order-independent broadening in the diffraction patterns [9].

To obtain a good fit between the theoretical and experimental XRD patterns, we also varied the values of the layer diameter assuming anisotropy of the CSD [72]. This enabled us to describe reasonably well the shape of the 220 diffraction peak (Fig. 4.12e). However, it is obvious that in this case the layer diameter is simply a fitting parameter, but at the moment we cannot attribute any clear physical sense to this parameter.

Some possibilities of our method for the study of systems with long-order distortion are outlined below:

1. In the framework of our model, the effective size of CSD responsible for the order-independent broadening of diffraction peaks is actually the long-order parameter, regardless of the type of structural defect. One can consider this parameter as an integral characteristic of the order–disorder degree in paracrystalline materials, which is especially important for the structural characterization of samples synthesized under variable preparation conditions. In our case, the effective particle size goes through a minimum, which indicates the process of homogenization of the solid solutions with an increase in time of mechanical activation of the alloys [72].
2. The concentration of the stacking faults can be calculated relatively easily, since these defects influence the position and shape of the diffraction peaks in a specific way, especially the 111 and 200 reflections.
3. Microstrains represented in the form of variable interplanar distances can be evaluated adequately, at least, for the directions similar to the layer normal in the layered structure model. In this sense, our method may be considered as a direct analog of the Warren–Averbach method for the separation of order-dependent and order-independent broadening effects, with the help of two reflection orders from the single system of lattice planes. Our method differs from that of Warren–Averbach only by the fact that we solve the inverse problem instead of the direct one.

4.6 Conclusion

Methods for the full profile analysis of X-ray diffraction patterns raise possibilities for the structural study of nanocrystalline materials. The Rietveld method is very useful for refinement of the average atomic structure of nanoparticles, but it suffers from serious problems in the description of anisotropically broadened XRD peaks from 1D systems. These problems can be solved in different ways; in particular, with the help of the proposed algorithm that allows us to refine the individual profile parameters of the X-ray diffraction peaks. With regard to the structural study of nanocrystalline materials, an approach involving the joint use of the modified Rietveld procedure and method of full profile analysis based on the simulation of partially disordered structures has surfaced. One of those methods, based on the diffraction theory for 1D disordered (lamellar) materials, is outlined in this chapter.

We would like to emphasize that our software works well not only in the case of lamellar systems. By using our software package, X-ray diffraction patterns of ensembles of crystallites with arbitrary syngonies can be calculated, taking account of the crystallite shape (cylinder, cube, or rectangular parallelepiped) and the crystallite size distribution. Upon simulation, it is possible to vary the vectors of the inter-layer translations in the directions parallel and perpendicular to the layers. Nonaccumulated fluctuations in the layer positions emulate static atom deviations that are analogous to the effect described by the Debye–Waller factor. Accumulated fluctuations in the layer positions are similar to strain fields of different types, including such particular cases as (i) uniform strain fields which result in a tangential scattering angle dependence of the integral breadth (positive correlated fluctuations) and (ii) strain fields existing in the paracrystalline lattice (independent fluctuations).

The Kakinoki–Komura matrix method, where a structural model is considered as a layer sequence with different degrees of layer ordering upon their alternation and superposition, is a powerful tool used in our software package. Owing to this approach, we can simulate a structure with randomly distributed stacking faults as well as 1D coherent nanosystems composed of thin alternating domains with different crystalline structures. It is important that our software enables us to calculate the scattering intensity at any point of the X-ray powder diffraction pattern rather than only in the Bragg positions, and near to them. It follows that diffusion scattering caused by the different correlations in the distribution of planar defects may be also analyzed.

At the moment, the main disadvantage of our method is the requirement of a scenario related to the case in which the planar defects lie parallel to each other in the same plane in every nanoparticle. In practice, many materials possess more complicated structures, with interacting stacking faults similar those found in magnesia–alumina spinel. We hope to solve such problems in future with the help of our developing algorithm.

Let us outline briefly the role of transmission electron microscopy when simulating a model of a partially disordered nanocrystal and interpreting the results of the full profile analysis of X-ray diffraction patterns. Firstly, TEM data appear to be the main criterion at the stage of qualitative choice of the structural model. The reader can see an example of such an approach in Sect. 4.5, which is devoted to the structural study of 1D coherent domains in Co metallic particles. The experimental diffraction pattern of this system could be considered as that of a mixture of particles that have different structures. However, the results of TEM analysis of this sample encouraged us to find another structural models one of those appear to be reliable. Lively discussions on the peculiarities of the shape of one diffraction peak or another might be wasteful without preliminary TEM information on the microstructure and types of the crystal structure disorders. But at the same time, XRD allows us to evaluate models based on TEM data and, more importantly, to provide quantitative nanostructural parameters.

In our opinion, the joint use of X-ray diffraction analysis and the method of transmission electron microscopy provide the possibility of adequate description of the features of the structural arrangement of nanocrystalline materials.

Acknowledgement

The authors are grateful to the organizers and sponsors of the Size-Strain III Conference for providing the possibility of publication of this material.

References

1. H.M. Rietveld: J. Appl. Cryst. **2**, 65 (1969)
2. G.K. Williamson and W.H. Hall: Acta Metall. **1**, 22 (1953)
3. B.E. Warren and B.L. Averbach: J. Appl. Phys. **23**, 497 (1952)
4. A.J.C. Wilson: Proc. Phys. Soc. **80**, 286 (1962)
5. M.S. Paterson: J. Appl. Phys. **23**, 805 (1952)
6. E. Bertaut: Acta Cryst. **3**, 14 (1959)
7. J.I. Langford: J. Appl. Cryst. **11**, 10 (1978)
8. J.I. Langford, R. Delhez, T.H. Keijser and E.J. Mittemeijer: Aust. J. Phys. **41**, 173 (1988)
9. J.G.M. van Berkum, R. Delhez, T.H. Keijser and E.J. Mittemeijer: Acta Cryst. A **52**, 730 (1996)
10. R.W. Siegel: *Proc. of NATO ASI.* Ed. by M. Nastasi, D.M. Parkin and H. Gleiter. Vol. 233 (Kluwer, Dordrecht Boston London 1993) p. 509
11. A. Makino, A. Inoue and T. Masumoto: Nanostruct. Mater. **6**, 985 (1995)
12. K. Yamauchi and Y. Yoshizawa: Nanostruct. Mater. **6**, 247 (1995)
13. A.I. Gusev: Uspekhi phyzicheskikh nauk **168**, 55 (1998) (in Russian)
14. F.C. Frank and W.T. Read: Phys. Rev. **79**, 722 (1950)
15. R.A. Andrievskii and A.M. Glezer: Fizika metallov i metallovedenie **88**, 50 (1999) (in Russian)

16. V.G. Gryaznov, I.A. Polonsky, A.E. Romanov and L.I. Trusov: Phys. Rev. B **44**, 42 (1991)
17. W.H. Straub, T. Gessman, W. Sigle, F. Phillipp, A. Seeger and H.E. Schaefer: Nanostruct. Mater. **6**, 571 (1995)
18. S. Wen and D. Yan: Ceramics Intern. **21**, 109 (1995)
19. D.H. Ping, D.X. Li and H.G. Ye: J. Mater. Sci. Letters. **14**, 1536 (1995)
20. G.A. Zenkovetz, V.A. Gavrilov, G.V. Tarasova and G.N. Kryukova: Kolloidni zhurnal **4**, 762 (1987) (in Russian)
21. S.V. Tsybulya, G.N. Kryukova, S.N. Goncharova and B.S. Bal'zhinimaev: J. Catalysis **154**, 194 (1995)
22. D.I. Kochubey, S.N. Pavlova, B.N. Novgorodov, G.N. Kryukova and V.A. Sadykov: J. Catalysis **161**, 500 (1996)
23. G.N. Kryukova: Cryst. Res. Technol. **33**, 1113 (1998)
24. R.Z. Valiev, A.V. Korznikov and R.R, Mulyukov: Mater. Sci. Engineering. A **168**, 141 (1993)
25. A.A. Nazarov, A.E. Romanov and R.Z. Valiev: Acta Metall. Mater. **4**, 94 (1994)
26. R.Z. Valiev: Nanostruct. Mater. **6**, 73 (1995)
27. R.Z. Valiev and I.V. Alexandrov: Nanostruct. Mater. **12**, 35 (1999)
28. M.M. Hall, V.G. Veeraraghavan, H. Rubin and P.G. Winchell: J. Appl. Cryst. **10**, 66 (1977)
29. R.A. Young and D.B. Wiles: J. Appl. Cryst. **15**, 430 (1982)
30. M. Ahtee, L. Unonius, M. Nurmeto and P. Suorti: J. Appl. Cryst. **17**, 352 (1984)
31. G. Caglioti, A. Paoletti and F.P. Ricci: Nucl. Instrum. Methods. **3**, 223 (1958)
32. A. Santoro: Godish. Jugosl. Cent. Kristallogr. **17**, 26 (1987)
33. P. Thompson, J.J. Really and J.M. Hastings: J. Less-Common Metals **129**, 105 (1987)
34. C. Lartigue, A. Le Bail and A.A. Percheron-Guegan: J. Less-Common Metals **129**, 65 (1987)
35. L. Lutterotti and P. Scardi; J. Appl. Cryst. **23**, 246 (1990)
36. T.B. Zunic, J. Dohrup: Powder Diffraction **14**, 203 (1999)
37. L. Solovyov: J. Appl. Cryst. **33**, 338 (2000)
38. N.C. Popa: J. Appl. Cryst. **31**, 176 (1998)
39. P. Stephens: J. Appl. Cryst. **32**, 281 (1999)
40. G.N. Kryukova, S.V. Tsybulya, L.P. Solovyeva, V.A. Sadykov and M.P. Andrianova: Mater. Sci. Engineering A **149**, 121 (1991)
41. S.V. Tsybulya: *Ph.D. Thesis.* Kishinev Institute of Applied Physics (1989)
42. S.V. Tsybulya, L.P. Solovyeva, G.N. Kryukova and E.M. Moroz: JSTCAM **32**, 121 (1991)
43. S.V. Tsybulya, L.P. Solovyeva: *Proceed. Int. Conf. Size-Strain-95.* Materials Structure in Chemistry, Biology, Physics and Technology **2**, 63 (1995)
44. V.I. Fadeeva, A.S. Kagan and L.S. Zevin: Acta Cryst. A **33**, 386 (1977)
45. B.E. Warren: *X-ray diffraction.* (Dover, New York, 1990)
46. L. Velterop, R. Delhez, T.H. de Keijser, E.J. Mittemejer and D. Reefman: J. Appl. Cryst. **33**, 296 (2000)
47. P. Scardi, M. Leoni and Y.H. Dong: Commission on powder diffraction. Newsletter **24**, 23 (2000)
48. V.A. Drits and C. Tchoubar: *X-ray Diffraction by Disordered Lamellar Structures.* (Springer, Berlin Heidelberg New York 1990) pp. 371

49. R.S. Reynolds, in *Modern Powder Diffraction.* Ed. by D.L. Bish and J.E. Post. (Reviews in Mineralogy; 20) (Mineralogical Society of America, Washington 1989) pp. 145–181
50. B.I. Nikolin, A.Y. Babkevich: Acta. Cryst. A **45**, 797 (1989)
51. A.Y. Babkevich and B.I. Nikolin: *Proceed. of the 3rd European Powder Diffraction Conf.* (Vienna, Austria 1993) pp. 57–62
52. R. Berliner and S.A. Werner: Phys. Rev. **34**, 3586 (1986)
53. A.I. Ustinov, in *Defect and Microsructure Analysis by Diffraction.* Ed. by R. Snyder, J. Fiala and H.J. Bunge. (Oxford Science Publication, New York 1999) pp. 264–317
54. J. Kakinoki and Y. Komura: J. Phys. Soc. Japan. **7**, 30 (1952)
55. J. Kakinoki and Y. Komura: J. Phys. Soc. Japan. **9**, 169 (1954)
56. M.S. Bartlett: *An introduction to stochastic process with special reference to methods and applications.* (Cambridge University Press, Cambridge 1960) pp. 312
57. H. Jagodzinski: Acta Cryst. **2**, 201 (1949)
58. B.A. Sakharov, A.S. Naumov and V.A. Drits: Dokladi Academii Nauk **265**, 339 (1982) (in Russian)
59. B.A. Sakharov, A.S. Naumov and V.A. Drits: Dokladi Academii Nauk **265**, 871 (1982) (in Russian)
60. A. Guinier: *Theorie et Technique de la Radiocristallographie.* (Dunod, Paris 1964) pp. 740
61. M.A. Krivoglaz: *Theory of Scattering of X-ray and Thermal Neutrons by Real Crystals.* (Fizmatgiz, Moscow 1967) pp. 336 (In Russian)
62. A.M. Hindeleh and R. Hosemann: J. Phys. C: Solid State Phys. **21**, 4155 (1988)
63. S.V. Cherepanova and S.V. Tsybulya: J. Molec. Catalysis. A **158**, 263 (2000)
64. R.W. Cheary and A.A. Coelho: *Size-Strain III Conference. 2–5 Dec. 2001. Trento, Italy*, Abstracts 21
65. R.E. Franklin: Acta Cryst. **4**, 253 (1951)
66. A.J.C. Wilson: Proc. Roy. Soc. A **180**, 277 (1942)
67. A.A. Khasin, T.M. Yur'eva, V.I. Zaikovskii, L.M. Plyasova and V.N. Parmon: Kinetics and Catalysis **39**, 400 (1998)
68. S.V. Tsybulya, S.V. Cherepanova, A.A. Khasin, V.V. Zaikovskii and V.N. Parmon: Dokladi Academii Nauk **366**, 216 (1999) (in Russian)
69. A.I. Ustinov: *Size-Strain III Conference. 2–5 Dec. 2001. Trento, Italy*, Abstracts 25–26
70. H. Yagodzinski: Acta Cryst. **7**, 17 (1954)
71. T.F. Grigorieva, A.P. Barinova, V.V. Boldyrev and E.Y. Ivanov: *Proc. Intern. Symp. on Metast. MA Nanostruct. Mat. ISMANAM 96. Rome, Italy 1996.* pp. 577–582
72. V.V. Boldyrev, S.V. Tsybuluya, S.V. Cherepanova, G.N. Kryukova and T.F. Grigorieva: Dokladi Academii Nauk **361**, 784 (1998) (in Russian)

5 Crystallite Size and Residual Strain/Stress Modeling in Rietveld Refinement

D. Balzar and N.C. Popa

We consider the problem of microstructure modeling in Rietveld refinement, in particular of crystallite size and residual strain/stress and propose the improvements to existing models. For the crystallite-size modeling, we consider the size-broadened profile given by the lognormal size distribution of spherical crystallites. We derive an analytical approximation in terms of Lorentz and Gauss functions for the size-broadened profile. Its advantage is that it can be analytically convolved with the strain-broadened and instrumental-broadened profiles, thus replacing time-consuming numerical convolutions that are commonly used. The method is tested on two CeO_2 powders; one shows "super-Lorentzian" profiles that we successfully model under the assumption of a broad lognormal size distribution. We show that the Voigt function, as an assumed size-broadened profile, fails for both very narrow and broad size distributions. In regard to the residual strain/stress modeling, we describe a novel approach to model diffraction line shifts caused by elastic residual or applied stresses in textured polycrystals. The model yields the complete texture-weighted strain and stress tensors as a function of crystallite orientations, the so-called weighted strain orientation distribution function. We also present the method to determine the average values of macroscopic strain and stress tensors that are frequently of interest. The model is particularly suitable for introduction in Rietveld-refinement programs, as the requirements on refinable parameters for all crystal Laue classes are outlined. The effects of sample symmetry are also included and conditions for strain invariance to both crystal and sample symmetries are discussed.

5.1 Introduction

Nowadays, Rietveld refinement [1] is a very mature set of procedures to refine complete structural information about crystalline materials. Since recently, there is an effort to include the determination of a variety of microstructural information. Notably, the inclusion of texture-determination methods through the generalized spherical harmonics [2,3], anisotropic line-broadening modeling for arbitrary crystal symmetry [4,5], and residual strain/stress determination [6,7] were very important extensions of the traditional Rietveld method into new areas. Although the main intention of these models is to allow for determination of important physical properties, the actual use is twofold; For instance, in order to obtain accurate structure factors and unit-cell dimensions, it is possible to only correct for "undesirable" effects (thus, they can be considered as systematic errors), whether it is change of relative line intensities (texture), line shifts (residual strain/stress) or anisotropic line broadening (crystallite-size and crystalline-defect related strain). In general, these new models are more robust and should be preferred over equivalent approaches that were used previously. As an example, March–Dollase [8] correction for preferred orientation proved to be less stable and even inadequate for strong sample textures, as opposed to the spherical-harmonics approach.

The aim here is to critically evaluate and present improvements in two areas: The first part of the chapter includes modeling of size-broadened profile in terms of the lognormal size distribution of spherical crystallites. This problem has recently received a considerable attention through studies of cubic ceria [9–12]. We focus on an analytical approximation of the size-broadened profile (Sect. 5.2.5), address deficiency of currently used analytical models based on the Voigt function and its approximation pseudo-Voigt (Sect. 5.2.6), and briefly outline an anisotropic extension to the model (Sect. 5.2.7). The second part discusses an implementation of residual strain/stress determination in Rietveld refinement. Our approach builds on recent efforts toward representation of strain and stress tensors in terms of generalized spherical harmonics [13–15]. Here, we present a modified approach that is based on a development of the texture-weighted strain tensor in a series of generalized spherical harmonics (Sect. 5.3.3) [16]. The approach is also extended for arbitrary crystalline symmetry (Sect. 5.3.4) and adopted specifically for implementation in Rietveld refinement programs.

5.2 Modeling of the Crystallite-Size Broadened Line Profile in Rietveld Refinement

5.2.1 Background

Diffraction line-broadening analysis (LBA) is the method to extract the crystallite size and inhomogeneous strain (strain associated with local distortions of the lattice due to defects) from powder diffraction data. Nowadays, there are two main phenomenological approaches: One is based on the integral breadth of the diffraction lines, and gives the volume-averaged apparent dimension in the direction normal to the reflecting planes (domain size), D_V (see, for instance, [17]). The other is based on the Fourier analysis of the line profile and gives the area-averaged apparent dimension (column length) in the direction normal to the reflecting planes, D_A [18]. These two crystallite-size dimensions are called "apparent" because they only relate to the real average crystallite dimension. The second derivative of the size Fourier coefficient is related to the column-length distribution function [19,20]. Presuming an identical shape for all crystallites, the column-length distribution function can be calculated as an integral with a variable limit over the crystallite distribution function [21]. Determination of the real crystallite size distribution then includes the third derivatives of the size Fourier coefficients, which further amplifies already large initial errors of the experimental Fourier coefficients. Additionally, Fourier coefficients for large harmonic numbers are unreliable because of approximations inherent to the size-strain separation approaches (see again [17, 20]).

An unbiased determination of crystallite size and strain can be undertaken only if the diffraction lines are not overlapped. Otherwise, a pattern fitting and decomposition must be performed first before any LBA can be undertaken. Gauss and Lorentz (Cauchy) analytical functions are used most often in the whole-pattern fitting with a rather loose association of strain broadening given by the former and size broadening by the latter. Therefore, it was generally accepted that the best profile is a convolution of the two, that is, the Voigt function. Especially, it was shown that modeling both size-broadened and strain-broadened profiles by Voigt functions [22, 23], is much more flexible and able to accommodate different sample types and deformations. Pearson VII and pseudo-Voigt functions were introduced as satisfactory approximations for the Voigt function, but much faster to evaluate, which is of utmost importance in Rietveld refinement. Hence, one of the most used function in the profile fitting is the Thompson, Cox and Hastings [24] pseudo-Voigt function, where the full widths at half maximum (FWHM) of the Gauss and Lorentz components are constrained to be the same and equal to the width of the pseudo-Voigt function itself.

There is no *a priori* reason to believe that a simple Voigt model can successfully describe all size and strain broadening-related effects because these profiles might correspond only to very special cases of size and strain broadening. A better approach might be the use of line profiles derived from physical models. Recently, Langford, Louër and Scardi [10] have discussed the influence of normal and lognormal size distributions of spherical crystallites on the diffraction line profile. An important conclusion of the authors was that the profile calculated from these physical models is intermediate between Gauss and Lorentz functions. However, they have not considered the case of a large dispersion. Here, by using the same model of spherical crystallites and the lognormal distribution, we show that at very large dispersion the Voigt function or its approximations fail to accurately describe the size-broadened profile. We show that in the case of a large dispersion the profile becomes "super-Lorentzian", which is normally assumed to be a result of a multimodal size distribution. We derive an analytical approximation that is a sum of up to three Gauss and/or Lorentz functions that can be convoluted analytically with the strain and instrumental profile, thus facilitating an easy introduction in the whole pattern fitting programs, such as Rietveld refinement. The gamma distribution is another bell-shaped distribution that was used to describe the distribution of crystallite sizes [25, 26] or crystalline defects [27]. We discuss the gamma distribution and its analytical approximation in a similar way elsewhere [11].

5.2.2 The Crystallite-Size Broadened Line Profile and Size Distribution

We first define the variables s and $\boldsymbol{h}$ in the following way:

$$s = \frac{2\sin\theta}{\lambda} - \frac{1}{d}, \quad \boldsymbol{h} = \frac{\boldsymbol{H}}{H}, \tag{5.1}$$

where 2θ is the scattering angle, λ is the wavelength and d is the interplanar distance corresponding to the reciprocal vector $\boldsymbol{H}$. When the crystallites are not of identical size, the peak profile $P(s, \boldsymbol{h})$, normalized to unit area, has to be averaged over the size distribution:

$$\bar{P}(s, \boldsymbol{h}) = \frac{\langle VP(s, \boldsymbol{h})\rangle}{\langle V\rangle}. \tag{5.2a}$$

The averaged peak profile can be written as follows:

$$\bar{P}(s, \boldsymbol{h}) = 2\int_0^\infty \bar{\psi}(r, \boldsymbol{h}) \cos 2\pi srdr, \tag{5.2b}$$

where $\bar{\psi}(r, \boldsymbol{h})$ is the average of the Fourier transform $\psi(r, \boldsymbol{h})$ representing the ratio between the volume common to the crystallite and its "ghost" displaced at the distance r in the direction $\boldsymbol{h}$ and the crystallite volume V.

5.2.3 Application to the Lognormal Size Distribution: Isotropic Case

The lognormal distribution for spherical crystallites is characterized by two parameters, the average radius $\bar{R}$ of the particles and the dispersion σ_R^2 (see [26, 28]. It is convenient to define the dimensionless ratio [11]

$$c = \frac{\sigma_R^2}{\bar{R}^2} \tag{5.3}$$

to characterize the distribution. Then the lognormal distribution can be written as follows:

$$f(R) = R^{-1}[2\pi\ln(1+c)]^{-1/2}\exp\left\{-\frac{\ln^2[R\bar{R}^{-1}(1+c)^{1/2}]}{[2\ln(1+c)]}\right\}, \tag{5.4}$$

One can calculate the volume- and area-averaged dimensions as [11]:

$$D_V = \frac{3\bar{R}(1+c)^3}{2}, \tag{5.5}$$

$$D_A = \frac{4\bar{R}(1+c)^2}{3}. \tag{5.6}$$

Unfortunately, the size-broadened profile cannot be calculated analytically either in direct or Fourier space. For the numerical computation, by using simple transformations of the integration variable, (5.2a) can be reduced to a standard quadrature formula:

$$\bar{P}(s) = \left(\frac{3\bar{R}}{2}\right)(1+c)^3\bar{\Phi}(2\pi s\bar{R}), \tag{5.7a}$$

$$\bar{\Phi}(\chi) = \pi^{-1/2}\int_{-\infty}^{\infty} dt\exp(-t^2) \times\Phi\left\{\chi(1+c)^{7/2}\exp\left[t(2\ln(1+c))^{1/2}\right]\right\}. \tag{5.7b}$$

Here, we assume an isotropic case with spherical crystallites with

$$\Phi(\chi) = \frac{(\chi^2+\sin^2\chi-\chi\sin 2\chi)}{\chi^4}, \tag{5.7c}$$

an interference function for a sphere. The integral (5.7b) can be computed by a standard Gauss–Hermite quadrature.

5.2.4 Determination of the Distribution Parameters

One can use D_V and D_A, obtained from the Fourier analysis of line profiles, to calculate values of $\bar{R}$ and c parameters by using (5.5) and (5.6) for the lognormal distribution [28]. However, except for materials with the highest crystallographic symmetry, line profiles commonly overlap, which makes it necessary to reconstruct overlapped profile tails, usually by profile fitting of simple analytical functions (Voigt or its approximations).

Another possibility to determine the size distribution parameters is from the parameters of the least-squares refined Fourier transform of size-broadened profile (5.7a) (the Fourier transform is given in [11, 26]). The result of the product of size Fourier coefficients and the distortion (strain) Fourier coefficients are fitted to the Fourier transform of measured diffraction profiles, corrected with the Fourier transform of the instrumental contribution [29]. However, in cases of significant profile overlapping, some way of pattern fitting also has to be used to resolve individual profiles, which is a requirement for performing the Fourier transform. Additionally, Fourier transforms of the experimental and instrumental profiles have to be calculated numerically, which introduces significant errors.

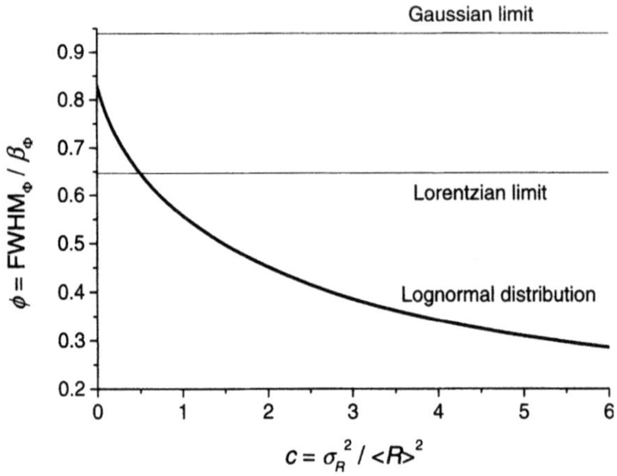

Fig. 5.1. The profile parameter ϕ as a function of c for spherical crystallites, distributed according to the lognormal distribution

To avoid the step of pattern fitting and decomposition with simple analytical functions, Langford, Louër and Scardi [10] proposed the whole-pattern fitting with individual profiles obtained by the numerical convolutions of the size, strain, and instrumental profiles in real space. Scardi and Leoni [26] proposed to calculate the profile in the whole pattern fitting by a numerical Fourier transform of size-broadened profile, multiplied by the strain and the instrumental Fourier coefficients. In this way multiple numerical integrations are avoided, all effects being accounted for in a single integral. Still, numerical convolution is computer-time intensive. As an illustration, we cite a recent example given by Scardi and Leoni [30]. They report on the whole pattern modeling (and fitting) by using numerical convolution through the Fourier transforms. By using a 1 GHz computer they report a time of 3 hours for the refinement of a pattern containing 7 peaks. With a 400 MHz computer, by using the analytical approximation for (5.7a)–(5.7c) [11], the time for the whole pattern fitting with 6 peaks was only 60–80 s. Therefore, it is advantageous, especially for the Rietveld refinement, to avoid any numerical integration or Fourier transform of the experimental pattern. This could be achieved if the exact size-broadened profile is accurately approximated by functions that can be analytically convolved with the strain and instrumental contributions, for instance, a combination of Gauss and Lorentz functions. We plot the profile parameter $\phi = (FWHM)_\Phi/\beta_\Phi$, defined as the ratio between the full width at half maximum of $\bar{\Phi}$ and its integral breadth, as a function of c in Fig. 5.1 for the lognormal distribution. $\bar{\Phi}\chi$ decreases with increasing χ much slower than a Lorentzian except for small values of c. But, for finite range of χ, the profile could be fairly described if more Lorentzian functions were added. This will be discussed in the next section.

5.2.5 The Analytical Approximation for $\bar{\Phi}(\chi)$

The function $\bar{\Phi}(\chi)$, given by (5.7b), has been calculated numerically in the range $1 \geq \bar{\Phi} > 10^{-2}$ by a Gauss–Hermite quadrature with 16 nodes for 53 values of c in the interval $0 \leq c \leq 6$. These exact profiles were least squares fitted by linear combinations of Gauss and Lorentz functions. The following combination was found to be a good approximation for c in this interval:

$$\bar{\Phi}(\chi) = \left(\frac{8}{3}\right)(1+c)^{-3} \times \left\{ \begin{array}{l} \eta_1 a_1^{-1}\left(1+\dfrac{4\chi^2}{a_1^2}\right)^{-1} + \eta_2 a_2^{-1}\left(1+\dfrac{4\chi^2}{a_2^2}\right)^{-1} \\ +(1-\eta_1-\eta_2)a_3^{-1} \left\{ \begin{array}{ll} \exp\left(\dfrac{-4\chi^2}{\pi a_3^2}\right) & \text{for } c \leq 1 \\ \left(1+\dfrac{4\chi^2}{a_3^2}\right)^{-1} & \text{for } c > 1 \end{array} \right. \end{array} \right\} . \tag{5.8a}$$

The parameters η_1, a_1, η_2, a_2 $(0 \leq \eta_i \leq 1)$ were freely adjusted during the fit but a_3 was constrained to conserve the integral breadth of the exact profile:

$$a_3 = \frac{(1-\eta_1-\eta_2)}{\left[\frac{3(1+c)^3}{8} - \frac{\eta_1}{a_1} - \frac{\eta_2}{a_2}\right]} . \tag{5.8b}$$

Further, the parameters η_1, a_1, η_2, a_2 were fitted by empirical analytical functions in c:

$$\eta_1(c) = 0.25631 + 0.018638c + 0.001155c^2 + 3.5671c\exp(-2.0467c^{0.93346}) \tag{5.9a}$$

$$a_1(c) = 4.02326\exp(-44.6429c) + 3.13982\exp(-7.01128c) + 0.580742\exp(-0.413958c) + 0.381245\exp(-1.10827c) \tag{5.9b}$$

$$\eta_2(c) = \begin{cases} 0 & \text{for } c \leq 0.4 \\ 0.59951 - 0.020058(c-0.4) - \dfrac{0.45347}{[1+3.3933(c-0.4)^2]} \\ -0.14604\exp[-0.49272(c-0.4)^2] & \text{for } c > 0.4 \end{cases} \tag{5.9c}$$

$$a_2(c) = 0.32781 \times [1 + 1.5399(c-0.4) - 0.21223(c-0.4)^2 + 0.18158(c-0.4)^3]^{-1} . \tag{5.9d}$$

From (5.9c), $\eta_2(c) = 0$ for $c \leq 0.4$. Therefore, from (5.8a) it follows that the pseudo-Voigt function is a satisfactory approximation for the size-broadened profile only for $c \leq 0.4$. For higher value of c a second Lorentz function must be added. Its weight increases with increasing c and the weight of the Gauss component decreases. For $c \approx 1$, the profile is well approximated by a sum of two Lorentz functions. A third Lorentz function must be added for $c > 1$. For $c > 6$, even three Lorentz functions are not enough for a satisfactory fit, but it is highly unlikely to find samples with such a large dispersion of crystallite sizes.

5.2.6 Limitations of the Common Analytical Approximations of the Size-Broadened Profile

From this discussion, it is evident that the size-broadened profile can be approximated by the pseudo-Voigt function, which is commonly used in Rietveld-refinement programs, only for samples that follow the lognormal size distribution of the crystallites with a limited dispersion in the region $0 \leq c \leq 0.4$. From Sect. 5.2.3, it follows that the ratio $D_V/D_A = (9/8)(1+c)$ can then take values in the range $(1.125, \infty)$. The size-broadened profile is very often modeled by the Voigt function. In order for the column-length distribution function to always be positive, it was found [23] that an assumed size-broadened Voigt function requires the ratio of volume-averaged and area-averaged domains to be in the range:

$$(2\pi e)^{1/2}\mathrm{erfc}\left(\frac{1}{2^{1/2}}\right) \approx 1.31 \leq \frac{D_V}{D_A} < 2\,. \tag{5.10}$$

The lower limit of this ratio constraints the dispersion parameter to $c \geq 0.164$. Therefore, the Voigt function appears to be an inadequate approximation for both very sharp and broad lognormal distributions of spherical crystallites! However, these additional restrictions are not placed on a pseudo-Voigt function, which indicates that the latter might be a better approximation for the size-broadened profile for samples with narrow crystallite-size distributions. As this appears as an apparent contradiction because pseudo-Voigt function is a good approximation to an exact Voigt function, we emphasize here that this is due to the mathematical formulation of these two functions. For the pseudo-Voigt approximation of $\Phi(\chi)$ the second derivative of the corresponding Fourier transform is always positive. This can be easily verified by using (5.9a)–(5.9d) with $c = 0$.

5.2.7 The Anisotropic Crystallite Shape

So far, we considered only spherical crystallites, that is, an isotropic crystallite shape. The size profiles (5.7a)–(5.7c) and (5.8a)–(5.8b) can be also applied to the nonspherical crystallites by using the phenomenological approach already reported by Popa [4]. In that paper, the size profile has been considered Lorentzian and the radius of the averaged "composite crystallite", dependent on the direction $\boldsymbol{h}$, was developed in symmetrized spherical harmonics specific to every Laue class. The "composite crystallite" was defined as an ensemble of identical crystallites with relative orientations given by the operation of the Laue point group. This model was introduced by Lutterotti [31] in his Rietveld program MAUD and successfully applied to refine the size anisotropy from an X-ray powder pattern measured on a sample of Y_2O_3 used in the Commission on Powder Diffraction (CPD) Round Robin [32]. The development of the crystallite radius in symmetrized spherical harmonics can be justified by invariance of the peak profile to the operations of the Laue point group. On this phenomenological basis, the extension of (5.7a)–(5.7c) and (5.8a)–(5.8b) to a general anisotropic case is made by replacing the parameters $\bar{R}$ and c by $\bar{R}_{\boldsymbol{h}}$ and $c_{\boldsymbol{h}}$ given by the following expressions:

$$\bar{R}_{\boldsymbol{h}} = \bar{R}_{00} + \sum_{l=2}^{\infty} \sum_{\mu=1}^{M(l)} \bar{R}_{l\mu} K_{l\mu}(\varphi, \beta) \,,$$

$$\bar{c}_{\boldsymbol{h}} = \bar{c}_{00} + \sum_{l=2}^{\infty} \sum_{\mu=1}^{M(l)} \bar{c}_{l\mu} K_{l\mu}(\varphi, \beta) \,. \tag{5.11}$$

Here $\bar{R}_{l\mu}$ and $\bar{c}_{l\mu}$ are refinable parameters in a profile fitting program, φ and β are the polar and azimuthal angles of $\boldsymbol{H}$ in an orthogonal crystallographic coordinate system and $K_{l\mu}(\varphi, \beta)$ are the symmetrized harmonics specific to the given Laue class. The symmetrized harmonics for all Laue classes were given by Popa [2].

5.2.8 Application to Two Cubic Ceria Samples

We tested the model of spherical crystallites distributed according to the lognormal size distribution on two CeO_2 powder samples. Sample 1 was prepared for the CPD Size-Strain Round Robin (see [7, 9]), and Sample 2 is a commercially available CeO_2 powder. A third sample was prepared from Sample 2 by annealing at 1300°C for 3 h, followed by slow cooling in the furnace, to determine the instrumental broadening. A comparative analysis with an NIST SRM660a LaB_6 powder, when used for correction of instrumental broadening, did not show significant changes (within a single standard deviation) in the final parameters of interest, that is, domain size and strain values. The X-ray diffraction patterns were collected using a commercial diffractometer in Bragg–Brentano geometry with Cu$K\alpha_{1,2}$ radiation. For every sample, three diffraction patterns were collected, at low (20–64.5° in 2θ, 0.01° step, 10 s/step for the annealed sample and 0.02° step, 65 s/step for broadened patterns), medium (64.5–102° in 2θ, 0.02° step, 25 s/step for the annealed sample and 0.04° step, 130 s/step for broadened patterns), and high (102–150° in 2θ, 0.02° step, 30 s/step for the annealed sample and 0.05° step, 160 s/step for broadened patterns) diffraction angles, to obtain comparable statistics.

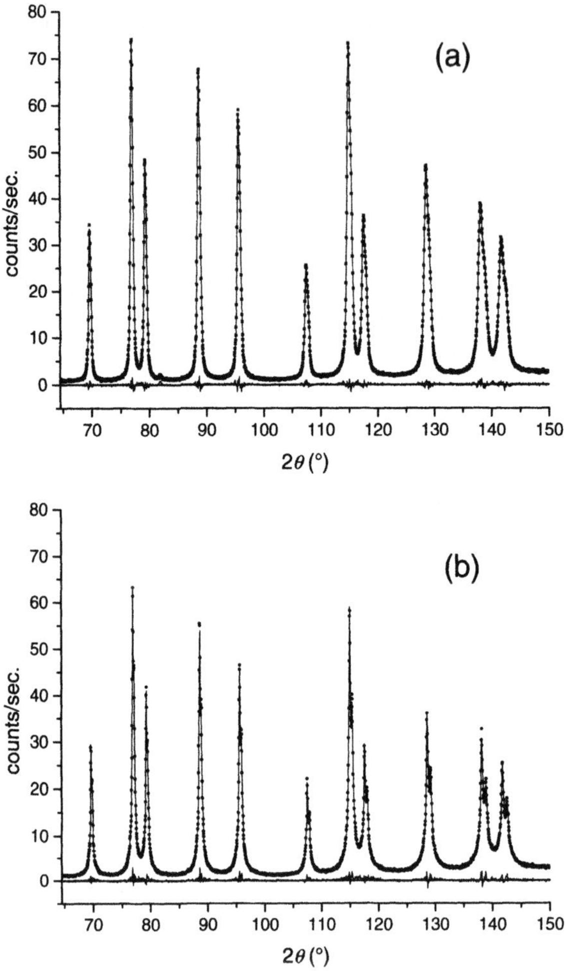

Fig. 5.2. The CeO_2 powder pattern fitted with the size-broadened profile calculated by (5.7a) and (5.8a) for the lognormal distribution: (**a**) Sample 1: $R_{wp} = 0.0390$; (**b**) Sample 2: $R_{wp} = 0.0477$

The diffraction patterns of the two samples show very different line profiles (see Fig. 5.2). Sample 1 shows "regularly" broadened lines with $K\alpha_1$ and $K\alpha_2$ overlapped at high angles, which can be successfully fitted by the Voigt function or its approximations, while Sample 2 exhibits "super-Lorentzian" profiles, with long tails and $K\alpha_1$ and $K\alpha_2$ separated at high angles.

Profiles of the annealed sample were corrected for the Lorentz-polarization factor and fitted by a sum of four Voigt functions resulting from the convolution of a Gaussian, modeling geometrical broadening, with four Lorentzians representing the spectral profile of the Cu $K\alpha_1$–$K\alpha_2$ line. The wavelengths and weights of four Lorentzians were taken from the literature [33]. The integral breadth, constant background, profile position and intensity were

Table 5.1. The results of the fit of an a priori lognormal size distribution for Sample 1 (S1) and Sample 2 (S2). The average radius $\bar{R}$, volume-averaged D_V, and area-averaged D_A apparent domain sizes are given in Å units

Sample	$\bar{R}$	c	D_V	D_A
S1	89.7(6)	0.181(4)	222(3)	167(2)
S2	16.8(2)	2.820(2)	1408(14)	328(3)

adjusted in the least-squares refinement. All profiles gave values of integral breadths within one standard deviation except for four lines at low angles, which are affected by asymmetry. To avoid systematic errors, the results from the first pattern at low diffraction angles were not included in the subsequent analysis.

Approximations (5.8a)–(5.8b) and (5.9a)–(5.9d) for the lognormal distribution were used to model the size-broadened profile. Although at least Sample 1 shows negligible strain broadening [9], we allowed for a small strain correction by convolving the size-broadened profile with a Gaussian and with the instrumental profile. Therefore, every line profile was a sum of up to 12 Voigt functions. The two diffraction patterns at medium and high diffraction angles were least-squares fitted independently. The free parameters were the intensity and position for every profile, and the global parameters included two background parameters, the size-distribution parameters $\bar{R}$, c and the Gaussian strain parameter. For Sample 1, the strain parameter was fixed to zero as we did not see any improvement in the fit. For Sample 2, the strain parameter, although very small ($\langle\varepsilon^2\rangle^{1/2} = 0.00033(1)$), improved the fit slightly.

The fit results for the lognormal distribution are shown in Fig. 5.2. For clarity, we plot both medium-angle and high-angle diffraction patterns together, although they were fitted independently. The numerical results are summarized in Table 5.1. The reported values are the average of the refined values obtained by independent fitting of the two diffraction patterns at medium and high diffraction angles. It is worthwhile to note that Sample 2 with "super-Lorentzian" profiles shows a good fit with large c parameter and consequently large difference between D_V and D_A ($D_V/D_A \approx 4.3$).

5.2.9 Conclusion

We show that the size-broadened line profiles can be satisfactorily modeled by an a priori size distribution. In particular, we discuss a widely used lognormal distribution. The line profiles of a commercial CeO_2 powder that showed tails falling off more slowly than the Lorentzian function (the so-called called "super-Lorentzian" profiles) were successfully fitted on the assumption of a broad lognormal size distribution. We avoid a need to numerically convolute either the size-broadened profile with the strain and instrumental profiles or numerical Fourier transforms by approximating the size-broadened profile by simple analytical functions that can be analytically convolved.

We found that the size-broadened profile can be approximated by the pseudo-Voigt function, which is commonly used in Rietveld-refinement programs, only for samples with the limited dispersion. Moreover, the Voigt function is not only inadequate for broad lognormal distribution of spherical crystallites, which is also true for the pseudo-Voigt function, but also for very sharp lognormal size distribution. The fact that the pseudo-Voigt and particularly Voigt function are found satisfactory in most cases probably indicates that most powders fall within a relatively narrow range of dispersion for the lognormal size distribution.

5.3 Modeling of Residual Strain/Stress in Rietveld Refinement

5.3.1 Background

A common approach for strain and stress determination uses the so-called $\sin^2 \Psi$ method [34,35], where the strain is derived from directional measurements of the interplanar spacing d as a function of the angle between the diffraction vector and an arbitrary direction in the specimen. An alternative approach to a traditional $\sin^2 \Psi$ method of strain determination was described by Ferrari and Lutterotti [6] and Balzar *et al.* [7]. It includes the refinement of strain and stress related parameters in Rietveld [1] refinement program. An advantage of this approach is that all available Bragg reflections are used simultaneously to obtain the strain tensor. Even if the strain/stress determination is not of interest, diffraction line shifts caused by residual stresses will generally be crystal-direction dependent. These diffraction line shifts should be corrected for, in order to carry out an accurate refinement using the Rietveld approach. However, for a successful application in Rietveld refinement, the challenge lies in the accurate modeling of strain and stress dependence on the crystallographic direction and the ability to handle arbitrary crystal symmetry. In a recent paper, Popa [36] presented a method for modeling of diffraction line shifts for all Laue classes in the frame of Voigt [37] and Reuss [38] approximations. The aim here is to propose an alternative method to accurately model diffraction line shifts in a Rietveld-refinement program for all Laue symmetries without making Voigt or Reuss approximations. This is accomplished by expanding strain and stress tensor components in series of spherical harmonics, similarly to the texture modeling, as described by Popa [2] and implemented into Rietveld-refinement program GSAS by Von Dreele [3]. Thus, strain and stress are determined by refinement of respective coefficients in the least-squares refinement procedure, simultaneously with other refinable parameters.

Recently, it was proposed that the strain/stress orientation distribution function (SODF), which is defined as a strain/stress tensor component as a function of crystallites orientations, be determined similarly to the crystallite orientation distribution function (CODF), through the expansion in series of generalized spherical harmonics [13–15]. That work assumes that the crystal symmetry acts on SODF in the same manner as on CODF. In this context, it means that all six components of the strain tensor in the sample coordinate system are invariant to the point-group symmetry operations. However, consider for instance, a polycrystalline assembly of non-spherical crystallites under the applied stress. A crystal symmetry operation on a crystallite could in general produce different physical strain states because of interactions with other crystallites in the aggregate, although two orientations are crystallographically equivalent. Therefore, strain/stress tensor elements are not invariant to the crystal symmetry operations in general. Contrarily, we postulate that the observable quantity measured by diffraction in a given sample direction, that is, the interplanar spacing d, averaged around the diffraction vector, is invariant to the point-group symmetry operations. Here, this invariance condition is used to derive the selection rules in SODF harmonic representation for all Laue classes.

5.3.2 The Measured Strain, the Average Strain and Stress Tensors

The diffraction method directly measures the interplanar spacing d along the direction of the diffraction vector, which must be parallel to a reciprocal lattice vector $\boldsymbol{H}$ for an (hkl) diffracting plane. Measured strain is then defined as an average change in the interplanar spacing from a reference value d_0:

$$\frac{\langle d \rangle}{d_0} - 1 = \frac{\langle \Delta H \rangle}{H} = -\langle \varepsilon_{\boldsymbol{h}}(\boldsymbol{y}) \rangle \,, \tag{5.12}$$

where the averaging is done by the rotation for ω around $\boldsymbol{h} = \boldsymbol{H}/H$, which is parallel to $\boldsymbol{y}$, the direction of the diffraction vector in the sample. If ε_i are the strain tensor elements in the crystallite coordinate system (in the condensed Voigt notation), (5.12) can be written as follows [36]:

$$\langle \varepsilon_{\boldsymbol{h}}(\boldsymbol{y}) \rangle = \sum_{i=1}^{6} E_i \frac{\int_0^{2\pi} d\omega f(\varphi_1', \Phi_0', \varphi_2') \varepsilon_i(\varphi_1', \Phi_0', \varphi_2')}{\int_0^{2\pi} d\omega f(\varphi_1', \Phi_0', \varphi_2')} \,. \tag{5.13}$$

Averaging is weighted by the CODF $f(\varphi_1, \Phi_0, \varphi_2)$. Here, $(\varphi_1, \Phi_0, \varphi_2)$ are the Euler's angles transforming the sample orthogonal coordinate system $\boldsymbol{y}_1, \boldsymbol{y}_2, \boldsymbol{y}_3$ into the crystallite orthogonal coordinate system $\boldsymbol{x}_1, \boldsymbol{x}_2, \boldsymbol{x}_3$, as defined by Bunge [39] (see Fig. 5.3). The integration in (5.13) is evaluated only

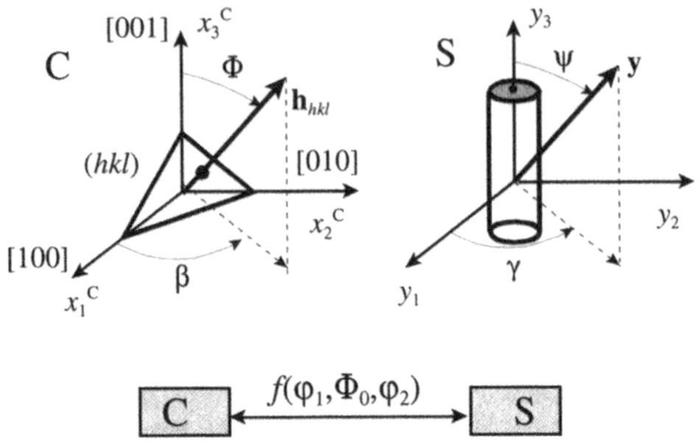

Fig. 5.3. Relationship between the sample coordinate system S and crystallite coordinate system C

over values of Euler's angles $(\varphi_1', \Phi_0', \varphi_2')$ that fulfill the condition $\boldsymbol{h} \parallel \boldsymbol{y}$. They depend on values of polar and azimuthal angles of $\boldsymbol{h}$ in $\boldsymbol{x}_1, \boldsymbol{x}_2, \boldsymbol{x}_3$, denoted as (Φ, β), of $\boldsymbol{y}$ in $\boldsymbol{y}_1, \boldsymbol{y}_2, \boldsymbol{y}_3$, denoted as (Ψ, γ), and on the rotation angle ω. The pairs of angles (Φ, β) and (Ψ, γ) give the direction cosines of $\boldsymbol{h}$ and $\boldsymbol{y}$ respectively, in their coordinate systems:

$$(A_1, A_2, A_3) = (\cos\beta \sin\Phi, \sin\beta \sin\Phi, \cos\Phi)\,, \tag{5.14a}$$

$$(B_1, B_2, B_3) = (\cos\gamma \sin\Psi, \sin\gamma \sin\Psi, \cos\Psi)\,. \tag{5.14b}$$

In (5.13) we used the following abbreviation:

$$(E_1 \ldots E_6) = (A_1^2, A_2^2, A_3^2, 2A_2A_3, 2A_1A_3, 2A_1A_2)\,. \tag{5.15}$$

The average strain and stress tensors in the sample coordinate system, $\bar{e}_i$ and $\bar{s}_i$, $(i = 1, 6)$, are defined as follows [36]:

$$\bar{e}_i = \left(\frac{1}{8\pi^2}\right) \int_0^{2\pi}\int_0^{\pi}\int_0^{2\pi} d\varphi_1 d\Phi_0 d\varphi_2 \sin\Phi_0 f(\varphi_1, \Phi_0, \varphi_2) e_i(\varphi_1, \Phi_0, \varphi_2)\,, \tag{5.16}$$

$$\bar{s}_i = \left(\frac{1}{8\pi^2}\right) \int_0^{2\pi}\int_0^{\pi}\int_0^{2\pi} d\varphi_1 d\Phi_0 d\varphi_2 \sin\Phi_0 f(\varphi_1, \Phi_0, \varphi_2) s_i(\varphi_1, \Phi_0, \varphi_2)\,. \tag{5.17}$$

If we denote by σ_i the stress tensor elements in the crystallite coordinate system, Hooke's law holds for the elastic part:

$$\sigma_i = \sum_{i=1}^{6} C_{ij} \varepsilon_j\,, \tag{5.18}$$

where C_{ij} are the monocrystal elastic stiffness moduli.

The following linear relations link the strain and stress tensor components in the two coordinate systems:

$$s_i = \sum_{j=1}^{6} P_{ij}\sigma_j \,, \tag{5.19a}$$

$$e_i = \sum_{j=1}^{6} P_{ij}\varepsilon_j \,, \tag{5.19b}$$

where the elements of the matrix $\boldsymbol{P}$ are sums of products of two Euler's matrix elements. Both the Euler's matrix and the matrix $\boldsymbol{P}$ were given explicitly by Popa [36].

5.3.3 The Strain Expansion in Generalized Spherical Harmonics

We now follow an approach similar to that of Wang [13,14] and Behnken [15], but with an important distinction that makes the problem of determination of components of the strain tensor equivalent to the texture problem and significantly simplifies the mathematical formalism. This is accomplished by replacing the product of SODF and CODF by the SODF weighted by texture (WSODF):

$$\varepsilon_i'(\varphi_1, \Phi_0, \varphi_2) = \varepsilon_i(\varphi_1, \Phi_0, \varphi_2) f(\varphi_1, \Phi_0, \varphi_2) \,. \tag{5.20}$$

The product (5.20) explicitly appears in (5.13) and implicitly in (5.16) and (5.17) through (5.18) and (5.19a)–(5.19b). This is an additional advantage in directly determining WSODF instead of SODF because the texture-weighted strain is actually measured in the diffraction experiment and required to calculate other properties of interest, in particular average strain and stress tensors.

With (5.20), the integral in the numerator of (5.13) becomes similar to the expression for calculating the pole distribution function $P_{\boldsymbol{h}}(\boldsymbol{y})$:

$$P_{\boldsymbol{h}}(\boldsymbol{y}) = \left(\frac{1}{2\pi}\right) \int_0^{2\pi} f(\varphi_1', \Phi_0', \varphi_2') d\omega \,. \tag{5.21}$$

Hence, the problem of strain becomes equivalent to the problem of texture. The measurable strain $\langle \varepsilon_{\boldsymbol{h}}(\boldsymbol{y}) \rangle$ is for WSODF what the pole distribution $P_{\boldsymbol{h}}(\boldsymbol{y})$ is for CODF. This is useful in implementation in Rietveld-refinement programs, because the texture and strain implementations become equivalent, only the selection rules being different.

Now, we follow the procedure used by Bunge [39] for texture, by developing ε_i' in generalized spherical harmonics:

$$\varepsilon_i'(\varphi_1, \Phi_0, \varphi_2) = \sum_{l=0}^{\infty} \sum_{m=-l}^{l} \sum_{n=-l}^{l} c_{il}^{mn} \exp(im\varphi_2) P_l^{mn}(\Phi_0) \exp(in\varphi_1) \,. \quad (5.22)$$

As ε_i' is a real quantity, the coefficients $c_{il}^{mn} = a_{il}^{mn} + ib_{il}^{mn}$ are complex numbers fulfilling the condition:

$$c_{il}^{-m,-n} = (-1)^{m+n} c_{il}^{mn*} \,, \quad (5.23)$$

and the functions P_l^{mn} and P_l^m are defined elsewhere [16, 39].

Then, for the integral over ω in the numerator of (5.13), we have (compare to Eq. 14.160 in [39]):

$$\int_0^{2\pi} d\omega \varepsilon_i'(\varphi_1', \Phi_0', \varphi_2') = \sum_{l=0} \frac{4\pi}{2l+1} \sum_{m=-l}^{l} \sum_{n=-l}^{l} \times c_{il}^{mn} \exp(-im\beta) P_l^m(\Phi) \exp(in\gamma) P_l^n(\Psi) \,. \quad (5.24)$$

We can rearrange (5.22) and consequently (5.13) in a more convenient form, with only positive indices m, n. Taking into account (5.15), (5.13) becomes:

$$\langle \varepsilon_{\boldsymbol{h}}(\boldsymbol{y}) \rangle P_{\boldsymbol{h}}(\boldsymbol{y}) = \sum_{l=0}^{\infty} \frac{2}{2l+1} I_l(\boldsymbol{h}, \boldsymbol{y}) \,, \quad (5.25)$$

where:

$$\begin{aligned} I_l(\boldsymbol{h}, \boldsymbol{y}) = {} & A_1^2 t_{1l}(\boldsymbol{h}, \boldsymbol{y}) + A_2^2 t_{2l}(\boldsymbol{h}, \boldsymbol{y}) + A_3^2 t_{3l}(\boldsymbol{h}, \boldsymbol{y}) \\ & + 2A_2A_3 t_{4l}(\boldsymbol{h}, \boldsymbol{y}) + 2A_1A_3 t_{5l}(\boldsymbol{h}, \boldsymbol{y}) + 2A_1A_2 t_{6l}(\boldsymbol{h}, \boldsymbol{y}) \end{aligned} \quad (5.26)$$

and:

$$t_{il}(\boldsymbol{h}, \boldsymbol{y}) = A_{il}^0(\boldsymbol{y}) P_l^0(\Phi) + \sum_{m=1}^{l} [A_{il}^m(\boldsymbol{y}) \cos m\beta + B_{il}^m(\boldsymbol{y}) \sin m\beta] P_l^m(\Phi) \,, \quad (5.27)$$

$$A_{il}^m(\boldsymbol{y}) = \alpha_{il}^{m0} P_l^0(\Psi) + \sum_{n=1}^{l} (\alpha_{il}^{mn} \cos n\gamma + \beta_{il}^{mn} \sin n\gamma) P_l^n(\Psi) \,, \quad (m = 0, l) \,, \quad (5.28)$$

$$B_{il}^m(\boldsymbol{y}) = \gamma_{il}^{m0} P_l^0(\Psi) + \sum_{n=1}^{l} (\gamma_{il}^{mn} \cos n\gamma + \delta_{il}^{mn} \sin n\gamma) P_l^n(\Psi) \,, \quad (m = 1, l) \,. \quad (5.29)$$

The coefficients α_{il}^{mn}, β_{il}^{mn}, γ_{il}^{mn}, δ_{il}^{mn} are obtained from the coefficients c_{il}^{mn} by the linear transformations, which are given in [16]. They can be directly refined in the Rietveld program to yield WSODF and the average strain tensor. The average stress tensor can also be determined if monocrystal elastic stiffness moduli C_{ij} are known. The required number of refined coefficients to achieve desired precision of WSODF, strain and stress tensors will depend on the crystal and sample symmetries, as well on the magnitude and gradient of strain and texture.

5.3.4 The Selection Rules for All Laue Classes

Equation (5.25) is a general formula for strain (diffraction-line shift) determination, valid for triclinic crystal symmetry. For a given value of l, the total number of the coefficients for every i is $(2l+1)^2$, where the number l takes only even values because of Friedel's law. If the crystal and sample symmetries are higher than triclinic, the number of coefficients α_{il}^{mn}, β_{il}^{mn}, γ_{il}^{mn}, δ_{il}^{mn} in (5.28) and (5.29) is reduced, some coefficients being zero and some being correlated. To find the selection rules for all Laue classes, we apply the invariance condition to the measured strain, $\langle \varepsilon_{\boldsymbol{h}}(\boldsymbol{y}) \rangle$, as outlined in Sect. 5.3.1.

We denote two operators for the crystal and sample point group by $\boldsymbol{X}$ and $\boldsymbol{Y}$, respectively. Because the terms in the sum (5.25) are independent, the invariance condition has to apply to every I_l in (5.25):

$$I_l(\boldsymbol{X} \cdot \boldsymbol{h}, \boldsymbol{Y} \cdot \boldsymbol{y}) = I_l(\boldsymbol{h}, \boldsymbol{y}) \,. \tag{5.30}$$

Selection Rules Imposed by the Sample Symmetry

For textured samples under stress, two sample symmetries must be distinguished: texture and stress/strain sample symmetry. Sometimes they are identical, but generally the strain sample symmetry can be lower than the texture sample symmetry. Furthermore, according to (5.25), the texture sample symmetry operations must form a supergroup of the strain sample symmetry point group because $P_{\boldsymbol{h}}(\boldsymbol{y})$ must be invariant to both symmetry operations. An example is the dependence of the observed strain on the direction in a sample with randomly oriented crystallites, derived in the Reuss [38] approximation by Popa [36]. Here the texture has a spherical symmetry but the dependence of $\langle \varepsilon_{\boldsymbol{h}}(\boldsymbol{y}) \rangle$ on $\boldsymbol{y}$, in general, shows triclinic symmetry and only for hydrostatic stress does it become independent of $\boldsymbol{y}$ (isotropic).

For strain sample symmetry higher than triclinic, the invariance relation (5.30) for any $\boldsymbol{h}$, with $\boldsymbol{X} = \boldsymbol{E}$ (where $\boldsymbol{E}$ is the identity operator), requires the invariance of the functions $A_{il}^{m}(\boldsymbol{y})$ and $B_{il}^{m}(\boldsymbol{y})$ in (5.27) independently for every m. In other words, the coefficients of different strain tensor components are independent. This implies that the selection rules that apply to the coefficients α_{il}^{mn}, β_{il}^{mn}, γ_{il}^{mn}, δ_{il}^{mn} are identical to those for the texture of the same sample symmetry. These rules were previously given by Popa [2] and are also summarized in [16]. Therefore, although strain and texture sample symmetries may be different, the selection rules for the same symmetry are identical, which greatly simplifies its implementation in the Rietveld refinement programs.

Selection Rules Imposed by the Crystal Symmetry

With $\boldsymbol{Y} = \boldsymbol{E}$ in (5.30), selection rules are obtained by solving the system of equations for crystal symmetry operators $\boldsymbol{X}$ in an arbitrary sample direction $\boldsymbol{y}$. A crystal symmetry operator $\boldsymbol{X}$ acts on both the functions $t_{il}(\boldsymbol{h}, \boldsymbol{y})$ and the coefficients $A_i A_j$ in (5.26). As a result, selection rules for the harmonic representation of WSODF are more complex than selection rules for the harmonic representation of CODF. Selection rules for all Laue classes were published by Popa and Balzar [16].

5.3.5 Determination of Average Strain and Stress Tensors

For the calculation of both average elastic strain and stress tensors, $\bar{e}_i$ and $\bar{s}_i$, only the coefficients α_{il}^{mn}, β_{il}^{mn}, γ_{il}^{mn}, δ_{il}^{mn} with $l = 0$ and $l = 2$ are needed. This is easily seen by combining (5.22) and (5.19b) into (5.16) for strain, and (5.22), (5.18), and (5.19a) into (5.17) for stress. The integrals of the terms with $l = 1$ and $l > 2$ are zero because the elements of the matrix $\boldsymbol{P}$ are sums of products of two Euler's matrix elements and the generalized harmonics are orthogonal. So, keeping in (5.22) only the terms with $l = 0$ and $l = 2$, and rearranging to have only positive indices m, n, in place of ε_i' we have the following:

$$\varepsilon_i''(\varphi_1, \Phi_0, \varphi_2) = \sum_{k=0}^{25} g_{ik} R_k(\varphi_1, \Phi_0, \varphi_2) . \tag{5.31}$$

The functions $R_k(\varphi_1, \Phi_0, \varphi_2)$ are linear combinations of $\cos(m\varphi_2 \pm n\varphi_1) \times Q_l^{mn}(\pm\cos\Phi_0)$ or $\sin(m\varphi_2 \pm n\varphi_1) Q_l^{mn}(\pm\cos\Phi_0)$ terms, where $Q_l^{mn} = P_l^{mn}$ for $m+n$ even and $Q_l^{mn} = iP_l^{mn}$ for $m+n$ odd. They were given by Popa and Balzar [16]. The vector $\boldsymbol{g}_i$ is defined as follows (the index t stands for transposed):

$$\boldsymbol{g}_i^t = \begin{pmatrix} \alpha_{i0}^{00}; \alpha_{i2}^{00}, \alpha_{i2}^{01}, \beta_{i2}^{01}, \alpha_{i2}^{02}, \beta_{i2}^{02}; \\ \alpha_{i2}^{10}, \alpha_{i2}^{11}, \beta_{i2}^{11}, \alpha_{i2}^{12}, \beta_{i2}^{12}; \\ \gamma_{i2}^{10}, \gamma_{i2}^{11}, \delta_{i2}^{11}, \gamma_{i2}^{12}, \delta_{i2}^{12}; \\ \alpha_{i2}^{20}, \alpha_{i2}^{21}, \beta_{i2}^{21}, \alpha_{i2}^{22}, \beta_{i2}^{22}; \\ \gamma_{i2}^{20}, \gamma_{i2}^{21}, \delta_{i2}^{21}, \gamma_{i2}^{22}, \delta_{i2}^{22} \end{pmatrix} . \tag{5.32}$$

By introducing (5.31) in (5.19b) and consequently, (5.19b) in (5.16), one obtains:

$$\bar{e}_i = \sum_{j=1}^{6} \sum_{k=0}^{25} \bar{w}_{ijk} g_{jk} \, . \tag{5.33}$$

Similarly, for stress one obtains after combining (5.31), (5.18), (5.19a), and (5.17), and rearranging the terms, the following:

$$\bar{s}_i = \sum_{j=1}^{6} \sum_{k=0}^{25} \bar{w}_{ijk} g'_{jk} \, , \tag{5.34}$$

where

$$g'_{jk} = \sum_{l=1}^{6} C_{jl} g_{lk} \, .$$

In (5.33) and (5.34), $\bar{\boldsymbol{w}}$ is

$$\bar{w}_{ijk} = \left(\frac{1}{8\pi^2}\right) \int_0^{2\pi} \int_0^{\pi} \int_0^{2\pi} d\varphi_1 d\Phi_0 d\varphi_2 \sin \Phi_0 P_{ij}(\varphi_1, \Phi_0, \varphi_2) R_k(\varphi_1, \Phi_0, \varphi_2) \, .$$

In principle, these integrals could be calculated analytically, but because there are 936 such integrals, we obtained components of $\bar{\boldsymbol{w}}$ by numerical integration, with better than 10^{-6} precision, by using Gauss quadrature over Φ_0 and Simpson quadrature over φ_1 and φ_2. Only 73 elements are different from zero and many are interrelated by an integer factor. The exact values of unique elements were calculated analytically and all nonzero elements were given by Popa and Balzar [16].

5.3.6 Conclusion

We present a method for the determination of texture-weighted strain orientation distribution function (WSODF) and average strain and stress tensors. The method is based on a development of texture-weighted strain tensor in a series of generalized spherical harmonics. It directly determines WSODF, which is in accord with the fact that a diffraction experiment yields the texture-weighted strain measure and that the average strain and stress tensors are texture weighted. We assume that only observable strain is invariant to the symmetry operations of a crystal point group, instead of each of six components of the strain tensor in the sample coordinate system [13–15]. Moreover, this approach is extended for arbitrary crystalline symmetry and adopted for implementation in Rietveld refinement programs.

5.4 Concluding Remarks

We presented two improvements to the Rietveld refinement that allow for the determination of crystallite size distribution and the state of residual strain/stress in the sample. Both models are based on physically sound parameters that can be compared with measurements obtained by other methods. However, as the Rietveld refinement becomes ever more powerful analysis tool it inevitably gains in complexity. It is inherent to Rietveld refinement, as a nonlinear least-squares refinement procedure, that some refined parameters might show a strong correlation. Furthermore, several physical effects will have similar manifestation; for instance, line broadening causes are numerous and besides small crystallite size include lattice defects (planar defects in particular also show diffraction-order independent line broadening), lattice parameter inhomogeneities, and others. Line shift is likewise influenced not only by residual strain but planar lattice defects in some crystal systems, lattice thermal expansion/contraction, etc. Therefore, a user of a particular Rietveld program has to pay careful attention to the physical connotation of refined parameters for the particular sample under investigation. It is always of advantage to have additional information about the sample acquired by complementary experimental techniques.

Acknowledgements

We gratefully acknowledge Nathalie Audebrand and Daniel Louër (University of Rennes) for preparing the CeO_2 powder and Commission on Powder Diffraction (CPD) of the International Union of Crystallography (IUCr) for supporting the Size-Strain Round Robin.

References

1. H. Rietveld: J. Appl. Cryst. **2**, 65 (1969)
2. N.C. Popa: J. Appl. Cryst. **25**, 611 (1992)
3. R.B. von Dreele: J. Appl. Cryst. **30**, 517 (1997)
4. N.C. Popa: J. Appl. Cryst. **31**, 176 (1998)
5. P.W. Stephens: J. Appl. Cryst. **32**, 281 (1999)
6. M. Ferrari and L. Lutterotti: J. Appl. Phys. **76**, 7246 (1994)
7. D. Balzar, R.B. von Dreele, K. Bennett and H. Ledbetter: J. Appl. Phys. **84**, 4822 (1998)
8. W.A. Dollase: J. Appl. Cryst. **19**, 267 (1986)
9. N. Audebrand, J.-P. Auffrédic and D. Louër: Chem. Mater. **12**, 1791 (2000)
10. J.I. Langford, D Louër and P. Scardi: J. Appl. Cryst. **33**, 964 (2000)
11. N.C. Popa and D. Balzar: J. Appl. Cryst. **35**, 338 (2002)
12. D. Balzar, N. Audebrand, M. Daymond, A. Fitch, A. Hewat, J.I. Langford, A. Le Bail, D. Louër, O. Masson, C.N. McCowan, N.C. Popa, P.W. Stephens, and B. Toby: J. Appl. Cryst. (2003) submitted.

13. Y.D. Wang, R.L. Peng and R.L. McGreevy: *Proceedings of the Twelfth International Conference on Textures of Materials ICOTOM-12, Canada, August 9–13.* (1999) pp. 553–559
14. Y.D. Wang, R.L. Peng, X.H. Zeng and R.L. McGreevy: Mater. Sci. Forum **347–349**, 66 (2000)
15. H. Behnken: Phys. Stat. Sol. (a) **177**, 401 (2000)
16. N.C. Popa and D. Balzar: J. Appl. Cryst. **34**, 187 (2001)
17. H.P. Klug and L.E. Alexander: *X-ray Diffraction Procedures.* 2nd edn. (Wiley, New York 1974) `Author: please give page-numbers.`
18. E.F. Bertaut: C.R. Acad. Sci. Paris **228**, 187 (1949)
19. A. Guinier: *X-ray Diffraction.* (Freeman, San Francisco 1963) p. 139
20. B.E. Warren: *X-ray Diffraction.* (Addison-Wesley, New York 1969) pp. 251–314
21. W.L. Smith: J. Appl. Cryst. **9**, 187 (1976)
22. J.I. Langford, in: *Accuracy in Powder Diffraction.* NBS Special Publication No. 567 (1980) pp. 255–269
23. D. Balzar and H. Ledbetter: J. Appl. Cryst. **26**, 97 (1993)
24. P. Thompson, D.E. Cox and J.B. Hastings: J. Appl. Cryst. **20**, 79 (1987)
25. B.R. York: Adv. X-ray Anal. **41**, 544 (1997)
26. P. Scardi and M. Leoni: Acta Cryst. A **57**, 604 (2001)
27. J.G.M. van Berkum: Ph.D. Thesis. (Delft University of Technology, 1994) p. 136
28. C.E. Krill and R. Birringer: Philos. Mag. A **77**, 621 (1998)
29. T. Ungár, J. Gubicza, G. Ribárik and A. Borbély: J. Appl. Cryst. **34**, 298 (2001)
30. P. Scardi and M. Leoni: Acta Cryst. A **58**, 190 (2002)
31. L. Lutterotti: `http://www.ing.unitn.it/ luttero/maud/index.html` (1998)
32. L. Lutterotti, S. Matthies and H.R. Wenk: IUCr CPD Newsletter **21**, (1999) pp. 14–15
33. G. Holzer, M. Fritsch, M. Deutsch, J. Hartwig and E. Forster: Phys. Rev. A **56**, 4554 (1997)
34. V. Hauk: Arch. für das Eisenhüttenwesen **23**, 353 (1952)
35. A.L. Christenson and E.S. Rowland: Transactions ASM **45**, 638 (1953)
36. N.C. Popa: J. Appl. Cryst. **33**, 103 (2000)
37. W. Voigt: *Lehrbuch der Kristallphysik.* (Teubner, Leipzig 1928)
38. A. Reuss: Z. Angew. Math. Mech. **9**, 49 (1929)
39. H.J. Bunge: *Texture Analysis in Material Science.* (London, Butterworth 1982)

6 The Quantitative Determination of the Crystalline and the Amorphous Content by the Rietveld Method: Application to Glass Ceramics with Different Absorption Coefficients

A.F. Gualtieri, A. Guagliardi, and A. Iseppi

The Rietveld method provides an accurate determination of the crystalline and the amorphous fractions in polyphase mixtures. To determine the content of the amorphous phase, the sample is diluted with an internal standard which is considered as a component itself and refined with the other phases. Although the method is accurate for systems containing crystalline and amorphous phases with an absorption coefficient comparable to that of the standard, it was not thoroughly tested for systems containing a weakly or highly absorbing amorphous phase. Commonly, this is the case for glass ceramics which may contain elements with fairly different atomic number.

X-ray powder diffraction data refined with the Rietveld method were utilized for the determination of the amorphous content in a number of systems containing glasses with different absorption coefficients (Li-, B-, Al-, P-, Zn-, Zr-, Ba-, Pb-glass) to check the accuracy of the method. The mixtures were prepared diluting the glass phase with corundum NIST 676 and hydroxyapatite NIST 2910 in fixed proportions. The refinements were carried out with GSAS and the new code QUANTO, especially developed for automatic quantitative phase analysis (QPA) of polycrystalline mixtures eventually including an amorphous component.

The glass weights estimated by the two programs are both in good agreement with the nominal value (45 wt%) although a minor dependency of the calculated amorphous fraction on the linear absorption coefficient of the glass was observed and attributed to microabsorption effects. The Brindley (1945) correction applied here according to the formulation of Taylor and Matulis (1991) revealed not to be fully satisfactory for mixture containing an amorphous component with a very different absorption coefficient, especially when highly accurate determinations are required. This is in concert with the general concern that the existing models for the correction of microabsorption effects should be revised as they are not adequate even for systems containing only crystalline phases (Madsen et al., 2001).

The method was successfully applied to a commercial glass ceramic indicating that it can be applied with a good accuracy also to glass ceramics with fairly different absorption coefficients.

6.1 Introduction

In the last decade, QPA using the Rietveld method [2–4] has been applied to systems of gradually increasing complexity (in the number of phases, their structure complexity and even their degree of structure order). One of the ultimate challenges is to achieve accurate weights in complex systems containing crystalline and amorphous phases as shown by the number of papers recently published in the specific literature (see for example [5–11] using an internal standard and [12,13] using empirical functions to fit the contribution of the amorphous phase). The interest in this area is predicted to increase if one considers that the appealing nano-science world describes crystal-size dependent properties of systems that cover a broad range between the crystalline and the amorphous state.

QPA using the Rietveld method, the weight fraction w_i of each ith crystalline component in the mixture are calculated from the corresponding refined scale parameter S_i, according to the equation

$$w_i = \frac{S_i M_i V_i}{\sum\limits_j S_j M_j V_j}\,, \tag{6.1}$$

with M_i and V_i, the unit cell mass and volume, respectively. This algorithm is based on the normalisation condition $\sum i = w_i = 1.0$. The scattering contribution of any amorphous component possibly existing in the sample is a part of the background and the above normalisation equation does not include it. The specific problem is usually solved using an internal standard, suitably chosen, which is added in known amount to the investigated mixtures and treated as a mixture component itself. At the end of the Rietveld process, the refined phase fractions are converted into weight fractions and rescaled to the values of the original mixture by the ratio between the refined $(X_{s,c})$ and the known (X_s) amount of added standard. Whenever an amorphous phase exists in the system, the values of the weight fractions are overestimated to satisfy the normalisation condition. The percentage of the amorphous phase X_a in the original mixture can be calculated directly from the weight of the internal standard according to the formula

$$X_a = \frac{100}{(100 - X_s)}\left(1 - \frac{X_s}{X_{s,c}}\right)\,. \tag{6.2}$$

Although in a recent paper [9] the major factors biasing the accuracy of the method (e.g., structure model, spike addition, choice of the thermal parameters, choice of the background polynomial, etc.) have been discussed, the absorption contrast of the amorphous phase vs. crystalline standard and microabsorption have not been elucidated.

The microabsorption effects describe the intensity reduction related to two main causes [14,15]: (i) the sample granularity [16]; (ii) the surface roughness of flat samples in the case of Bragg–Brentano geometries.

The former causes the underestimation of the highly absorbing phases in multiphase systems; The latter causes an angle-dependent attenuation of the intensity which becomes stronger with decreasing scattering angle [17–19,29]. These effects can be minimised by reducing the particle size or by using a more penetrating radiation, otherwise they should be corrected.

It is worth here to give a short overview of the three current models available to correct microabsorption effects keeping in mind that in practice only the first two methods have been applied to the quantitative phase analysis of polyphase mixtures:

1. The *Brindley correction* [21] applies to heterogeneous multiphase systems constituted by medium-coarse sized materials whose constituents have different absorption coefficients. The correction requires the *absorption particle factor* τ:

$$\tau = \frac{1}{A_i}\int_0^{A_i} \exp[-(\mu_i - \bar{\mu})x]dA_i \,, \tag{6.3}$$

with A_i, the particle volume of i and $\bar{\mu}$, the mean linear absorption coefficient of the solid matrix. Brindley tabulated the τ values for spherical particles of radius R_i versus $(\mu_i - \bar{\mu})R_i$. After the success of the Rietveld approach to quantitavive phase analysis, Taylor and Matulis [22] suggested a simple modification of (6.1) to exploit the Brindley particle absorption factor, by

$$w_i = \frac{S_i M_i V_i/\tau_s}{\sum\limits_j (S_j M_j V_j/\tau_j)} \,. \tag{6.4}$$

It is important to mention here that missing knowledge of the particle size of each phase can be a dramatic source of error in practice.

2. The *Hermann and Ermrich correction* [14,23] applies to both monophase and polyphase samples. The correcting function $P(\theta)$ (to be applied to the integrated intensities) is now described in terms of bulk porosity P_o and surface roughness P_s:

$$P(\theta) = P_0 + P_S = 2\mu\beta(1-\alpha_0)\left[1 - \frac{(1-\alpha_0)}{4}\right] + 2\mu\beta\alpha_0\left(\frac{t}{\sin\theta}\right)\left(1 - \frac{\tau}{\sin\theta}\right) \,, \tag{6.5}$$

with $\beta = \bar{l}\ln\alpha_0$, where $\bar{l}$ is the mean chord length of the powder particles, α_0 = packing fraction of the bulk material, and τ the degree of surface roughness. There are no dramatic effects if α_0 and θ are not too small. In the case of multiphase powders, Hermann and Ermrich [24] demonstrated that the Brindley [21] formalism, suggesting that the microabsorption

effect is essentially governed by the size of the particle and the absorption contrast (μ_i-$\bar{\mu}$) between the linear absorption coefficient μ_i of the i-type phase and the mean linear absorption coefficient ($\bar{\mu}$) of the solid material forming the powder, is not in all cases a correct measure for the absorption contrast. An *effective absorption coefficient* should be introduced: $\mu_i' = \mu/[1-2(\mu_i-\bar{\mu})b_i]$ considering that $\mu = \sum_j c_j\mu_j$ with c_j = concentration of phase j and $b_i = 2B_i/3$ with B_i = mean breadth of the i-type particle. The limiting value for very fine powders is $(|\mu_i - \bar{\mu}|b_i \to 0)$ is $\mu_i' = \bar{\mu}$.
3. The *correction using the measured fluorescence intensity reduction with* 2θ [15,19,20]. It is a linear correction which represents an empirical statement for strong absorption:

$$P(\theta) = P + C(2\theta)\,, \tag{6.6}$$

with P and C refinable parameters.

Several examples of applications of the first two corrections are found in literature. The earlier work by Taylor and Matulis [22] reported gross errors in the weights of mixtures of LiF and $Pb(NO_3)_2$ if the Brindley correction was not applied. According to Liu and Kuo [25], the incorporation of the *Hermann–Ermrich* correction in the Rietveld code can give much higher accuracy to the QPA results. On the same line, Winburn et al. [26] recently obtained accurate results using rutile as internal standard for the QPA of NIST fly ash standards with 66.3–77.8 wt% glass contents. DuPont's 0.3 µm Ti-Pure R900 Rutile with a 94.4 wt% crystalline rutile was selected because it has a linear absorption coefficient in the midrange of the various fly ash phases and thus helps to minimize an intensity error due to microabsorption. Moreover, Winburn [27] and Winburn et al. [28] reported a post-refinement Brindley correction (see also [22]) to achieve more accurate results in the quantifica hematite, magnetite and other phases in the fly ash samples. At last, Madsen et al. [29] in the outcomes of the Round Robin on Quantitative Phase Analysis, stated that: "*...the correction models for microabsorption in current use are not adequate and need to be revised as a matter of high priority*". Although microabsorption in systems containing only crystalline phases may be somehow corrected, little is known about microabsorption effects in systems containing crystalline and amorphous phases. An earlier work by Mandile and Hutton [30] on the QPA of organic matter rich rocks containing an organic amorphous phase pointed out the problem. The authors used a combined method with a calibration curve obtained by calculating the organic hump vs. organic amorphous fraction content and the Brindley correction for the remaining crystalline phases stressing the fact that low X-ray absorption coefficient of the organic matter may introduce microabsorption effects.

Of course, absorption and microabsorption effects are crucial issues also in the Rietveld method which uses a spiking for the determination of the amorphous phase. The method proved to be accurate for mixtures of natural silicates as zeolite tuffs or traditional ceramics diluted with corundum NIST SRM 676[1] [9,31–35] but these systems invariably contain an amorphous phase (a silica glass) with a linear absorption coefficient comparable to that of corundum and the other silicates in the mixtures. It is then interesting to test the accuracy of the method in the case of mixtures containing an amorphous phase with a different absorption coefficient. To this aim, a number of glasses used for the production of glass ceramics were selected as test candidates for the study. Glass ceramics represent a relatively new class of ceramic materials with different applications. They are polycrystalline solids prepared by controlled crystallisation of glasses [36]. The glasses which are used to promote crystallisation may have very different technological properties according to their composition, which is extremely variable and may include oxides of elements with a large atomic number such as Zr, Cd, Sn, Ba, and Pb. The understanding of the physical and technological properties of the glass ceramics requires a full chemical and mineralogical characterisation and of course QPA is of great importance to this aim.

The accuracy of the method was tested on Li-, B-, Al-, P-, Zn-, Zr-, Ba-, Pb-glasses. Ternary mixtures were prepared using the NIST standards corundum and hydroxyapatite. Corundum was selected as internal standard and hydroxyapatite was selected as an additional standard crystalline phase with an absorption coefficient different from that of corundum. A commercial glass ceramic with a high absorption coefficient was also investigated in order to apply the method to a real case. In this work, only copper radiation was considered as it is the most widely used in the academic and industrial lab sources.

6.2 Experimental

6.2.1 Sample Selection and Preparation

Glasses with different compositions (Table 6.1) were selected for the study. A number of mixtures with different absorption due to the different composition or to the different particle size (mixtures with the same number but different letter in the label) were prepared. The complete list is reported in the Table 6.2. The mixtures were obtained by carefully weighing the samples and homogenise the powder by hand grinding in agate mortar for the X-ray data collection. The seven mixtures of the group 1 (1,1b–1f) were prepared to test the accuracy of the quantitative results with different amount

[1] This approach has been referred by some authors as the *Rietveld and R.I.R. method* to indicate that the spike added to the mixture for the determination of the amorphous phase is specifically the *α-Al_2O_3 corundum*, in concert with the original definition of the Reference Intensity Ratio given by Visser and De Wolff [37].

Table 6.1. Chemical analyses of the glasses selected for the study

oxides (wt%)	1 (Si-Pb)*	2 (Si-Ca)	3 (Si-B)	4 (Si-Ca-Zr)	5 (Si-Zn-Pb)	6 (Ca-Si-P)	7 (Si-Al-Ca-Li)	8 (Pb-Si-Ba)	9 (B-Si-Li)
SiO_2	21.0	54.2	37.0	53.7	36.0	38.0	70.3	20.72	21.0
Al_2O_3	–	7.0	5.0	1.3	3.0	2.0	11.0	0.02	–
CaO	–	22.0	11.0	30.0	4.5	47.0	11.0	0.02	–
MgO	–	11.7	–	3.0	2.0	–	–	–	–
BaO	–	–	7.0	–	–	–	–	8.09	–
K_2O	–	3.2	–	–	–	1.5	1.2	0.02	–
Na_2O	–	1.9	3.0	–	3.0	1.5	–	–	–
B_2O_3	–	–	37.0	–	3.0	–	–	–	70.0
ZrO_2	–	–	–	12.0	–	–	–	–	–
PbO	79.0	–	–	–	23.5	–	–	71.13	–
ZnO	–	–	–	–	25.0	–	–	–	–
Li_2O	–	–	–	–	–	–	6.5	–	9.0
P_2O_3	–	–	–	–	–	10.0	–	–	–
TOT	100.0	100.0	100.0	100.0	100.0	100.0	100.0	100.0	100.0

* chemical species in parenthesis are the major elements contained in the investigated glasses

Table 6.2. List of the sample mixtures refined in the study

Sample	glass no.	glass (wt%)	corundum NIST (wt%)	Hydroxyapatite NIST (wt%)	Ceria NIST (wt%)
1	1	45.0	10.0	45.0	–
1b	1	20.0	10.0	70.0	–
1c	1	70.0	10.0	20.0	–
1d	1	35.0	30.0	35.0	–
1e	1	25.0	50.0	25.0	–
1f	1	15.0	70.0	15.0	–
2	2	45.0	10.0	45.0	–
3	3	45.0	10.0	45.0	–
3b	3	45.0	10.0	–	45.0
4	4	45.0	10.0	45.0	–
5	5	45.0	10.0	45.0	–
6	6	45.0	10.0	45.0	–
7	7	45.0	10.0	45.0	–
7b	7	45.0	10.0	45.0	–
7c	7	45.0	10.0	45.0	–
8a	8	45.0	10.0	45.0	–
8b	8	45.0	10.0	45.0	–
9a	9	45.0	10.0	45.0	–
9b	9	45.0	10.0	45.0	–
10	–	–	18.18	81.82	–

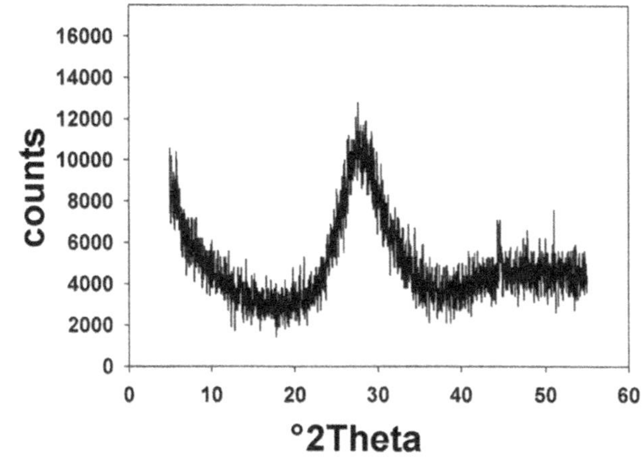

Fig. 6.1. Example of qualitative X-ray powder diffraction analysis of glass 1. The tiny peak at 44.5° 2θ is a contribution of the Aluminum sample holder

of glass and different spike additions. Indirectly, this group enables us to control possible source of errors from the crystalline phases (structural model, residual amorphous fraction, microabsorption correction and others) as the effects are expected to be enhanced with increasing the amount of the corresponding crystalline phase; to this aim, also a mixture of "corundum + hydroxyapatite" only was prepared (Sample 10). All the sample mixtures in Table 6.2 with proportions 45 : 10 : 45 of glass, corundum and hydroxyapatite, respectively, were prepared to investigate the specific influence of the different kind of glasses on the QPA results. The number of the sample mixture indicates the correspondence with the number glass in Table 6.1. Some mixtures have the same number (e.g., 7, 7b, 7c and 8a, 8b and 9a, 9b) but different mean size of the glass particles and were prepared to rule out any influence of the size. Finally, the Sample 3b was analysed to check the influence of the highly absorbing cerium oxide with respect to the hydroxyapatite. The presence of crystalline phases in the glasses was checked by X-ray powder diffraction. None of the selected glasses showed diffraction peaks indicating a complete amorphous state (as an example, Fig. 6.1 reports the experimental X-ray powder pattern of Glass 1).

One of the problems in the quantification of the amorphous phase using the internal standard methods is that the standard may not be fully crystalline. Corundum NIST 676[2] was annealed to increase the crystallinity to 100 wt%. The value of amorphous phase in the treated standard was reasonably assumed to be about 0 wt% by checking the peak ratio of the peak

[2] According to earlier reports [41, 42] Corundum NIST 676 contained 1.6 wt% of amorphous fraction; according to the more recent reports [43] it contains 4.97 wt% of amorphous fraction.

intensities before and after the treatment with the intensities of an internal standard (fluorite 311 and 422) although it is not possible to exclude that a minor quantity of amorphous alumina may still be present in the standard. Similarly, hydroxyapatite NIST 2910 was annealed for 1 d at 250 °C to eliminate the certified amorphous fraction of 7.5(4.1) (see Sect. 6.3).

6.2.2 Calculation of the Linear Absorption Coefficient

The linear absorption coefficient μ of each glass was calculated using (6.7) [38]:

$$\mu = \rho \sum_i g_i \left(\frac{\mu}{\rho}\right)_i , \tag{6.7}$$

where g_i is the mass fraction of the atoms of the ith species for which the mass absorption coefficient is $(\mu/\rho)_i$. The density ρ of each glass powder was obtained using the picnometer method [39] and divided by the density calculated for a known volume (apparent density = density of the particles + air in the pores) to calculate the packing fraction α_0 of the powder.

The linear absorption coefficient of the glasses was also independently cross-checked for each sample mixture according to the Beer–Lambert equation in transmission mode [40]:

$$I_x = I_0 \exp(-\mu x) , \tag{6.8}$$

where I_x is the intensity of the transmitted beam; I_0 is the intensity of the incident beam; x is the thickness of the sample; μ is the linear absorption coefficient. I_0, the blank collection (the sample holder and air scattering) and I_x (the sample holder + powder and air scattering) were measured at the Dipartimento di Scienze della Terra (The University of Milano, Italy) with the goniometer at $2\theta = 0°$ and a primary beam attenuator in a conventional Bragg–Brentano (BB) parafocusing geometry with a Cu$K\alpha$ monochromatized radiation. The thickness of the sample holder was 0.1 cm.

To rule out any influence of the grain size of the glasses on the absorption and consequently the calculated amorphous phase, the grain size of each powder was measured using a laser diffraction system Malvern SB.0B. Some grain size distributions were also independently checked by TEM using a Philips JEM 2010 instrument operating at 200 kV.

All the physical measured characteristics of the amorphous powders are reported in Table 6.3.

6.2.3 Data Collection for the QPA

The powders of each sample mixture (see Table 6.2) were mounted on a 2 mm thick flat Al holder and the data collection was performed using a Philips PW 1710 instrument with a conventional Bragg–Brentano (BB) parafocusing geometry equipped with a copper tube and a secondary beam pyrolitic graphite

Table 6.3. Measured and calculated physical characteristics of each glass phase

Sample	Density D_r (g/cm^3)	Apparent density D_a (g/cm3)	Powder packing fraction $\alpha_0(\%)^{a)}$	Calculated linear absorption coefficient μ (cm^{-1})	Measured specific surface area (m^2/g)	Measured average grain diameter (μm)
1	11.00	5.51	50.1	1976	4.9	35.0
2	5.31	2.60	48.9	298	3.2	45.0
3	4.60	2.25	48.9	154	3.0	15.0
4	8.50	4.29	50.5	594	4.5	35.0
5	5.79	2.92	50.4	489	3.9	30.0
6	5.12	2.52	49.2	411	4.7	18.0
7	5.13	2.85	55.5	224	3.6	42.0
7b	5.13	1.55	30.2	224	3.4	10.0
7c	5.13	2.29	44.6	224	1.6	25.0
8a	10.10	5.10	50.5	1852	5.1	45.0
8b	10.10	4.94	48.9	1852	3.5	10.0
9a	5.05	2.52	49.9	73	0.3	100.0
9b	5.05	2.01	39.8	73	2.5	25.0
corundum	3.98 [b)]	1.757	44.0	124 [b)]	n.d.	1.0
hydroxyapatite	3.15 [b)]	2.103	67.0	262 [b)]	18.3 [c)]	1.5

[a)] $P(\%)=100^*(D_a/D_r)$
[b)] theoretical values calculated using theoretical densities and calculated values of the packing fraction (0.44 for corundum; 0.67 for hydroxyapatite; 0.52 for quartz)
[c)] certificate of Analysis NIST 2910

crystal monochromator. The data for the QPA refinements were collected in the angular range 8–100°2ϑ with steps of 0.02° and 10 s/step, a divergence slit of 1° and a receiving slit of 0.2 mm.

6.2.4 Rietveld Refinements

Each data set was refined using GSAS [44] and QUANTO [45]. The starting structure models for the sample mixtures and the industrial glass ceramic were: corundum from Ishizawa et al. [46], hydroxyapatite from Sudarsanan and Young [47], quartz from Le Page and Donnay [48], zircon from Hazen and Finger [49], spinel from Peterson, Lager and Hitterman [50], and ceria from Blumental et al. [51]. The LP correction used in the refinements with both programs was $[P_h + (1 - P_h)\cos^2 2\theta]/[2\sin^2\theta\cos\theta]$ with an independently refined P_h value of 0.55. The atomic displacement parameters $U_{\rm iso}$ were set to the single crystal values for every phase and kept fixed during the refinement procedure. The equivalent $U_{\rm iso}$ derived from the $U_{\rm aniso}$ were used for hydoroxyapatite. The GSAS refinements were carried out according to the following features: (a) the background was successfully fitted with a

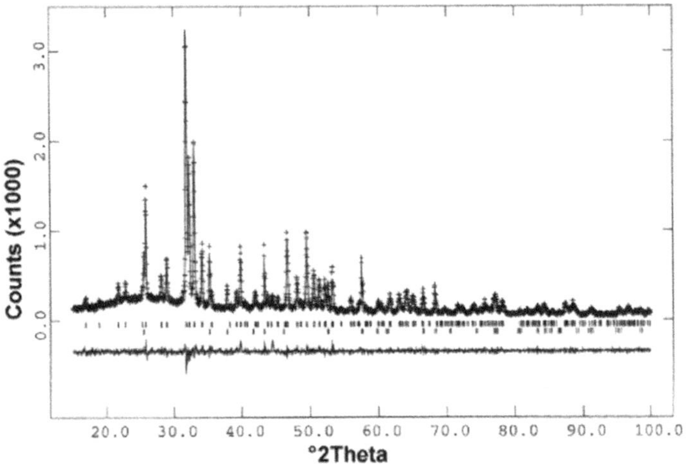

Fig. 6.2. Example of the Rietveld observed (*crosses*), calculated (*continuous line*) and difference curve (*bottom line*) of the refined pattern of Sample 3. Legend for the markers: first line from bottom = corundum; second line from bottom = hydroxyapatite

Chebyshev function with a variable number of coefficients depending on its complexity; (b) the peak profiles were modelled using a pseudo-Voigt function with one Gaussian and one Lorenzian component; c) The lattice constants, the phase fractions and the coefficients corresponding to the sample displacement and asymmetry were also refined. The cut-off value for the calculation of the peak profiles in all refinements was 0.05%. The QUANTO program was run in the automatic mode, according to the following features: (a) Pearson VII as a profile shape function; (b) Chebyshev polynomials for background modeling (the number of coefficients is automatically set by the program).

Figure 6.2 is an example of the observed, the calculated and the difference curve of the refined pattern of Sample 3. Figure 6.3 is the observed (*crosses*), calculated (*continuous line*) and difference curve (*bottom line*) of the refined pattern of the industrial glass ceramic. The full data sets of the GSAS experimental files of the refinements are available upon request to the corresponding author.

6.3 Results and Discussion

The refined weights of the crystalline phases and relative estimated amorphous fraction of each sample mixture are reported in Tables 6.4 and 6.5 for GSAS and QUANTO refinements, respectively, together with the agreement profile indices (R_{wp} and R_p) and the additional statistical indicators supplied by the two programs (χ_2 and $R(F^2)$ for GSAS, Gof and $R(|F|)$

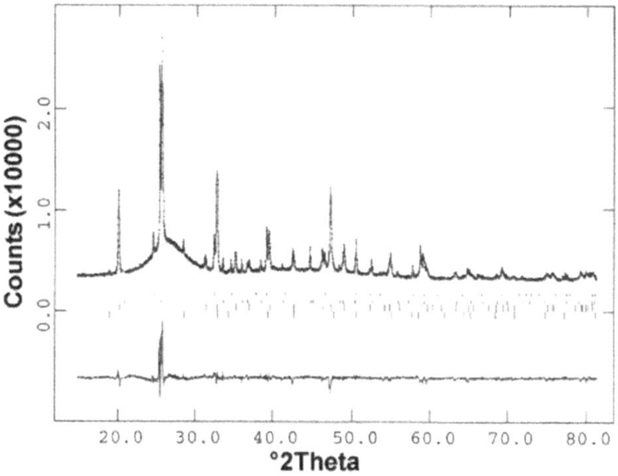

Fig. 6.3. The Rietveld observed (*crosses*), calculated (*continuous line*) and difference curve (*bottom line*) of the refined pattern of the industrial glass ceramic. Legend for the markers: first line from bottom = zircon; second line from bottom = spinel; third line from bottom = quartz; fourth line from bottom = corundum

for QUANTO). First of all, the QPA results obtained with the two Rietveld packages are highly comparable indicating a good accuracy of almost all the estimated glass fractions, the maximum absolute error being less than 3.5% (excluding mixture 3b, for which a microabsorption correction must be applied).

The mixtures 1, 1b, 1c show comparable accurate amorphous fractions with respect to the increasing amount of hydroxyapatite, while the mixtures 1d, 1e, 1f with increasing added fraction of corundum, always present overestimated amorphous percentages. This result could reveal a residual amorphous fraction for corundum, despite the previous check. In any case, considering that the investigated mixtures contain only 10 wt% of corundum and considering the results obtained for the mixture 10 with and without the Brindley correction in QUANTO giving a fraction of 18.31 wt% (see Table 6.5) vs. a theoretical value of 18.18 wt%, the calculated amorphous fraction in the mixture 10 is 0.71 wt%. Thus, the contribution of this residual to the total estimated glass fraction of the investigated mixtures with 0.45 wt% glass is $0.71 \times 10.0/18.18 = 0.39$ wt%. This residual (0.39 wt%) can be eventually subtracted to the values of amorphous fraction reported in Table 6.6 (columns 1 and 3) to obtain a final content of the amorphous fraction corrected for the amorphous phase contained in the crystalline standards.

In systems with a silica glass, it was earlier observed that increasing amounts of the internal standard result in an underestimation of the glass fraction (see [9], Fig. 2). Oppositely, the results for the system with Pb glass reported in Tables 6.4–6.6 and Fig. 6.4 indicate that such relationship is

Table 6.4. Refinement statistics and results: GSAS refinements

Sample	Rwp (%)	Rp (%)	$R(F^2)$ (%)	χ_2	Corundum refined wt%	Hydroxyapatite refined wt%	Ceria wt%	Calculated amorphous fraction wt%
1	5.46	4.12	3.95	1.72	18.28(16)	81.72(7)	–	45.29(23)
1b	7.57	5.83	3.22	2.11	12.49(11)	87.51(3)	–	19.93(14)
1c	5.00	3.83	7.29	1.32	33.36(23)	66.64(2)	–	70.02(25)
1d	6.08	4.55	4.15	2.04	46.78(11)	53.22(13)	–	35.87(24)
1e	6.59	4.85	3.45	2.41	67.64(6)	32.36(14)	–	26.08(20)
1f	7.60	5.78	3.56	3.36	83.01(3)	16.99(12)	–	15.67(15)
2	8.45	5.86	4.68	3.12	18.37(11)	81.63(6)	–	45.56(17)
3	8.78	6.77	4.41	1.43	19.24(15)	80.76(6)	–	48.02(21)
3b	10.47	7.65	1.71	6.23	20.54(27)	–	79.46(3)	51.31(30)
4	6.49	4.89	3.58	1.92	18.38(11)	81.62(4)	–	45.59(15)
5	5.50	4.23	3.76	1.79	18.51(13)	81.49(5)	–	45.97(18)
6	8.75	6.35	4.16	2.05	18.58(15)	81.42(6)	–	46.18(21)
7	7.91	5.82	3.53	1.94	18.69(13)	81.31(5)	–	46.49(18)
7b	9.33	6.94	4.28	1.66	19.31(13)	80.69(5)	–	48.21(18)
7c	9.27	7.11	4.39	1.53	18.89(15)	81.02(6)	–	47.31(21)
8a	7.55	5.59	4.07	1.79	18.42(16)	81.58(7)	–	45.71(23)
8b	7.29	5.67	3.59	1.20	18.23(2)	81.77(10)	–	45.14(12)
9a	8.94	6.93	3.52	1.36	18.67(14)	81.33(5)	–	46.44(19)
9b	8.54	6.58	3.59	1.37	19.13(13)	80.87(5)	–	47.72(18)
10	10.95	7.45	4.63	3.32	18.40(10)	81.60(9)	–	1.19(19)

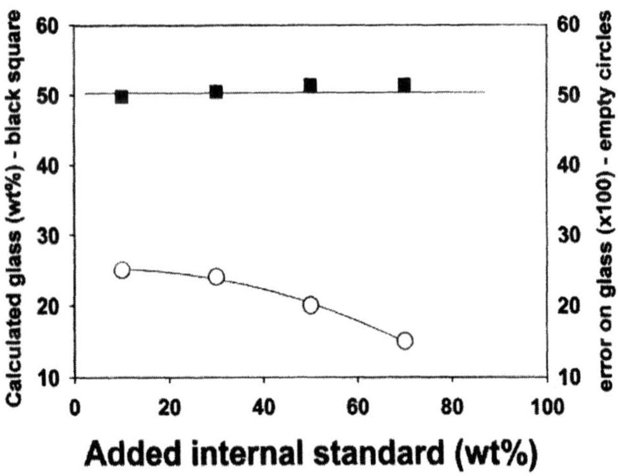

Fig. 6.4. Plot of the amount of added internal standard and the calculated glass fraction for the system with the Pb glass (Sample 1)

Table 6.5. Refinement statistics and results: QUANTO refinements

Sample	Rwp (%)	Rp (%)	GoF	Corundum refined wt%	R(|F|) (%) α_1/α_2	Hydroxy apatite refined wt%	Ceria wt%	R(|F|) (%) α_1/α_2	Calculated amorphous wt%
1	5.53	4.27	1.33	18.45(15)	4.43/4.16	81.55(6)	–	5.36/6.70	45.80(21)
1b	7.46	5.55	1.43	12.58(14)	2.91/2.24	87.42(6)	–	4.01/4.79	20.53(20)
1c	5.39	4.25	1.23	33.54(31)	7.30/5.34	66.46(30)	–	10.59/12.47	70.18(61)
1d	5.94	4.45	1.39	47.56(14)	2.55/2.15	52.44(11)	–	5.28/6.41	36.92(25)
1e	6.22	4.67	1.46	68.33(8)	1.89/1.59	31.67(12)	–	5.29/6.26	26.83(20)
1f	6.79	5.09	1.63	83.55(4)	1.85/1.87	16.45(10)	–	5.74/7.01	16.22(14)
2	7.94	5.34	1.66	18.05(12)	2.85/2.71	81.95(5)	–	4.29/5.10	44.60(17)
3	6.87	5.32	1.23	18.83(11)	8.82/8.59	81.18(5)	–	4.57/5.66	46.88(16)
3b	10.28	7.51	1.33	20.78(36)	24.57/24.76	–	79.22(4)	0.96/1.24	51.89(40)
4	6.36	4.72	1.36	18.29(11)	2.42/2.26	81.71(4)	–	4.09/4.86	45.33(15)
5	5.32	4.03	1.30	18.35(16)	2.74/2.11	81.65(5)	–	4.29/5.11	45.49(21)
6	8.60	6.18	1.41	18.30(18)	2.88/3.27	81.70(6)	–	5.00/5.71	45.35(22)
7	7.68	5.60	1.35	18.62(12)	2.64/2.07	81.38(5)	–	4.35/5.01	46.29(17)
7b	9.09	6.70	1.24	19.35(15)	2.71/2.31	80.65(8)	–	4.63/5.48	48.31(23)
7c	8.62	6.62	1.16	18.51(13)	2.17/2.56	81.49(7)	–	4.56/6.28	45.97(20)
8a	7.48	5.57	1.33	18.45(19)	2.86/3.36	81.55(9)	–	5.22/6.11	45.80(28)
8b	7.65	6.05	1.14	18.46(23)	5.06/5.46	81.54(13)	–	7.83/9.78	45.39(36)
9a	8.86	6.82	1.15	18.56(18)	2.10/2.83	81.44(5)	–	3.78/4.53	46.12(23)
9b	8.38	6.43	1.14	19.08(13)	2.42/2.87	80.92(5)	–	3.94/4.74	47.60(18)
10	9.05	6.40	1.78	18.46(9)	2.20/2.10	81.54(3)	–	4.00/4.87	1.52(12)
				18.31[a]		81.69[a]			0.71

[a] with the application of the Brindley correction

not found here as the amount of calculated glass fraction is nearly constant with added corundum. Although the refinements were repeated several times and the same result was obtained, it is possible that the data reported in Gualtieri [9] are somehow affected by an unknown bias.

As far as the mixtures with fixed proportions of glass, corundum and hydroxyapatite (45 : 10 : 45) are concerned, the calculated amorphous fractions in the range 45.15–48.21 wt% for GSAS and 44.60–48.31 wt% for QUANTO, respectively, are in good agreement with the nominal value, indicating that the Rietveld method can be successfully applied to systems containing an amorphous phase with a different linear absorption coefficient with respect to the crystalline phases (see also the good results on the commercial glass ceramic in Table 6.7), although the accuracy of the estimation seems to be influenced by the absorption properties of the glass. Indeed, the estimated amorphous fractions reveal a minor trend versus the linear absorption coefficient μ of the glass. Figure 6.5 (where the mean value of GSAS and QUANTO estimations is reported for each mixture) shows that the deviation from the

Table 6.6. Refinement statistics and results: final results after the microabsorption correction in QUANTO

Sample	Calculated amorphous wt% with the Brindley correction in QUANTO considering the glass absorption	Calculated linear absorption coefficient of the mixture considering the glass absorption μ (cm^{-1})	Calculated amorphous wt% with the Brindley correction in QUANTO not considering the glass absorption
1	46.59	1033.13	45.41
2	43.88	264.12	44.11
3	46.94	196.84	46.46
3b	47.67	885.77	46.90
4	45.10	398.63	44.85
5	45.21	351.41	45.08
6	45.00	315.66	44.88
7	45.84	250.53	45.85
7b	47.86	229.77	47.87
7c	45.50	230.63	45.51
8a	46.49	976.40	45.34
8b	46.52	976.86	45.37
9a	45.54	160.94	45.67
9b	47.03	158.16	47.14

Table 6.7. Theoretical, refined and calculated weights of the crystalline and amorphous phase present in the investigated industrial glass ceramic

Phase	Theoretical wt%	Refined wt%	Rescaled wt%
Corundum NIST	10.0	22.8(3)	10.0
Quartz	10.0	24.1(2)	10.6
Zircon	20.0	48.1(2)	21.1
Spinel	3.0	5.0(2)	2.2
Pb-Ba-Si Glass	57.0	–	56.1

true amorphous fraction increases with the decreasing μ value. This trend is enhanced if the mean value of GSAS and QUANTO estimations is plotted against the μR values in Fig. 6.6, where μ is the absorption coefficient and R the mean particle radius of the glass contained in the mixture.

This behaviour is unsatisfactory explained in terms of microabsorption effects involving corundum ($\mu = 124\,cm^{-1}$) and hydroxyapatite ($\mu = 262\,cm^{-1}$) mixed to a considerable amount of amorphous material with a fairly different μ which determines the mean absorption coefficient ($\bar{\mu}$) of the mixture. To this aim, the Brindley correction, already implemented in QUANTO according to the formalism suggested by Taylor and Matulis [22] has been adapted in order to take the linear absorption coefficient of the glass into account for

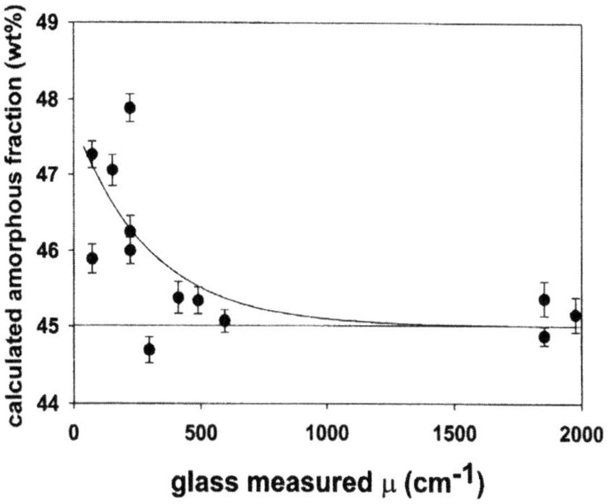

Fig. 6.5. Plot of the calculated amorphous (glass) weights (with the subtracted 0.39 wt% amorphous fraction of hydroxyapatite) versus the linear absorption coefficient m of the glass (see text for details)

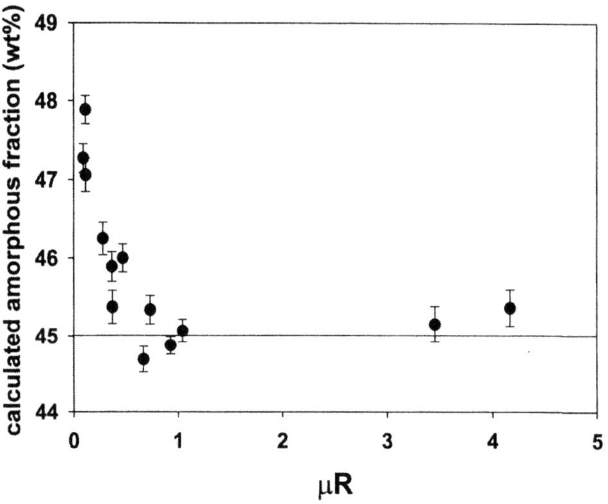

Fig. 6.6. Plot of the calculated glass weights (with the subtracted 0.39 wt% amorphous fraction of hydroxyapatite) versus μR (see text for details)

computing the $\bar{\mu}$ of the mixture. A similar application of the Brindley correction has been used by Orlhac et al. [11] to a borosilicate glass ceramic with an amorphous content > 95 wt%; in this application the glass is the phase with the smaller μ value.

The amorphous fractions supplied by QUANTO after the Brindley correction are reported in Table 6.6, together with the corresponding $\bar{\mu}$ value of the mixture. For comparison, the glass fractions were also computed exclud-

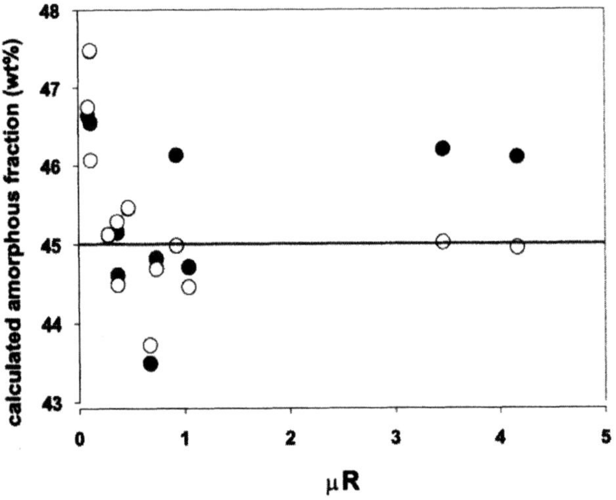

Fig. 6.7. Plot of the calculated glass weights corrected for Brindley (*black dots*: column 1 of Table 6.6; *white dots*: column 3 of Table 6.6) and subtracted 0.39 wt% amorphous fraction of hydroxyapatite versus μR for the data (see text for details)

ing the glass contribution to the values of $\bar{\mu}$ of the mixture[3] for the Brindley correction of crystalline phase fractions (last column of Table 6.6). Compared to the results shown in Table 6.5, the results in Table 6.6 clearly show that when the values of μ of the glass is not very different from those of crystalline phases, the correction supplies very accurate results (as for Samples 4, 5 and 6; indeed, in this case excluding the amorphous contribution from the calculation of $\bar{\mu}$ is not critical). The error is reduced but still evident for glasses with small μ (or μR) values (as for Samples 3, 7 and 9), while it is remarkably increased for highly absorbing glasses (Fig. 6.7). In this case the correction excluding the amorphous contribution seems to be more suitable. It is worthwhile noting the comparable overestimation of the amorphous fraction for both 3 and 3b mixtures, despite the stronger microabsorption problems due to the cerium oxide.

These results seem to indicate that the Brindley correction, required whenever an absorption contrast among the crystalline phases of the systems exists, should be revised for mixtures containing an amorphous component with a very different absorption coefficient, at least in the formulation suggested by Taylor and Matulis [22] for the integration with the QPA Rietveld method, especially when highly accurate determinations are required. In particular, the results of the present chapter outline some difficulties to correctly define the mean absorption coefficient of the mixture to be introduced in the correction (with or without the glass contribution) taking into account the fact that the normalization condition of (6.4) includes crystalline phases only.

[3] These values are in the range $\bar{\mu} = 235$–$237\,\mathrm{cm}^{-1}$ for all the mixtures but the 3b.

6.4 Conclusions and Future Perspectives

The Rietveld method has been applied for the calculation of the amorphous fraction in a number of glasses reproducing glass ceramic systems (Li-, B-, Al-, P-, Zn-, Zr-, Ba-, Pb-glass) with different absorption coefficients with respect to the added internal standard. In addition to GSAS, a new code (QUANTO) especially developed for the quantitative phase analysis (QPA) has been used and the results of the refined amorphous fractions are highly comparable. The calculated glass weights show a moderate scatter around the nominal value of 45 wt%. A minor dependency of the calculated amorphous fraction on the linear absorption coefficient of the glass is found, probably related to some microabsorption phenomena. The Brindley [21] correction applied in QUANTO according to the formulation of Taylor and Matulis [22] revealed to need some revisions for mixtures containing an amorphous component with a very different absorption coefficient, especially when highly accurate determinations are required. This corroborates the general feeling that the existing models for the correction of microabsorption effects should be revised as they appear to be not adequate even for simple systems containing only crystalline phases [29].

The method has also been successfully applied to a commercial glass ceramic containing Pb and Ba. Previously reported observation that the addition of the internal standard results in an underestimation of the glass fraction [9] is not found here as the amount of calculated glass fraction is nearly constant with the amount of added corundum in the system containing the Pb glass.

Acknowledgments

The glasses were kindly prepared at the research laboratories of ICC FERRO Italia. Thanks to M. Dapiaggi for the measurements of the absorption coefficient performed in Milan.

References

1. H.M. Rietveld: J. Appl. Cryst. **2**, 65 (1969)
2. D.L. Bish and S.A. Howard: J. Appl. Cryst. **21**, 86 (1988)
3. D.L. Bish and J.B. Post: Am. Min. **78**, 932 (1993)
4. R.J. Hill: Powder Diff. **6(2)**, 74 (1991)
5. G. Artioli, A. Kern, M. Marchi and M. Dapiaggi: *Collected Abstracts, Proc. EPDIC-8.* (The 8th European Powder Diffraction Conference, Uppsala (SW) 23-26 May 2002) p. 202
6. A.G. De La Torre, S. Bruque and M.A.G. Aranda: J. Appl. Cryst. **34**, 196 (2000)
7. A.G. De la Torre, I. Pajares, S. Martinez-Ramírez, S. Bruque and M.A.G. Aranda: *Collected Abstracts.* (Proc. EPDIC-8, The 8th European Powder Diffraction Conference, Uppsala (SW) 23-26 May 2002) p. 201

8. F.R. Feret and D. Roy: Spectrochimica Acta B **57**, 551 (2002)
9. A.F. Gualtieri: J. Appl. Cryst. **33**, 267 (2000)
10. F. Guirado, S. Galí and S. Chinchón: Cement and Concrete Res. **30**, 1023 (2000)
11. X. Orlhac, C. Fillet, P. Deniard, A.M. Dulac and R. Brec: J. Appl. Cryst. **34**, 114 (2000)
12. P. Riello, P. Canton and G. Fagherazzi: J. Appl. Cryst. **31**, 78 (1998a)
13. P. Riello, G. Fagherazzi and P. Canton: Acta Cryst. A **54**, 219 (1998b)
14. W. Pitschke, N. Mattern and H. Hermann: Powder Diff. **8(1)**, 74 (1993a)
15. W. Pitschke, N. Mattern and Hermann: Powder Diff. **8(4)**, 223 (1993b)
16. R.J. Harrison and A. Paskin: Acta Cryst. A **43**, 401 (1964)
17. P.M. De Wolff: Acta Cryst. **9**, 682 (1956)
18. P. Suortti: J. Appl. Cryst. **5**, 325 (1972)
19. C.J. Sparks, K. Kumar, E.D. Specht, P. Zschack and G.E. Ice: Adv. X-ray Anal. **35**, 57 (1991)
20. R. Kumar, C.J. Sparks, T. Shiraishi, E.D. Specht, P. Zschack, G.E. Ice and K. Hisatune: Mat. Res. Soc. Symp. Proc. **213**, 369 (1991)
21. G.W. Brindley: Philos. Mag. **36**, 347 (1945)
22. J.C. Taylor and C.E. Matulis: J. Appl. Cryst. **24**, 14 (1991)
23. H. Hermann and M. Ermrich: Acta Cryst. A **43**, 401 (1987)
24. H. Hermann and M. Ermrich: Powder Diff. **4(4)**, 189 (1989)
25. H. Liu and C. Kuo: Mat. Letters **26**, 171 (1996)
26. R.S. Winburn, D.G. Grier, G.J. McCarthy and R.B. Peterson: Powder Diff. **15(3)**, 163 (2000b)
27. R.S. Winburn: Ph.D. dissertation (North Dakota State University, Fargo USA 1999a)
28. R.S. Winburn, S.L. Lerach, B.R. Jarabek, M.A. Wisdom, D.G. Grier and G.J. McCarthy: Adv. X-ray Anal. **42**, 387 (2000b)
29. I.C. Madsen, N.V.Y. Scarlett, L.M.D. Cranswick and L. Thaung: J. Appl. Cryst. **34**, 409 (2001)
30. A.J. Mandile and A.C. Hutton: Intern. J. Of Coal Geology **28**, 51 (1995)
31. M. Bellotto and C. Cristiani: Proc. EPDIC1, Munich 1991, **64** (1991) p. 64
32. G. Artioli, G. Alberti, G. Cagossi and M. Bellotto, in: *Atti 1° Convegno Italiano di Scienza e Tecnologia delle Zeoliti.* Ed. by C. Colella (De Frede, Napoli 1991) p261–270
33. A.F. Gualtieri and G. Artioli: Powder Diff. **10(4)**, 269 (1995)
34. A.F. Gualtieri: Powder Diff. **11[2]**, 97 (1996)
35. A.F. Gualtieri and M. Zanni: Mat. Science Forum **278–281**, 834 (1997)
36. P.W. MacMillan: *Glass-ceramics.* (Academic Press, London 1979) pp. 283
37. J.W. Visser and P.M. De Woff: *Report 641.109.* (Technisch Physische Dienst, Delft 1964)
38. D.C. Creagh. In: *Int. tables for Crystallography.* Vol. C, IUCr, (Kluver Academic Publishers, 1995) p. 182–189
39. A.L. Tisdall: Australian J. of Agricultural Res. **2**, 349 (1951)
40. D.L. Bish and R.C. Reynolds Jr.: Mineralogy **20**, 73 (1989)
41. NIST SRPM (1999). SRM 676, 100 Bureau Drive, Gaithersburg, MD 20899-2322 USA
42. J.P. Cline and R.B. Von Dreele: *The Denver X-ray Conference.* (Colorado Springs, CO 1998) (unpublished)

43. J.P. Cline and R.B. Von Dreele: *Late contribution at the Size-Strain III Conference. December 2-5, 2001* (Trento, Italy 2001)
44. A.C. Larson and R.B. Von Dreele: *Los Alamos National Laboratory Report LAUR 86-748.* (1994)
45. A. Altomare, M.C. Burla, C. Giacovazzo, A. Guagliardi, A.G. Moliterni, G. Polidori and R. Rizzi: J. Appl. Cryst. **34**, 392 (2001)
46. N. Ishizawa, T. Miyata, I. Minato, F. Marumo and S. Iwai: Acta Cryst. B **36**, 228 (1980)
47. K. Sudarsanan and R.A. Young: Acta Cryst. B **25**, 1534 (1969)
48. Y. Le Page and G. Donnay: Acta Cryst. B **32**, 2456 (1976)
49. R.M. Hazen and L.W. Finger: Am. Min. **79**, 196 (1979)
50. R.C. Peterson, G.A. Lager and R.L. Hitterman: Am. Min. **76**, 1455 (1991)
51. N. Blumental, F.S. Brugner and J.E. Garner: J. Electrochem. Soc. **120**, 1230 (1973)
52. B.L. Davis and D.K. Smith: Powder Diff. **3(4)**, 205 (1988)
53. H.P. Klug and L.E. Alexander: *X-Ray diffraction procedures.* (Wiley, New York 1974) pp. 716
54. C.R. Ward J.C. Taylor, C.E. Matulis and L.S. Dale: Int. J. of Coal Geology **46**, 67 (2001)
55. P.G. Weidler, J. Luster, J. Schneider, H. Sticher and A.U. Gehring: Europ. J. of Soil Science **49**, 95 (1998)

7 Quantitative Analysis of Amorphous Fraction in the Study of the Microstructure of Semi-crystalline Materials

P. Riello

In this chapter a Rietveld procedure that is useful in performing the quantitative phase analysis of semi-crystalline materials without using an internal standard is described and some applications are reported. Although the applicability of this method is not as wide as the internal standard one, it represents a valid alternative when the standard cannot be added to the sample, e.g. in bulk materials. This method can only be applied if the chemical composition of the amorphous phase or, as an alternative, of the whole sample is known. This limitation can be a serious problem in the study of natural materials, but it usually does not constitute a problem in the study of synthetic products the composition of which is well known.

7.1 Introduction

A large variety of materials that are very interesting both from the technological and scientific point of view are semi-crystalline: polymers, ceramic glasses, metallic glasses, catalysts, composites (metal matrix, polymer matrix) and others. Many of these materials are constituted of nanocrystalline phases embedded in an amorphous matrix.

Their chemical and/or physical properties are usually dependent on the relative amount of the amorphous to crystalline phases, so the studies of these materials should require the determination of the amorphous/crystalline content.

Powder diffraction is commonly used to perform the quantitative phase analysis of polycrystalline materials and it is also possible to quantify an amorphous phase by adding a known amount of a well-crystallised standard to the sample. If combined with the Rietveld method this procedure gives very good results.

This is undoubtedly one of the most used methods for the quantitative phase analysis of semi-crystalline materials but other methods that do not employ the internal standard have been developed.

7.2 A Short Historical Introduction to the Standardless Quantitative Methods

Analytical methods that do not use standards were historically developed to study the degree of crystallization of semi-crystalline polymers. The earliest methods gave only a "crystallinity index" that is useful in the comparison of different samples belonging to same homogeneous group [1].

Later, other procedures with a more solid theoretical basis were developed. Since these new methods had also been developed to study semi-crystalline polymers they were tailored to analyse materials whose crystalline and amorphous phases have the same chemical composition.

These methods are based on a fundamental property of the kinematical theory of diffraction that is a consequence of the Fourier transform theorems and of the atomicity hypothesis [2]. In fact if we consider the global intensity diffracted by N atoms in the reciprocal space V_s we can verify that it is independent of the atomic spatial order and it is only related to the number and to the species of the diffracting atoms:

$$\int_{V_s} I^{\mathrm{cor}}(\boldsymbol{s})d\boldsymbol{s} = \int_0^\infty I^{\mathrm{cor}}(s)4\pi s^2 ds \propto \sum_{i=1}^{N} \int_0^\infty [|f_i(s)|^2 + I_i^{\mathrm{ic(s)}}]4\pi s^2 ds \,, \quad (7.1)$$

where the used symbols are: $|\boldsymbol{s}| = s = 2\sin(\theta)/\lambda$, $I^{\mathrm{cor}}(\boldsymbol{s})$ = diffracted intensity (corrected for absorption and polarization), N = number of the diffracting atoms, $f_i(s)$ = atomic scattering factor of the atom i, $I_i^{\mathrm{inc}}(s)$ = incoherent (Compton) scattering of atoms i. So if we are able to split the diffraction pattern $I^{\mathrm{cor}}(s)$ of a semicrystalline material into its crystalline, $I_{cr}^{\mathrm{cor}}(s)$, and amorphous, $I_{am}^{\mathrm{cor}}(s)$, components (7.2a) we can use this invariant to determine the crystalline degree X_{cr} by means of

$$I^{\mathrm{cor}}(s) = I_{cr}^{\mathrm{cor}} + I_{am}^{\mathrm{cor}}(s) \quad (7.2a)$$

$$X_{cr} = \frac{\int_0^\infty I_{cr}^{\mathrm{cor}}(s)s^2 ds}{\int_0^\infty [I_{cr}^{\mathrm{cor}}(s) + I_{am}^{\mathrm{cor}}(s)]s^2 ds} \,. \quad (7.2b)$$

The measured sample is assumed to be isotropic.

In all these equations the superscript "cor" means that the intensities are corrected for absorption and polarization.

This equation is valid only if the crystalline phase has the same chemical composition of the whole sample but as we will see later a generalisation is also possible. The use of this approach first requires the solution of these two problems:

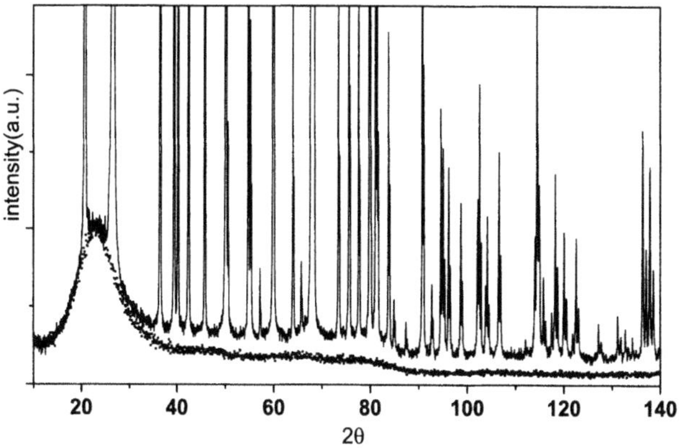

Fig. 7.1. Comparison of the diffraction pattern of the mixture α-quartz (80 wt.%) + amorphous silica (20 wt.%) (*continuous line*) and the pattern of the amorphous silica (*points*)

1. The separation of the scattering belonging to the crystalline phase from the amorphous one.
2. The evaluations of integrals in (7.2b) that have infinite limits of integration.

In Fig. 7.1 the pattern of a semi-crystalline sample constituted by a mixture of amorphous silica and quartz is reported. This example explains the origin of the first problem.

The comparison of the pattern of the semicrystalline sample with the pattern of the amorphous silica immediately evidences the fact that we cannot attribute the whole smooth background of the semicrystalline sample to the amorphous phase because this would result in an overestimation of the amorphous content. In fact an important fraction of the whole smooth background belongs to the crystalline phase; it is principally due to

- Incoherent (Compton) scattering of the atoms of the crystalline phase.
- Disorder and thermal diffuse scattering (TDS) (both depend on the thermal factors B)
- Fluorescence of the atoms of the crystalline phase (which can be avoided with a careful choice of the X-Ray wavelength).

Some authors [3] tried to avoid both the problems limiting the analysis to the 2θ range which corresponds to the first halo of the amorphous pattern. Studying many samples with the same chemical composition but with a different degree of crystallization and plotting the global areas of the Bragg peaks, in the chosen 2θ range, versus the areas of the smooth background they obtained a calibration line. This line intercepts the abscissa and the ordinate in two points corresponding to a completely amorphous ($X_{cr} = 0$) and to a completely crystalline ($X_{cr} = 1$) sample respectively. The other points of this line correspond to intermediate value of X_{cr} that are linearly interpolated between the two extreme values. Obviously the use of this calibration line is limited to the set of selected samples and allows only the definition of a "crystallinity index" useful to compare different samples.

Ruland [4] and Vonk [5] introduced a different approach: after the subtraction of the incoherent scattering from the experimental data (this is possible only if the chemical composition of the sample is known) they used a modified expression for X_{cr} (7.3) that takes into account the disorder and TDS.

$$X_{cr} = \frac{\int_0^{Sp} I_{\mathrm{Bragg}}^{\mathrm{cor}}(s) s^2 ds}{\int_0^{Sp} [I(s) - I^{\mathrm{inc}}(s)] s^2 ds} \; \frac{\int_0^{Sp} \langle f^2 \rangle s^2 ds}{\int_0^{Sp} \langle f^2 \rangle \exp(-(B/2) s^2) s^2 ds} . \quad (7.3)$$

The new introduced symbols are: $I_{\mathrm{Bragg}}^{\mathrm{cor}}$ = scattering belonging to the Bragg peaks; $\langle f^2 \rangle$ = average squared atomic scattering factor; I^{inc} = global incoherent scattering; B = overall thermal factor; $s_p = 2\sin(\theta_p)/\lambda$, where $2\theta_p$ is the upper measured scattering angle.

The new equation requires the contemporary evaluation of X_{cr} and B. Ruland evaluated them with a trial and error procedure that determines the thermal factor B that gives a constant X_{cr} for different choices of the upper integration limit s_p. The same equation was used by Vonk that introduced a computerized method able to determine both B and X_{cr}.

All the above methods always analyse materials where the amorphous and crystalline phases have the same chemical composition. Vonk et al. [6] introduced a procedure that takes into account the differences in the chemical composition between the crystalline and amorphous phases. Other approaches that are essentially based on the same ideas were developed by other authors and applied both to polymeric and glass ceramic materials [7,8].

The intrinsic limitations of the cited methods are:

- The crystalline phases are always treated as a whole: the global intensity belonging to Bragg peaks does not take into account the fact that different crystalline phases may be present.
- Usually the amorphous and the crystalline phases must have the same chemical composition. Only Vonk et al. [6] introduced a correcting factor in order to overcome this problem.
- Some methods require the pattern of a completely amorphous sample.
- Other approaches require different samples belonging to the same homogeneous group in order to trace out a calibration line.

7.3 A Rietveld Based Solution

The main problem, as discussed in the previous section, is the separation of the scattering of the amorphous phase from the global smooth background of the pattern of the semicrystalline sample (see Fig. 7.1).

This problem can be solved using the Rietveld method if we take into account the smooth contribute due to TDS, disorder and Compton scattering in the expression of the intensity diffracted by a crystalline phase [9].

By using the hypothesis of uncorrelated thermal disorder, the new expression is:

$$I_{cr} = KAP\left\{\sum_{k=1}^{n_B} J_k L_k O_k |F_k|^2 \Phi(2\theta - 2\theta_k) + Y^{bk}\right\}, \qquad (7.4)$$

$$Y^{bk} = \frac{(16\pi^2 V_c)}{(180\lambda^3)}\left\{\sum_{i=1}^{n} I_i^{\text{inc}} + \left[1 - \exp\left(\frac{2B\sin^2(\theta)}{\lambda^2}\right)\right]\sum_{i=1}^{n}|f_i|^2\right\}, \qquad (7.5)$$

where K is the scale factor; n_B is the number of Bragg reflections; n is the number of atoms in the unit cell; A is the correction factor for absorption; P is the polarisation factor (depending also on the monochromator set-up); J_k is the multiplicity factor; L_k is the Lorentz factor $1/\sin^2(\theta_k)\cos(\theta_k)$; O_k is the preferred orientation correction; $|F_k|^2$ is the squared structure factor for the k-th reflection, including the Debye–Waller factor $\exp(2B\sin^2(\theta)/\lambda^2)$; $\Phi(2\theta^\circ - 2\theta_k^\circ)$ is the normalised profile function (in 2θ degree), including asymmetry correction; B is the thermal factor (the same into the Debye–Waller factor); V_c is the cell volume; I_i^{inc} is the incoherent (or Compton) scattering from atoms i; f_i is the atomic scattering factor of the i-th atom. In this expression we do not consider the intensity due to fluorescence.

The incoherent scattering of each atom can be evaluated by using the analytical expression and the relevant parameters published by Smith, Thakkar and Chapman [10]. It must, however, be corrected for the Breit–Dirac factor, for specific absorption effects and for the band-pass function of the monochromator if used in the diffracted beam [11, 12].

All these corrections are very important especially the band-pass function that takes into account the angle-dependent attenuation of the incoherent scattering due to the monochromator when it is placed in the diffracted beam.

In spite of the crude hypothesis of uncorrelated disorder this expression describes very well the pattern of a fully crystallised sample (including its background), see Fig. 7.2, without introducing any ad hoc functions or parameters.

It can be seen that the new terms depend only on the global scale, the cell volume and the thermal factor B (the same thermal parameter used in the Debye–Waller factor) thus its shape and intensity depend only on the usual parameters refined by the Rietveld procedure.

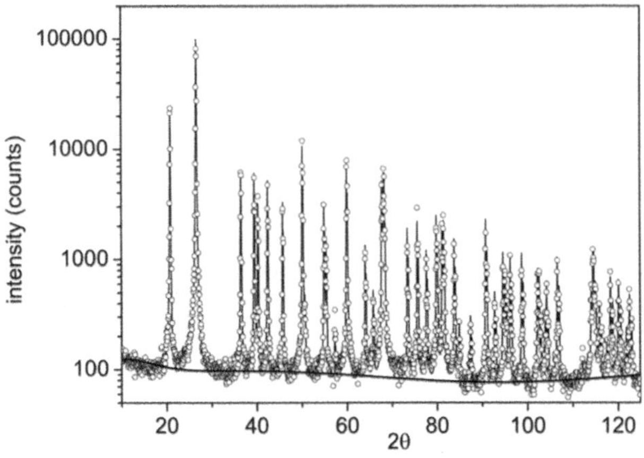

Fig. 7.2. Fitted pattern of α-quartz: the observed intensities are fitted by using (7.4) and (7.5). The background is evaluated taking into account the Compton and to the disorder scattering

Moreover it is important to observe that the shape of the function describing the background depends only on the thermal factor B: if B increases we have a shift of the intensity from the Bragg peaks to the smooth background while the global scattered intensity is conserved. This reduces the correlation between the thermal parameter and the global scale K that, as we will show in the following discussion, depend only on the global intensity integrated on the reciprocal space.

The overall pattern of a semi-crystalline material can now be described with the function:

$$
\begin{aligned}
I(2\theta) - I^{\text{air}}(2\theta) &= \sum_{l=1}^{N} I_l^{cr}(2\theta) + I^{\text{am}}(2\theta) \\
&= \left\{ \sum_{l=1}^{N} K_l A P \left[\sum_{k=1}^{n_B} J_k L_k O_k |F_k|^2 \Phi(2\theta - 2\theta_k) + Y_l^{bk} \right] \right\} \\
&\quad + I^{\text{am}}(2\theta)\,,
\end{aligned}
\tag{7.6}
$$

where I^{am} and I^{air} are respectively the amorphous and air scattering (see Fig. 7.3). This last term can easily be measured and subtracted from the measured pattern $I(2\theta)$ [13].

The amorphous contribution I^{am} can be described either by using the air corrected scattering of a pure amorphous material Y^{am} multiplied by a scale factor K^{am}:

$$
I^{\text{am}} = K^{\text{am}} Y^{\text{am}} \tag{7.7}
$$

or by using some ad hoc chosen analytical functions able to fit the amorphous pattern.

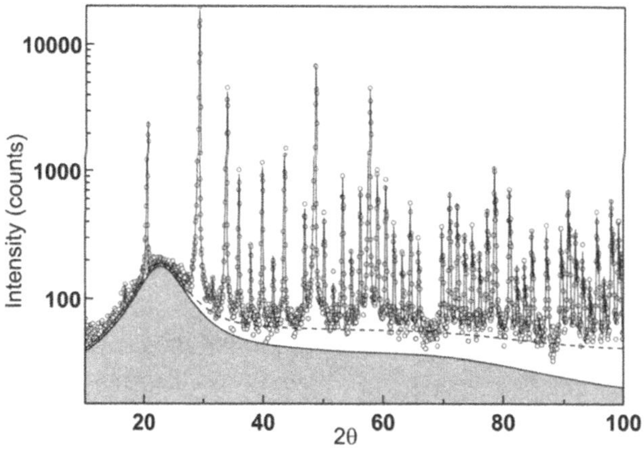

Fig. 7.3. Fit of the air corrected pattern of the mixture Y_2O_3 (60 wt.%) + amorphous silica (40 wt.%). The *shaded area* corresponds to the smoothed experimental pattern of the silica. The *dashed line* is the global background of the semicrystalline pattern. It is possible to appreciate the background due to the disorder and Compton scattering belonging to the crystalline Y_2O_3 phase

It is important to observe that the smooth background belonging to the crystalline phases is already taken into account in (7.4), (7.5) and (7.6) so the amorphous contribution can be fairly well described by using a polynomial or other more suited functions.

Once the pattern has been split into its crystalline and amorphous components the quantitative analysis can then be carried out [15]. Equation (7.1) can be applied to each phase in the pattern. Since each phase can be considered constituted of $N = Mn$ atoms, where M is the number of composition units of n atoms each, Eq. (7.1) becomes

$$\begin{aligned}\int_0^\infty I^{\mathrm{cor}}(s)4\pi s^2 ds \equiv \int_0^\infty \frac{I(s)}{AP} 4\pi s^2 ds \\ \propto \sum_{i=1}^{N} \int_0^\infty \left[|f_i^0|^2 + I_i^{\mathrm{inc}}\right] 4\pi s^2 ds \\ \propto M \sum_{i=1}^{n} \int_0^\infty \left[|f_i^0|^2 + I_i^{\mathrm{inc}}\right] 4\pi s^2 ds \,. \end{aligned} \quad (7.8)$$

Thus, the number of composition units can be evaluated if the chemical composition is known:

$$CM = \frac{\int_0^\infty \frac{I(s)}{AP}(s) 4\pi s^2 ds}{\sum_{i=1}^{n} \int_0^\infty \left[|f_i^0|^2 + I^{\mathrm{inc}}\right] 4\pi s^2 ds} . \tag{7.9}$$

The proportionality constant $\boldsymbol{C}$ in Eq. (7.9) is the same for all the phases in the pattern, when the microabsorption correction can be ignored, otherwise $\boldsymbol{C} = \tau^{-1}$, where τ is the particle absorption factor given by

$$\tau = \frac{1}{A_0} \int_{A_0} \exp[-(\mu - \bar{\mu})] dA ,$$

where A_0 is the particle volume and $\mu, \bar{\mu}$ are respectively the linear absorption coefficient of the considered phase and the mean linear absorption coefficient of the solid matrix [15, 16].

The calculation of the number of composition units by the use of the ratio in Eq. (7.9) obviously requires the knowledge of the chemical composition.

The evaluation of Eq. (7.9) by using the analytical expression (7.4) of the intensity diffracted by a crystalline material gives (see [2] for details):

$$\begin{aligned} \int_0^\infty I_{cr}^{\mathrm{cor}} 4\pi s^2 ds \equiv \int_0^\infty \frac{I_{cr}}{AP} 4\pi s^2 ds \\ = K \frac{16\pi^2 V_c}{(180\lambda^3)} \sum_{i=1}^{n} \int_0^\infty \left[|f_i^0|^2 + I_i^{\mathrm{inc}}\right] 4\pi s^2 ds , \end{aligned} \tag{7.10a}$$

$$\boldsymbol{CM} \propto \frac{16\pi^2}{180\lambda^3} K V_c , \tag{7.10b}$$

where the unit cell is assumed as compositional unit.

If for example, there is a doublet with intensity ratio 0.5, as is usual in laboratory apparatus, (7.10b) must be modified:

$$\boldsymbol{CM} \propto \frac{16\pi^2 K V_c}{180} \left(\frac{1}{\lambda_1^3} + \frac{1}{2\lambda_2^3} \right) . \tag{7.10c}$$

This result proves that equations (7.4) and (7.5) are consistent with the general integral property (7.1) and that the number of composition units (cells) M is proportional to the factor KV_c used in the conventional Rietveld quantitative analysis of crystalline materials. This proves that the described method, when wholly crystalline materials are considered, is equivalent to the usual Rietveld quantitative analysis.

The originality of this approach is that we can evaluate the number of composition units from the diffraction pattern of unknown or amorphous phases if we know their chemical composition. We can also apply this method to the diffraction pattern of the whole sample, i.e. without distinguishing between crystalline and amorphous phases, as described in the example on the crystallisation of glass ceramic material. In such example we evaluate the number of composition units of the whole sample. The necessary steps of the quantitative analysis are:

1. The separation of the crystalline and of the amorphous phases by using the described Rietveld procedure.
2. a) The chemical composition of the amorphous phase is known. In this case, the evaluation of the number of the composition units of all the phases (amorphous included) can be done by using the integral expression (7.9) [for the crystalline phases it is possible to the use the simpler formula (7.10b)]. The weight fraction of each phase is thus given by:

$$W_l = \frac{CM_l \sum_{i=1}^{n_l} w_l^i}{\sum_{l=1}^{N} CM_l \sum_{i=1}^{n_l} w_l^i}, \qquad (7.11)$$

   where w_l^i is the atomic weight of the i-th atoms of the composition unit of the l-th phase and n_l is the number of atoms in the composition unit. If we use (7.10b) for the crystalline phases we have to remember that the composition unit is the chosen elemental cell.

   b) The overall chemical composition of the material is known: In this case the evaluation of the number of the composition units of each crystalline phase can be carried out as in point 2a). The number of the composition unit CM_G of the whole sample is obtained by using the integral expression (7.9). In this way the weight fractions W_l^{cr} of the crystalline phases is evaluated by

$$W_l^{cr} = \frac{CM_l \sum_{i=1}^{n_l} w_l^i}{CM_G \sum_{i=1}^{n_G} w_l^i}, \qquad (7.12)$$

   where n_G the number of atoms in the composition unit of the whole sample. The amount of the amorphous phase is then obtained by $W_{am} = 1 - \sum W_l^{cr}$.

In any case we have to evaluate the number of composition units by using (7.9). The main difficulty is obviously related to the infinite integral limits required by the Fourier theorem when instead the intensity readings are collected in a limited angular range $s_{\min} < s < s_{\max}$.

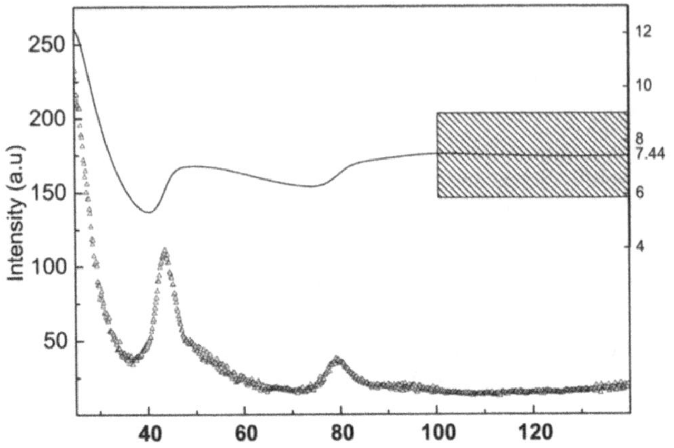

Fig. 7.4. Evaluation of the number of composition units. The symbols represent the pattern of the amorphous carbon collected with Cu$K\alpha$ radiation while the continuous line is the function $CM(s_p)$ (*right hand scale*). The asymptotic value (7.44) of the function $CM(s_p)$ is obtained as the average value of the function in the *shaded region*

In order to evaluate (7.7) we introduce the function $[CM](s_p)$ of the upper limits of integration:

$$[CM](s_p) = \frac{\int\limits_{s_{\min}}^{s_p} \frac{I(s)}{AP}(s) 4\pi s^2 ds}{\sum\limits_{i=1}^{n} \int\limits_{s_{\min}}^{s_p} \left[|f_i^0|^2 + I^{\text{inc}}\right] 4\pi s^2 ds} . \tag{7.13}$$

Now we can approximate $CM \simeq \lim_{(sp \to \infty)} [CM](s_p)$ if $s_{\min}$ is low enough. Since, for large values of s_p, this function oscillates around its asymptotic value it is possible to approximate the limit as the average value of the function in the range $s > 1$.

For example, with Cu radiation, the measures are collected in the $5 < 2\theta < 140$ range so that the asymptotic value is evaluated as the mean value of $CM(s_p)$ in the range $100 < 2\theta_p < 140$ (i.e. $1 < s_p < 1.2$). The problem of the choice of $s_{\min}$ is not so critical because of the weighting factor s^2 in the integrals.

Figure 7.4 illustrates the evaluation of the number of composition units for an amorphous material (this is the active carbon of the first illustrated examples). The asymptotic value is approximated with the average value of the function $CM(s_p)$ (continuous line) in the hatched region. This function is evaluated using the intensity reported in the same figure.

7.4 Examples

7.4.1 Example 1: Study of a Supported Metal Catalyst

The goal in the preparation of this kind of material is to obtain very dispersed metal particles. In fact the use of small particles means that a large fraction of the atoms of the active phase is exposed at the surface of particles where the chemical reactions take place. It is also known that the nanometric size changes the electronic structure and enhances the specific catalytic efficiency (selectivity and/or activity).

A special effort is made to obtain catalysts with small, size controlled particles so it is very important to develop analytical methods able to check the obtained particle size distributions both from the qualitative and quantitative point of view. In this example, concerning a carbon supported palladium catalyst the results obtained with a careful use of powder diffraction are described.

When the particles have a size of about 2 nm their diffraction peaks are so broad that it is very difficult to separate their contribution from the background, especially if these particles are supported and the amount of the loaded metal is low. In this case, structured background and bad signal to noise ratio make the separation particularly difficult.

Very often, only a small fraction of the high-dispersed metal gives origin to diffraction peaks. Furthermore, the majority of the metal is dispersed in such fine particles that their very broad diffraction peaks are hidden in the background. In this case, line profile analysis, that is a common tool in the characterization of these materials, must be combined with the quantitative phase analysis in order to know if the obtained microstructural information are characteristic of the total loaded metal.

Figure 7.5 illustrates this problem: all the patterns belong to the same catalyst that was annealed at different temperatures in order to promote the growth of the metal particles that in this way become "visible". If line profile analysis is not coupled to the quantitative phase analysis one could think that the measured sizes of the particles are characteristic of all the metal in the catalyst while, as often happens, they are only due to a small fraction of it. Moreover, if we consider that catalysis is a surface localized phenomenon, often the measured size is due to a fraction of particles (the larger ones) that are irrelevant to the comprehension of the performance of the catalyst. The study of these systems requires therefore, a careful data collection and analysis in order to obtain an optimal separation of the background. The optimum data treatment is possible only if we can compare the pattern of the catalyst with the pattern of the support but if this is not possible we can still judge the limits of our results by using the quantitative analysis.

These ideas are clarified in the following example. In the first place, one of the characteristics of the highly porous support used is a high hygroscopicity,

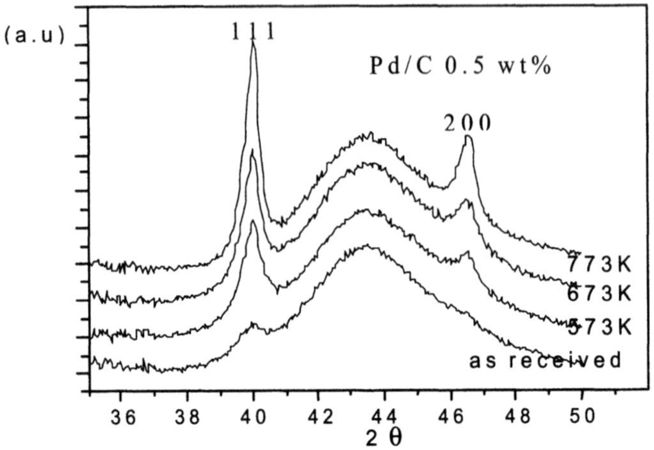

Fig. 7.5. Diffraction patterns of thermally treated carbon supported Pd catalysts. The catalyst was thermally treated at three different temperatures for 4 h and then measured at room temperature. The curves are vertically shifted in order to improve the readability of the plot

which makes the shape of the background strongly dependent on the water content.

In order to remove this uncontrolled effect the supporting carbon is previously fluxed for a couple of hours with N_2 and successively the data collection is carried out, while the sample continues to undergo the same fluxing, in an ad hoc developed sample holder [17].

The catalyst samples however, previously undergo an analogous treatment with H_2 for a few minutes. This treatment allows the elimination of the palladium oxide, which has the notable property that it can be easily reduced to metallic palladium at room temperature when the particles are very small.

In Fig. 7.6 there is the comparison of the pattern of the dried carbon with that of the catalyst (Pd/C 0.66 wt.%) before and after gas fluxing. The difference between the patterns of the catalyst before and after the gas treatment is evident. The fluxing with N_2 allows the collection of the diffraction data on reproducible dried samples. In fact, without drying the samples, the untreated patterns of the catalyst and of the carbon usually do not fit in the angular range $25 < 2\theta < 35$ degree (see Fig. 7.6) because of a different content of adsorbed water.

Since we have the experimental pattern of the carbon (i.e. the amorphous phase) we used (7.7) to perform the Rietveld analysis (Fig. 7.7). Pseudo-Voigt functions were used to fit the Bragg peaks of Palladium and the Fourier analysis of the 111 fitted peak was carried out evaluating the analytical Fourier transform of the fitted profile. The instrumental broadening, evaluated by using a reference polycrystalline α-quartz standard, was also taken into account even if the introduced correction was very small. The line broadening was ascribed only to size effects.

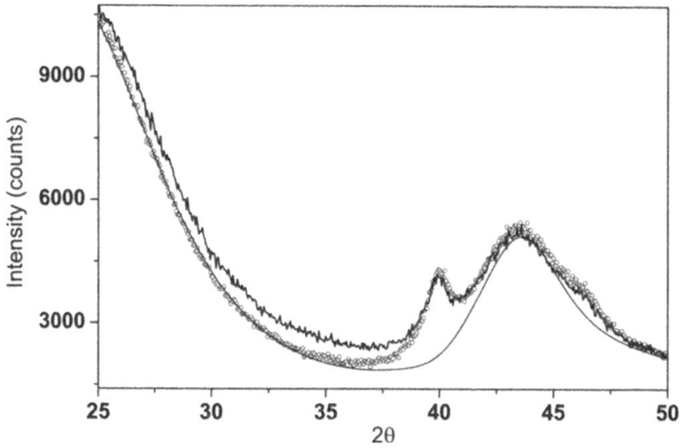

Fig. 7.6. Comparison of the carbon pattern (*smoothed continuous line*) that was fluxed with N_2, with the catalyst pattern before (*continuous line*) and after (symbols) the H_2 and N_2 flow. The pattern of the carbon was fitted with a combination of polynomial and pseudo-Voigt functions in order to smooth the data before the Rietveld fitting (see Fig. 7.7)

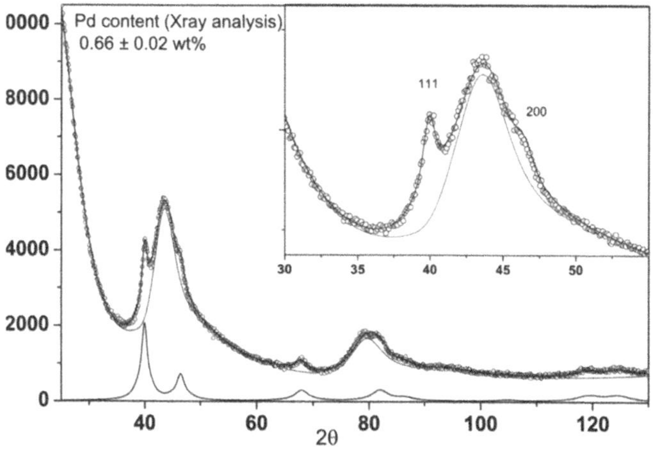

Fig. 7.7. Pd/C (0.66(3) wt.%) fluxed with H_2 for ten minutes and successively fluxed with N_2. The fit is obtained by using the described Rietveld procedure and the smoothed pattern of the carbon, that underwent the same gas treatment, was used to fit the amorphous phase. The fitted Palladium pattern (*bottom*) shows the long tails of the Bragg peaks. In the *inset* it is possible to see an enlargement of the region with the main two Palladium peaks

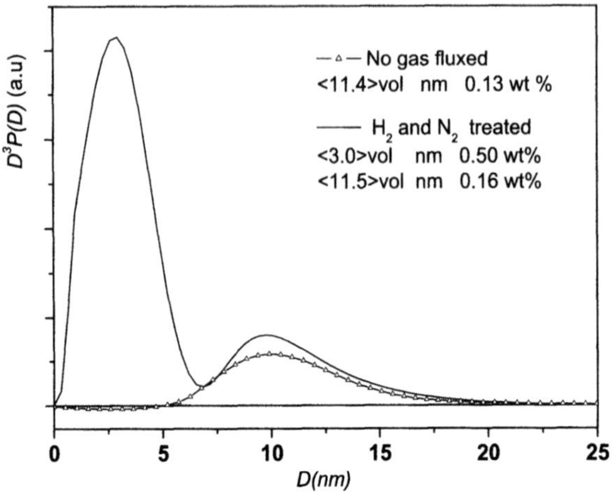

Fig. 7.8. Volumetric particles distributions of the palladium as obtained by using the Fourier analysis of the fitted 111 peaks corrected for the instrumental broadening. Continuous line is the distribution obtained studying the gas-fluxed sample. Symbols shows the results obtained analysing the as received sample (without gas flowing) using a polynomial function to fit the background. The weight fractions, that are proportional to the peak areas, are obtained with the procedure described in the main text

Usually the Fourier analysis is employed to obtain the column distribution that is independent of the shape of the crystallites [18], but in this case, we have preferred to consider the spherical particle distribution given by [19]

$$P(D) = -\frac{d}{dD}\left[\frac{1}{D}\frac{d^2A(D)}{dD^2}\right], \tag{7.14}$$

where D is the diameter of the particles and $A(D)$ is the Fourier transform of the profile corrected for the instrumental broadening. This choice is supported by the observation that the palladium particles are nearly spherical, as proved by the TEM images of this catalyst.

We have also carried out the line broadening analysis by fitting a Pseudo Voigt function to the 111 Pd peak on the as received sample (without H_2 and N_2 fluxing). In such analysis we used a polynomial function to fit background as often done in the XRD studies of supported catalyst. The imprecise description of the background line underestimates the long tails of the peak of the metal.

Figure 7.8 shows the volumetric particle distributions obtained by using the two described procedures. The areas under the peaks are proportional to the volumetric fraction of the palladium and in order to compare the two distributions we have set their areas to be proportional to the area of the fitted 111 peaks. It is clear that the gas-fluxed sample studied by the described Rietveld analysis evidences an important group of particles with very small sizes (3 nm ca.) that is completely lost when the study is done with the standard methods.

The Rietveld quantitative phase analysis applied to the fluxed catalyst gives a Pd content equal to 0.66 ± 0.02 wt.% thus detecting all the metal in the sample. Furthermore the particle size distribution shows two well defined groups of particles with average sizes $\langle D \rangle = 3$ nm (0.5 wt.%) and $\langle D \rangle = 11.5$ nm (0.16 wt. respectively. The comparison of the size distributions of the treated and of the untreated catalyst indicates that in the latter distribution only a small fraction of the loaded metal is detected.

7.4.2 Example 2: Nucleation of Glass Ceramic Materials in the System Li_2O–Al_2O_3–SiO_2

This example concerns the study of the nucleation of glass ceramic materials in the Lithium–Aluminium–Silicon oxides system [20].

These glasses are very interesting because it is possible to obtain transparent semi-crystalline materials with a very large fraction of crystalline phases that provide good mechanical characteristics. Moreover, the glass ceramics of the system Li_2O–Al_2O_3–SiO_2 develop crystalline phases with very low thermal expansion parameters, which give the glass a high resistance to the thermal shocks.

These glasses are transparent because during the crystallisation process very small crystals, with sizes that are much smaller than the visible light wavelength, develop in the glassy matrix. In order to obtain such a controlled crystallisation other nucleating oxides (TiO_2 and ZrO_2) are added to the main composition. The nucleating oxides develop a large number of nuclei, that will then be the seeds of the main crystallisation (Fig. 7.9), when a thermal treatment is carried out at a temperature which is lower than the temperature necessary for the main crystallisation. A large number of nuclei means that a large number of very small crystallites will develop during the crystallisation treatment and a transparent material with a high fraction of crystalline phases is obtained.

The control of the nucleation step in the manufacturing glass ceramic is very important so it is necessary to know both the mean size of the nuclei and their concentration. The peaks in Fig. 7.10, bottom curve, correspond to $ZrTiO_4$ that constitutes the nucleated phase.

In this case the quantitative analysis is carried out taking into account that the global composition of the glass is perfecly known and that does not change with the thermal treatment so the method 2b described in the Sect. 7.3 is employed. This method was also used to study the main cristallisation process of the glass [15].

From the analysis of the peak broadening it is possible to obtain the average size of the crystallites (5 nm) that, in this case, as the TEM image shows, corresponds to the average size of the particles, thus they are generally single domain (Fig. 7.11).

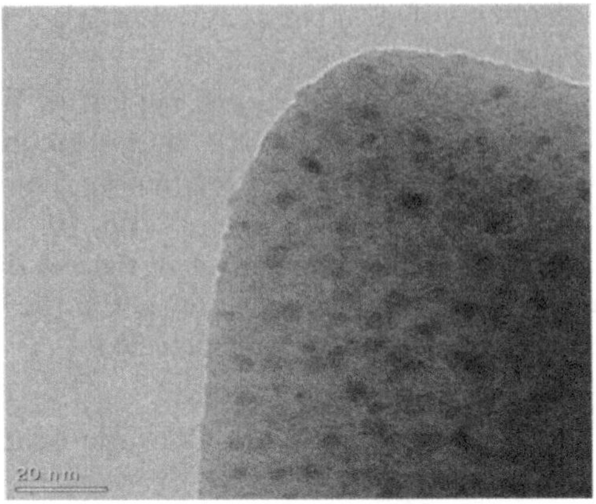

Fig. 7.9. Digitized TEM images of the nucleated glass (953 K for 20 h). A large number of nuclei are distinctly visible. H. Hofmeister and S. Hopfe obtained this image at the Max Planck Institute of Microstructure Physics, Halle, Germany

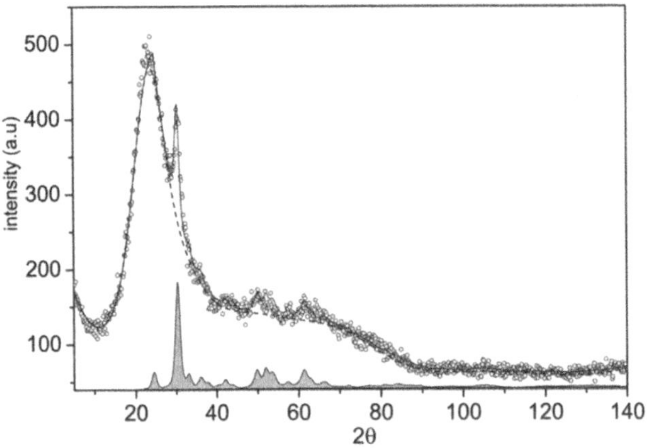

Fig. 7.10. Air corrected XRD pattern of the nucleated glass. The *dashed line* is the global background that is the sum of the amorphous, the thermal and the incoherent scattering. The fitted diffraction pattern of the nucleated $ZrTiO_4$ phase is shown at the bottom of the graph

Once the weight fraction m of the nucleated phase is evaluated it is easy, knowing the glass density, to evaluate the concentration of the nuclei:

$$C_n \left(\frac{\text{nuclei}}{\text{cm}^3} \right) = \frac{m\rho_G}{V_n \rho_N} .$$

In the above equation ρ_G and ρ_N are respectively the mass density of the glass and of the nucleated phase and V_n is the average volume of the nuclei that are assumed spherical. It was so possible to characterize in a satisfactory

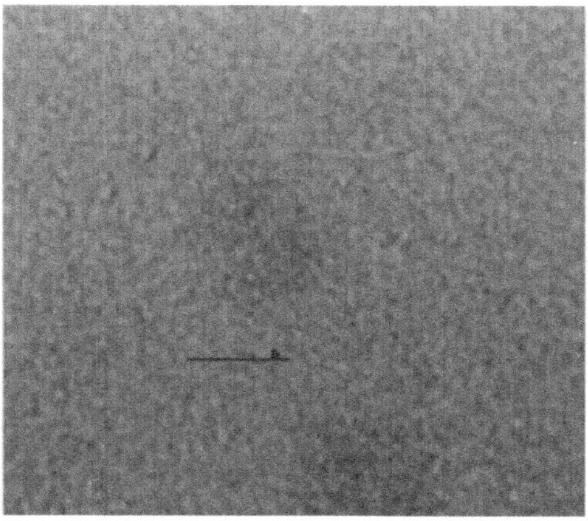

Fig. 7.11. Digitised HRTEM image of a spherical crystalline nucleus of $ZrTiO_4$ in a glassy matrix. The bar corresponds to 5 nm. H. Hofmeister and S. Hopfe obtained the image at the Max Planck Institute of Microstructure Physics, Halle, Germany

way the studied glass obtaining $m = 4.3\,\text{wt.\%}$ that corresponds to $C_n = 7.8 \times 10^{16}\,\text{nuclei/cm}^3$ (the TEM evaluated number is $9.6 \times 10^{16}\,\text{nuclei/cm}^3$).

7.5 Conclusion

The applicability of the method for quantitative phase analysis used in this article is quite general even if it is related to the knowledge of the chemical composition of the studied material. When we study synthetic materials we usually know their chemical composition but this knowledge is obtained with difficulty in the study of natural ones (i.e. rocks, sediments and other).

However this analytical method can be applied to a large group of technologically important materials and often it is the only possible method when the internal standards cannot be employed.

The reported examples show that the study of the microstructure of complex materials cannot ignore quantitative analysis that is very useful to understand their chemical-physical behaviour and to develop new materials.

Another important aspect of the described Rietveld method is the possibility of tracing the background line by using a physical model that, even though in a approximate way, allows a better and physically consistent separation of the diffraction peaks. This is particularly important when the peaks are very broad. By using this approach it was possible to obtain quantitative information on particles with size of 3 nm ca. supported on an amorphous phase even if their content was less than 1 wt.%.

References

1. J.H. Wakelin, H.S. Virgin and E. Crystal: J. Appl. Phys. **30**, 1654 (1959)
2. P. Riello, G. Fagherazzi, P. Canton: Acta Crystallogr. A **54**, 219 (1998)
3. P.H. Hernabbs and A. Weidinger: Makromol. Chem. **44–46**, 24 (1961)
4. W. Ruland: Acta Cryst. **14**, 1180 (1961)
5. C.G. Vonk: J. Appl. Cryst. **6**, 148 (1973)
6. C.G. Vonk and G. Fagherazzi: J. Appl. Cryst. **16**, 274 (1983)
7. R. Gehrke and H.G. Zachmann: Makrom. Chem. **182**, 627 (1981)
8. S. Polizzi, G. Fagherazzi, A. Benedetti, M. Battagliarin, T. Asano: J. Appl. Crystallogr. **23**, 359 (1990)
9. P. Riello, G. Fagherazzi, D. Clemente, P. Canton: J. Appl. Crystallogr. **28(2)**, 115 (1995)
10. V.H. Smith Jr., A.J. Thakkar and D.C. Chapman: Acta Cryst. A **31**, 391 (1975)
11. W. Ruland: Br. J. Appl. Phys. **15**, 1301 (1964)
12. P. Riello, P. Canton and G. Fagherazzi: Powder Diffraction **12**, 160 (1997)
13. S. Ottani, P. Riello and S. Polizzi: Powder Diffraction **8**, 149 (1993)
14. P. Riello, P. Canton and G. Fagherazzi: J. Appl. Crystallogr. **31(1)**, 78 (1998)
15. T.C. Taylor and C.E. Matulis: J. Appl. Crystallogr. **24(1)**, 14 (1991)
16. G.W. Brindley: Philos. Mag. **36**, 347 (1945)
17. P. Canton, P. Riello, A. Furlan, A. Minesso, L. Bertoldo, F. Pinna, A. Benedetti: Catal. Lett. **69(1,2)**, 17 (2000)
18. A. Guinier: *X-Ray Diffraction.* (Freeman, San Francisco 1963) p. 139
19. S. Ciccariello, G. Fagherazzi, A. Benedetti: Acta Crystallogr. A **46**, 187 (1990)
20. P. Riello, P. Canton, N. Comelato, S. Polizzi, M. Verita, G. Fagherazzi, H. Hofmeister and S. Hoppe: J. Non-Cryst. Solids **288(1-3)**, 127 (2001)

Part III

Crystallite Size and Shape

8 A Bayesian/Maximum Entropy Method for the Certification of a Nanocrystallite-Size NIST Standard Reference Material

N. Armstrong, W. Kalceff, J.P. Cline, and J. Bonevich

A Bayesian/Maximum Entropy (MaxEnt) method for determining crystallite size distributions and morphologies from size-broadened X-ray line profiles is presented. This method will be used in certifying a nanocrystallite-size standard reference material (SRM) being developed at the National Institute of Standards and Technology (NIST). The proposed SRM will assist in ensuring that uniform procedures in quantifying the microstructure of nanocrystallites from X-ray line profile data are developed. This will become increasingly important as emerging nanotechnology applications begin to call for crystallites designed to have particular morphology and size distributions.

The Bayesian/MaxEnt method is presented as a way of overcoming the inherent difficulties in determining the size distribution from the experimental X-ray line profile data. These difficulties pertain to solving the inverse problem of the governing integral equation for crystallite size-broadening. It is proposed that the principles of inductive reasoning be used to determine the most probable size distribution, given our knowledge of the experimental data and equipment, while employing any suitable a priori information. These principles are embodied in Bayesian theory, with the entropy function used to assign values to the size distribution constrained by the experimental data, and resulting in a size distribution (or solution) with minimal assumptions.

The application of the Bayesian/MaxEnt method to X-ray diffraction data gathered from a CPD/IUCr round-robin cerium oxide (CeO_2) specimen is presented. Our analysis shows that the CeO_2 specimen has a near spherical morphology and a well defined size distribution, comparing satisfactorily with transmission electron microscopy data.

8.1 Introduction

In this chapter we present a general Bayesian/maximum entropy (MaxEnt) method for determining the crystallite size distribution and morphology from size-broadened X-ray line profiles. Motivation for this work has been the ongoing development of a nanocrystallite-size Standard Reference Material (SRM) by the National Institute of Standards and Technology (NIST), Gaithersburg. This SRM will enable any laboratory to assess its experimental and analytical techniques for quantifying size-broadened X-ray profiles. It will also ensure that uniform standards are developed in quantifying the microstructure of nanocrystallites from X-ray line profile data.

Two nanocrystallite-size SRMs are under development. The first is a chemically inert cerium oxide (CeO_2) specimen with a spherical morphology in the 10–20 nm size range; it offers a basic morphology that can be analysed using the standard line profile methods (for a detailed review of various methods see [1]), while its cubic symmetry results in well separated reflections. The second size SRM will be a zinc oxide (ZnO) specimen in the 40–60 nm crystallite size range, with a hexagonal-prism morphology [2], which can be approximated by a cylindrical morphology [3–5]. This SRM will require more sophisticated models and analysis techniques, as the low symmetry of the lattice also results in overlapping line profiles. However, the generality of the Bayesian/MaxEnt method is such as to allow these additional complications to be accommodated.

The development of this SRM and the Bayesian/MaxEnt method is timely, given the growing nanotechnology industry. Recent developments in determining the nanocrystallite-size distribution from X-ray line profiles have seen these methods compared to existing direct techniques such as transmission electron microscopy (TEM) [6–8]. Using the size-broadened line profile as a starting point, the underlying nanocrystallite-size distribution and morphology can be determined and compared with the TEM size distribution. This is a significant development in that it makes X-ray line profile analysis and TEM methods complementary. However, each method also offers distinct advantages. For example, while TEM techniques can be applied to miniature quantities of a material specimen (perhaps insufficient for X-ray diffraction analysis), X-ray line profile methods ensure that suitably large numbers of crystallite are "sampled" to produce representative statistics and significant deductions about its microstructure.

In determining the size distribution from the observed line profiles, an inverse problem must be solved. This immediately raises issues concerning the existence of a solution, and also questions the uniqueness of the crystallite size distribution. The Bayesian/MaxEnt method resolves both questions by applying the principles of inductive reasoning and determining the most probable or plausible distribution, given any suitable a priori knowledge about the observed line profile, instrumental effects and statistical noise. It is this fundamental characteristic of the Bayesian/MaxEnt method that ensures its suitability for certifying a nanocrystallite-size SRM in that it preserves the fundamental properties of the size distribution, tests any underlying assumptions in the presence of experimental data and quantifies the uncertainties in the resulting size distribution. In applying the Bayesian/MaxEnt method to size broadening analysis, we have followed the alogrithm of Bryan [9] and Jarell and Gubernatis [10] for evaluating the probabilities and treating the Lagrangian multiplier, α.

The background theory for this discussion is developed by investigating crystallite size broadening for the ideal and realistic cases (see Sect. 8.2.1).

This derivation enables physical properties to be quantified, including the size distribution and crystallite shape. It also demonstrates that these properties are encoded into the specimen line profile. It is shown that the observed line profile can be expressed in terms of the size distribution, while the size-broadening and instrumental kernel are "wrapped" into a single kernel, $K(s, \boldsymbol{D})$, where s and $\boldsymbol{D}$ are identified in Sect. 8.2. The resulting integral equation (of the first kind), suggests that the underlying crystallite size distribution and morphology can be determined from the experimental data. It also establishes that the size distribution and line profile are described by a particular class of distributions – called *positive* and *additive* distributions.

The general functional properties of the integral equations for size-broadening are discussed and it is shown that the kernel, $K(s, \boldsymbol{D})$, is intrinsically (near) singular, which questions the uniqueness of the size distribution (Sect. 8.3). Based on these observations, the principles of inductive reasoning are presented in terms of probabilities and Bayesian analysis, and proposed as an approach to determine the most plausible size distribution, given the experimental data and a priori information (see Sect. 8.4). The theoretical development towards a Bayesian/MaxEnt method is presented in Sect. 8.5; this discussion outlines how the probability distributions can be evaluated. In addition, the fundamental properties of the size distribution and line profiles (i.e. *positivity* and *additivity*) are used to argue that the entropy function is the *only* function that can consistently assign values to these distributions, while taking into consideration the experimental and a priori information. In Sect. 8.5.6 it is shown that the Langford et al. method [7] is a special and limited application of Bayes' theorem, where the a priori distributions for the parameters of the size broadening are set as uniform distributions and the likelihood function is minimised. Generally, the values of the size broadening are determined using non-linear least squares. However, a Bayesian/Markov chain Monte Carlo (MCMC) method is proposed as an alterative numerical technique for implementing this method.

The Bayesian/MaxEnt method is applied to a specimen of cerium oxide (CeO_2) which was prepared for a CPD/IUCr crystallite size round-robin [11] (see Sect. 8.6). The results for this analysis are generally very satisfactory and there exists agreement between the Bayesian/MaxEnt results and the TEM distribution. These results are also compared with the analysis from "standard approaches" such as the Krill and Birringer method [6] and the Langford et al. method [7] embodied in the Bayesian/MCMC approach. Although the comparison is favorable, it is argued that the Bayesian/MaxEnt method addresses the problem of determining the size distribution in a fundamental way.

8.2 Nanocrystallite-Size Broadening of X-ray Line Profiles

8.2.1 Size-Broadened Profiles

Ideal Specimen

Consider the diffraction of a monochromatic beam of X-rays from a specimen which has a perfect atomic arrangement and is infinite in size. The phase of the scattered beam will vary continuously from 0 to 2π due to the path difference defined over the thickness of the crystallite. This results in destructive interference of the scattered beam when the Bragg condition[1] is not met, and constructive interference when it is. The resulting diffraction spot (in reciprocal space) is an intensity distribution (in three dimensions) which is infinitesimally narrow and of infinite intensity. After integrating across the surface normal of a sphere which intersects the diffraction spot at s, the line profile, I_p, from this specimen can be expressed as

$$\begin{aligned} I_p(s) &\approx A\,\frac{\sin^2(2\pi D s)}{(2\pi d_{\mathrm{hkl}} s)^2}\,, \\ &= A\,\delta(s)\,, \quad \text{as} \quad D \to \infty\,, \end{aligned} \tag{8.1}$$

where D refers to the dimensions of a crystallite of finite size; d_{hkl} is the interplanar spacing; $s = \frac{2}{\lambda}(\sin\theta - \sin\theta_0)$; λ is the X-ray wavelength, and $2\theta_0$ is the Bragg angle. In (8.1), $\delta(s) = \infty$ when $s = 0$ (i.e. $\theta = \theta_0$) and is zero otherwise. Also in (8.1), the complete line profile from this ideal specimen will be characterised by a set of delta-functions, $\delta(s)$, centered about $\frac{2}{\lambda}\sin\theta_0$. These profiles will have integrated intensity A, since $\int_{-\infty}^{+\infty} \delta(s)\,\mathrm{d}s = 1$. The width of the line profile will be $\sim 1/D \to 0$, as the dimension of the crystallite parallel to the diffraction vector $D \to \infty$; that is, the broadening of the line profile is infinitesimally small.

Imperfect Specimen

Now consider the line profile from an "imperfect" specimen consisting of nanocrystallites which are finite in size ($\lesssim 100\,\mathrm{nm}$) and oriented in all directions. The crystallites also have well-defined boundaries and the same shape and size, defined by $\boldsymbol{D} = \{D_1, D_2, D_3\}$, where D_i is the length of the crystallite along the x_i axis as shown in Fig. 8.1. In the simplest case of a specimen

[1] Given that the incident beam is defined by the wave vector $\boldsymbol{S}_i$ and the final or diffracted wave is given be $\boldsymbol{S}_f$, the Bragg condition requires $\boldsymbol{S}_f - \boldsymbol{S}_i = \vec{\boldsymbol{g}}$, where $|\vec{\boldsymbol{g}}| = 1/d_{\mathrm{hkl}}$ and d_{hkl} is the inter-atomic d-spacing for a set of (hkl) planes. By setting $|\boldsymbol{S}_i| = |\boldsymbol{S}_f| = 1/\lambda$ and observing that the angle between $\boldsymbol{S}_f$ and $\boldsymbol{S}_i$ is 2θ, Bragg's equation follows. Throughout this discussion we use $\vec{\boldsymbol{g}}$ to denote the diffraction vector and $\boldsymbol{g}$ to denote the observed line profile.

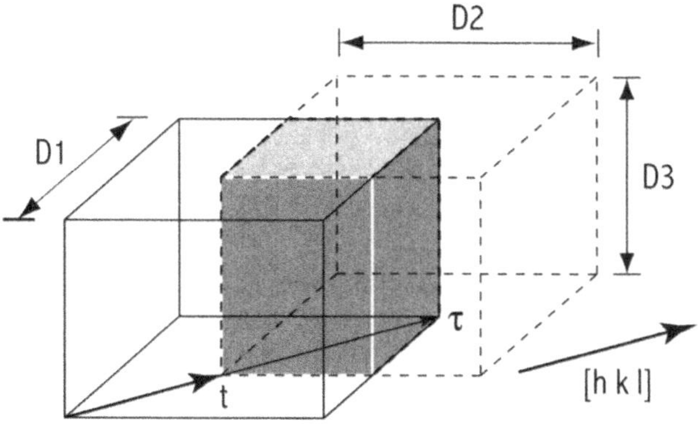

Fig. 8.1. The crystallite (*solid line*) and its "ghost" (*dashed line*) shifted a distance t in the direction of the scattering vector $\vec{g} = [\text{hkl}]$. The crystallite and ghost have dimensions, $\boldsymbol{D} = \{D_1, D_2, D_3\}$. The shaded region represents the common volume between the crystallite and ghost. The maximum thickness of the crystallite in the direction $[hkl]$ is τ

consisting of spherical crystallites, the principal dimension is the diameter, D; the more general case of a specimen consisting of cylindrical crystallites, would have the principal dimensions of height, H, and diameter, D.

The line profile and broadening from this specimen arises from crystallites diffracting independently of each other ([12] 1950, 1952). Further, only a fraction of the total number of crystallites will have planes oriented in the diffracting condition. As was the case above, the thickness of the crystallites parallel to $\vec{g}$ will cause a phase difference in the scattered beam, but this will not cause complete destructive interference due to the finite size of the crystallites; constructive interference will still result from the diffraction condition being met.

The corresponding diffraction spot in reciprocal space will now be an intensity distribution which has finite width and is smeared or "spread out" in space [13]. In addition, second order and third order diffraction peaks may be detected as having considerably diminished intensity. The corresponding line profile from crystallites of dimension $\boldsymbol{D}$, $I_p(s, \boldsymbol{D})$, can be expressed as [14,15]

$$I_p(s, \boldsymbol{D}) = \int_{-\tau}^{+\tau} V(t, \boldsymbol{D}) \exp\left[-2\pi \imath s t\right] \mathrm{d}t \tag{8.2a}$$

$$= \int_{-\tau}^{+\tau} V(t, \boldsymbol{D}) \cos 2\pi s t \, \mathrm{d}t \tag{8.2b}$$

where $\imath = \sqrt{-1}$; t defines the displacement of a crystallite shifted parallel to the diffraction vector, $\vec{g}$; $V(t, \boldsymbol{D})$ is the common-volume (or "ghost") function of the crystallite and represents the volume common (or shared) when the "ghost" crystallite is shifted a distance t parallel to $\vec{g}$. The dimension τ defines the maximum thickness of the crystallite parallel to $\vec{g}$ and is a function of the crystallite dimensions, $\boldsymbol{D}$. Figure 8.1 is a schematic representation of the crystallite and ghost shifted a distance t parallel to the diffraction vector. The common-volume function is an even function and will result in $I_p(s, \boldsymbol{D})$ also being an even function, which accounts for (8.2b). The common-volume function has the boundary conditions $V(0, \boldsymbol{D}) = V_0$ and $V(\pm\tau, \boldsymbol{D}) = 0$. As $t \to \tau$, $V(t, \boldsymbol{D}) \to 0$.

It is important to note that diffraction spots correspond to a Fourier transform of the crystallite. This is a direct consequence of diffraction taking place in the far-field approximation. (The same reasoning can also be applied to account for the line profile in (8.1) i.e. the Fourier transform of an infinite array of lattice points is a $\delta(s)$ function.) Given this, (8.2b) is the result of integrating out two of the three dimensions that define the diffraction spot. This allows (8.2a) to be interpreted as the Fourier transform of the common-volume function. That is, the Fourier transform of the crystallite, given by $V(t, \boldsymbol{D})$ represents the crystallite-size broadened line profile.

The crystallite size-broadening can be quantified in terms of the Fourier coefficients and integral breadth of (8.2b). Following on from the above discussion and (8.2a), the Fourier coefficients, $A(t)$, are given as (normalised for unity)

$$A(t) = \frac{V(t, \boldsymbol{D})}{V(0)}, \quad \forall t \in [0, \tau], \tag{8.3}$$

where $V(0)$ is the volume of the crystallite (see Fig. 8.1). In quantifying the crystallite size-broadening in terms of the Fourier coefficients, an area-weighted size can be defined as

$$\langle t \rangle_a^{-1} = \left. \frac{\mathrm{d}\, A(t)}{\mathrm{d}t} \right|_{t=0}. \tag{8.4}$$

It should be pointed out that $\langle t \rangle_a$ does not represent the physical dimension of the crystallite, but the apparent "thickness" (i.e. area-weighted size) of the crystallite in the direction of the diffraction vector, $\vec{g}$.

The integral breadth of (8.2b) is a direct consequence of the additivity of I_p and can be defined in reciprocal space units as

$$\beta = \frac{1}{I_{p,max}} \int_{-\infty}^{+\infty} I_p(s, \boldsymbol{D})\, \mathrm{d}s \tag{8.5a}$$

$$= \left[2 \int_0^{\tau} A(t)\, \mathrm{d}t \right]^{-1} \tag{8.5b}$$

where $I_{p,\max}$ is the maximum intensity of the line profile (8.2b). Also as (8.3) is used in (8.5b), β can be defined in terms of $V(t, \boldsymbol{D})$. From (8.5b), another dimension, the volume-weighted size, $\langle t \rangle_v$ can be defined as

$$\langle t \rangle_v = \beta^{-1}. \tag{8.6}$$

Like $\langle t \rangle_a$, $\langle t \rangle_v$ is also an "apparent" thickness of the crystallite in the direction of the diffraction vector, $\vec{g}$. These quantities are referred to as apparent sizes because a physical shape of the crystallite is needed to interpret them. This generally requires determining the Scherrer constant in order to relate $\langle t \rangle_a$ and $\langle t \rangle_v$ to the actual physical thickness of the crystallite, τ, parallel to $\vec{g}$. Given a diffraction pattern consisting only of size-broadened line profiles, each hkl defines the thickness of the crystallite in a particular direction. Using this information, the shape of the crystallite can be developed [1–5, 16, 17].

Real Specimen

We consider a real specimen to consist of nanocrystallites of the same shape, but with a distribution of sizes, $P(\boldsymbol{D})$. Thus, in a given specimen the crystallites have a range of dimensions $\boldsymbol{D}$ to $\boldsymbol{D} + \mathrm{d}\boldsymbol{D}$. The size-broadening, $f(s)$, in this case will consist of a superposition of line profiles, $I_p(s, \boldsymbol{D})$, having a specific dimension, $\boldsymbol{D}$, as given in (8.2b) but weighted by $P(\boldsymbol{D})$

$$f(s) = \int_0^\infty I_p(s, \boldsymbol{D})\, P(\boldsymbol{D})\, \mathrm{d}\boldsymbol{D}, \quad \forall\, s \tag{8.7}$$

where the measure, $\mathrm{d}\boldsymbol{D}$, has been used and is dependent on a specific co-ordinate system. The common-volume function for (8.7) does not define the common-volume for a single crystallite, as was the case in (8.2b), but involves the contribution of crystallites in the $\boldsymbol{D}$ to $\boldsymbol{D} + \mathrm{d}\boldsymbol{D}$ size-range,

$$\tilde{V}(t) = \int_t^\infty V(t, \boldsymbol{D})\, P(\boldsymbol{D})\, \mathrm{d}\boldsymbol{D}\,, \quad \forall\, t \in [0, \infty]\,, \tag{8.8}$$

where $\tilde{V}(t)$ defines a *modified* common-volume function. The Fourier coefficients are now given by

$$A(t) = \frac{1}{\tilde{V}(0)} \int_t^\infty V(t, \boldsymbol{D})\, P(\boldsymbol{D})\, \mathrm{d}\boldsymbol{D}\,, \tag{8.9}$$

where

$$\tilde{V}(0) = \int_0^\infty V(0, \boldsymbol{D})\, P(\boldsymbol{D})\mathrm{d}\boldsymbol{D}\,. \tag{8.10}$$

From (8.9) amd (8.10), the area- and volume-weighted sizes can also be defined. Further, each of these dimensions has a corresponding distribution $p_a(t)$ and $p_v(t)$, respectively. These distributions can also be determined (for example see [18]), but as they are only apparent dimensions it is physically more useful and meaningful to focus on determining the crystallite size distribution, $P(\boldsymbol{D})$.

We have assumed that the size-broadened line profile consists of crystallites of the same shape, but having a distribution of sizes, $P(\boldsymbol{D})$. This information is "encoded" into the specimen line profile, $f(s)$. The shape, size, and distribution of crystallites are important in designing nanomaterials, such as polishing media and catalysts. If this information could be extracted from the X-ray line profile data, it would provide a relatively simple method for determining the microstructural properties of crystallites. It would also complement other direct methods of determining the shape and size distribution of crystallites, such as TEM, SEM and AFM. However, the overall advantage of X-ray line profile methods is that they can generate representative statistics about the specimen, and can be sampled quickly and efficiently.

8.2.2 Observed Line Profiles

The difficulty in determining $P(\boldsymbol{D})$ from the observed line profile, $g(s)$, is that there are other contributions which must be removed. These include instrumental broadening or "blurring" due to the imperfect optics in the diffractometer, background counts from diffuse scattering, and statistical noise from imperfect counting of the detector and other sources. We can express the observed line profile $g(s)$ in reciprocal space units, s as

$$g(s) = \int_{-\infty}^{+\infty} k(s - s')\, f(s')\, \mathrm{d}s' + b(s) + n(s)\,, \tag{8.11}$$

where k is the instrument line profile, b is the background level and n is the noise contribution. In practice, $k(s)$, can be modelled from a specimen that contains few microstructural defects and large crystallites, while n generally follows a Poisson distribution, which for large counts ($\gg 10$) can be approximated by a Gaussian distribution.

Although it is useful to have $g(s)$ expressed in terms of the specimen profile, $f(s)$ (see (8.7)), it does not explicitly show the dependency of $g(s)$ on the size distribution $P(\boldsymbol{D})$. This can be achieved if (8.7) is substituted into (8.11) and the order of integration changed such that

$$g(s) = \int_{0}^{\infty} K(s, \boldsymbol{D})\, P(\boldsymbol{D})\, \mathrm{d}\boldsymbol{D} + b(s) + n(s)\,, \quad s \in [-\infty, \infty]\,, \tag{8.12}$$

where the new kernel "wraps up" the instrumental and line-broadening contribution into a single result,

$$K(s, \boldsymbol{D}) = \int_{-\infty}^{\infty} k(s - s')\, I_p(s, \boldsymbol{D})\, \mathrm{d}s'\,. \tag{8.13}$$

The explicit dependency of $g(s)$ on $P(\boldsymbol{D})$ is given in (8.12) and (8.13), highlighting the problem of determining $P(\boldsymbol{D})$ from the experimental data, $g(s)$; it also allows us to explore how to determine $P(\boldsymbol{D})$ from $g(s)$, given our knowledge of $K(s, \boldsymbol{D})$, $b(s)$ and $n(s)$.

The properties of $g(s)$, $f(s)$, $P(\boldsymbol{D})$ and $A(t)$ are important in developing a method for determining the specimen profile and size distributions, which are all both *positive* and *additive* distributions. In developing such a method, it is expected that positivity will be preserved without having to make additional assumptions concerning, say, the functional form of the distributions. The additivity property ensures that quantities such as integral breadth, (8.5b), can be determined, and is critically important for characterising $P(\boldsymbol{D})$. In other words, positivity ensures that unrealistic negative probabilities can not be obtained, while additivity ensures $P(\boldsymbol{D})$ can be normalised and its moments can be determined. These two fundamental properties allow a particular class of function to be defined without making additional assumptions concerning the shape and functional form of $P(\boldsymbol{D})$; also they are important in solving the inverse problem (see Sect. 8.3.2) and assigning values (i.e. real positive numbers) to $P(\boldsymbol{D})$ and $f(s)$ (see Sects. 8.4 and 8.5). Expressed differently, the solution set of $P(\boldsymbol{D})$ and $f(s)$ is restricted to those solutions which comply with the positivity and additivity conditions.

8.3 Developing the Inverse Problem

8.3.1 Forward Mapping

Equations (8.12) is an integral equation of the first kind [19] and defines the mapping of a size distribution, $P(\boldsymbol{D})$, into another continuous function, given by the observed line profile, $g(s)$. $P(\boldsymbol{D})$ is continuous on $[0, \infty]$, $K(s, \boldsymbol{D})$ is continuous on $[0, \infty] \times [-\infty, \infty]$, and the observed line profile is continuous on $[-\infty, \infty]$.

Alternatively (8.12) can be viewed as an *operator equation* [19]

$$g = \mathcal{K}P\,, \tag{8.14}$$

where the operator defines the mapping $\mathcal{K} : \mathcal{X} \to \mathcal{Y}$, and $\mathcal{X}$ and $\mathcal{Y}$ are sets of the domain and range, respectively. For the moment, the influences of $b(s)$ and $n(s)$ in (8.12) and (8.13) are ignored, since these distributions are additive, while the functional properties of $g(s)$, $K(s, \boldsymbol{D})$ and $P(\boldsymbol{D})$ are critical in understanding the forward and inverse problems. In (8.14), for every $P \in \mathcal{X}$, the mapping $\mathcal{K}$ defines a unique element $g \in \mathcal{Y}$.

Figure 8.2a is a schematic of the mapping $\mathcal{K} : \mathcal{X} \to \mathcal{Y}$, such that $P \in \mathcal{X}$ and $g \in \mathcal{Y}$. We notice that this mapping is singular-valued, in that there exists a unique size distribution which gives rise to a unique line profile, $g(s)$. The breadth, tails and characteristics of $g(s)$ are a signature of the underlying microstructure of the specimen defined by $P(\boldsymbol{D})$.

8.3.2 Inverse Problem

The existence and uniqueness of the size distribution, $P(\boldsymbol{D})$, is dependent on the existence of the *inverse operator*, $\mathcal{K}^{-1}$ [19]. In the case of (8.12) or (8.14),

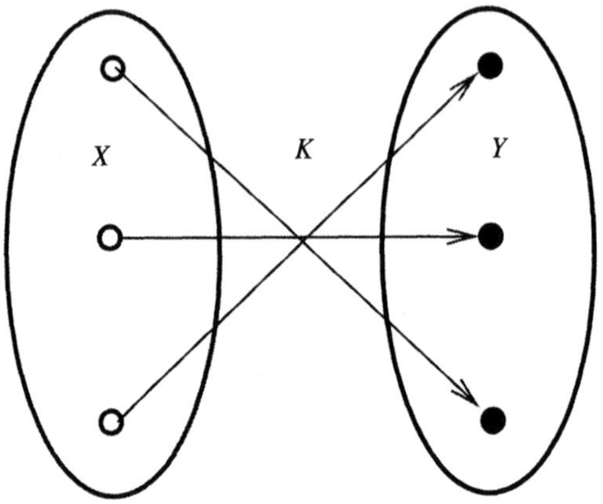

(a)

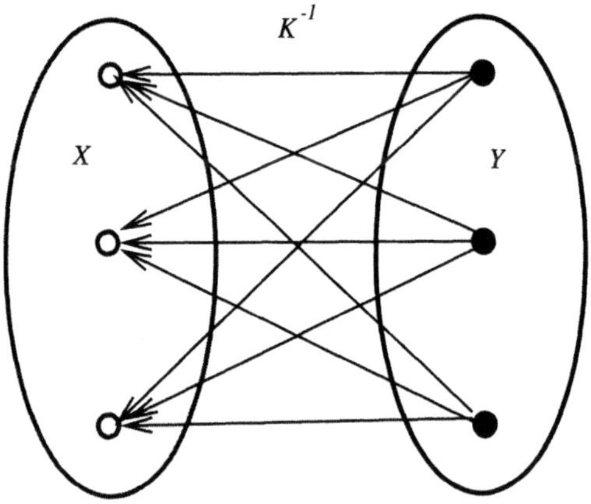

(b)

Fig. 8.2. Representing the "forward" and "inverse" problems. (**a**) The mapping $\mathcal{K} : \mathcal{X} \to \mathcal{Y}$ schematically represents measurement of the observed line profile in terms of the *singular-value mapping* from $\mathcal{X}$ (*open circle*) to $\mathcal{Y}$ (*solid circle*). (**b**) The singular-value mapping shown in (**a**) breaks down due to the (near) singular nature of $\mathcal{K}$. The mapping $\mathcal{K}^{-1} : \mathcal{Y} \to \mathcal{X}$ results in a *one-to-many mapping* from $\mathcal{Y}$ (*solid circle*) to $\mathcal{X}$ (*open circle*). As a consequence there may exist many plausible solutions for $P(\boldsymbol{D})$. Determining the most plausible $P(\boldsymbol{D})$ from the set of solutions $\{P(\boldsymbol{D})\}$ requires inductive reasoning. (Adapted from [22])

the inverse kernel or operator may be near-singular or at worst may not exist at all (i.e. singular). We can explore this issue from a practical point of view.

Initial efforts in finding a $P(\boldsymbol{D})$ may involve removing the background level, defining the inverse of the kernel and acting on both sides of (8.12) with this inverse kernel, to reproduce the original $P(\boldsymbol{D})$. However, the results from such an approach are generally ill-conditioned. The solution $P(\boldsymbol{D})$ will usually have large nonphysical oscillations, and the positivity and additivity conditions of $P(\boldsymbol{D})$ are violated [20]. A direct consequence of the ill-conditioning is that the uniqueness of the solution $P(\boldsymbol{D})$ is also compromised, so there may exist many possible solutions that fit the data.

Figure 8.2b schematically illustrates the mapping for the inverse problem and the effect of the (near) singular nature of $\boldsymbol{K}$. The mapping of the inverse problem is defined as $\mathcal{K}^{-1} : \mathcal{Y} \to \mathcal{X}$, such that $g \in \mathcal{Y}$ and $P \in \mathcal{X}$. We notice that the one-to-one mapping given in Fig. 8.2a breaks down, producing a mapping from one-to-many, with many possible solutions. Each solution may be plausible in that it may reproduce the observed line profile within some defined goodness-of-fit.

Why does this ill-conditioning arise? It essentially arises from amplification of noise in the observed line profile, due to the near singular kernel, $K(s, \boldsymbol{D})$. That is, if we attempt to decompose $K(s, \boldsymbol{D})$ into a set of basis functions, it is found that at least one of the basis functions is (near) linearly dependent [20]. This can also be demonstrated if $K(s, \boldsymbol{D})$ is expressed as a matrix and its determinant evaluated, such that det $\boldsymbol{K}^T\boldsymbol{K} \sim 0$ [8].

Given this situation, the best we can do is determine the most probable or plausible $P(\boldsymbol{D})$, using available information about the noise $n(s)$, background level, $b(s)$, and $K(s, \boldsymbol{D})$, while having the option to incorporate any existing a priori information that arises from other data or theoretical models. This approach is called *inductive* or *plausible reasoning* [21–23]. In addition, the fundamental properties of $P(\boldsymbol{D})$, namely the positivity and additivity, should be preserved, while making the least assumptions [8].

Mathematically speaking, the general class of methods used to determine $P(\boldsymbol{D})$ are called *inverse* or *deconvolution* methods. These types of problems can be expressed as matrix/vector equations using a suitable quadrature [19, 24]. Determining the specimen profile, $f(s)$, from (8.11) is such a problem, requiring the same inductive reasoning to determine the most plausible specimen line profile given our information about the noise $n(s)$, background level, $b(s)$, and instrument kernel, $k(s)$.

In order to apply inductive reasoning, we must be able to quantify our belief that a particular size distribution, $P(\boldsymbol{D})$ from a set of plausible solutions $\{P(\boldsymbol{D})\}$ is true.

8.4 Axioms for Inductive Reasoning

8.4.1 Two Basic Axioms

The axioms of inductive logic are based on probability. Cox [21] using the rules of Boolean algebra, essentially showed that inductive reasoning follows the rules for manipulating probabilities. We quantify our belief in a proposition by a real number mapped into the range [0, 1]. This enables the propositions to be ranked in a descending (or ascending) order according to their numerical value.

The first basic axiom says that given some background information, $\mathcal{I}$, and $\Pr(x|\mathcal{I})$ our belief that proposition x is true, our belief that the same proposition is false must also be defined, given by $\Pr(\bar{x}|\mathcal{I})$. Further, the sum of these probabilities must be equal to unity [22],

$$\Pr(x|\mathcal{I}) + \Pr(\bar{x}|\mathcal{I}) = 1\,, \tag{8.15}$$

where $\bar{x}$ in (8.15) reads "not x" or "x is false"; $\Pr(x|\mathcal{I})$ reads as the probability of x conditional on $\mathcal{I}$, where $\mathcal{I}$ defines the background information or assumptions concerning x. Thus, "|" means *conditional on* or *given*, and anything to the left of "|" is conditional on anything to the right of "|". The axiom in (8.15) asserts that $\Pr(x|\mathcal{I})$ and $\Pr(\bar{x}|\mathcal{I})$ obey the additivity condition for probabilities.

The second axiom defines the product of propositions x and y. Consider that the proposition x is true *given* that proposition y is also true i.e. $\Pr(x|y,\mathcal{I})$; supposing our belief that proposition y is true, and $\Pr(y|\mathcal{I})$ has also been defined, our belief that propositions x *and* y are both true, $\Pr(x,y|\mathcal{I})$, can be stated. In $\Pr(x,y|\mathcal{I})$ the comma reads as "and". This axiom can be expressed concisely as [22]

$$\Pr(x,y|\mathcal{I}) = \Pr(x|y,\mathcal{I})\,\Pr(y|\mathcal{I})\,, \tag{8.16}$$

where $\Pr(x,y|\mathcal{I})$ is often called the *joint probability* of x and y.

Axioms (8.15) and (8.16) define the *sum* and *product* rules, respectively, and form the basis for manipulating probabilities. Using these rules, additional important results can be developed, which are critical to developing a method for determining the most plausible size distribution, $P(\boldsymbol{D})$.

8.4.2 Bayes' Theorem

Bayes' theorem of probabilities enables conditional probabilities to be manipulated. This theorem follows directly from (8.16).

Using commutations, $\Pr(x,y|\mathcal{I})$ can also be expressed as

$$\Pr(x,y|\mathcal{I}) = \Pr(y|x,\mathcal{I})\ \Pr(x|\mathcal{I})\,, \tag{8.17}$$

and on combining (8.16) and (8.17),

$$Pr(x|y,\mathcal{I})\ \Pr(y|\mathcal{I}) = \Pr(y|x,\mathcal{I})\ \Pr(x|\mathcal{I})\,.$$

On dividing both sides by $\Pr(y|\mathcal{I})$, Bayes' theorem follows,

$$Pr(x|y,\mathcal{I}) = \frac{\Pr(y|x,\mathcal{I})\ \Pr(x|\mathcal{I})}{\Pr(y|\mathcal{I})}\,, \tag{8.18}$$

where $\Pr(x|\mathcal{I})$ is defined as the a priori probability, $\Pr(y|x,\mathcal{I})$ as the likelihood probability, and for the present $\Pr(y|\mathcal{I})$ is considered as a normalisation constant.

More specifically, the likelihood probability $\Pr(y|x, \mathcal{I})$ defines how likely it is that y is true given that proposition x is true, while the a priori probability, $\Pr(x|\mathcal{I})$, defines the probability of x conditional on the assumptions $\mathcal{I}$. In practice, this may represent a guess based on some notion. However, in defining $\Pr(x|y, \mathcal{I})$, x is being tested, given our knowledge of y. Hence, $\Pr(x|y, \mathcal{I})$ revises our knowledge of x in the light of $\Pr(y|x, \mathcal{I})$.

Bayes' theorem, (8.18), is important in that it combines our experimental data with our hypothesis or "guess" [22]. As will be shown later (see Sect. 8.5), it enables us to combine our understanding of the observed line profile (i.e. experimental data) with the proposition about the size distribution (i.e. hypothesis). The normalisation term in (8.18), $\Pr(y|\mathcal{I})$, also plays an important role. It is usually called the *evidence* and becomes important in the selection of different size distribution models [25] and different models for crystallite shape. The use of inductive reasoning allows our hypothesis to be tested and revised in the light of experimental data. This is the basis of scientific reasoning.

8.4.3 Marginalisation or "Integrating Out"

The process of marginalisation or *integrating out* variables is another important tool in developing a method for determining $P(\boldsymbol{D})$. It is a direct consequence of (8.15) and (8.16).

Suppose the proposition x and the array of propositions $\boldsymbol{y} = \{y_1, y_2, y_3, \cdots, y_n\}$ are given. Further, if y_j are continuous variables and given the joint probability $\Pr(x, \boldsymbol{y}|\mathcal{I})$, marginalisation allows us to integrate out $\boldsymbol{y}$,

$$\Pr(x|\mathcal{I}) = \int_Y \Pr(x, \boldsymbol{y}|\mathcal{I})\, \mathrm{d}\boldsymbol{y}\,, \tag{8.19}$$

where $\mathrm{d}\boldsymbol{y} = \mathrm{d}y_1 \mathrm{d}y_2 \mathrm{d}y_3 \ldots \mathrm{d}y_n$; Y defines the parameter space for $\boldsymbol{y}$ and the normalisation condition $\int_Y \Pr(\boldsymbol{y}|x, \mathcal{I})\mathrm{d}\boldsymbol{y} = 1$ has been applied. Hence, the marginalisation enables unwanted or "nuisance" variables to be integrated out from the joint probability.

8.5 Developing a "Method" for Determining $P(\boldsymbol{D})$

8.5.1 Bayes' Theorem for $P(\boldsymbol{D})$

We now investigate how to determine the most plausible $P(\boldsymbol{D})$, by applying (8.15), (8.16), (8.18) and (8.19) to formulate Bayes' theorem for $P(\boldsymbol{D})$.

From (8.18), Bayes' theorem for the size distribution, given the observed line profile $\boldsymbol{g}$, an a priori model $\boldsymbol{m}$, statistical noise σ, the kernel $\boldsymbol{K}$, and a positive constant α, can be expressed as

$$\Pr(\boldsymbol{P}|\boldsymbol{g}, \boldsymbol{m}, \boldsymbol{K}, \sigma, \alpha, \mathcal{I}) = \frac{\Pr(\boldsymbol{P}|\boldsymbol{m}, \alpha, \mathcal{I}) \Pr(\boldsymbol{g}|\boldsymbol{P}, \boldsymbol{K}, \sigma, \mathcal{I})}{\Pr(\boldsymbol{g}|\boldsymbol{K}, \boldsymbol{m}, \sigma, \mathcal{I})}\,, \tag{8.20}$$

where bold face type denotes matrix/vector notation (for discussion concerning notation in this and related equations, see [8]).

Equation (8.20) has the same form as (8.18). It involves the a priori probability distribution, $\Pr(\boldsymbol{P}|\boldsymbol{m},\alpha,\mathcal{I})$, which defines our hypothesis concerning $\boldsymbol{P}$ conditional on a model, $\boldsymbol{m}$. The a priori model, $\boldsymbol{m}$, may incorporate some additional data concerning the characteristics of $\boldsymbol{P}$ or a theoretical model pertaining to the functional form of $\boldsymbol{P}$. The likelihood probability distribution, $\Pr(\boldsymbol{g}|\boldsymbol{P},\boldsymbol{K},\sigma,\mathcal{I})$ expresses how probable it is to reproduce $\boldsymbol{g}$, conditional on our hypothesis $\boldsymbol{P}$, the kernel $\boldsymbol{K}$ (which includes our knowledge of the crystallite shapes), and the statistical noise, σ. Also, in (8.20) we treat $\Pr(\boldsymbol{g}|\boldsymbol{K},\boldsymbol{m},\sigma,\mathcal{I})$ as a normalisation constant.

On combining $\Pr(\boldsymbol{P}|\boldsymbol{m},\alpha,\mathcal{I})$ and $\Pr(\boldsymbol{g}|\boldsymbol{P},\boldsymbol{K},\sigma,\mathcal{I})$ in (8.20), the axioms of inductive reasoning are being applied and our hypothesis concerning $\boldsymbol{P}$ in the light of the experimental data, $\boldsymbol{g}$, is being tested.

8.5.2 Likelihood and Entropy Functions

We have seen that Bayes' theorem for $P(\boldsymbol{D})$ combines our hypothesis with the experimental data. However, it does not tell us how to explicitly assign values (i.e. real numbers) to the a priori and likelihood probability distributions given in (8.20). In order to do this, the *likelihood* and *entropy* functions must be defined.

Likelihood Function

The likelihood probability distribution quantifies the probability of an observed line profile, $\boldsymbol{g}$, given the size distribution $\boldsymbol{P}$, the kernel $\boldsymbol{K}$, and the statistical noise, σ. In defining the likelihood distribution, we make an assumption that the intensity (or counts) in $\boldsymbol{g}$ are large ($\gg 10$). On applying the *central limit theorem*, the likelihood distribution can be expressed as a Gaussian distribution

$$\Pr(\boldsymbol{g}|\boldsymbol{P},\boldsymbol{K},\sigma,\mathcal{I}) = Z_L^{-1}(\sigma)\exp\left[-\frac{1}{2}L(\boldsymbol{P},\boldsymbol{g},\boldsymbol{K})\right], \tag{8.21}$$

where the likelihood function, L is given by

$$L(\boldsymbol{P},\boldsymbol{g},\boldsymbol{K},\sigma) = \sum_{i=1}^{M} \frac{\left(g_i - \sum_{j=1}^{N} K_{ij}P_j\right)^2}{\sigma_i^2}, \tag{8.22}$$

and $Z_L(\sigma)$ is the normalisation term, which is dependent on the statistical error,

$$Z_L(\sigma) = \prod_{i=1}^{M} \sqrt{2\pi\sigma_i^2}\,. \tag{8.23}$$

In (8.22), the likelihood function compares the trial data, given by $\boldsymbol{KP}$, with the observed line profile $\boldsymbol{g}$, while incorporating the statistical uncertainty, σ. In defining the statistical uncertainty, both the counting statistics and uncertainty in estimating the background level must be taken into account. For large counts, the variance in the counting statistics is given by $\sigma_{c\,i}^2 = g_i$, while the variance in the background level estimation is given by $\sigma_{bi}^2 = b_i^{\text{est}}$. Combining these two contributions results in the overall variance, $\sigma_i^2 = g_i + b_i^{\text{est}}$. If the peak-to-background ratio is large, then counting statistics will be the dominant source of uncertainty, while if the peak-to-background ratio is small, the uncertainty in background level will be significant. Also, care should be taken in correctly transforming σ from 2θ-space into s-space, such that $\sigma(s) = \sigma(2\theta)|[\mathrm{d}(2\theta)]/(\mathrm{d}s)|$.

Entropy Function

The basic properties of $\boldsymbol{P}$, its positivity and additivity, define the class of functions that can be used to assign values to it. It can be shown that the *only* function producing consistent solutions with the fewest assumptions is the entropy function (see [23, 26–32]),

$$S(\boldsymbol{P}, \boldsymbol{m}) = \sum_{j=1}^{N} P_j - m_j - P_j \ln(P_j/m_j)\,. \tag{8.24}$$

Using (8.24), the a priori probability distribution can be expressed as

$$\Pr(\boldsymbol{P}|\boldsymbol{m}, \alpha, \mathcal{I}) = Z_S^{-1}(\alpha) \exp\left[\alpha S(\boldsymbol{P}, \boldsymbol{m})\right]\,, \tag{8.25}$$

where the normalisation term, $Z_S(\alpha)$ in (8.25) is given by

$$Z_S(\alpha) = \left(\frac{2\pi}{\alpha}\right)^{\frac{N}{2}}\,. \tag{8.26}$$

The entropy function quantifies our hypothesis that the size distribution and specimen profile are positive and additive. Further, the entropy function is a measures of the "distance" between the a priori model $\boldsymbol{m}$, and the size distribution $\boldsymbol{P}$ in the presence of the observed line profile. That is, the entropy function measures the plausibility of the a priori model, $\boldsymbol{m}$, in the light of the observed line profile. For example, if the size distribution $\boldsymbol{P}$ were found to "lie close" to the a priori model (i.e. there is little change in the entropy function), when determined in the presence of the observed line profile, it could be inferred that the underlying size distribution closely resembles the a priori model. However, if it were found that there is a considerable difference between $\boldsymbol{P}$ and $\boldsymbol{m}$ (i.e. large change in the entropy), it could be inferred that $\boldsymbol{m}$ is a poor choice for $\boldsymbol{P}$.

8.5.3 Combining S and L

Using (8.21)–(8.23) and (8.24)–(8.26) in (8.20), the probability for the size distribution, $\boldsymbol{P}$, can be expressed as

$$\begin{aligned}\Pr(\boldsymbol{P}|\boldsymbol{g},\boldsymbol{K},\boldsymbol{m},\sigma,\alpha,\mathcal{I}) &\propto Z_S^{-1}(\alpha)Z_L^{-1}(\sigma)\exp\left[\alpha S(\boldsymbol{P},\boldsymbol{m}) - L(\boldsymbol{P},\boldsymbol{g},\boldsymbol{K},\sigma)\right] \\ &\propto Z_S^{-1}(\alpha)Z_L^{-1}(\sigma)\exp\left[Q(\boldsymbol{P},\alpha)\right] , \qquad (8.27)\end{aligned}$$

where the exponent combines the entropy and likelihood functions into a single function,

$$Q = \alpha\, S - L\,. \qquad (8.28)$$

In (8.27) and (8.28), Q is defined solely as a function of $\boldsymbol{P}$ and α, since they are the only unknown quantities, and the $\boldsymbol{g}$, $\boldsymbol{K}$, $\boldsymbol{m}$ and σ are known from the data. The result given by (8.27) can be interpreted as a Lagrangian function and the positive parameter α as an unknown Lagrangian multiplier.

Thus from (8.27) and (8.28) the steps for determining the most plausible or probable size distribution, $\boldsymbol{P}$, become clear. This requires maximising (8.27), which in turn requires that (8.28) be a maximum. This can only occur if $\nabla Q = \alpha\nabla S - \nabla L = 0$. In other words, we want to maximise S with respect to the experimental data, given by L. This is referred to as the maximum entropy (MaxEnt) method [33]. However, in order to develop a MaxEnt method, it is important to understand the role of α and how to treat it in determining $\boldsymbol{P}$.

8.5.4 What to Do with α?

In developing (8.28), a "coupling" between S and L has been defined, given by α. This moderates the interaction of the hypothesis, defined by S, with the experimental data, defined by L, to produce an optimum $\boldsymbol{P}$, and is schematically represented in Fig. 8.3. In the case of small α, the influence of the experimental data becomes dominant (i.e. strong coupling) as depicted by the "stiffer spring" in Fig. 8.3a attaching L to $\boldsymbol{P}$ and a "weaker spring" attaching S to $\boldsymbol{P}$. In this case, there is a tendency to "over fit" $\boldsymbol{P}$ to the observed line profile. This results in features, such as statistical noise and uncertainty in the background level estimation, filtering into $\boldsymbol{P}$ [22, 25, 34].

On the other hand, when α is large, the influence of the hypothesis becomes dominant and the experimental data becomes less significant. Again, this is depicted in Fig. 8.3b, where the "stiffer spring" attached between S and $\boldsymbol{P}$ and the "weaker spring" attaches L to $\boldsymbol{P}$. In this case $\boldsymbol{P}$ defaults to the a priori model, $\boldsymbol{m}$ [22]. In determining $\boldsymbol{P}$, an optimum condition must be found which reflects the influence of S and L.

Traditionally, the approach for dealing with α is to determine a single value of α which results in $\nabla Q = 0$ [33, 35, 36]. However, in the view of Bayesian theory (8.20), α can also be treated as "Bayesian variable". Using the tools of (8.16) and (8.19), a probability distribution for α can be determined [25] as

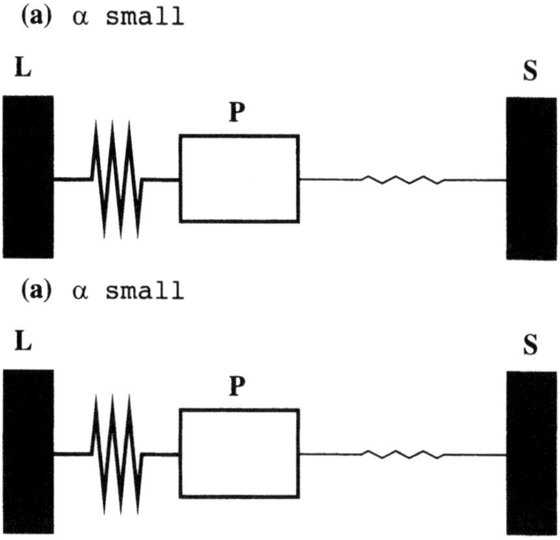

Fig. 8.3. Schematic representation of the coupling between L, $\boldsymbol{P}$ and S through α. (**a**) In the case where α is small, the likelihood function, L is dominant and the resulting $\boldsymbol{P}$ will be "over fitted" to $\boldsymbol{g}$. (**b**) In the case were α is large, S is dominant and $\boldsymbol{P}$ defaults to $\boldsymbol{m}$, with very little effect from the experimental data, L. In these schematic diagrams S and L are considered to be fixed

$$\Pr(\alpha|\boldsymbol{g},\boldsymbol{K},\boldsymbol{m},\sigma,\mathcal{I}) = \int \Pr(\alpha|\mathcal{I})\Pr(\boldsymbol{P}|\boldsymbol{g},\boldsymbol{K},\boldsymbol{m},\sigma,\alpha,\mathcal{I})\mathcal{D}\boldsymbol{P} \tag{8.29a}$$

$$= \int \Pr(\boldsymbol{P},\alpha|\boldsymbol{g},\boldsymbol{K},\boldsymbol{m},\sigma,\mathcal{I})\mathcal{D}\boldsymbol{P}\,, \tag{8.29b}$$

where the measure $\mathcal{D}\boldsymbol{P} = \prod_{j=1}^{N} P_j^{-\frac{1}{2}}\mathrm{d}P_j$ accounts for the intrinsic curvature of the entropy function and $\Pr(\alpha|\mathcal{I})$ in (8.29a) is the a priori probability distribution for α. Applying (8.16) produces $\Pr(\alpha|\boldsymbol{g},\boldsymbol{K},\boldsymbol{m},\sigma,\mathcal{I})$. Also, in (8.29,b) $\boldsymbol{P}$ has been integrated out of the joint probability using the rule of marginalisation in (8.19). The distribution for α is conditional on $\boldsymbol{g}$, $\boldsymbol{K}$, $\boldsymbol{m}$ and σ; thus all the available information has been used to determine α.

At a practical level, we use the algorithm developed by Bryan [9] and the approach outlined by Jarrell and Gubernatis [10]. This involves starting with a large α and stepping towards $\alpha \approx 0$. At each α, a $\boldsymbol{P}$ is determined using an algorithm developed by Bryan [9]. Over the range of α-values, a set of solutions for the size distribution, $\{\boldsymbol{P}(\alpha)\}$, which fit the experimental data is formed and parameterised by $\alpha \in [\alpha_{\min}, \alpha_{\max}]$. This demonstrates the difficulties associated with uniqueness discussed in Sect. 8.3.2 and shown in Fig. 8.2b. It also highlights the fundamental difficulties faced by other methods for determining the size distribution from X-ray line profile data [6,7].

In order to determine a single size distribution, the result in (8.29,b) can be used to determine the average size distribution, $\langle \boldsymbol{P} \rangle$,

$$\langle \boldsymbol{P} \rangle = \int_{\alpha_{\min}}^{\alpha_{\max}} \mathrm{d}\alpha \boldsymbol{P}(\alpha) \Pr(\alpha|\boldsymbol{g}, \boldsymbol{K}, \boldsymbol{m}, \sigma, \mathcal{I}) , \tag{8.30}$$

where $\int_{\alpha_{\min}}^{\alpha_{\max}} \mathrm{d}\alpha \Pr(\alpha|\boldsymbol{g}, \boldsymbol{K}, \boldsymbol{m}, \sigma, \mathcal{I}) = 1$, and $\alpha \in [10^{-3}, 10^5]$ has been found to be a sufficient range.

8.5.5 Determining $f(2\theta)$

Traditionally a picture of the microstructure of a specimen has been developed by first determining the specimen profile, $f(2\theta)$ or $f(s)$. Subsequently, qualitative methods such as Williamson–Hall and Fourier analyses of $f(2\theta)$ were carried out to identify the nature of the specimen broadening (i.e. dislocation-broadening and/or nanocrystallite size-broadening). However, the problem with this approach of first determining $f(2\theta)$, is that it is also an inverse problem and the discussion presented in Sect. 8.3 applies.

Determining the specimen profile, $f(2\theta)$ provides an intermediate approach for analysing the specimen line broadening. For example, the Krill and Birringer method [6] is dependent on determining the specimen profile (see Sect. 8.6.3). However, the conventional deconvolution approach, such as Stokes [37] and the early MaxEnt method (e.g. see [33]) have the problem that "ringing effects" appear in the specimen profile. This is a result of noise effects being amplified in the observed line profile and producing overfitting of $\boldsymbol{g}$, as depicted in Fig. 8.3a. However, these effects can be reduced if local smoothing is introduced into $f(2\theta)$ [8, 22]. Following the suggestion of Sivia [38], we introduce the spatial correlation length scale, ω, and decompose $f(2\theta)$ in terms of a set of basis functions that are dependent on ω,

$$f(2\theta) = \sum_{l=1}^{N} \psi(2\theta; 2\theta_l, \omega) a_l , \tag{8.31}$$

where a_l are the set of coefficients which define the amplitude of $f(2\theta)$; $\psi(2\theta; 2\theta_l, \omega)$ is the basis function located at the lth position with a width, ω. In the limit $\omega \to 0$ we expect that $\psi(2\theta; 2\theta_l, \omega) \to \delta(2\theta; 2\theta_l)$. Physically this corresponds to the case where there is no noise in the experimental data. A simple choice for ψ is a Gaussian function [8, 22]. Other functions could be used, but the Gaussian is a good choice as it mimics a band-limited function whose tails decrease rapidly to zero.

In applying (8.31), the Bayesian/MaxEnt method can be used to determine ω and a_l. That is, by applying (8.18) and (8.19), the a posteriori distribution for ω conditional on the observed data can be determined. The optimum choice of ω is the value which maximises the a posteriori distribution. This defines the *fuzzy pixel width*, $\hat{\omega}$. The a posteriori distribution for ω can be expressed [8] as

$$\Pr(\omega|\boldsymbol{g},\boldsymbol{m},\boldsymbol{\sigma},\mathcal{I}) = \int \mathcal{D}\boldsymbol{a} \int \mathrm{d}\alpha \Pr(\boldsymbol{a},\alpha,\omega|\boldsymbol{g},\boldsymbol{m},\boldsymbol{\sigma},\mathcal{I}) \tag{8.32a}$$

$$= \int \mathcal{D}\boldsymbol{a} \int \mathrm{d}\alpha \Pr(\alpha|\mathcal{I}) \Pr(\omega|\mathcal{I}) \times \Pr(\boldsymbol{a}|\boldsymbol{g},\boldsymbol{m},\boldsymbol{\sigma},\alpha,\omega,\mathcal{I})\,. \tag{8.32b}$$

Once $\hat{\omega}$ has been determined, $\psi(2\theta; 2\theta_l, \omega)$ can be evaluated and the MaxEnt method can be applied to determine the coefficients, a_l for $l \leq N$.

Using the Bayesian/MaxEnt/fuzzy pixel approach, the specimen profile and broadening contributions can be identified, which allows a strategy for determining the microstructure of the specimen to be developed.

8.5.6 Derivation of Langford et al. (2000)

It is worth discussing the Langford et al. [7] approach and placing it into a Bayesian context since it has direct consequence to methods widely used in line profile analysis. The theoretical foundation of this method corresponds to a special and limited case of Bayes' theorem, where the a priori distribution is uniform. Unlike the Bayesian/MaxEnt method discussed in the previous section, Langford et al. [7] explicitly assume the functional form of the crystallite size distribution. For example, the principal parameters of the log-normal distribution are the median diameters, D_0, and the log-normal standard deviation, σ_0. The full Bayesian derivation takes into consideration any suitable a priori information about D_0 and σ_0. However, in setting these a priori distributions as uniform, any additional information about these parameters is ignored and the likelihood distribution is *only* maximised to determine the value of the parameters.

On applying (8.16) and (8.18), the joint a posteriori probability for D_0 *and* σ_0 conditional on the experimental data, $\boldsymbol{g}$, kernel, $\boldsymbol{K}$ and statistical noise, σ can be expressed as

$$\Pr(D_0,\sigma_0|\boldsymbol{g},\boldsymbol{K},\sigma,\mathcal{I}) = \frac{\Pr(D_0|\mathcal{I}) \Pr(\sigma_0|\mathcal{I}) \Pr(\boldsymbol{g}|D_0,\sigma_0\boldsymbol{K},\sigma,\mathcal{I})}{\Pr(\boldsymbol{g}|\boldsymbol{K},\sigma,\mathcal{I})}\,, \tag{8.33}$$

where $\Pr(D_0|\mathcal{I})$ and $\Pr(\sigma_0|\mathcal{I})$ define the a priori distribution for D_0 and σ_0, respectively. The likelihood distribution, $\Pr(\boldsymbol{g}|D_0,\sigma_0\boldsymbol{K},\sigma,\mathcal{I})$, is also defined in terms D_0 and σ_0, while also taking into consideration the kernel and statistical noise.

Noting that (8.33) is a joint probability and using the rules for marginalisation (8.19), the a posteriori probability for each individual parameter becomes

$$\Pr(D_0|\boldsymbol{g},\boldsymbol{K},\sigma) = \int_1^\infty \mathrm{d}\sigma_0 \Pr(D_0,\sigma_0|\boldsymbol{g},\boldsymbol{K},\sigma,\mathcal{I}) \tag{8.34}$$

and

$$\Pr(\sigma_0|\boldsymbol{g},\boldsymbol{K},\sigma) = \int_0^\infty \mathrm{d}D_0 \Pr(D_0,\sigma_0|\boldsymbol{g},\boldsymbol{K},\sigma,\mathcal{I})\,, \tag{8.35}$$

where any correlation between the D_0 and σ_0 parameters is integrated out.

Supposing the a priori distribution in (8.33) were set as a uniform distribution, such that $\Pr(D_0|\mathcal{I}) = 1/(D_{0\max} - D_{0\min})$ and $\Pr(\sigma_0|\mathcal{I}) = 1/(\sigma_{0\max} - \sigma_{0\min})$, then (8.33) would reduce to maximising the likelihood distribution (or minimising the likelihood function). This is theoretically equivalent to Langford et al. [7], expressed as

$$\log \Pr(D_0, \sigma_0 | \boldsymbol{g}, \boldsymbol{K}, \sigma, \mathcal{I}) = c + \log \Pr(\boldsymbol{g} | D_0, \sigma_0, \boldsymbol{K}, \sigma, \mathcal{I}) , \tag{8.36}$$

where c is a constant that takes into consideration the magnitude of the a priori distributions, since they were defined as uniform distributions. However, maximising the likelihood distribution does not take into consideration any a priori information about the parameters, D_0 and σ_0 and is dependent on the experimental data for inferring the values of the parameters [39]. In addition, simply maximising the likelihood distribution does not limit the number of parameters in the size distribution model. That is, more complex size distributions will generally produce better fits. This issue is of particular importance in model selection; for example, determining the modal characteristic of the size distribution is dependent on the a priori distribution [22].

Defining the a priori distribution in terms of the entropy function (8.24) overcomes these limitations. No explicit assumption concerning the distribution is made while the a priori model for $P(\boldsymbol{D})$ is being introduced.

There are several numerical approaches in which (8.36) can be implemented, such as nonlinear least squares. However, as demonstrated in Sect. 8.6.3, using the Markov Chain Monte Carlo (MCMC) method is a numerical method which produces equivalent results with the additional benefit that an a priori distribution for D_0 and σ_0 can be readily introduced and (8.34), (8.35) are explicitly evaluated (see [39]).

8.6 Analysis of CeO_2 X-ray Diffraction Data

A cerium oxide (CeO_2) specimen was used to demonstrate the application of the Bayesian/MaxEnt methods. This specimen was prepared for the CPD/IUCr size round-robin [11].

8.6.1 XRD Details

Diffraction patterns were collected on a Siemens D500[2] diffractometer that was equipped with a focusing Ge incident beam monochromator, sample spin-

[2] Certain commercial materials, equipment and software are identified in order to adequately specify the experimental procedure. Such identification does not imply a recommendation or endorsement by NIST, nor does it imply that the materials or equipment or software are necessarily the best available for the purpose.

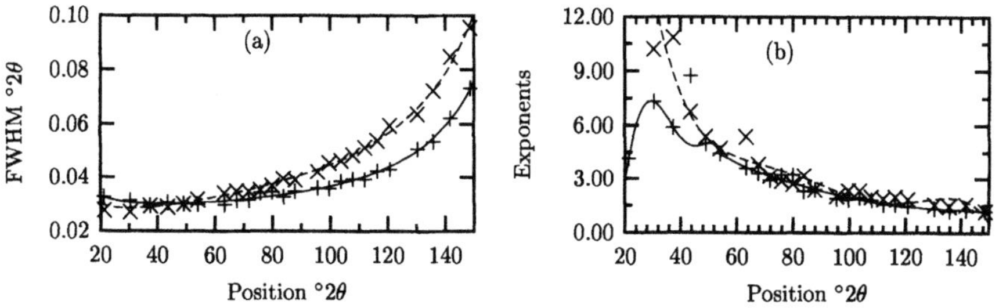

Fig. 8.4. Calibration plots for the split-Pearson VII line profiles, generated from the SRM 660a (LaB_6) diffraction pattern and used to model the instrument function, $K(2\theta)$. (**a**) The FWHM vs 2θ for low- (+ and *solid line*) and high- (× and *dashed line*) angle sides of the peak. (**b**) The exponents vs 2θ for low- (+ and *solid line*) and high- (× and *dashed line*) angle sides of the peak

ner and a scintillation detector. Copper $K_{\alpha 1}$ radiation with a wavelength $\lambda = 0.15405945\,\text{nm}$ was used. The divergence slit was 0.67°, while the receiving optics included a slit of 0.05° and 2° Soller slits. Data were collected in discrete regions straddling the maxima of each profile, with the step and scan width of each region being varied in correspondence with the FWHM. Count times were varied so as to obtain an approximately constant total number of counts for each scan region. The instrument profile was determined using a split-Pearson VII shape function fitted to 22 reflections collected from SRM 660a (LaB_6). Figure 8.4, shows the FWHMs and exponents for the split-Pearson VII profile function. The low- and high-FWHMs were fitted [40] using

$$\mathrm{FWHM}^2 = A\tan^2\theta + B\cot^2\theta + C\tan\theta + D\,, \tag{8.37}$$

while the low- and high-side exponents were fitted using a fifth-order polynomial.

The count times for the CeO_2 data were adjusted so that the error was kept in the range 1–3% for all peaks in the CeO_2 pattern. The scan ranges for the CeO_2 data were considerably wider, in proportion to the FWHM, than those used for the data collection from SRM 660a. This ensured a reasonable determination of the tails of the profiles and background levels.

8.6.2 TEM Details

Particle agglomerates were gently crushed in ethanol using a mortar and pestle. A portion of the dilute slurry was dispersed on a holey carbon film and left to dry. Once in the TEM, a series of micrographs of particles were taken at a fixed magnification of 200k×. In the *preliminary* examination reported here, these negatives were scanned and analysed by manually approximating the particle size with an oval. The oval's major and minor axes were adjusted so as to tangentially intersect the particle surface facets.

There are several sources of error in the measurements: TEMs typically have a 5% error in length scale measurements; also, imaging the particle clusters means that particles are at different heights, which results in Fresnel fringes around the particles making it harder to identify particle edges. Further, larger particles give better contrast and it is easier to detect their edges, so it is possible to inadvertently preferentially choose larger particles over smaller ones.

A frequency histogram for about 853 particles is shown in Fig. 8.13b. This figure also shows the Bayesian/MaxEnt size distributions (discussed in Sect. 8.6.4) determined from the non-overlapped *hkl* profiles of the CeO_2 diffraction pattern. A TEM micrograph of the CeO_2 crystallites is shown in Fig. 8.14. It can be seen from the larger particles that they have a spherical-like morphology.

8.6.3 Identifying Specimen Broadening

Using the MaxEnt/Fuzzy Pixel Method to Determine $f(2\theta)$

In this first part of the analysis, the specimen profile, $f(2\theta)$, is determined using the MaxEnt/fuzzy pixel method. This approach identifies the specimen broadening contribution and the necessary information needed for the Bayesian/MaxEnt method in Sect. 8.6.4.

The non-overlapped lines from the round-robin CeO_2 diffraction pattern were identified; they include the 111, 200, 220, 400, 422, 511 and 531 line profiles. In each case the instrumental line profile was modelled using a split-Pearson-VII function (see [1]). Figure 8.5 shows the 200 line profile for this specimen. In Fig. 8.5a the observed line profile, $g(2\theta)$, is given before the background level is removed, while Fig. 8.5b shows (on a logarithmic scale) the effect of background level estimation on $g(2\theta)$. In this case the background level was modelled using a linear function and its characteristics were determined using the Fourier coefficients of $g(2\theta)$ (see [8]).

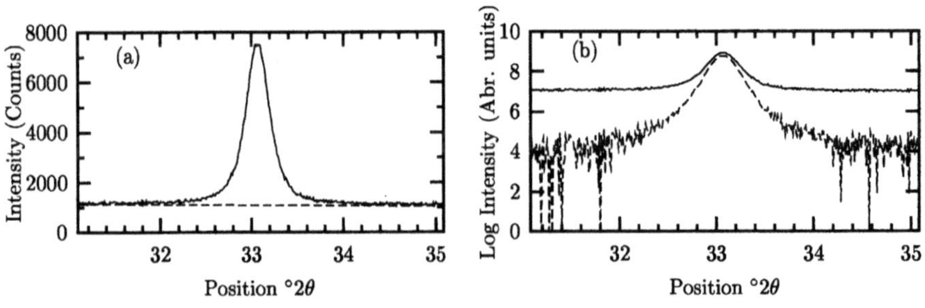

Fig. 8.5. Experimental CeO_2 200 profile, $g(2\theta)$. (**a**) The observed 200 profile (*solid line*) and estimated background level (*dashed line*) over $(2\theta_0 \pm 2)°2\theta$, the range over which the analysis was carried out. (**b**) Logarithm of the 200 measured profile before (*solid line*) and after background removal (*dashed line*)

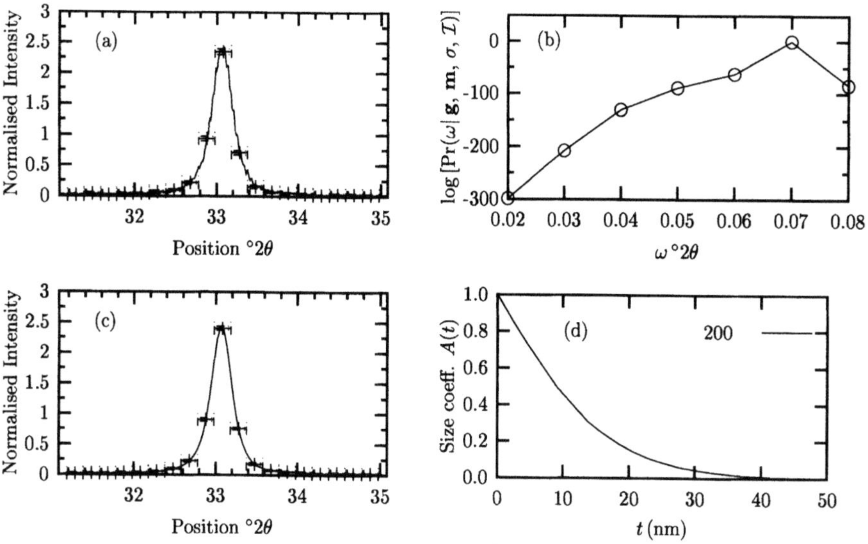

Fig. 8.6. Specimen profiles from the "old" MaxEnt method and fuzzy pixel/MaxEnt method for the measured CeO_2 200 line. (**a**) "Old MaxEnt" specimen profile (*solid line + error bars*). (**b**) $\log \Pr(\omega|\boldsymbol{g}, \boldsymbol{m}, \sigma, \mathcal{I})$ distribution to determine the optimum fuzzy pixel width, $\hat{\omega} \approx 0.07°2\theta$. (**c**) MaxEnt/Fuzzy Pixel specimen profile (*solid line + error bars*). (**d**) MaxEnt/Fuzzy Pixel Fourier coefficients (*solid line*)

Figure 8.6 compares the 200 CeO_2 line profile from the "old" MaxEnt method (see [18]) with the Bayesian/MaxEnt/fuzzy pixel method [8,22]. Figure 8.6a illustrates the "old" MaxEnt specimen profile, showing that the line profile has been "over fitted" and the characteristics of the statistical noise in $g(2\theta)$ have filtered into the 200 line profile. This result emphasises the effect of α on determining the optimum f; when α is small (see Fig. 8.3a), the influence of the likelihood function on f is dominant. For the results in Fig. 8.6, $\alpha \approx 0.01$ (i.e. small).

The MaxEnt/fuzzy pixel 200 line profile results are given in Fig. 8.6b–d, in contrast to the results in Fig. 8.6a. This approach imparts local smoothing on f by decomposing it into a set of basis functions, with a specified *fuzzy pixel width*, ω. This width acts as a filter and defines the region of local smoothing as the length of ω. However, in order to determine the optimum ω the tools of Bayesian analysis are used to formulate the probability distribution, $\Pr(\omega|\boldsymbol{g}, \boldsymbol{m}, \sigma, \mathcal{I})$ (see (8.32b) and Sect. 3.7 in [8]. Figure 8.6b shows $\Pr(\omega|\boldsymbol{g}, \boldsymbol{m}, \sigma, \mathcal{I})$ on a logarithmic scale, where it can be seen clearly that the optimum fuzzy pixel width is $\hat{\omega} \approx 0.07°2\theta$. Once $\hat{\omega}$ has been determined, the MaxEnt method is used to reconstruct the specimen profile, while simultaneously removing the instrumental broadening effects. Thus, Fig. 8.6c shows a considerable improvement in f. The over-fitting given in Fig. 8.6a has been replaced by the local smoothing of ω. In addition, there has been some reduction in the error bars, with an $\sim$ 8–28% decrease. In the case of the fuzzy pixel results, $\alpha \approx 0.1$, which yields a better choice of f in that neither the S nor L functions dominate. Figure 8.6c clearly demonstrates that the specimen profile can be determined without making assumptions about its functional form, while still preserving its positivity and additivity properties. Figure 8.6d shows the Fourier coefficients for the 200 line. This is typical of the Fourier coefficients for the non-overlapping lines in this specimen. In addition, when all the Fourier coefficients are plotted on the same axis (not shown here) no dependency on hkl is evident. This suggests that nanocrystallite size-broadening is the dominant broadening mechanism and it is also independent of the direction of scattering, $\boldsymbol{g}$ or [hkl]. Furthermore, this result also suggests that the maximum thickness of the crystallites is $\sim$ 50–60 nm, since this is where $A(t) \to 0$ (with $A(t) = V(t)/V(0)$, see Fig. 8.1).

Krill and Birringer (1998) Method

The MaxEnt/fuzzy pixel results were used to determine the integral breadths, using (8.5,b), and a Williamson–Hall plot was developed (see Fig. 8.7a). It is evident from Fig. 8.7a that the integral breadth is independent of d^*_{hkl}. The slope of the linear fit is zero, implying there are no dislocation broadening effects. It can be concluded that the dominant source of specimen broadening in f is due to size effects. This also suggests that the crystallites are near-spherical in shape, since there is no evident anisotropic scattering of β. From the Fourier coefficients and the Williamson–Hall plot the average area- and volume- weighted sizes can be determined using (8.4) and (8.6), respectively. These results are given in Table 8.1. From the Fourier coefficient data and the Williamson–Hall plot, it can be argued that the crystallites are (near) spherical, with a maximum thickness of $\sim$ 50–60 nm.

Table 8.1. Summary of CeO_2 data analysis using the MaxEnt/fuzzy pixel and Krill and Birringer methods [6]. The area- and volume-weighted sizes were determined from the specimen profile of the MaxEnt/fuzzy pixel method. The $\langle t\rangle_a$ and $\langle t\rangle_v$ results were determined directly from f using (8.4) and (8.6), respectively. The area- and volume-weighted diameters were determined using (8.38a,b), while the log-normal parameters were determined according to Krill and Birringer [6] and using (8.41)

hkl	$\langle t\rangle_a$ (nm)	$\langle D\rangle_a$ (nm)	$\langle t\rangle_v$ (nm)	$\langle D\rangle_v$ (nm)	D_0 (nm)	σ_0	$\langle D\rangle_{KB}$ (nm)	$\sigma^2_{\langle D\rangle}$ (nm^2)
111	19.21 ± 0.05	28.81 ± 0.07	22.88 ± 0.03	30.50 ± 0.04	25.0 ± 0.2	1.270 ± 0.007	25.7 ± 0.2	39 ± 2
200	16.04 ± 0.06	24.06 ± 0.08	22.22 ± 0.05	29.63 ± 0.06	14.3 ± 0.2	1.578 ± 0.007	15.9 ± 0.2	58 ± 2
220	18.92 ± 0.04	28.38 ± 0.06	23.24 ± 0.04	31.00 ± 0.05	22.8 ± 0.2	1.345 ± 0.006	23.8 ± 0.2	52 ± 2
400	15.76 ± 0.06	23.64 ± 0.09	22.03 ± 0.11	29.4 ± 0.2	13.7 ± 0.3	1.59 ± 0.01	15.3 ± 0.3	56 ± 3
422	15.45 ± 0.08	23.2 ± 0.1	21.5 ± 0.1	28.6 ± 0.1	13.7 ± 0.3	1.58 ± 0.01	15.2 ± 0.3	54 ± 3
511	15.91 ± 0.07	23.9 ± 0.1	21.9 ± 0.1	29.2 ± 0.2	14.4 ± 0.3	1.57 ± 0.01	16.0 ± 0.3	57 ± 3
531	15.04 ± 0.04	22.55 ± 0.07	21.9 ± 0.1	29.20 ± 0.16	11.8 ± 0.2	1.66 ± 0.01	13.5 ± 0.2	53 ± 3
Average	16.6 ± 0.2	24.9 ± 0.2	22.2 ± 0.2	29.6 ± 0.3	16.5 ± 0.6	1.51 ± 0.03	17.9 ± 0.7	53 ± 7

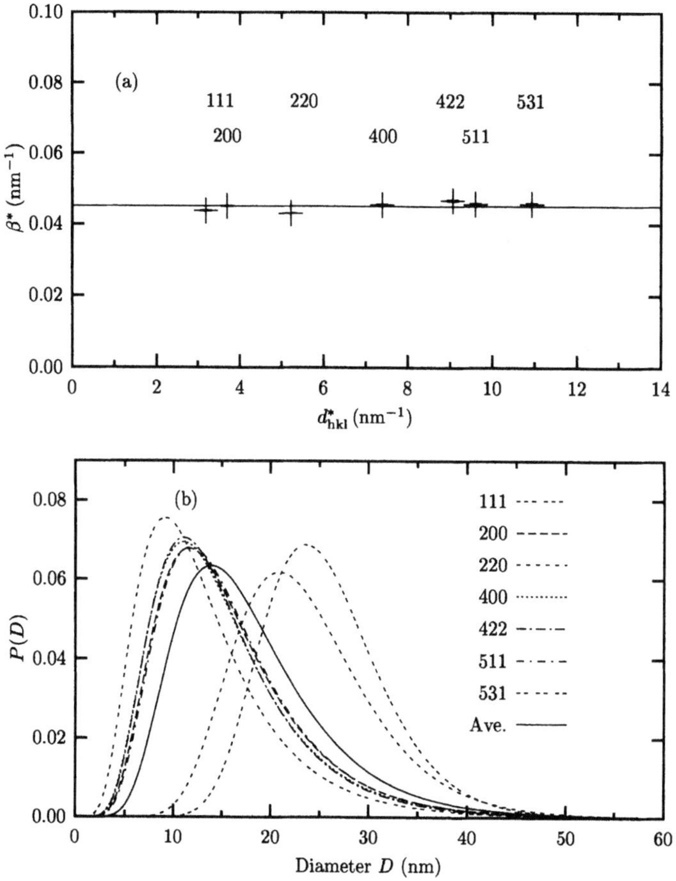

Fig. 8.7. Volume- and area-weighted sizes from the MaxEnt/Fuzzy Pixel line profiles for CeO_2 data. (**a**) Classical Williamson–Hall plot of the integral breadths, showing no dependency on hkl. This suggests that the crystallites are spherical in morphology. (**b**) Log-normal size distribution using the Krill and Birringer method [6] (also see (8.41) and Table 8.1)

The area- and volume-weighted sizes can be related to a spherical nanocrystallite model, by determining the corresponding area- and volume-weighted diameters [7],

$$\langle D\rangle_a = \frac{3}{2}\langle t\rangle_a \tag{8.38a}$$

$$\langle D\rangle_v = \frac{4}{3}\langle t\rangle_v \,. \tag{8.38b}$$

The ratio of the area to volume-weighted sizes also gives a qualitative indicator as to the presence of a distribution of crystallite diameters. That is, if $\langle t\rangle_a/\langle t\rangle_v = \langle D\rangle_a/\langle D\rangle_v = 8/9 \approx 0.89$, for spherical crystallites, then it can be inferred that the size distribution is a δ-function about D_0, $P(D) = \delta(D)$, such that $\delta(D) = \infty$ at $D = D_0$, and $\delta(D) = 0$, $\forall D \neq D_0$ and $\int_0^\infty \delta(D) = 1$. In other words, the crystallites all have the same size. Using the $\langle t \rangle_a$ and $\langle t \rangle_v$ values in Table 8.1, $\langle t \rangle_a / \langle t \rangle_v = (16.6 \pm 0.2)/(22.2 \pm 0.2) \approx 0.75 \pm 0.01$. This ratio suggests that the nanocrystallites are not uniform and need a size distribution to describe them.

The log-normal size distribution is a model for spherical crystallites which has wide application in the nanoscience literature and has produced some debate (see [6,7,41,42]). On applying the results of Krill and Birringer (see (6)–(8), p. 625 in [6]), the $\langle D \rangle_a$ and $\langle D \rangle_v$ values in Table 8.1 can be used to determine a median diameter, D_0, and log-normal variance, σ_0^2. In turn, these results can be related to the average diameter and variance. That is, the log-normal size distribution for spherical crystallites can be written as

$$N(D) = \frac{1}{\sqrt{2\pi D^2 \ln^2 \sigma_0}} \exp\left[-\frac{1}{2} \left(\frac{\ln(D/D_0)}{\ln \sigma_0} \right)^2 \right], \tag{8.39}$$

where D_0 is the median size and σ_0^2 is the log-normal variance. The average diameter, $\langle D \rangle_{model}$, and variance, $\sigma^2_{\langle D \rangle}$, of the distribution are related to these quantities by [8]

$$\langle D \rangle_{\text{model}} = D_0 e^{\ln^2 \sigma_0 / 2} \tag{8.40}$$

and

$$\sigma^2_{\langle D \rangle} = D_0^2 e^{\ln^2 \sigma_0} \left(e^{\ln^2 \sigma_0} - 1 \right). \tag{8.41}$$

The values for D_0, σ_0^2, $\langle D \rangle_{\text{model}}$ and $\sigma^2_{\langle D \rangle}$ for the different hkl are given in Table 8.1, which shows the average crystallite diameter was found to be $\langle D \rangle_{KB} = (17.9 \pm 0.7)$ nm and the variance $\sigma^2_{\langle D \rangle} = (53 \pm 7)\,\text{nm}^2$.

The CeO_2 size distribution using the method proposed by Krill and Birringer [6] is presented in Fig. 8.7b. It can be seen that the majority of hkl lines produce similar log-normal size distributions, with the exception of the 111 and 220 lines. For these size distributions, difficulties in determining the area-weighted size were encountered. A graphical representation of the crystallites can be developed using the diameter results in Table 8.1 and Fig. 8.7b. This is shown in Fig. 8.8, where $\langle D \rangle_a$, $\langle D \rangle_v$ and $\langle D \rangle_{KB}$ have been plotted, dependent on [hkl] relative to the [010]-direction. From Fig. 8.8, the $\langle D \rangle_a$ results are disappointing. The scattering of the $\langle D \rangle_a$ from the 111 and 220 lines highlights the difficulties of determining the area-weighted size from (8.4). That is, in evaluating $\mathrm{d}A(t)/\mathrm{d}t$ as $t \to 0$, small changes in $\mathrm{d}A(t)/\mathrm{d}t$ are produced, and on using (8.4) there are large changes in $\langle t \rangle_a$ and consequently $\langle D \rangle_a$. The results for $\langle D \rangle_v$ demonstrate the suitability of the spherical crystallite model, with a close fit between the model and $\langle D \rangle_v$ values. However, on combining these results to determine $\langle D \rangle_{KB}$, a large scatter is produced. This is largely due to the variation in the $\langle D \rangle_a$ values for the 111 and 220 lines; if these lines are not considered, the expected trend for the remaining hkl can be seen in Fig. 8.8.

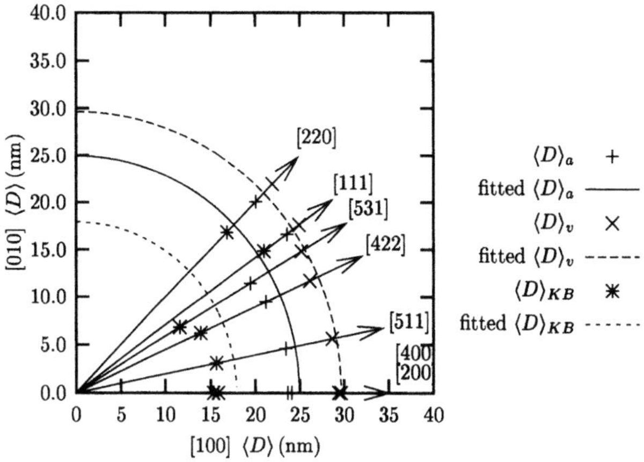

Fig. 8.8. A graphical representation using the area- and volume-weighted diameters, (8.38b), and log-normal diameters, (8.40) (also see Table 8.1). Each diameter is plotted according to its [hkl] direction, relative to the [010] direction. The [001] direction is out of the page

Bayesian/Markov Chain Monte Carlo (MCMC) Method

As pointed out in Sect. 8.5.6, a Bayesian/Markov Chain Monte Carlo (MCMC) method can be developed to sample $\Pr(D_0|\boldsymbol{g}, \boldsymbol{K}, \sigma, \mathcal{I})$ and $\Pr(\sigma_0|\boldsymbol{g}, \boldsymbol{K}, \sigma, \mathcal{I})$ assuming a log-normal size distribution and unform distribution of the a priori probability. The uniform distribution for D_0 and σ_0 were defined over the range $D_0 \in [7, 18]$ nm and $\sigma_0 \in [1.1, 3.0]$, respectively. The analysis of the CeO_2 data applied the Metropolis–Hastings algorithm [43,44] for the Markov Chain Monte Carlo method. The algorithm discussed in [45] provided the basis for the numerical implementation, while Gibbs sampling of the D_0 and σ_0 parameters was employed (see [39, 46]).

Gibbs sampling essentially involves initialising the starting values, $\{D_0^{\{0\}}, \sigma_0^{\{0\}}\}$ and randomly drawing D_0 and σ_0 in a cyclical manner from their conditional probabilities, expressed as

$$\begin{aligned}
\text{1st iteration:}\quad D_0^{\{1\}} &\leftarrow \Pr(D_0|\,\sigma_0^{\{0\}},\, \boldsymbol{g},\, \boldsymbol{K},\, \sigma,\, \mathcal{I})\\
\sigma_0^{\{1\}} &\leftarrow \Pr(\sigma_0|\,D_0^{\{1\}},\, \boldsymbol{g},\, \boldsymbol{K},\, \sigma,\, \mathcal{I})\\
\text{2nd iteration:}\quad D_0^{\{2\}} &\leftarrow \Pr(D_0|\,\sigma_0^{\{1\}},\, \boldsymbol{g},\, \boldsymbol{K},\, \sigma,\, \mathcal{I})\\
\sigma_0^{\{2\}} &\leftarrow \Pr(\sigma_0|\,D_0^{\{2\}},\, \boldsymbol{g},\, \boldsymbol{K},\, \sigma,\, \mathcal{I})\\
&\ \vdots\\
\text{nth iteration:}\quad D_0^{\{n\}} &\leftarrow \Pr(D_0|\,\sigma_0^{\{n-1\}},\, \boldsymbol{g},\, \boldsymbol{K},\, \sigma,\, \mathcal{I})\\
\sigma_0^{\{n\}} &\leftarrow \Pr(\sigma_0|\,D_0^{\{n\}},\, \boldsymbol{g},\, \boldsymbol{K},\, \sigma,\, \mathcal{I})
\end{aligned}$$

where $D_0^{\{n\}}$ and $\sigma_0^{\{n\}}$ represent the nth iteration.

Table 8.2. Summary of CeO_2 data analysis using a log-normal distribution in the Markov Chain Monte Carlo (MCMC) method. The values for $\langle D \rangle_{mcmc}$ and $\sigma^2_{\langle D \rangle}$ were determined from (8.41)

hkl	D_0 (nm)	σ_0	$\langle D \rangle_{mcmc}$ (nm)	$\sigma^2_{\langle D \rangle}$ (nm^2)
111	16.861 ± 0.002	1.50728 ± 0.00007	18.341 ± 0.002	61.63 ± 0.03
200	16.922 ± 0.003	1.5027 ± 0.0001	18.386 ± 0.003	60.99 ± 0.04
220	16.556 ± 0.002	1.51243 ± 0.00008	18.035 ± 0.002	60.61 ± 0.04
400	14.06 ± 0.006	1.5834 ± 0.0002	15.637 ± 0.007	57.45 ± 0.08
422	13.881 ± 0.004	1.5846 ± 0.0002	15.433 ± 0.004	56.2 ± 0.05
511	14.086 ± 0.004	1.5809 ± 0.0002	15.643 ± 0.005	57.10 ± 0.06
531	13.255 ± 0.004	1.6098 ± 0.0002	14.846 ± 0.004	56.08 ± 0.06
Average	15.1 ± 0.6	1.55 ± 0.02	16.6 ± 0.6	58.6 ± 0.9

The Bayesian/MCMC method was applied to all non-overlapping hkl lines in the CeO_2 data. It was found that 1.5×10^4 iterations were needed in order to sufficiently sample $\Pr(D_0|\boldsymbol{g}, \boldsymbol{K}, \sigma, \mathcal{I})$ and $\Pr(\sigma_0|\boldsymbol{g}, \boldsymbol{K}, \sigma, \mathcal{I})$. In addition, the first $k \sim 500$–2000 iterations were rejected in order for the Markov chain to "thermalise". The value of k that produced a small autocorrelation value ($\lesssim 0.1$) was used. This value also defined an appropriate sampling interval which was used to minimise the correlation between sampled values and from which the mean and standard error for D_0 and σ_0 parameters were determined. Different starting values for D_0 and σ_0 were also used to ensure that the Bayesian/MCMC method had converged on to the correct $\Pr(D_0|\boldsymbol{g}, \boldsymbol{K}, \sigma, \mathcal{I})$ and $\Pr(\sigma_0|\boldsymbol{g}, \boldsymbol{K}, \sigma, \mathcal{I})$ distributions. In developing this approach, using both simulated and experimental CeO_2 size broadened data, it was found that the Bayesian/MCMC method was sensitive to small changes in the background estimation ($\ll 10\%$). The background that produced "small" changes in the $\Pr(D_0|\boldsymbol{g}, \boldsymbol{K}, \sigma, \mathcal{I})$ and $\Pr(\sigma_0|\boldsymbol{g}, \boldsymbol{K}, \sigma, \mathcal{I})$ distribution was selected.

The results from the Bayesian/MCMC analysis for all non-overlapping hkl are given in Table 8.2, while the distributions for each hkl line are shown in Fig. 8.9. Overall the comparison between the individual hkl is good. The "scatter" between the distributions is less than the results from the Krill and Birringer method [6] (compare Fig. 8.7b with Fig. 8.9). However, it is interesting to point out that the 111, 200 and 220 lines are all shifted slightly forward relative to the remaining hkl (see Fig. 8.7b). These include the 111 and 220 which were also found in the Krill and Birringer method [6] to be shifted forward relative to the other hkl. This may suggest the background level effects have been minimised. The Bayesian/MCMC method incorporates this observed line profile and statistical noise in determining D_0 and σ_0, while the Krill and Birringer method [6] relies on accurately determining the specimen profile and not on the observed line profile. In this analysis

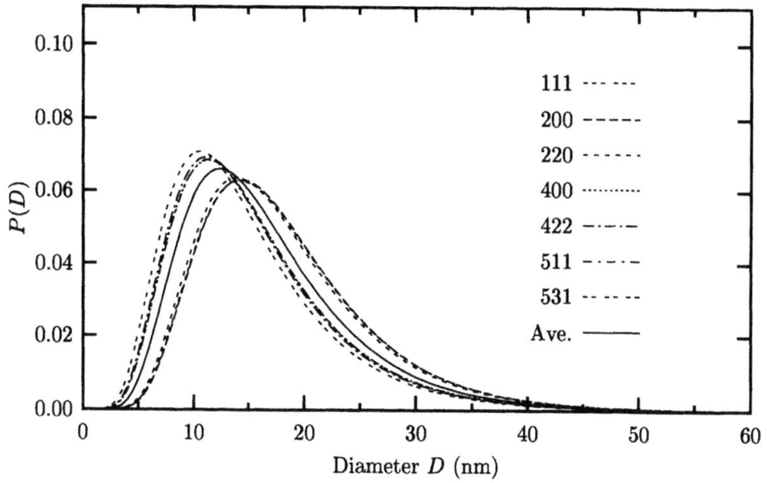

Fig. 8.9. CeO_2 size distributions from the Markov Chain Monte Carlo (MCMC) method. In this method the size distribution was assumed to be a log-normal function

uniform distributions for D_0 and σ_0, were used. However, if additional information that specified an average value for D_0 and σ_0 were available, then this information could be included in (8.33).

These results show that the Bayesian/MCMC method provides useful and consistent information concerning the parameter values for a specific size distribution.

8.6.4 Determining the P(D) Using the Bayesian/MaxEnt Method

In this section the full Bayesian/MaxEnt method is used to determine the crystallite size distribution for all the non-overlapping line profiles observed with the CeO_2 data set. In carrying out this analysis, two models are presented, namely a uniform crystallite size distribution model and a log-normal crystallite size distribution model, which uses the results from Sect. 8.6.3.

Determining $P(D)$ Using a Uniform Model

As we have seen, Fourier coefficients for the 200 line, given in Fig. 8.6d, suggest that the nanocrystallite-size is in the range $D \in [0, 60]$ nm. In order to test whether there exists any "structure" in the line profile data a uniform model can be used. This model essentially defines our ignorance concerning the size distribution. That is, it contains no information about the shape and functional properties of the size distribution. In this sense it defines our worst case, against which any other model can be compared.

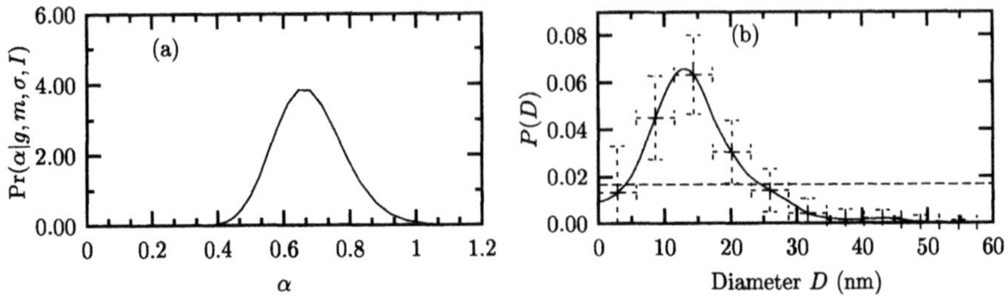

Fig. 8.10. CeO_2 Bayesian/MaxEnt crystallite-size distribution using a uniform a priori model over $D \in [0, 60]$ nm. (**a**) The $\Pr(\alpha|\boldsymbol{g}, \boldsymbol{m}, \sigma, \mathcal{I})$ distribution, (8.29), used to average over the set of solutions, $\{\boldsymbol{P}(\alpha)\}$ for the 200 line profile; (**b**) Bayesian/MaxEnt crystallite-size distribution (solid line + error bars) and the a priori model over $D \in [0, 60]$ nm

A uniform model was used in the entropy function, (8.24) given by

$$m(D) = \frac{1}{D_{\max} - D_{\min}}, \tag{8.42}$$

where $D_{\min} = 0$ nm and $D_{\max} = 60$ nm. For this range, the kernel $\boldsymbol{K}$ was determined using the results in [8].

The α-range used in these calculations started at $\alpha = 10^5$ and was reduced to $\alpha = 10^{-3}$, with each new α determined from the old one by $\alpha_{\text{new}} = 0.95\alpha_{\text{old}}$. For each new α, a solution $\boldsymbol{P}(\alpha)$ was determined giving a set of solutions $\{\boldsymbol{P}(\alpha)\}$. Also, in accordance with (8.29), $\Pr(\alpha|\boldsymbol{g}, \boldsymbol{K}, \boldsymbol{m}, \sigma, \mathcal{I})$ was determined and used to average over the set of solutions using (8.30), where the a priori α distribution needed in (8.29) was defined as $\Pr(\alpha|\mathcal{I}) = 1/(\alpha_{\max} - \alpha_{\min})$.

The Bayesian/MaxEnt size distributions using a uniform model are given in Fig. 8.10. The α probability distribution, using a uniform a priori model, is given in Fig. 8.10a. The range of α-values shown in Fig. 8.10a suggests that the influence of L in (8.28) is dominant (see Fig. 8.3a). This α probability distribution was used to average over $\{P(D)\}$ to produce the single $P(D)$ for the 200 line in Fig. 8.10b. The magnitudes of the error bars in Fig. 8.10b indicate no useful information has been imparted to $P(D)$ by the a priori model. That is, as $m(D) \to P(D)$ the error bars in $P(D)$ are reduced to the uncertainty defined by the observed line profile. There are also physical difficulties with the size distribution in Fig. 8.10b, in that for $D = 0$, $P(D) > 0$, suggesting that there exist a finite number of crystallites with $D = 0$. However, both the positivity and additivity of the solution $P(D)$ have been preserved.

The size distributions, $P(D)$, for all the non-overlapping hkl lines are given in Fig. 8.11a. Theoretically $P(D)$ should be identical for all hkl. However, the results in Fig. 8.11a, although showing "structure" in the size distribution, are generally disappointing in that there is considerable "scatter" in the solution

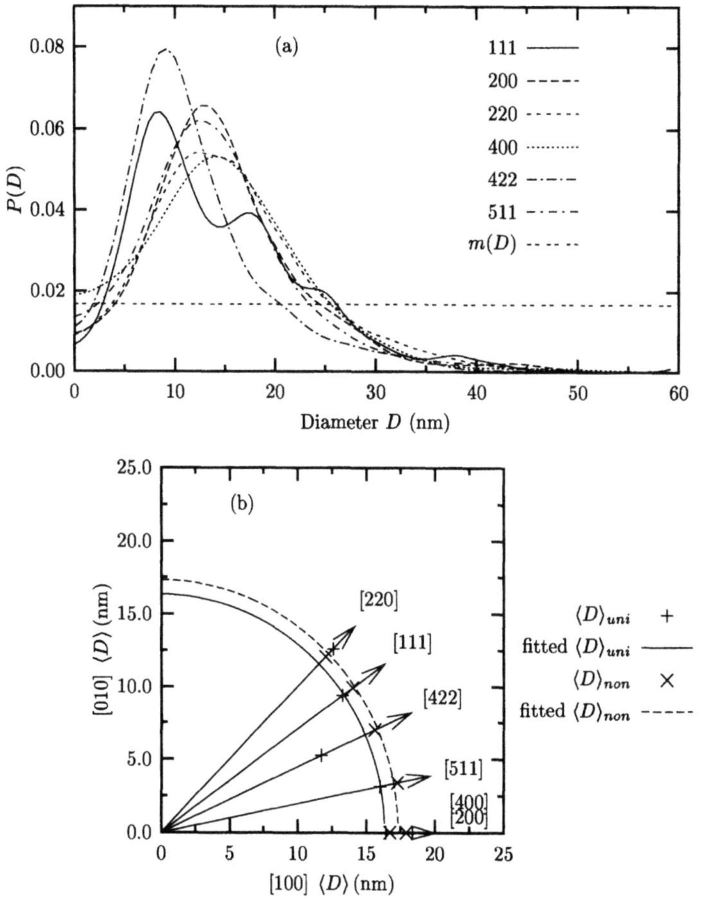

Fig. 8.11. CeO_2 Bayesian/MaxEnt crystallite-size distribution for all hkl lines using a uniform a priori model over $D \in [0, 60]$ nm. (**a**) Bayesian/MaxEnt crystallite-size distributions for the various CeO_2 hkl profiles using a uniform a priori model; (**b**) a graphical representation using the Bayesian/MaxEnt diameters for a uniform a priori model, $\langle D \rangle_{\text{uni}}$ and a log-normal (non-uniform) a priori model, $\langle D \rangle_{\text{non}}$ (also see Tables 8.3 and 8.4, respectively), according to their [hkl] direction and relative to the [010] direction. The [001] direction is out of the page

$P(D)$ over the different hkl lines. This scatter is also seen in the $\langle D \rangle_{\text{uni}}$ results in Fig. 8.11b, which provides a graphical representation of the crystallites according to the [hkl] directions. These results were fitted to a log-normal distribution, (8.41), and are shown in Table 8.3; they indicate the average crystallite, $\langle D \rangle_{\text{uni}} = (14 \pm 1)$ nm is of the same order as the results given in Tables 8.1 and 8.2.

Table 8.3. Size distribution results using a uniform a priori model in the Bayesian/MaxEnt method. The Bayesian/MaxEnt size distributions given in Fig. 8.11a were fitted with a log-normal size distribution and the following parameters determined

hkl	D_0 (nm)	σ_0	$\langle D\rangle_{uni}$ (nm)	$\sigma^2_{\langle D\rangle}$ (nm^2)
111	13.1 ± 0.4	1.9 ± 0.1	16.3 ± 0.8	140 ± 43
200	14.7 ± 0.3	1.61 ± 0.03	16.4 ± 0.4	69 ± 7
220	15.5 ± 0.4	1.70 ± 0.08	17.8 ± 0.6	104 ± 25
400	15.8 ± 0.8	1.7 ± 0.1	18 ± 1	120 ± 48
422	11.1 ± 0.2	1.728 ± 0.007	12.8 ± 0.2	58 ± 2
511	14.3 ± 0.3	1.66 ± 0.05	16.3 ± 0.4	78 ± 15
Average	14 ± 1	1.7 ± 0.2	16 ± 2	95 ± 71

Determining $P(D)$ Using a Log-Mormal Model

In these calculations the $P(D)$ was determined using a log-normal a priori model. The same steps were followed as in the case of the uniform a priori model, $m(D)$.

For these calculations $m(D) = N(D)$, where the values $D_0 = (16 \pm 2)$ nm and $\sigma_0 = 1.7 \pm 0.2$ were used from Table 8.1. The average results from Bayesian/MCMC (see Table 8.2) could also be used as the a priori model. In other words, the size distribution $P(D)$ is *not* explicitly defined as $N(D)$, in (8.41), but the a priori model, $m(D)$, is used to test whether a log-normal distribution is a valid size distribution to represent the crystallite diameters. Referring to the discussion in Sect. 8.5.2, it was argued that S in (8.24) was a measure that can be used to qualitatively assess the applicability of $m(D)$. In the above discussion for the uniform a priori model, it was shown that $P(D)$ is considerably different from (8.42) and can be thought of as a poor choice for $m(D)$.

However, by setting $m(D) = N(D)$, useful information is imparted into the size distribution, $P(D)$. This is evident from the results given in Fig. 8.12. In Fig. 8.12a, the $\Pr(\alpha|\boldsymbol{g}, \boldsymbol{m}, \sigma, \mathcal{I})$ distribution for the 200 line profile is given. It can be seen that the range of α-values that defines this distribution is a factor 10–20 times greater than that of Fig. 8.10a. The corresponding size distribution for the 200 line is given in Fig. 8.12b. These results show a considerable improvement upon the results in Fig. 8.10b. The error-bars have been reduced markedly, while at $D = 0$, $P(D) = 0$. Also, the $m(D)$ and $P(D)$ in Fig. 8.12b are similar, suggesting that the log-normal distribution is a valid size distribution model to represent the crystallites, while being constrained by the experimental data. If we have $m(D) = P(D)$, this implies that S is dominant and $\alpha \sim 10^5$ (large) and thus experimental data would have no influence in determining $P(D)$ (see Fig. 8.3b). However, the fact that

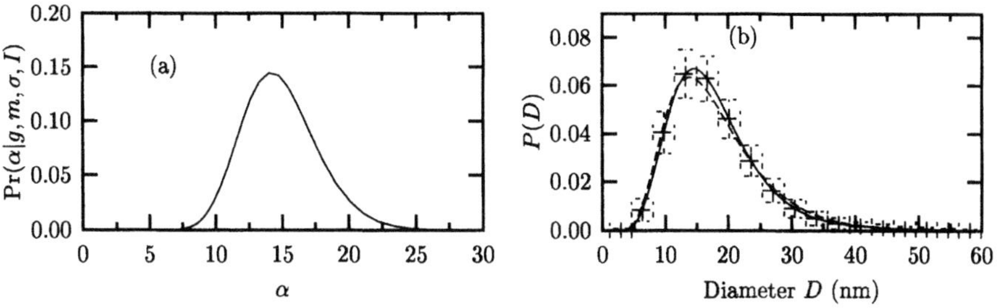

Fig. 8.12. The 200 CeO_2 Bayesian/MaxEnt crystallite-size distribution using a log-normal a priori model over $D \in [0, 60]$ nm. (**a**) The $\Pr(\alpha|\boldsymbol{g}, \boldsymbol{m}, \sigma, \mathcal{I})$ distribution, (8.29), used to average over the set of solutions $\{\boldsymbol{P}(\alpha)\}$, for the 200 line. (**b**) Bayesian/MaxEnt crystallite-size distribution (solid line + error bars) and the a priori model (dashed line) over $D \in [0, 60]$ nm

Table 8.4. Size distribution results using a log-normal (non-uniform) a priori model in the Bayesian/MaxEnt method. The Bayesian/MaxEnt size distributions given in Fig. 8.13a were fitted with a log-normal size distribution and from these the following parameters were determined

hkl	D_0 (nm)	σ_0	$langleD\rangle_{\rm non}$ (nm)	$\sigma^2_{\langle D\rangle}$ (nm^2)
111	15.9 ± 0.2	1.505 ± 0.008	17.2 ± 0.2	54 ± 2
200	16.64 ± 0.04	1.469 ± 0.005	17.91 ± 0.04	51 ± 1
220	15.68 ± 0.06	1.502 ± 0.008	17.04 ± 0.08	52 ± 2
400	15.48 ± 0.01	1.480 ± 0.005	16.72 ± 0.03	46 ± 1
422	15.86 ± 0.07	1.4799 ± 0.0002	17.12 ± 0.08	48.7 ± 0.5
511	16.21 ± 0.07	1.497 ± 0.002	17.58 ± 0.08	54.7 ± 0.6
531	16.32 ± 0.07	1.500 ± 0.005	17.72 ± 0.09	56 ± 1
Average	16.0 ± 0.2	1.49 ± 0.01	17.3 ± 0.3	52 ± 3

there is some difference between $m(D)$ and $P(D)$ and α is in the range of 5–25 (see Fig. 8.12), suggests that S has not been dominant and the $P(D)$ results in Fig. 8.12b–c are physically valid crystallite size distributions for the CeO_2 specimen.

The $P(D)$ for all non-overlapping hkl in Fig. 8.13a further supports the theoretical assertion that the size distribution for all hkl should be identical. Compared to the Krill and Birringer [6] and Bayesian/MCMC results, there is very little "scatter" in the non-overlapping hkl lines. Furthermore, by including a priori information with structure, i.e. non-uniform distribution, the difficulties encountered with the Bayesian/MaxEnt results for uniform model have been resolved.

The results in Fig. 8.13a were fitted to a log-normal model (8.41) and are shown in Table 8.4. They further indicate that the Bayesian/MaxEnt $P(D)$ is similar to a log-normal size distribution. In addition, a graphical

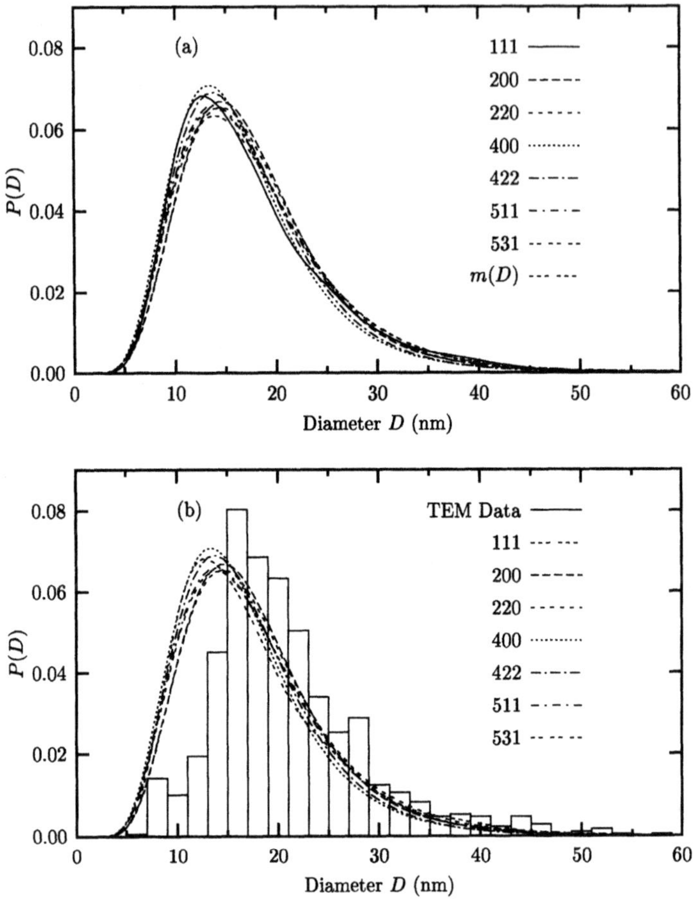

Fig. 8.13. CeO_2 Bayesian/MaxEnt crystallite-size distribution for all the non-overlapping hkl lines using a log-normal a priori model over $D \in [0, 60]$ nm. (**a**) Bayesian/MaxEnt crystallite-size distributions for the non-overlapped line profiles. (**b**) Comparison of TEM and Bayesian/MaxEnt size distributions. Due to difficulties in de-aggregating the CeO_2 particles and identifying smaller crystallites, the TEM size distribution is poorly defined for $D \leq 15$ nm. However, for $D \geq 15$ nm, the comparison is reasonable

representation of the crystallites using uniform and log-normal a priori models is given in Fig. 8.11b. This figure shows the considerable improvement that is gained in going from a uniform a priori model to a log-normal (nonuniform) a priori model. That is, useful information in being imparted to $P(D)$. The results in Fig. 8.11b clearly demonstrate that the CeO_2 crystallites have a near-spherical morphology.

8.6.5 TEM Size Distribution

The real test for the Bayesian/MaxEnt $P(D)$ comes in comparing it with the size distribution obtained from a TEM analysis of CeO_2 crystallites, given in Fig. 8.13b. The TEM distribution was obtained from manual size measurements from 853 crystallites (see [8]). In Fig. 8.13b, the TEM sizes were binned into 2 nm widths. This width was approximated by estimating the number of bins $\sim \sqrt{853} \approx 30$, while keeping in mind the size range, $D \in [0, 60]$ nm [47].

On comparing the TEM and Bayesian/MaxEnt $P(D)$, we notice for $15 \lesssim D \lesssim 60$ nm, there is biasing towards large crystallites sizes relative to the Bayesian/MaxEnt $P(D)$, but on the whole there is a reasonably good agreement between them. However, for $D \lesssim 15$ nm, there is considerable discrepancy between the two sets of results. The main reason for this difference is the difficulty in identifying smaller crystallites. The processing history of the specimen is unknown to us, but the TEM micrographs (see Fig. 8.14) reveal that the CeO_2 crystallites have formed aggregates. A direct consequence of this is that aggregates of small crystallites can be mistaken as large crystallites [8]. This accounts for the bias of the TEM distribution to larger D-values and the discrepancy between the TEM and Bayesian/MaxEnt distribution for $D \lesssim 15$ nm. Several attempts to de-aggregate CeO_2 specimen have been unsuccessful. It is also noticeable from Fig. 8.14 that there is considerable overlaying between crystallites, adding to the difficulty in discerning between groups of smaller crystallites and large ones. On the other hand, Fig. 8.14 shows that the crystallites are generally spherical in shape, confirming an earlier result; they also lie in the size range specified by Fig. 8.13b.

Fig. 8.14. A TEM micrograph of the CeO_2 specimen taken at a magnification of 200 k×. The crystallites appear to have near-spherical shapes and sizes that are in the range predicted by the crystallite-size analysis presented here

The above-mentioned difficulties with the CPD/IUCr CeO_2 specimen highlights an additional consideration in selecting a specimen for the purpose of a NIST nanocrystallite-size SRM. That is, the nanocrystallite size SRM must consist of a dispersed material, such that all the crystallites in a TEM micrograph are countable and a $P(D)$ can be determined using these techniques. This will ensure that the X-ray line profile and electron microscopy techniques are complementary.

It can be concluded, therefore, that the underlying CeO_2 size distribution resembles a log-normal distribution with a near-spherical morphology having $\langle D \rangle = (17.3 \pm 0.3)\,\mathrm{nm}$ and $\sigma_{\langle D \rangle} = (52 \pm 3)\,\mathrm{nm}^2$.

8.6.6 Comparison of Methods

The average distribution from the Krill and Birringer [6], Bayesian/MCMC and full Bayesian/MaxEnt methods can be compared with the TEM distribution. A comparison of these methods is given in Fig. 8.15, where uncertainty for the TEM distribution has been included. The uncertainty was assumed to follow Poisson statistics with the square-root of the counts taken and scaled accordingly when the distribution was normalised for unit area.

The size distributions for these methods compare closely with each other. There is some difference between the Bayesian/MCMC and Bayesian/MaxEnt method. However, given that the Bayesian/MCMC calculations used uniform priors for the D_0 and σ_0 parameters, and the Bayesian/MaxEnt method incor-

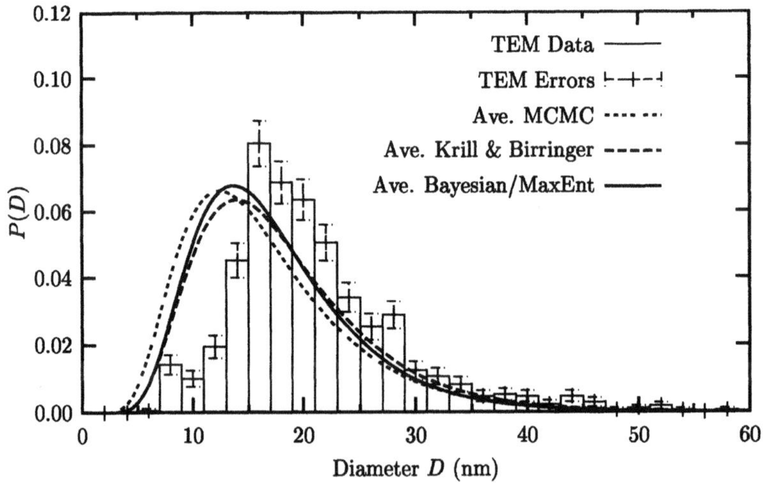

Fig. 8.15. A comparison between the TEM distribution, average Krill and Birringer method (see Table 8.1), average MCMC method (see Table 8.2), and average Bayesian/MaxEnt method (see Table 8.4). The error-bars for the TEM data were assumed to be the square-root of the counts. The TEM histogram was normalised to unit area and the error-bars scaled accordingly

porated non-uniform a priori information about the distributions, differences between the results are expected. Also keeping in mind the difficulties with the experimental (TEM) distribution, especially for $D \lesssim 15\,\mathrm{nm}$, each method compares reasonably with the TEM distribution for $D \gtrsim 15\,\mathrm{nm}$.

An obvious question that arises from this comparison is: *Why apply the Bayesian/MaxEnt method if the other methods (e.g. [6], Bayesian/MCMC or [7]) produce comparable results?*

The critical difference between these methods and the Bayesian/MaxEnt method is that the Bayesian/MaxEnt method makes an inference as to the best size distribution, given the a priori information and experimental data. Consequently, a minimal number of assumptions concerning the size distribution have been made. That is, the hypothesis is tested in the presence of the experimental data (see Sect. 8.4). It is only after the Bayesian/MaxEnt solution has been determined that it is fitted to a size distribution, such as the log-normal function, and the parameters are determined. The full benefits of this approach will become more apparent when applied to cases involving more complex distribution, such as non-spherical crystallites and bimodal distributions. On the other hand, methods like those of Krill and Birringer [6] and Langford et al. [7] are embodied in the Bayesian/MCMC approach; they assume a particular size distribution and estimate the values of the parameters that best fit the data. In other words, a specific assumption is made, but not tested. Hence, for the purpose of parameter estimation, methods like Krill and Birringer [6] or Langford et al. [7] are sufficient. These methods provide useful a priori information. However, for the purpose of a NIST nanocrystallite-size SRM, a size distribution based on the fewest assumptions is required, while taking into consideration any suitable a priori distribution. The theoretical basis and generality of the Bayesian/MaxEnt method ensures that this requirement is satisfied.

8.7 Discussion

The motivation for this research work has been the development of a general method for quantifying the nanocrystallite-size distribution and morphology from X-ray line profile data. This method will play an integral part in the development and certification of a nanocrystallite-size standard reference material (SRM) to be produced by NIST. In addition, this work has applications in the nanotechnology industry, where nanocrystallites will be designed to have a particular morphology and size distribution.

In Sect. 8.2–8.5, size broadening of line profiles was introduced and it was shown how this can be treated as an inverse problem. The necessary physics was developed (see Sect. 8.2), outlining how the shape and size information is encoded in the diffraction profile, but also necessary mathematical insights were introduced to define the inverse problem (see Sect. 8.3-8.5). It was shown how "standard" approaches like the Langford et al. method [7] can be interpreted in the wider Bayesian context (see Sect. 8.5.6). Although the issues discussed in Sect. 8.2–8.5 are dispersed in the line profile and mathematics literature, we have attempted to present them in a unified and integrated manner and relate them directly to size broadening analysis.

The application of the Bayesian/MaxEnt method to a cerium oxide specimen for a CPD/IUCr size round-robin were presented in Sect. 8.6. This method was also compared with other (existing) methods, namely the Krill and Birringer [6] and Langford [7] methods, where the latter was embodied in the Bayesian/Markov Chain Monte Carlo analysis. Two steps were taken in the analysis. Firstly, the MaxEnt/Fuzzy Pixel method was used to determine the specimen line profile, $f(2\theta)$, from which the specimen broadening could be identified (see Sect. 8.6.3). This intermediate analysis enabled the Krill and Birringer [6], as well as the Williamson–Hall and Fourier methods to be applied. Secondly, the Bayesian/MCMC and full Bayesian/MaxEnt methods were used to determine the crystallite size distribution from the observed line profiles (see Sect. 8.6.3 and 8.6.4).

The specimen broadening was identified as due to nanocrystallite-size effects, and it was shown that the morphology of the crystallites was near-spherical in the 0–60 nm size range. This analysis demonstrated that the specimen line profile and underlying broadening contribution can be determined without making assumptions concerning the functional form of specimen profiles or the nature of the specimen broadening. This intermediate step in the analysis has an important role in identifying broadening effects and in the case of size-broadening, a suitable crystallite shape model.

The Krill and Birringer [6] and Bayesian/MCMC methods assume the size distribution as a log-normal function. On the whole, these method produced consistent results (compare Tables 8.1 and 8.2). However, these methods do differ markedly, in that the [6] method is dependent in the specimen profile, and statistics of the observed profile are only indirectly taken into consideration. The Langford et al. method [7], as embodied in the Bayesian/MCMC approach, does determine the D_0 and σ_0 parameters from the observed line profile.

Information from the Krill and Birringer [6] or the Langford et al. [7] methods can be used as the non-uniform a priori model in the full Bayesian/MaxEnt method. This information is tested in the light of the observed line profile, statistical noise and kernel (see Sect. 8.6.4). That is, the size distribution was not explicitly defined as a log-normal distribution, rather the a priori model was defined as a log-normal distribution and this model (or hypothesis) was tested against the experimental data. This defines the critical difference between Krill and Birringer [6] and Bayesian/MCMC methods and the Bayesian/MaxEnt methods, in that the Krill and Birringer [6] and Bayesian/MCMC approaches are a consequence of "parameter estimation", while the Bayesian/MaxEnt method makes an inference of the size distribution. This feature become important in solving more complex problems such as determining the modal characteristics of distribution and determining the size distribution from non-spherical crystallites.

On applying the full Bayesian/MaxEnt method to a non-uniform model, the average crystallite diameter and variance were found to be $\langle D \rangle = (17.3 \pm 0.3)$ nm and $\sigma^2_{\langle D \rangle} = (52 \pm 3)\,\mathrm{nm}^2$, respectively. These results were also found to closely fit a log-normal distribution.

The Bayesian/MaxEnt results were compared to the TEM distribution and micrographs, from which it was evident that the crystallites have a near-spherical morphology and the size distribution is in good agreement in the range of $15 \lesssim D \lesssim 60$ nm, while difficulties in identifying smaller crystallites account for the differences between the TEM and Bayesian/MaxEnt results for $D \lesssim 15$ nm (see Figs. 8.13 and 8.14). The TEM analysis of the CPD/IUCr specimen highlights an additional difficulty in developing a nanocrystallite size SRM, in that the material must be sufficiently dispersed so that TEM techniques can be successfully applied. This is important if both electron microscopy and X-ray diffraction are to provide independent sources of data and be considered as complementary techniques. However, ensuring that an SRM is sufficiently disperse in large quantities (~ 1 kg scale) presents unique technical difficulties, which at present are unresolved.

Finally, the analysis has demonstrated that the CPD/IUCr round-robin specimen is not suitable as a nanocrystallite-size SRM. The development of the Bayesian/MaxEnt method, which is a consequence of the SRM project, clearly demonstrates that it is suited to certifying the Standard Reference Material.

References

1. J.I. Langford and D. Louër: Rep. Prog. Phys. **59**, 131 (1996)
2. R. Vargas, D. Louër and J.I. Langford: J. Appl. Cryst. **16**, 512 (1983)
3. J.I. Langford and A.J.C. Wilson: J. Appl. Cryst. **11**, 102 (1978)
4. J.I. Langford and D. Louër: J. Appl. Cryst. **15**, 20 (1982)
5. J.I. Langford: NIST Spec. Pub. **846**, 110 (1992)
6. C.E. Krill and R. Birringer: Phil. Mag. **A77(3)**, 621 (1998)
7. J.I. Langford, D. Louër and P. Scardi: J. Appl. Cryst. **33**, 964 (2000)
8. N. Armstrong, W. Kalceff, J.P. Cline and J. Bonevich: J. Res. Nat. Inst. Stand. Techn. 2001 (Submitted); *Proceedings of Accuracy in Powder Diffraction III, 22–25 April 2001, NIST, Gaithersburg, USA*
9. R.K. Bryan: Eur. Biophys. J. **18**, 165 (1990)
10. M. Jarrell and J.E. Gubernatis: Phys. Reports **269**, 133 (1996)
11. D. Louër and N. Audebrand (University of Rennes, 2001) For details see links at `www.iucr.org/iucr-top/comm/cpd/projects/index.html`
12. E.F. Bertaut: Acta Cryst. **3**, 14 (1950)
13. A. Guinier: *X-ray diffraction in crystals, imperfect crystals and amorphous bodies.* (Freeman, San Francisco 1963) Reprinted by Dover (New York), 1994
14. A.R. Stokes and A.J.C. Wilson: Proc. Camb. Phil. Soc. **38**, 313 (1942)

15. P. Scardi and M. Leoni: Acta Cryst. A **57**, 604 (2001)
16. D. Louër, J.P. Auffrédic and J.I. Langford: J. Appl. Cryst. **16**, 183 (1983)
17. J.I. Langford, A. Boultif, J.P. Auffrédic and D. Louër: Appl. Cryst. **26**, 22 (1993)
18. N. Armstrong: *Application of the maximum entropy method to X-ray profile analysis.* PhD thesis (Department of Applied Physics, University of Technology, Sydney, Australia 1999)
19. R. Kress: *Linear integral equations.* 2nd edn. (Springer, Berlin Heidelberg New York 1999)
20. N. Armstrong and W. Kalceff: J. Appl. Cryst. **31**, 453 (1998)
21. R.T. Cox: Amer. J. Phys. **14(1)**, 1 (1946)
22. D.S. Sivia: *Data Analysis: A Bayesian Tutorial.* (Oxford Science Pub., Oxford 1996)
23. K. Knuth: *Inductive logic: From data analysis to experimental design.* Presented at MaxEnt2001, APL John Hopkins Uuniversity, August 4–9, 2001. (URL: `xxx.lanl.gov/list/physics.data-an/recent/0204068`, 2002)
24. G.B. Arfken and H.J. Weber: *Mathematical Methods for Physicists.* 4th edn. (Academic Press, San Diego 1996)
25. S.F. Gull: *Developments in maximum entropy data analysis.* In: Maximum Entopy and Bayesian Methods, ed. by J. Skilling (Kluwer Acad. Publ., Netherlands 1989) pp. 53–71
26. J.E. Shore and R.W. Johnson: IEEE Trans. IT **26(1)**, 26 (1980)
27. R.W. Johnson and J.E. Shore: IEEE Trans. IT **26(6)**, 942 (1983)
28. Y. Tikochinsky, N.Z. Tishby and R.D. Levine: Phys. Rev. Lett. **52(16)**, 1357 (1984)
29. S.F. Gull and J. Skilling: IEE Proc. **F131**, 646 (1984)
30. J. Skilling: *The eigenvalues of mega-dimensional matrices.* In: Maximum Entropy and Bayesian Methods, ed. by J. Skilling (Kluwer Acad. Pub., Netherlands 1989) pp. 455–566
31. J. Skilling: *Quantified Maximum Entropy.* In: Maximum Entropy and Bayesian Methods, ed. by P.F. Fougère (Kluwer Acad. Pub., Netherlands 1990) pp. 341–350
32. F. Goambo and E. Gassiat: Annals Stat. **25(1)**, 328 (1997)
33. J. Skilling and R.K. Bryan: Mon. Not. R. astr. Soc. **211**, 111 (1984)
34. D.M. Titterington: Astron. Astrophys. **144**, 381 (1985)
35. E.T. Jaynes: Phys. Rev. **106(4)**, 620 (1957)
36. B.R. Frieden: J. Opt. Soc. Am. **62(4)**, 511 (1972)
37. A.R. Stokes: Proc. Phys. Soc. London A **61**, 382 (1948)
38. D.S. Sivia, (Private communication 1999)
39. J.J.K. O'Ruanaidh and W.J. Fitzgerald: *Numerical Bayesian methods applied to signal processing.* Springer, Berlin Heidelberg New York 1996)
40. R.W. Cheary and J.P. Cline: Adv. X-ray Anal. **38**, 75 (1995)
41. J.A. Blackman, B.L. Evans and A.I. Maaroof: Phys. Rev. B **49(19)**, 13863 (1994)
42. L.B. Kiss, J. Söderland, G.A. Niklasson and C.G. Granqvist: NanoStruct. Mater. **12**, 327 (1999)
43. N. Metropolis, A.W. Rosenbluth, M.N. Rosenbluth, A.H. Teller and E. Teller: J. Chem. Phys. **21**, 1087 (1953)
44. W.K. Hastings: Biometr. **57**, 97 (1970)

45. S.E. Koonin and D.C. Meredith: *Computational physics: Fortran version.* (Wesly, New York 1990)
46. S. Geman and D. Geman: IEEE Trans. Pattern Anal. Mach. Intell. **6(6)**, 721 (1984)
47. L. Kirkup, Private communication, Department of Applied Physics, UTS 11 July, 2002

9 Study of Submicrocrystalline Materials by Diffuse Scattering in Transmitted Wave

R. Kužel, V. Holý, M. Čerňanský, J. Kuběna, D. Šimek, and J. Kub

A new method of crystallite size determination has been applied to submicrocrystalline materials prepared by high-pressure torsion deformation. The method is based on the measurement of the small-angle diffuse scattering in the transmitted wave and requires thin foil samples and well-defined incoming plane wave. The physical information is encoded in the wave modification due to primary extinction in the crystallites.

Two methods were used for the evaluation of the diffuse scattering: calculation of the autocorrelation function of the crystallite shape by the transformation of (1) the measured signal or (2) by the direct fitting. In the former case, the normalized shape function is compared with that calculated for spherical blocks and crystallite size estimated from the simulation. In the latter procedure, a log-normal distribution of spheres is included and both the mean crystallite size and variance of the distribution are refined.

The agreement between experimental and calculated values is not always satisfactory when using a single distribution of crystallite size. In these cases a bimodal block size distribution was assumed. This assumption was supported by classical film method as well as by the observation of long tails in the conventional diffraction profiles.

9.1 Introduction

Materials with a mean grain size of hundreds of nanometers are called submicro-crystalline or ultra fine-grained materials. These materials are attracting a growing interest due to a number of unusual physical properties. It has been shown mainly by X-ray powder diffraction and by positron annihilation spectroscopy [1–4] that in addition to small crystallite size, such metals contain a large number of inhomogeneously distributed dislocations and also microvoids. The addition of a small amount of Al_2O_3 stabilizes the fine-grained highly defected structure to much higher temperatures. Significant changes are observed above 400°C in comparison with about 150°C for pure copper. So far, studies of these materials were mainly performed by conventional XRD line profile analysis, for instance in terms of the simplified analysis by means of the modified Williamson–Hall plots. The results of conventional line profile analysis can be summarized as follows. In all samples investigated, the lattice parameters are close to the "standard" PDF-2 values for bulk or large grained materials. In most samples, a [111] texture was found. The main feature of the powder diffraction pattern was the anisotropic broadening due to dislocations, typical of f.c.c. materials. The size broadening was relatively weak and did not allow precise determination of the mean crystallite size, that was estimated above 100 nm. In this work, however, the main attention is devoted to testing small-angle diffuse scattering in the transmitted wave for the estimation of crystallite size [5].

A plane monochromatic wave incident upon a small diffracting crystallite produces a divergent diffracted wave (diffuse X-ray scattering). Information about the shape and size of the crystallite can be obtained from the distribution of the scattered intensity in the reciprocal space. We can measure the reciprocal space distribution of the diffusely scattered intensity in a vicinity of a non-zero reciprocal lattice point hkl, this intensity is denoted as diffuse component of the diffracted wave. Similarly, the diffusely scattered intensity concentrated around the origin 000 of the reciprocal lattice represents the diffuse component of the transmitted (non-diffracted) wave.

The diffuse scattering in the transmitted wave is originated by two different mechanisms. The first mechanism is based on inhomogeneities of the electron density of the sample (i.e. changes of refractive index) and refers to the classical small-angle scattering. It is the dominating effect in powder and porous samples, owing to the large differences of refractive index, and it does not occur in compact polycrystalline samples. The second mechanism is caused by the primary extinction of X-rays. If a crystallite is in a diffraction position, the intensity of the wave transmitted through this crystallite is reduced due to the diffraction in the crystallite. This effect results in a non-homogeneous space distribution of the intensity in the transmitted wave and from this inhomogeneity the mean size of the crystallites can be deduced.

This mechanism has significant role in dense compact samples – bulk materials, thin films, thin foils – and it is sensitive to the size of coherently scattering domains (crystallites).

Conventional method for the diffraction determination of crystallite size and size distribution is XRD line profile broadening analysis (e.g. [6–8]. The analysis has been well developed during last decades [9] and still new approaches are suggested [10–15]. The scope of this subject is so broad and interesting that conferences devoted to the investigations of the real structure of materials (called "Size-Strain") are regularly held in recent years (August 1995, Liptovsky Mikulas, Slovakia; November 1998, Freiberg, Germany; December 2001, Trento, Italy). However, profile analysis has significant difficulties. Major issues concern a correct evaluation of instrumental broadening, and a proper consideration of the simultaneous influence of two main effects – size broadening, that is order independent (the same for all points in the reciprocal space), and strain broadening, proportional to the modulus of the diffraction vector. An independent determination of one of the components would be desirable.

The diffuse scattering in the transmitted wave does not depend on the internal structure of the crystallite – strains, dislocation structure etc. Being sensitive to domain size is an advantage in this context, even if the technique is more sensitive to larger particles in comparison with the classical profile analysis.

9.2 Theoretical Background

The method of small-angle diffuse scattering (e.g. [16,17]) is based on the loss of coherence of the primary beam due to extinction. This is not caused by phase difference as in the scattering from cavities or optically denser grains, but on the amplitude decay of the wave transmitted through a grain in diffraction position with respect to the part of wave-plane passing by. This results in a distortion of the wave-plane similar to the case of phase distortion.

Let us consider the wave in a grain, which is close to the exact diffraction position for an arbitrary hkl reflection. This happens when the hkl planes and the incident beam direction include angle θ satisfying the Bragg condition $n.\lambda = 2d_{hkl}\sin\theta$. Then the exponential decay of the wave amplitude (if the grains are small enough for the kinematical approximation to be applicable) can be described as follows:

$$E(\boldsymbol{r}) = E_0(\boldsymbol{r})e^{-z.p(\eta)}\,. \tag{9.1}$$

Here $E_0(\boldsymbol{r})$ is the amplitude of the wave passing through the same grain not in diffraction position (i.e. the incident wave reduced by absorption) and $p(\eta)$ is the extinction coefficient depending on the angular orientation η. Under the assumption of weak extinction, we can expand the exponential in (9.1) and write an expression for the change of wave amplitude after passing the sample:

$$E(\boldsymbol{x}) = E_0(\boldsymbol{x})\left[1 - \sum_j p_j(\eta_j)l(\boldsymbol{x}-\boldsymbol{x}_j)\right] \equiv E_0(\boldsymbol{x}) + \Delta E(\boldsymbol{x})\,, \tag{9.2}$$

where $l(\boldsymbol{x}-\boldsymbol{x}_j)$ is the length of the wave-path through the j-th crystallite in the lateral position $\boldsymbol{x}$ (lateral means perpendicular to the incident beam). $\boldsymbol{x}_j$ is the lateral position of the j-th crystallite. The mutual coherence function [18] of the transmitted wave is then:

$$\Gamma(\boldsymbol{x},\boldsymbol{x}') \equiv \langle E(\boldsymbol{x})E^*(\boldsymbol{x}') = \Gamma_{\mathsf{coh}}(\boldsymbol{x},\boldsymbol{x}') + \Gamma_{\mathsf{diff}}(\boldsymbol{x},\boldsymbol{x}')\,. \tag{9.3}$$

The first term denotes the mutual coherence function of the coherent part of the transmitted wave

$$\Gamma_{\mathsf{coh}}(\boldsymbol{x},\boldsymbol{x}') \equiv \Gamma_0(\boldsymbol{x},\boldsymbol{x}') + E_0(\boldsymbol{x})\langle\Delta E(\boldsymbol{x}')\rangle^* + E_0^*(\boldsymbol{x}')\langle\Delta E(\boldsymbol{x})\rangle\,, \tag{9.4}$$

where $\Gamma_0 \equiv \langle E_0(\boldsymbol{x})E^*(\boldsymbol{x}')\rangle$ and the second term describes the diffuse (incoherent) part

$$\Gamma_{\text{diff}} = \langle \Delta E^*(\boldsymbol{x})\Delta E(\boldsymbol{x}')\rangle . \tag{9.5}$$

The coherent part does not contribute to the diffusely scattered wave. Therefore, only the incoherent part will be discussed below. The averaging in (9.5) can be done assuming that the random positions $\boldsymbol{x}_j$ of the crystallites are not statistically correlated to their random angular orientations. Then, the incoherent mutual coherence function is

$$\Gamma_{\text{diff}}(\boldsymbol{x},\boldsymbol{x}') = \Gamma_0(\boldsymbol{x},\boldsymbol{x}')\,|\langle p\rangle|^2 \left\langle \sum_{j,k}(\boldsymbol{x}-\boldsymbol{x}_j)l(\boldsymbol{x}'-\boldsymbol{x}_k)\right\rangle . \tag{9.6}$$

The diffuse part Γ_{diff} represents an additional diffuse scattering due to extinction. According to the Wiener–Khintchine theorem [18] its angular dependence (the distribution in reciprocal space) is given by the Fourier transform of G_{diff}:

$$I(\boldsymbol{q}) = \int \Gamma_{\text{diff}}(\boldsymbol{x},\boldsymbol{x}')e^{i\boldsymbol{q}(\boldsymbol{x}-\boldsymbol{x}')}d^2(\boldsymbol{x}-\boldsymbol{x}') . \tag{9.7}$$

Here, $\boldsymbol{q}$ denotes the component of the wave vector of the scattered wave perpendicular to the direction of the primary wave. Its magnitude is proportional to the rotation angle of the analyzing crystal. The transformation can be considered two-dimensional only, because for small scattering angles, q_z is almost zero. After some rearrangement (9.2), (9.5), (9.7), we obtain finally

$$\begin{aligned} I(\boldsymbol{q}) &= I_0 \left\langle |p|^2\right\rangle \left\langle |l^{FT}(\boldsymbol{q})|^2\right\rangle \left\langle \sum_{j,k} e^{i\boldsymbol{q}(\boldsymbol{x}_j-\boldsymbol{x}_k)}\right\rangle \\ &\cong N I_0 \left\langle |p|^2\right\rangle \left\langle |l^{FT}(\boldsymbol{q})|^2\right\rangle , \end{aligned} \tag{9.8}$$

where

$$l^{FT}(\boldsymbol{q}) = \int l(\boldsymbol{x})e^{i\boldsymbol{q}\cdot\boldsymbol{x}}d^2\boldsymbol{x} \equiv \int V(\boldsymbol{r})e^{i\boldsymbol{q}\cdot\boldsymbol{r}}d^3\boldsymbol{r} \tag{9.9}$$

is the Fourier transform of the mean path length $l(\boldsymbol{x})$. The last equality in (9.8) stands for non-correlated orientation of neighboring grains, so that crystallites with the same orientation (indices j and k) are far enough and uniformly distributed over the whole irradiated volume with density N (see below). The second equality in (9.9) is due to $q_z \approx 0$, $V(\boldsymbol{r})$ is the shape function of the crystallite (equal to one inside, zero elsewhere).

For averaging in (9.8), the well-known Debye formula can be used, i.e.

$$I(\boldsymbol{q}) = I_0 \left\langle |p|^2\right\rangle \left\langle |l^{FT}(\boldsymbol{q})|^2\right\rangle N \left[1 - \frac{N}{S}\int_0^\infty [1-P(x)]J_0(qx)dx(2\pi x)\right] . \tag{9.10}$$

S is the area of the irradiated sample surface, N is the number of crystallites in the diffraction position, J_0 is the Bessel function of 0th order,

$$P(|x_j - x_k|) = p_{j,k}\left(\frac{S}{N}\right)^2 ,$$

and $p_{j,k}d^2\boldsymbol{x}_j d^2\boldsymbol{x}_k$ is the probability of finding a projection of the crystallite in the elementary area $d^2\boldsymbol{x}_j$ and another crystallite in the area $d^2\boldsymbol{x}_k$.

The second term on the right side of (9.10) expresses the influence of the correlations of the crystallite positions on the scattered intensity. If the probability of finding a crystallite in the diffraction position is sufficiently small, the value of N/S is small too and this term can be neglected. Within this approximation, we assume that the scattering polycrystalline sample is *diluted*, i.e. the mean distance between two crystallites in the diffraction position is much larger than the crystallite size.

Then the scattered intensity can be expressed as a Fourier transform of the autocorrelation function of the crystallite:

$$I(\boldsymbol{q}) = A\left\langle \int F(\boldsymbol{r})e^{i\boldsymbol{q}\cdot\boldsymbol{r}}d^3\boldsymbol{r}\right\rangle , \tag{9.11}$$

where

$$\begin{aligned} F(\boldsymbol{r}) &= V(\boldsymbol{r})^*V(-\boldsymbol{r}) = \int V(\boldsymbol{r}')V(\boldsymbol{r}'-\boldsymbol{r})d^3\boldsymbol{r}' , \\ A &= I_0\left\langle |p|^2\right\rangle N . \end{aligned} \tag{9.12}$$

The averaging is performed over all orientations of the crystallite. A straightforward calculation yields

$$I(q) = \frac{A}{4\pi q}\int_0^\infty r\langle F(r)\rangle \sin(qr)dr , \tag{9.13}$$

where

$$\langle F(r)\rangle = \frac{1}{4\pi}\int F(r)d\Omega$$

is the orientational average of the autocorrelation function.

According to (9.13), the scattered intensity distribution is given by the Hankel transformation of 3rd order of the average $\langle F(r)\rangle$. The inverse Hankel transformation gives

$$\langle F(r)\rangle = \frac{A}{2\pi^2 r}\int_0^\infty qI(q)\sin(qr)dq . \tag{9.14}$$

In principle, this formula allows determination of the autocorrelation function directly from the measured data. Then it can be compared with a theoretical function, for example calculated for spherical crystallites. The autocorrelation function of a spherical crystallite with radius R is given by (see [19], among others)

$$F_R = \langle F(r) \rangle = \frac{4}{3}\pi R^3 \left[1 - \frac{3}{4}\frac{r}{R} + \frac{1}{16}\left(\frac{r}{R}\right)^2\right] H(2R - r) , \qquad (9.15)$$

where $H(x)$ is the Heaviside function ($H(x) = 1$ for $x \geq 1$ and $H(x) = 0$ for $x < 0$).

However, in practice, the influence of vertical divergence must be taken into account. This is much larger in the experiment than the horizontal one. Then, the measured signal Φ is given by an integral of the intensity $I(\boldsymbol{q})$ over the vertical component q_y of the wave vector transfer. Since the vertical angular window of the experiment is much larger than typical angles of small-angle scattering, the integration can be extended to infinity,

$$\Phi(q_x, q_z) = \frac{1}{K} \int_{-\infty}^{\infty} I(q) dq_y , \qquad (9.16)$$

where $I(\boldsymbol{q})$ is given by (9.11). Similarly to (9.13) we can get

$$\Phi(q_{xz}) = \frac{4\pi^2 A}{K} \int_0^{\infty} r \langle F(r) \rangle J_0(q_{xz} r) dr ; \quad q_{xz} = \sqrt{q_x^2 + q_z^2} , \qquad (9.17)$$

where J_0 is the Bessel function of order 0. The measured signal $\Phi(q_1)$ is expressed as a Fourier–Bessel transformation of the averaged shape function $\langle F(r) \rangle$. Then the shape function can be expressed as follows:

$$\langle F(r) \rangle = \frac{K}{8\pi^2 A} \int_{-\infty}^{\infty} |q_1| \Phi(q_1) | J_0(q_1 r) dq . \qquad (9.18)$$

It is useful to introduce a normalized shape function

$$P(r) = \frac{\langle F(r) \rangle}{\langle F(0) \rangle} . \qquad (9.19)$$

$P(r)$ can be evaluated from the measured curves. For spherical blocks, this function equals (see (9.15))

$$P_R(r) = 1 - \frac{3}{4}\frac{r}{R} + \frac{1}{16}\left(\frac{r}{R}\right)^3 . \qquad (9.20)$$

In order to characterize the block size, the *effective block radius* R_{eff} can be defined as the radius of a spherical block having the same volume as the mean volume of the blocks. R_{eff} can be obtained from $P(r)$ by using the formula

$$R_{\text{eff}} = \left[3 \int_0^{\infty} r^2 P(r) dr \right]^{1/3} . \tag{9.21}$$

9.3 Experimental Setup and Evaluation

In order to measure the diffuse scattering, a high-intensity source is required, and the profile corresponding to the incoming wave should be narrow and with very short tails. This condition was realized using a Rigaku rotating anode generator and by means of several reflections of the beam (channel-cut monochromators and crystal analyzer). The experimental setup consisted of a primary germanium monochromator with 111 reflection set active, a slit selecting the Cu$K\alpha_1$ wavelength, after which there was a three-reflection silicon channel-cut (111). Then the incident beam was quite well monochromatic with a narrow angular divergence of 6 arc seconds (FWHM), moreover the intensity of the incident beam decreases 20 arc seconds from the maximum down to 10–4, which is more important, than the FWHM parameter. Angular properties of the beam after the first channel-cut were measured by means of an additional 4-reflection channel-cut on the scanning axis before the detector (see Fig. 9.1).

The sample was placed either between both channel-cuts (position A) or between the second channel-cut and the detector (position B). In position B, the detector gains all the scattered intensity (coherent and diffuse) passing through the sample. The crucial point is, that the angular dependence of the measured intensity should be the angular profile of the primary beam convoluted with the analyzer angle aperture function. The same profile should be observed in the coherent part of the wave passing through the sample in position A. Thus the diffuse part of the wave can be separated by a simple subtraction of an appropriate multiple of the measured intensity in position B from the intensity in position A.

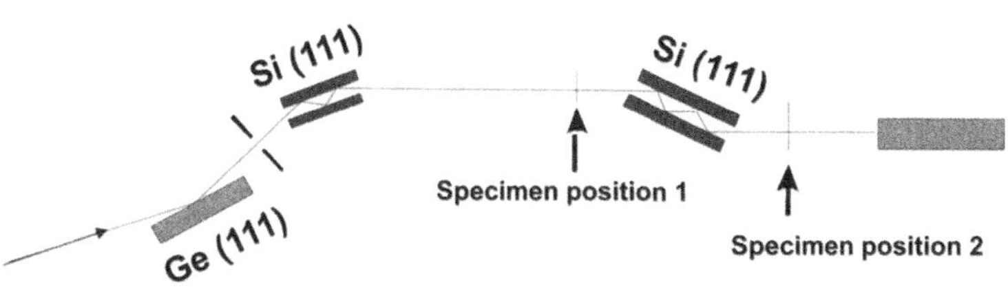

Fig. 9.1. Experimental setup. Rigaku rotating anode, Ge and Si monochromator, Bede detector. Sample is measured in positions A and B, respectively

9.3.1 Determination of the Autocorrelation Function by Transformation of Measured Data

All the above formulas and evaluation procedure could be easily performed for example in the Mathcad environment. Usually, it is not realistic to assume that the materials are made of grains with the same size. Therefore, experimental results (9.19) can be simulated by summation of several components (9.20) or, better, by considering a suitable size distribution. In many cases (e.g. [12, 14]) a log-normal distribution seems to be appropriate:

$$f(R) = \frac{1}{\sqrt{2\pi}\sigma R} \exp\left(-\frac{\ln^2(R/m)}{2\sigma^2}\right), \tag{9.22}$$

where m is the median and σ the log-normal variance of the distribution $(0 < s < 1)$.

9.3.2 Evaluation by the Fitting of Measured Data

The second possibility of evaluation is direct fitting of experimental data by using the following relation:

$$I_{\exp}(\boldsymbol{q}) = \int_{-\infty}^{+\infty} I(\boldsymbol{q}) dq_y = B\left\langle \int_0^{\infty} F_R(r) J_0(q_r r) r dr \right\rangle \tag{9.23}$$

including (9.22). Note that $F_R(r) = (R/R_0)^3 F_{R_0}(rR_0/R)$. The final expression for experimental small-angle diffuse scattering is

$$I_{\exp}(q) = \int_0^{\infty} f(R) \Phi_\lambda\left(\frac{qR}{\lambda}\right)\left(\frac{R}{\lambda}\right)^5 dR\,; \tag{9.24}$$

$$\Phi_\lambda(\xi) = \int_0^{\infty} F_\lambda(r) J_0(\xi r) r dr\,, \tag{9.25}$$

where $q = 4p/\lambda.\sin(\theta)$, and θ is half of the scattering angle (angle of the scanning axis). m and σ can then be determined by fitting the above expression to the data by means of a linearised least squares procedure.

Following the above theory, we have two physical parameters describing the sample – the log-normal distribution parameters m and σ. The median can be related to the mean grain size M as follows

$$M = m.\exp(0.5\sigma^2)\,, \tag{9.26}$$

where σ describes the sharpness of the distribution. The median is less sensitive to long tails of size distribution than the mean. If a single size distribution is not sufficient, an additional log-normal distribution can be added. Then three more parameters are introduced: median, variance and a weight parameter relating the two distributions. It is formally possible to introduce more distributions, but the procedure tends to be unreliable as more fitting parameters are introduced.

The experimental conditions require some more parameters. These are the scanning axis zero (i.e. the maximum position) and the shift of the angular zeros in the sample position A and B. The latter are necessary because the stability of the high precision experiment over the long time necessary for collecting the data is not perfect. Further parameters are the appropriate scaling parameter for the intensity measured in B position with respect to the data collected in position A (scaling two subsequent measurements), background and scale factor (primary intensity).

The data are simulated by least squares and a residual function is calculated using the noise-adjusted weight, which is inversely proportional to measured intensity. The differentials are calculated numerically. Both integrations in (9.24) and (9.25) have to be performed numerically. The Hankel transformation (9.25) is performed only once for a wide enough range and its values for (9.24) are then evaluated by interpolation. Corresponding computer programs were written in the *Matlab* environment.

9.4 Results and Discussion

Samples of pure copper and copper with aluminum oxide were subjected to high-pressure (6 GPa) torsion deformation. The samples were prepared at the Ufa Aviation State University [20] in the form of thin-foils of diameter of about 1 cm and thickness of about 0.25 mm. Some of them were thinner.

It follows from the description of the experiment that only thin foil samples are suitable for transmission measurements. In previous experiments [5] with powder samples, it was shown that small-angle scattering on inhomogeneities is dominating and prevents the application of the method for crystallite size determination. The same is true for highly porous polycrystalline samples. Therefore compact foils are necessary. Even in that case, however, appropriate thickness is required because if samples are too thick there is no measurable signal and if too thin the effect of diffuse scattering is too weak to be measured. It was found that the optimal thickness range for copper is between about 0.1 mm and 0.15 mm. Of course, this information is strongly connected with the used arrangement and especially the 18 kW rotating anode. Synchrotron radiation would offer much greater flexibility, both because of the higher intensity of incident beam and because of the possibility of selecting the proper wavelengths for low sample absorption.

Due to the mentioned limitations only a few measured samples could be analysed. Some of the samples were etched in order to get smaller thickness.

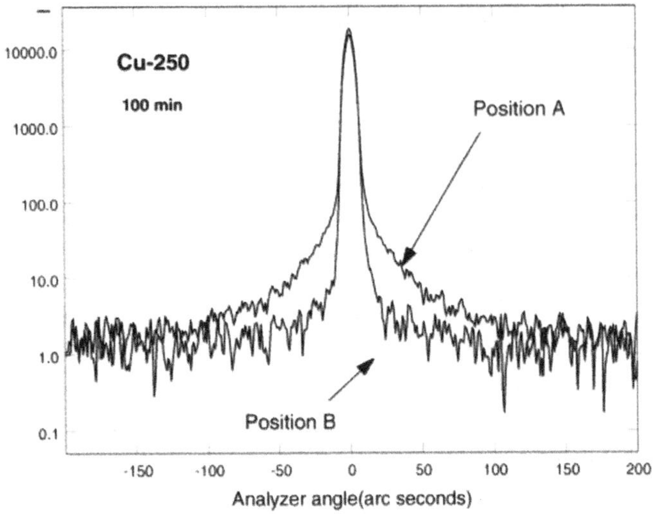

Fig. 9.2. Measured profiles (collected in two positions, see Fig. 9.1) for a Cu sample deformed under 6 GPa pressure by torsion plastic deformation and then annealed at 250°C for 100 min

Finally, the reasonable and interesting results were obtained for the deformed copper samples annealed at 250°C for 4 s (250–4) and 100 min (250–100), respectively. The short annealing is used for the investigations of recovery dynamics.

Typical results are shown in Fig. 9.2. Experimental profiles were measured at positions A and B, respectively. Intensities are normalized and shown in logarithmic scale. The following step is the separation of diffuse scattering. This cannot be done correctly in the central part of the profile where the diffuse scattering is much less than the transmitted radiation. Fortunately, final results are nearly completely determined by the tails. Central part is determined mainly by the instrumental and spectral component and by large particles which cannot be studied by the method. Normalized shape of the diffuse scattering is shown in Fig. 9.3 for two samples and corresponding normalized shape function P (9.19) is plotted in Fig. 9.4. This is critical point of the evaluation procedure, as the result is strongly affected by profile truncation and especially sensitive to a correct determination of the background level. See for comparison two curves calculated with different truncation of the experimental profile.

As expected, calculation of the shape functions for spheres showed a poor agreement with the experimentally obtained P. The sum of two or three components can give a better agreement (Fig. 9.5a,b). In any case, there is an indication of a broad size distribution, that we represented as a log-normal function 9.22. Fitting results are shown in Fig. 9.6. For the sample annealed for longer time the agreement is not good, and for the other annealed sample

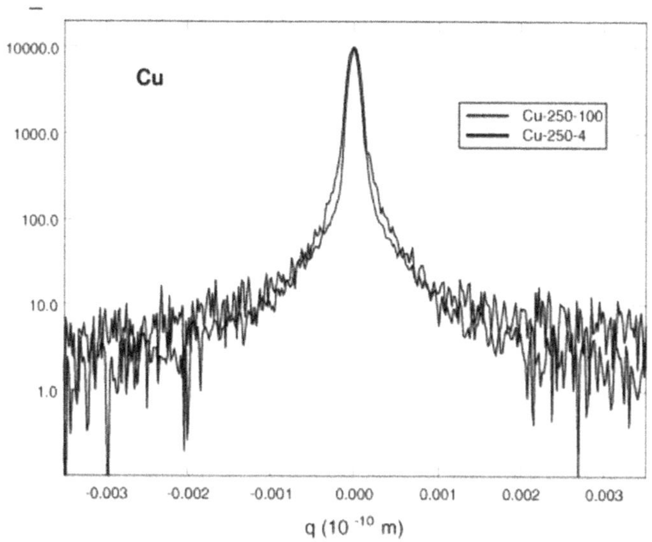

Fig. 9.3. Diffuse scattering in the transmitted wave obtained from measurements shown in Fig. 9.2 plotted against $\boldsymbol{q}$-vector (thin) and compared with the scattering measured on another sample annealed for 4 s

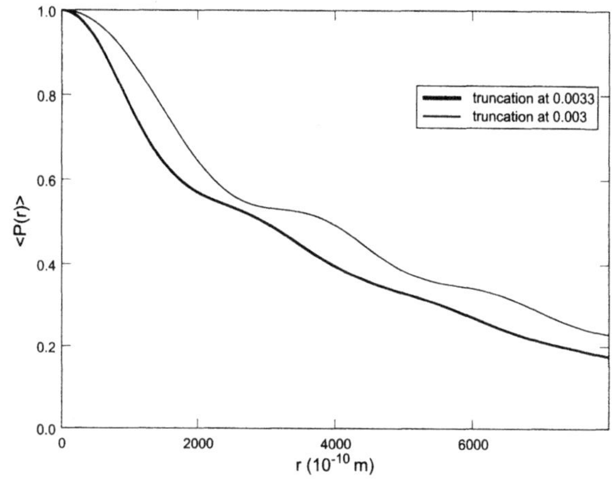

Fig. 9.4. Normalized shape function calculated from the diffuse scattering (Fig. 9.3, sample annealed for 100 min) with two different truncations of the profile

the shape function cannot be described by a single log-normal distribution at all (Fig. 9.7). Then at least two distributions must be introduced.

The same indication was obtained by the second method – fitting of the measured profile. Therefore, a bimodal block size distribution was assumed. Fitting result with two log-normal distributions is shown in Fig. 9.8a,b. Often, one component was rather sharp, as shown in Fig. 9.9. A markedly nonuni-

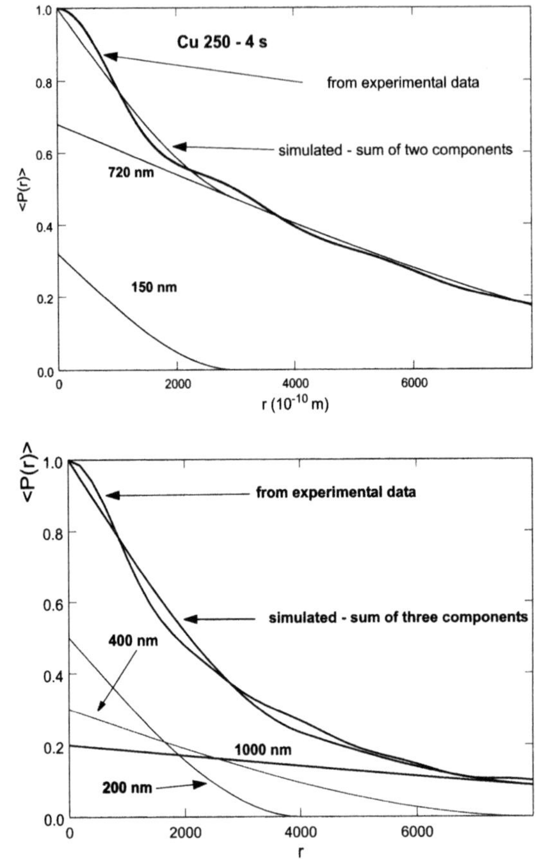

Fig. 9.5. Normalized shape functions obtained from the diffuse scattering shown for samples annealed for 4 s (**a**) and 100 min (**b**). They are compared with the simulated curves calculated for the sum of contributions from spheres of two (**a**) and three (**b**) different radii, respectively, as indicated in the graphs

form distribution was also confirmed by the classical film method (Fig. 9.10), as well as by the long tails (super-Lorentzian) of conventional diffraction profiles.

The results are in agreement with previous TEM studies showing that after short annealing, when the strains are partially relaxed, some of the grains grow rapidly.

The large dots in some of the back-reflection photos indicate the presence of very large grains, of the size of tens of micrometers. Such grains cannot be studied neither by the present diffuse scattering method. Therefore, the classical profile analysis, diffuse scattering in the transmitted wave and back-reflection photos are three methods which are – to some extent – complementary.

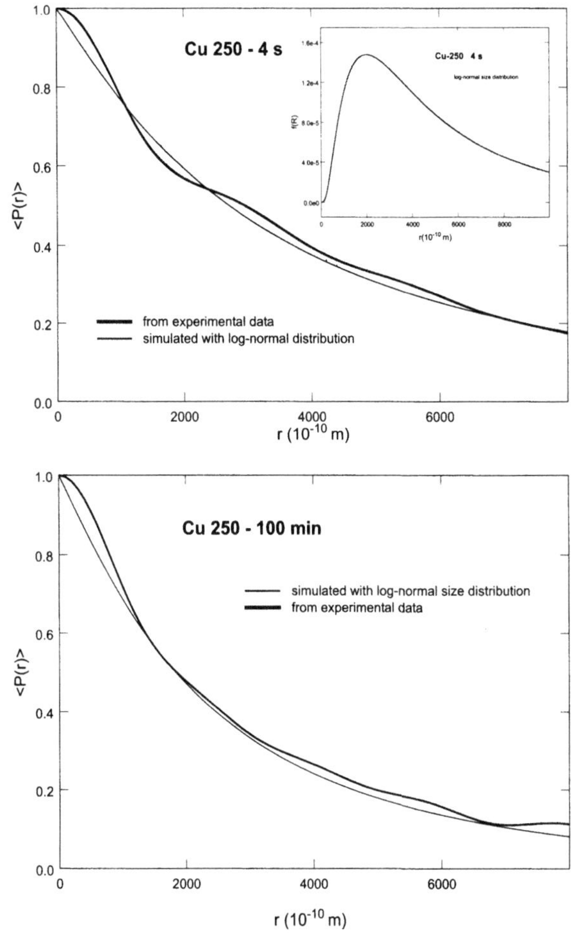

Fig. 9.6. Normalized shape functions obtained from the diffuse scattering shown for samples annealed for 4 s (**a**) and 100 min (**b**). They are compared with the log-normal distributions of sphere radii ($m = 450$ nm, $s = 0.9$ for 250–4; $m = 300$ nm, $s = 0.8$ for 250–100). The very broad distribution for the sample 250–4 (**a**) is shown in the inset

The comparison between two annealed samples investigated cannot be done well. The effective values of the block size are very close for the short-annealed sample (490 nm) and the annealed one (250–100, 498 nm). However, the distribution is broader for the former, which can be surprising. Nevertheless, better sample statistics should be achieved before drawing definite conclusions. Moreover, as found in previous studies of thin foils, the method "sees" larger crystallites than profile analysis [5]. On the other hand, the method is not sensitive to very large ones (tens of µm) which are clearly more present in the long-annealed sample, as shown by the back-reflection method.

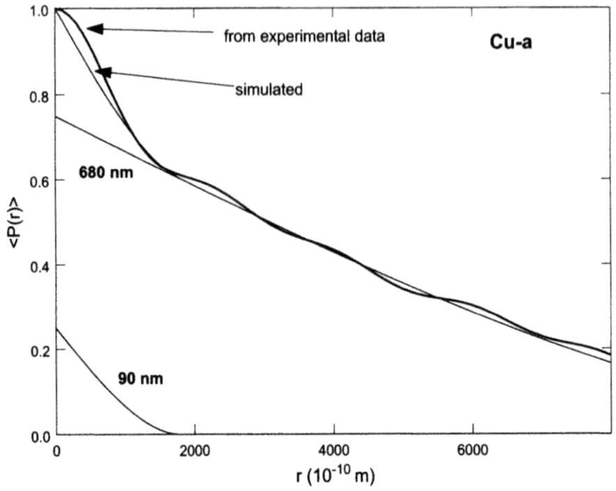

Fig. 9.7. Normalized shape function obtained from the diffuse scattering for an annealed sample different from that shown in Figs. 9.2–9.6. It is compared with the simulated curves calculated for the sum of contributions from spheres of two different radii. The bimodal distribution is a more appropriate model than in previous cases: the shape function cannot be modeled by a single log-normal distribution

It must be emphasized that these studies on submicrocrystalline materials are preliminary. Unfortunately, only a few samples could be studied up to now mainly because of the very narrow range of thicknesses that can be investigated. It would be desirable to measure compact samples with really small crystallites, below 100 nm, where the comparison with classical line profile analysis can be made.

In contrast to a conventional powder diffraction method, the method presented here uses an analyzing crystal for the determination of the direction distribution of the scattered wave. The angular resolution of an analyzer crystal is determined by the width of its intrinsic dynamical diffraction curve; if we neglect the wave dispersion, the angular resolution can reach few arc sec. Therefore, the method presented here is suitable for the measurement of diffuse scattering from large crystallites. The high sensitivity to larger crystallites can be seen also from relation (9.24), where the R^5 dependence is shown.

From the two applied methods – Hankel transformation of measured profile and direct profile fitting, the latter seems to be less sensitive to the noise and to a correct background determination. On the other hand, direct profile fitting is more time consuming. Additional fit parameters like zero-position corrections do not correlate with the experimental ones, since no asymmetry is given by the theory. The background level correlation is removed by measurement in a wide angular range, where the intensity of diffuse scattering drops to zero. Both the background and the scale factors are linear parameters, which make no complication for the least squares method. The

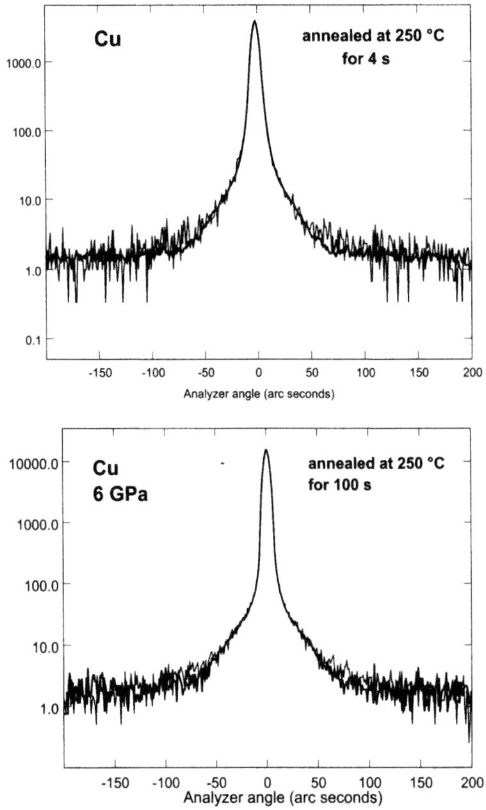

Fig. 9.8. Experimental profiles measured in sample position A for the samples 250–4 (**a**) and 250–100 (**b**), respectively. They are compared with the curves reconstructed from the fitting (23), (24) procedure

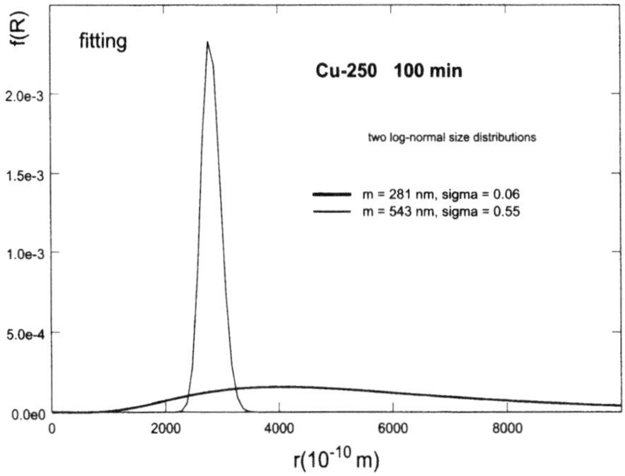

Fig. 9.9. Two sphere radius distributions obtained by the fitting procedure for sample 250–100

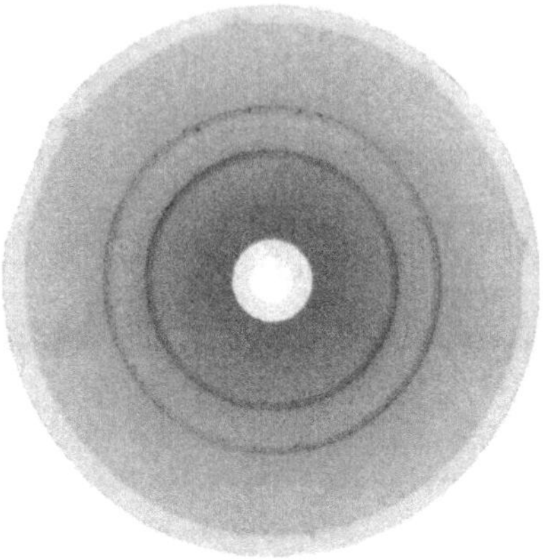

Fig. 9.10. Typical back-reflection photo for the annealed Cu samples. Clear dots on the background of continuous Debye rings indicate the presence of larger grains, leading to a broad (in some cases bi-modal) crystallite size distribution

only troublesome may be the proper scale correction between the data measured in positions A and B, respectively. Since the coherent wave can be also simulated by the diffuse scattering on very large grains, the scale correction can drop to a value smaller than one (it should be comparable with 1). This is the experimental limit caused by the non-zero angular divergence of the primary beam. The possible long tail of the distribution (9.22) representing very large grains cannot be determined well in principle and therefore it may be omitted in integration (9.24).

9.5 Conclusions

This chapter presents a new method of crystallite size-distribution determination from small angle scattering experiments on the basis of primary extinction in diffracting grains causing diffuse small angle scattering. Thereby it can be applied to massive polycrystalline specimens where no large differences in electron density occur. The method as presented relies on two main assumptions: (i) it is assumed that the positions of the crystallites are randomly distributed and (ii) the average distance between two diffracting crystallites should be much larger than the average crystallite size.

When the experimental setup is prepared and carefully aligned, the method is extremely simple and fast. A measurement on a typical sample can be done in a few minutes. The main problem is the high sensitivity of the high-resolution setup to thermal and mechanical instabilities. A further disadvantage is that this type of analysis can be used only for specimens having the form of thin compact foils of a certain thickness. In powders and porous materials the scattering by inhomogeneities is dominating, and it does not allow the determination of crystallite size. However, the method can be useful and deserves attention at least as a complementary technique to the traditional procedures.

Acknowledgement

This work was partially supported by The Grant Agency of Czech Republic (106/02/1521) and Grant Agency of Charles University (187/2001).

References

1. J. Čížek, I. Procházka, R. Kužel, R.K. Islamgaliev: Monatshefte für Chemie, Vol. **133**, No. 6, pp. 873–887, (2002)
2. R. Kužel, J. Čížek, I. Procházka, F. Chmelík, R.K. Islamgaliev, N.M. Amirkhanov: Materials Science Forum **378–381** (2), 463 (2001)
3. J. Čížek, I. Procházka, O. Melikhova, G. Brauer, W. Anwand, R. Kužel, M. Cieslar, R.K. Islamgaliev: Applied Surface Science, Vol. **194**, No. 1–4, pp. 140–144, (2002)
4. J. Čížek, I. Procházka, M. Cieslar, R. Kužel, J. Kuriplach, F. Chmelík, I. Stulíková, F. Bečvář, O. Melikhova, R.K. Islamgaliev: Phys. Rev. B **65**, 094106 (2002)
5. M. Čerňanský, V. Holý, J. Kub, J. Kubena: *Materials Structure.* Vol. 5, Special Issue B, (1998) p. 201 and to be published
6. B.E. Warren: *X-ray diffraction.* (Reading, Massachusetts. 1969)
7. H.P. Klug, L.E. Alexander: *X-ray diffraction procedures for polycrystalline and amorphous materials.* (John Wiley and Sons, New York 1974)
8. M.A. Krivoglaz: *X-ray and neutron diffraction in nonideal crystals.* (Springer, Berlin Heidelberg New York 1996)
9. *Defect and Microstructure Analysis by Diffraction.* Eds. R. Snyder, J. Fiala and H. Bunge (Oxford University Press 1999)
10. D. Breuer, P. Klimanek, W. Pantleon: J. Appl. Cryst. **33**, 1284 (2000)
11. T. Ungár, I. Dragomir, A. Révész, A. Borbély: J. Appl. Cryst. **32**, 992 (1999)
12. T. Ungár, J. Gubicza, G. Ribárik, A. Borbély: J. Appl. Cryst. **34**, 298 (2001)
13. G. Ribárik, T. Ungár, J. Gubicza: J. Appl. Cryst. **34**, 298 (2001)
14. P. Scardi, M. Leoni: Acta Crystallographica, A **58**, 190 (2002)
15. P. Scardi, M. Leoni, Y.H. Dong: Eur. Phys. J. B **18**, 23 (2000)
16. A. Guinier, G. Fournet: *Small-angle scattering of X-rays.* (John Wiley, New York 1955)
17. *Small angle X-ray scattering.* Ed. O. Glatter and O. Kratky (Academic Press, London 1982)
18. M. Born and E. Wolf: *Principles of Optics: Electromagnetic Theory of Propagation, Interference and Diffraction of Light.* 7th edn (Cambridge University Press, Cambridge 1999)
19. L.V. Azaroff, R. Kaplow, N. Kato, R.J. Weiss, A.J.C. Wilson, and R.A. Young: *X-Ray Diffraction* (McGraw-Hill, New York 1974)
20. R.Z. Valiev: Mat. Sci. Eng. A **234–236**, 59 (1998)

Part IV

Dislocations and Stacking Faults

10 Determining the Dislocation Contrast Factor for X-ray Line Profile Analysis

N. Armstrong and P. Lynch

Dedicated to Professor R.W. Cheary
for teaching and inspiring us to tackle this problem

An important step in quantifying dislocation broadening of a line profile is the evaluation of the *dislocation contrast factors*. In this way the elastic properties of the material, displacement field of the dislocations and geometric orientation of the diffraction vector relative to the dislocation slip-system are incorporated into the analysis.

We present a detailed description of how to carry out a first-principle evaluation of the contrast factors for a range of industrially important materials. This requires an understanding of the displacement fields and slip systems of dislocations in elastically anisotropic materials. This approach enables the contrast factors and Fourier coefficients for single crystal, textured and poly crystalline specimens to be determined.

In this discussion, we also present detailed examples of the evaluation of the contrast factors for a range of face-centred and body-centred cubic materials. An atypical case is also presented which demonstrates the need of a first-principles approach to determining the contrast factors.

10.1 Introduction

The shape and broadening of X-ray line profiles provides microstructural information about material specimens which have important industrial applications (e.g. nanopowders, catalysts and computer chip interconnectors). This information includes: the shape, size and distribution of crystallites in a 5–100 nm range; lattice faulting/twinning; the type, spatial arrangement and density of dislocations in the material specimen. This microstructural information is generally convoluted or "blurred" together in the recorded line-profile and can be grouped into *size-* and *distortion*-broadening contributions [1]. Using the recorded line profile as the starting point, line profile analysis attempts to "work backwards" to quantify the microstructural state of the material specimen.

Modern theoretical developments in X-ray line profile analysis have seen a shift from a phenomenological description of line profile broadening [2–6] to theories which explain the broadening in terms of the microstructural state of the material specimen [7–14]. In turn, methods have been developed to relate the size-broadening of line profiles to the size, shape and distribution of nanocrystallites [15–23]. Also, dislocation broadening has been related to the type, spatial arrangement and density of dislocations in the material specimen [20,21,23–29]. Consequently, these developments have made X-ray line profile methods comparable to electron microscopy methods (i.e. TEM and SEM) in terms of the microstructural information they can provide.

The following discussion concentrates specifically on dislocation broadening and on how to evaluate the *dislocation contrast factor* for face-centred- and body-centred-cubic structures. Evaluating this factor is critical in determining the dislocation density from dislocation broadened line profiles. That is, this factor accounts for the influence of dislocations in broadened line profiles by including the elastic properties of the material, displacement fields and slip-system of the dislocations [8–10,24,28,30–35]. The combination of these components results in the contrast factor describing the "visibility" or broadening of the recorded line profile due to a dislocation. Furthermore, it accounts for the anisotropic broadening in the Fourier coefficients and integral breadths of dislocation broadened line profiles, by including the elastic anisotropic nature of most real materials. In turn, this observation enables the Fourier coefficients and integral breadth of the diffraction profile to be "unscrambled" by defining the scaling parameter $d^{*2}_{\mathrm{hkl}}\langle C_{\mathrm{hkl}}\rangle$ for the Fourier coefficients and $d^{*}_{\mathrm{hkl}}\sqrt{\langle C_{\mathrm{hkl}}\rangle}$ for the integral breadths [24].

Although there exist parameterised methods for determining the average contrast factor for cubic and hexagonal lattices [28,34,35], we demonstrate that a first-principles approach requires an understanding of dislocations in real (i.e. elastically isotropic and anisotropic) materials. Our evaluation of the contrast factors for a range of cubic materials follows the approach developed by Klimanek and Kužel [31] and Kužel and Klimanek [32], who carried out the contrast factor calculations for hexagonal materials. This approach is general and can be applied to any lattice symmetry.

Specifically, we consider the first-principles calculations of the contrast factor, C_{hkl}, for a single slip-system. This approach is critical in analysing the dislocation line-broadening of textured materials [29,36] and can be used to determine the average contrast factor value, $\langle C_{\mathrm{hkl}}\rangle$ for powder specimens. We present examples of the evaluation of the elastic and geometrical components of the contrast factor values for the $\{1\bar{1}1\}\langle 110\rangle$ and $\{110\}\langle\bar{1}11\rangle$ slip-systems for real (i.e. elastically anisotropic) fcc and bcc materials, respectively. The elastic component describes the distortion due to the dislocation, while the geometric component defines the orientation of the scattering vector in the dislocation slip system.

Our discussion begins with an overview of X-ray line profiles (Sect. 10.2), focusing on the need to determine the contrast factor in order to quantify the dislocation broadening of line profiles (see Sect. 10.2.2). In Sect. 10.3, the theoretical background for determining the C_{hkl} is presented by introducing the elastic and geometric components of the contrast factors, defining the important slip-systems for fcc and bcc materials (see Sect. 10.3.1). In

Sect. 10.3.2, a detailed discussion of the properties of screw and edge dislocations is presented for elastically anisotropic fcc and bcc materials. Specific numerical calculations for the $C_{\rm hkl}$ in fcc and bcc materials are presented in Sect. 10.4. The numerical calculations concentrate on determining the $C_{\rm hkl}$ and $\langle C_{\rm hkl}\rangle$ values, for both pure screw and edge dislocations in gold (an fcc example) and in α-iron (a bcc example), while the $C_{\rm hkl}$ and $\langle C_{\rm hkl}\rangle$ value, for both pure screw and edge dislocations, are also given for Ag, Pt, Mo and Nb (Sects. 10.4.1 and 10.4.2). To highlight the importance and advantages of the first-principles approach for determining the contrast values, the atypical case of Cu_2O is presented [37] (see Sect. 10.4.3).

10.2 X-ray Line-Profile Broadening

10.2.1 Quantifying the Line-Broadening

The experimental X-ray diffraction measurement of a material containing nanocrystallites, lattice faulting/twinning and dislocations results in broadened and asymmetrical line profiles, $f(s)$. Let us assume a diffraction intensity (in reciprocal space) is given by the function, $\bar{I}(\boldsymbol{q})$, defined in terms of a set of Cartesian coordinates $\boldsymbol{q} = [q_1, q_2, q_3]$ and centred at the diffraction vector, $\boldsymbol{g}$, such that $|\boldsymbol{g}| = d^*_{\rm hkl} = 1/d_{\rm hkl}$. The coordinate system is orientated such that q_1 is parallel to $\boldsymbol{g}$ [38]. The recorded line broadening profile (not including instrumental effects) can be expressed as [38,39]

$$f(s) = \int \bar{I}(\boldsymbol{q})\,\mathrm{d}A^s \tag{10.1a}$$

$$\approx \int \bar{I}(\boldsymbol{q})\,\mathrm{d}q_2\mathrm{d}q_3\,, \tag{10.1b}$$

where $\mathrm{d}A^s$ is the area element of a sphere of radius $d^*_{\rm hkl}+s$ about the origin of diffraction and can be approximated be the area element of a plane, $\mathrm{d}q_2\mathrm{d}q_3$, normal to $\boldsymbol{g}$. The reciprocal space units, s, are defined as

$$\begin{aligned} s &= |\boldsymbol{g}-\boldsymbol{q}| - |\boldsymbol{g}| \\ &= \frac{2}{\lambda}(\sin\theta - \sin\theta_0)\,, \end{aligned} \tag{10.1c}$$

where λ is the X-ray wavelength and $2\theta_0$ is the Bragg angle.

The shape and breadth of $f(s)$ characterises the microscopic state of the material specimen. The three broadening mechanisms, viz. lattice faulting/twinning, crystallite size and dislocation broadening contributions, are generally convoluted together

$$f(s, d^*_{\rm hkl}) = \int f^{ft}(s-s', d^*_{\{\rm hkl\}}) \int f^{\rm size}(s'-s'') f^{\rm disl}(s'', d^*_{\rm hkl})\,\mathrm{d}s''\,\mathrm{d}s'\,, \tag{10.2}$$

where $f^{ft}(s, d^*_{\{hkl\}})$ is the "faulting/twinning profile" for a set of {hkl}, $f^{\rm size}(s)$ is the "crystallite size profile" and $f^{\rm disl}(s, d^*_{\rm hkl})$ is the "dislocation profile" for a given $d^*_{\rm hkl}$. Faulting/twinning is dependent on the set of $\{hkl\}$, and on the faulting/twinning probabilities (see [1]). This component will result in an effective domain size broadening (i.e faulting) and line profile asymmetry (i.e twinning) (see [1]). The crystallite size broadening is $\sim 1/\langle t\rangle$ in s-space or $\sim \lambda/\langle t\rangle \cos\theta_0$ in 2θ-space, where $\langle t\rangle$ is the average thickness of the crystallite in the direction of $\boldsymbol{g}$ [2]. Also, crystallite size broadening is order-independent. On the other hand, the dislocation broadening is order-dependent and varies as $\sim \sqrt{\langle\epsilon^2_{\rm hkl}\rangle}\tan\theta$ in 2θ-space, where $\sqrt{\langle\epsilon^2_{\rm hkl}\rangle}$ is the root-mean-square strain in the direction $\boldsymbol{g}$ [3]. In quantifying the broadening of the line profile, we consider the integral breadth and the Fourier coefficients.

The integral breadth $\beta(d^*_{\rm hkl})$ of $f(s)$ (defined in s-space) is given as,

$$\beta(d^*_{\rm hkl}) = \frac{1}{f_{max}} \int_{-\infty}^{+\infty} f(s, d^*_{\rm hkl})\,{\rm d}s\,, \tag{10.3}$$

where $f_{\rm max}$ is the maximum intensity for the profile $f(s, d^*_{\rm hkl})$ for a given hkl. The Fourier coefficients represent the Fourier transform of $f(s)$.

In Fourier space, the convolution in (10.2) is expressed as [1,23]

$$A(t, d^*_{\rm hkl}) = A^{\rm size}(t)\, A^{\rm disl}(t, d^*_{\rm hkl})\, A^{ft}(t, d^*_{\{\rm hkl\}})\,. \tag{10.4}$$

The separation of these broadening contributions is generally treated by the Warren–Averbach and Williamson–Hall methods [4–6]. However, these methods and the various derivatives of them (e.g. [24]) work over a small range of $t \lesssim 10\,$nm. Recent developments have seen whole pattern modelling, using (10.4), while fitting the calculated pattern to the experimental pattern [20, 21, 23, 40]. However, the development of a general method that determines each contribution for all orders of t and over the whole pattern, while addressing the uniqueness issues, remains one of the outstanding theoretical problems in X-ray line profile analysis.

In providing an overview of the broadening contributions, the discussion now turns to quantifying the dislocation broadening in the line profiles.

10.2.2 Quantifying the Dislocation Broadening

The dislocation broadening of X-ray line-profiles is dependent on $\boldsymbol{g}\cdot\boldsymbol{u}$, where $\boldsymbol{u}$ is the displacement field of the dislocations. This is analogous to the visibility condition of dislocations in transmission electron microscopy (TEM) [41]. The "visibility" in the case of X-ray diffraction is the dislocation broadening of the line-profiles.

Based on the previous discussion, broadening can be quantified using integral breadth and Fourier coefficients. The integral breadth arising from dislocations can be expressed as

$$\beta(d^*_{\mathrm{hkl}}) \propto b\, d^*_{\mathrm{hkl}} \sqrt{C_{\mathrm{hkl}}\rho}\,, \tag{10.5}$$

where C_{hkl} is the contrast factor of the dislocation, and $b = |\boldsymbol{b}|$ is the length of the Burgers vector; ρ is the *total* dislocation density. We note in (10.5) there exist differing definitions for the proportionality term (e.g. compare [8, 10]). The dislocation Fourier coefficients, $A^{\mathrm{disl}}(t, d^*_{\mathrm{hkl}})$, for a single-slip system can be expressed as [8, 9],

$$A^{\mathrm{disl}}(t, d^*_{\mathrm{hkl}}) = \exp\left[-\frac{\pi}{2}\, d^{*2}_{\mathrm{hkl}}\, b^2 C_{\mathrm{hkl}} \rho_j t^2 f^*(\eta)\right], \tag{10.6}$$

where t is the Fourier length parallel to the diffraction vector $\boldsymbol{g}$, R_e is the "outer cut-off radius" and ρ_j is the jth dislocation density, such that the total dislocation density, ρ is defined by $\rho = \sum_{j=1}^{N} \rho_j$. The $f^*(\eta)$ term in (10.6) ensures that the Fourier coefficients can be evaluated for all t. A full discussion and derivation of the $f^*(\eta)$-function is given in Wikens [8] (see Appedix 1, A.8, p. 1213) and the definition of η in Wilikens [9] (see (2.9), p. 361).

From (10.5) and (10.6), we see that the principal term to be evaluated in order to quantify the dislocation broadening in terms of ρ, is the dislocation contrast factor, C_{hkl}. This term describes the visibility of the dislocation with respect to the displacement field, $\boldsymbol{u}$, the slip-coordinate system of the dislocation, and the diffraction vector, $\boldsymbol{g}$. The evaluation of C_{hkl} requires an understanding of the displacement fields, $\boldsymbol{u}$, the types of dislocation and their slip-system for a given lattice. We now examine each of these terms and show by example how the calculations are performed.

10.3 Dislocation Contrast Factors, C_{hkl}

10.3.1 Evaluating the Contrast Factor, C_{hkl}

The contrast factor for screw and edge dislocations in elastically isotropic materials can be determined analytically and expressed as a function of the Miller indices[1] h, k, l and Poisson ratio, ν (for edge dislocations) [8, 25, 31]. In this case the standard solutions for edge and screw dislocations can be used (see [42]). However, for the case of elastically anisotropic materials, the contrast factors need to be numerically evaluated [30–32]. This evaluation can be split into two components: the geometric component, $\mathbf{G} \equiv [G_{KL}]$, which defines the orientation of $\boldsymbol{g}$ in the slip-system of the dislocation; and the elastic component, $\mathbf{E} \equiv [E_{KL}]$, which defines the distortion due to a dislocation. Taking these two components into consideration, the contrast factor for a single slip-system is defined as [31, 32]

$$C_{\mathrm{hkl}} = \sum_{K,L=1}^{6} G_{KL} E_{KL}\,. \tag{10.7}$$

[1] For notational reasons, the Miller indices are given by h, k and l, while the indices used in the equations are given by i, j, k and l.

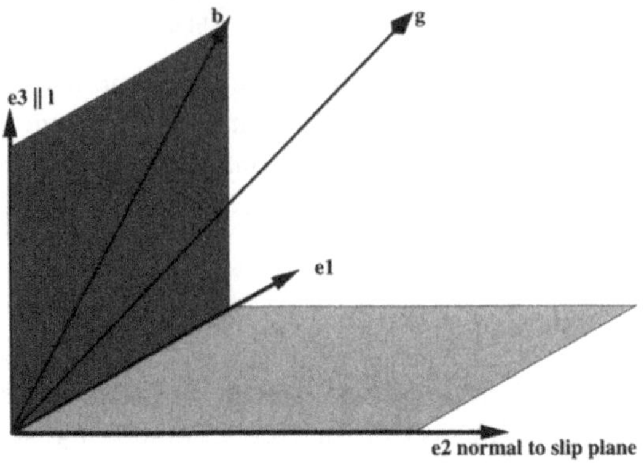

Fig. 10.1. A slip system defined by the unit vectors $\boldsymbol{e}_1$, $\boldsymbol{e}_2$ and $\boldsymbol{e}_3$ and the orientation of the scattering vector, $\boldsymbol{g}$. $\boldsymbol{e}_3$ is parallel to the line vector of the dislocation, $\boldsymbol{l}$. $\boldsymbol{e}_2$ is the unit vector normal to the slip plane. The Burgers vector, $\boldsymbol{b}$, is defined in the plane formed by $\boldsymbol{e}_2$ and $\boldsymbol{e}_1$. The orientation of the Burgers vector defines the cases for screw (10.13) and edge (10.14) dislocations. (Adapted from [31])

In this result a single slip-system is defined by three unit vectors, $\boldsymbol{e}_1$, $\boldsymbol{e}_2$ and $\boldsymbol{e}_3$, and their orientation with respects to the Burgers vector, $\boldsymbol{b}$, line-direction of the dislocation, $\boldsymbol{l}$, and the normal to the slip-plane. Figure 10.1 schematically illustrates a single-slip system.

Important qualitative information can be gained by expressing the contrast factor in the form of (10.7). For example, if we consider an elastically isotropic material of a given cubic lattice containing pure screw dislocations, the geometric component is dominant, while the elastic components will be equal to unity, such that $E_{33} = E_{66} = 1$ [31]. Conversely, for the same lattice and elastically anisotropic form of the same material, the geometric component will remain unchanged, while the elastic component introduces the anisotropy which in turn influences the contrast factor. Translating this to the integral breadths and Fourier coefficients of the line profiles, the elastic anisotropy causes these quantities to be "scrambled" as d^*_{hkl} increases, while for the isotropic case the integral breadths and Fourier coefficients will vary systematically over the same d^*_{hkl} range.

In the case of a crystal lattice consisting of a single-slip system, such as an fcc material, the contrast factors for all the slip-systems can be averaged [31, 32],

$$\langle C_{\mathrm{hkl}} \rangle = \frac{1}{N} \sum_{K,L=1}^{6} \sum_{i=1}^{N} G^{i}_{KL} E_{KL} \tag{10.8a}$$

$$= \sum_{K,L=1}^{6} \langle G_{KL} \rangle E_{KL}\,, \tag{10.8b}$$

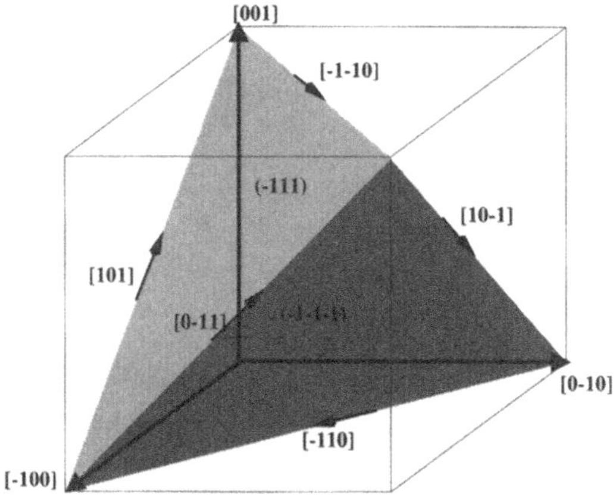

Fig. 10.2. A tetrahedron embedded in a cubic lattice illustrates the $\{1\bar{1}1\}\langle 110\rangle$ slip-system in fcc lattices

where G^i_{KL} is the geometric component of the ith slip-system and is averaged over the total number of slip-systems, N. The result (10.8b) can also be expressed as a function of the hkl. In the case of cubic lattices, (10.8b) is a linear equation [3, 33],

$$\langle C_{\mathrm{hkl}}\rangle = \mathcal{A} + \mathcal{B}\,\Gamma(\mathrm{h,k,l}) \,, \tag{10.9}$$

where $\Gamma(\mathrm{h,k,l}) = (\mathrm{h^2k^2 + h^2l^2 + k^2l^2})/(\mathrm{h^2 + k^2 + l^2})^2$. In (10.9), $\mathcal{A}$ and $\mathcal{B}$ are constants that depend on the elastic and geometric components of the contrast factor. In the case of $\Gamma(\mathrm{h},0,0) = 0$, $\langle C_{\mathrm{h00}}\rangle = \mathcal{A}$. It is interesting to note that Ungár et al. [28] express (10.9) in the form of $\langle C_{\mathrm{hkl}}\rangle = \langle C_{\mathrm{h00}}\rangle\,[1 - q\Gamma(\mathrm{h,k,l})]$, where $q = \mathcal{B}/\langle C_{\mathrm{h00}}\rangle$. This representation allows $\langle C_{\mathrm{hkl}}\rangle$ to be parameterized by q, c_{44}/c_{12}-ratio and by the anisotropy factor or Zener constant, A [43]

$$A = \frac{2c_{44}}{c_{11} - c_{12}} \,. \tag{10.10}$$

The number of slip-systems in an fcc lattice, for example, is 12 (i.e. 4 slip-planes × 3 slip-directions per plane). Figure 10.2 shows a tetrahedron embedded in a cubic lattice. Each face of the tetrahedron represents a slip-plane for dislocations, while its edges represent the possible slip directions.

Equation (10.8a,10.8b) also defines the contrast factor over a selected number of slip-systems corresponding to a particular slip-plane, such as the $\{\bar{1}\bar{1}\bar{1}\}$ slip-plane, where there are the $[\bar{1}10]$, $[0\bar{1}1]$ and $[10\bar{1}]$ slip-directions [29, 36]. This plane is also depicted in Fig. 10.2. In (10.8a,10.8b), the elastic component, $\mathbf{E}$, is invariant over all N slip-systems, for a particular dislocation type.

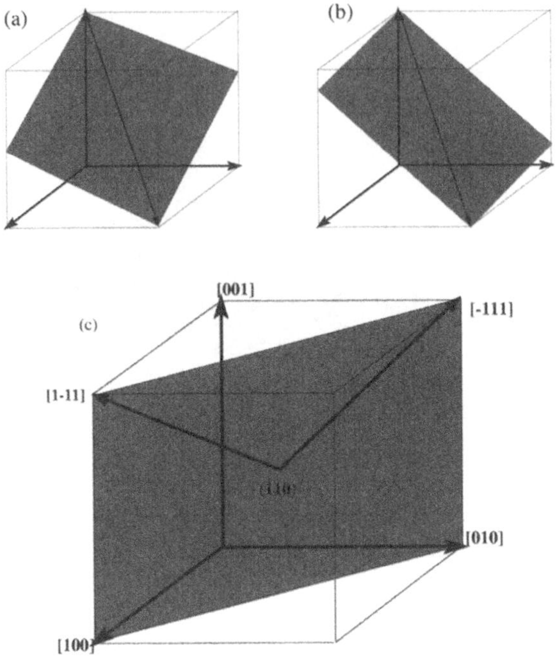

Fig. 10.3. Schematic representation of the multiple slip-systems for bcc lattices. (**a**) The $\{211\}\langle\bar{1}11\rangle$ slip-system. (**b**) The $\{321\}\langle\bar{1}11\rangle$ slip-system. (**c**) The $\{110\}\langle\bar{1}11\rangle$ slip-system which is commonly observed in bcc lattices

The most general case for the contrast factor involves a lattice with multiple slip-systems, such as bcc[2] or hcp lattices (see [31]). In this case, the contrast factor value is weighted by the dislocation density for each slip-system [31, 32]

$$\langle C_{\mathrm{hkl}}\rangle = \sum_{j=1}^{M} \sum_{K,L=1}^{6} \frac{\rho_j}{\rho} \langle G_{KL}\rangle^j E^j_{KL} , \tag{10.11}$$

where ρ_j is the dislocation density for the jth slip-system, ρ is the total dislocation density, $\langle G_{KL}\rangle^j$ and E^j_{KL} are the average geometric and elastic components for the jth slip-system, respectively. Figure 10.3 shows a schematic of the multiple slip-systems for a bcc lattice. Figures 10.3a and b show the $\{211\}\langle\bar{1}11\rangle$ and $\{321\}\langle\bar{1}11\rangle$ systems, respectively. Figure 10.3c shows the $\{110\}\langle\bar{1}11\rangle$ system, where there is a total of 12 slip-systems (6 planes $\times$ 2 slip-direction per plane) and will be of principal concern in this discussion.

The above discussion for the contrast factors has assumed that the dislocations are either pure edge or pure screw, and has not considered mixed

[2] For bcc there are 12 slip-systems for the $\{211\}\langle\bar{1}11\rangle$ and 24 for the $\{312\}\langle\bar{1}11\rangle$ slip-systems. Here we only consider the $\{110\}\langle\bar{1}11\rangle$ slip-system.

dislocations. Generally this case is treated by a weighed sum of the edge and screw contrast factors (see [24]), given as

$$C_{\mathrm{hkl}} = \xi C_{\mathrm{hkl}}^{\mathrm{edge}} + (1-\xi) C_{\mathrm{hkl}}^{\mathrm{screw}} , \tag{10.12}$$

where $\xi \in [0, 1]$ defines the "relative population" of dislocations. This approach assumes that the screw and edge dislocations are independent and do not interact.

Wilkens [30] showed that for the case of mixed dislocations there exists an *interaction* contrast factor for the edge and screw components. Essentially, this is dependent on the angle between the Burgers vector, $\boldsymbol{b}$, and dislocation line, $\boldsymbol{l}$ [30]. The formulation for C_{hkl} in (10.7) can be extended to mixed dislocation. However, care must be taken in determining the $\mathbf{E}$ component, since it depends on the edge and screw components of $\boldsymbol{b}$ (see [41–44]).

The results for the contrast factor, given in (10.7)–(10.12) cover the full range of possibilities. The calculations outlined here correspond to the average contrast factor for the $(\bar{1}1\bar{1})$ slip-plane and its three slip-directions in an fcc lattice. Calculations are also presented for the (110) slip-plane and its two slip-directions in a bcc lattice. These calculations will be compared with the contrast factor values averaged over all the slip-systems in the corresponding cases.

Geometrical Component, $[G_{KL}]$

The geometrical component describes the orientation of the diffraction vector, $\boldsymbol{g}$, with respect to the Burgers vector, $\boldsymbol{b}$, slip-plane and line-direction of the dislocation, $\boldsymbol{l}$ (see Fig. 10.1). In other words, the diffraction vector is transformed into the slip coordinate system defined by the basis vectors, $\{\boldsymbol{e}_i;\ i = 1, 2, 3\}$. However, the type of dislocation defines the specific axes of the coordinate system. In the calculations presented here, we assume pure screw and edge dislocations.

For screw dislocations, the slip coordinate system is given as (compare with Fig. 10.1)

$$\boldsymbol{e}_1 = \boldsymbol{e}_2 \times \boldsymbol{e}_3 , \quad \boldsymbol{e}_2 \perp \text{slip-plane} ; \quad \text{and} \quad \boldsymbol{e}_3 \| \boldsymbol{l} \| \boldsymbol{b} \tag{10.13}$$

and for edge dislocations,

$$\boldsymbol{e}_1 \| \boldsymbol{b} , \quad \boldsymbol{e}_2 \perp \text{slip-plane} ; \quad \boldsymbol{e}_3 = \boldsymbol{e}_1 \times \boldsymbol{e}_2 , \quad \text{such that} \quad \boldsymbol{e}_3 \| \boldsymbol{l} . \tag{10.14}$$

The geometrical component of the contrast factor, G_{KL}, is defined in terms of the cosine directions of the diffraction vector, $\boldsymbol{g}$, with respect to the axes defined in the slip-system. In order to determine G_{KL} for a single slip-system, we must evaluate a 4-rank tensor $\hat{G}_{ijkl}$, given as [31,32]

$$\hat{G}_{ijkl} = \gamma_i \gamma_j \gamma_k \gamma_l \quad \text{for} \quad i, k = 1, 2, 3 \quad \text{and} \quad j, l = 1, 2 , \tag{10.15}$$

where γ_i are the direction cosines given in (10.37). The 4-rank tensor can be expressed as a [6 × 6] matrix, G_{KL}, i.e. $\hat{G}_{ijkl} \to G_{KL}$, by using the following algorithm: $K = i$ for $j = 1$; $K = i+3$ for $j = 2$; $L = k$ for $l = 1$; $L = k+3$ for $l = 2$ [31]. Essentially, the scattering vector, $\boldsymbol{g}$ is transformed into the slip-coordinates defined by (10.13) for screw dislocations and by (10.14) for edge dislocations. The geometrical component can be expressed as a symmetric [6 × 6] matrix, $\mathbf{G} = [G_{KL}]$.

Elastic Component, $[E_{KL}]$

The elastic term of the contrast factor, $[E_{KL}]$, also requires the evaluation of a 4-rank tensor and the explicit evaluation of the dislocation displacement field, $\boldsymbol{u} = [u_i(x_1, x_2); i = 1, 2, 3]$. In Sect. 10.3.2 the evaluation of the displacement fields is outlined in the context of fcc and bcc lattices, while Appendix B outlines the general field equations for planar strain in anisotropic materials. Assuming $u_i(x_1, x_2)$ is known, we define the elastic distortion due to the dislocation as [31,32]

$$D_{ij}(x_1, x_2) = \frac{2\pi r}{b} \partial_j u_i(x_1, x_2) \quad \text{for} \quad i = 1, 2, 3 \quad \text{and} \quad j = 1, 2\,. \tag{10.16}$$

On transforming (x_1, x_2) into polar coordinates (r, φ), where $r^2 = x_1^2 + x_2^2$, $x_1 = r\cos\varphi$ and $x_2 = r\sin\varphi$. This results in the transformation $D_{ij}(x_1, x_2) \to \hat{D}_{ij}(\varphi)$ i.e. $\hat{D}_{ij}$ becomes solely a function of φ. Using the results for $\hat{D}_{ij}(\varphi)$, the 4-rank elastic component of the contrast factor is given by [31,32]

$$\hat{E}_{ijkl} = \frac{1}{\pi} \int_0^{2\pi} \hat{D}_{ij}(\varphi) \hat{D}_{kl}(\varphi) \mathrm{d}\varphi\,, \tag{10.17}$$

where the integral can be evaluated numerically in the anisotropic case. Using the above algorithm (10.17), we can transform $\hat{E}_{ijkl}$ into $\mathbf{E}$, a symmetrical [6 × 6] matrix. Essentially, (10.17) represents an averaging of the distortion.

10.3.2 Determining the Dislocation Displacement Fields

Dislocations in Anisotropic Materials

In this discussion, the numerical evaluation of the displacement fields for a given type of dislocation is based on the representation developed by Teodosiu [44]. This representation places no restrictions on the orientation or type of dislocations and elastic properties of the material.

The analysis of dislocations in elastically anisotropic materials provides both qualitatively and quantitatively different results when compared with dislocations in elastically isotropic materials [43]. In this analysis, the dislocations are assumed to be straight and infinite in an elastically anisotropic continuum.

A Cartesian coordinate system $\{x_1, x_2, x_3\}$ for the dislocation is defined, such that the x_3-axis is parallel to the dislocation line, **l**. This allows the displacement vector, $\boldsymbol{u}$, to be dependent only on x_1 and x_2. The dislocation core is not considered in this discussion. More specifically, a cylindrical region, Σ_0, of radius r_0 is defined, with its axis parallel to x_3, to be the boundary of the dislocation core. The generalised field equations (outlined in Appendix B) define the planar strain outside Σ_0 [44]. The displacement vectors are expressed in terms of a complex potential function, $f_\alpha(z_\alpha)$ [44]

$$u_k(x_1, x_2) = 2\mathrm{Re} \sum_{\alpha=1}^{3} A_{k\alpha} f_\alpha(z_\alpha) \tag{10.18a}$$

$$= \frac{1}{\pi} \mathrm{Im} \sum_{\alpha=1}^{3} A_{k\alpha} D_\alpha \ln z_\alpha \,, \text{ for } k = 1, 2, 3 \,, \tag{10.18b}$$

where

$$f_\alpha(z_\alpha) = \frac{D_\alpha}{2\pi\imath} \ln z_\alpha \tag{10.18c}$$

and $z_\alpha = x_1 + p_\alpha x_2$, p_α, **A**, **L** and **D** are quantities that depend on the elastic constants of the material and will generally be complex. In evaluating the displacement fields of dislocations in fcc and bcc materials, careful attention must be given to determining the characteristic roots, p_α. In addition, an outline of the steps needed to determine **A**, **L** and **D** are given in Appendix B.

Dislocations in Elastically Anisotropic fcc Materials

The characteristic roots, p_α, incorporate the symmetry of the slip system. They are solutions to the sextic polynomial (10.45e), whose roots are always complex due to the fact that the strain energy is positive definite [45]. In fcc materials, the primary slip system for dislocations is $\{1\bar{1}1\}\langle 110\rangle$.

For screw dislocations, the line and Burgers vectors will be parallel to the $\langle 110\rangle$ direction[3], and taking $\boldsymbol{e}_2$ normal to the slip plane, the $\boldsymbol{e}_1$ axis can be fixed as $\langle\bar{1}12\rangle$ (see (10.13)) i.e. $\boldsymbol{e}_3 \| \boldsymbol{b} \| \boldsymbol{l} \| \langle 110\rangle$ has a twofold rotational symmetry, while the $\boldsymbol{e}_2 \| \langle 1\bar{1}1\rangle$ has a three fold symmetry axis.

Using these basis vectors, the elastic constants can be transformed from crystallographic axes into the slip coordinate system for screw dislocations (see 10.13, 10.33–10.36). The general structure of the transformed elastic constants matrix is given by (also see Tables 1.1 and 1.2, p. 7 of [43])

[3] In this section and the next we use $\langle \cdots \rangle$ to denote the *set* of vector.

$$c' = \begin{pmatrix} c'_{11} & c'_{12} & c'_{13} & c'_{14} & 0 & c'_{16} \\ & c'_{22} & c'_{12} & 0 & 0 & 0 \\ & & c'_{11} & -c'_{14} & 0 & c'_{16} \\ & & & c'_{44} & -c'_{16} & 0 \\ & & & & \frac{1}{2}(c'_{11} - c'_{13}) & c'_{14} \\ & & & & & c'_{44} \end{pmatrix} . \tag{10.19}$$

Given that the dislocation line, $\boldsymbol{l}$, is parallel to a twofold symmetry axis, the sextic polynomial (10.45e) can be decoupled into quartic and quadratic equations [44, 46]. A pure screw dislocation corresponds to a quadratic case having only one displacement component. That is, (10.45e) reduces to $l(p) = l_2(p)l_4(p)$, where the $l_2(p)$ corresponds to a pure screw dislocation[4]. The ordering of the roots can be achieved by observing that $l_2(p_1) \neq 0$, $l_2(p_2) \neq 0$ and $l_2(p_3) = 0$; $l_4(p_1) = l_4(p_2) = 0$ and $l_4(p_3) \neq 0$. Generally, the characteristics roots, p_α, for the screw dislocation in this slip-system will be distinct, $p_\alpha = a_\alpha + \imath\, c_\alpha$, where $\{a_\alpha, c_\alpha\} \in \mathbb{R}$, $\forall\, \alpha = 1, 2, 3$. In this case the roots corresponding to the $l_2(p)$ terms are needed (p_3).

The slip direction for pure screw dislocation has the Burgers vector, $\boldsymbol{b} = [0, 0, b]$. Furthermore, the coefficients A_{33} and D_3 from (10.47a), (10.47b) and (10.48) will also be nonzero, resulting in the displacement vector having the components

$$\boldsymbol{u} = \left[0,\ 0,\ \frac{1}{\pi}\mathrm{Im} A_{33} D_3 \ln z_3\right] \tag{10.20a}$$

$$= \left[0,\ 0,\ \frac{-b}{2\pi}\arctan\left(\frac{c_3 x_2}{x_1 + a_3 x_2}\right)\right] , \tag{10.20b}$$

where $a_3 = \mathrm{Re}\{p_3\}$, $c_3 = \mathrm{Im}\{p_3\}$, $z_3 = x_1 + p_3 x_2$ and $\ln z_3 = \ln r_3 + \imath\theta_3$, such that $r_3 = \sqrt{(x_1 + a_3 x_2)^2 + (c_3 x_3)^2}$ and $\tan\theta_3 = c_3 x_3/(x_1 + a_3 x_2)$. In (10.20b) the $\ln r_3$ term is dropped since it corresponds to the real-part of z_3. In the limit of a screw dislocation for an isotropic fcc material, the characteristic roots will have the values, $p_\alpha = \imath$, $\forall\alpha$, and the usual displacement for screw dislocations, $[0, 0, \frac{-b}{2\pi}\arctan(x_2/x_1)]$ results (see Sect. 3.2 in [42]).

Although the results for screw dislocations in anisotropically and isotropically elastic fcc materials are similar, the differences are also clear. In (10.20), the elastic anisotropy is expressed in the "stretching" of the x_2 coordinate with respect to x_1, while there is no stretching in the isotropic result.

For edge dislocations in real fcc materials, $\boldsymbol{e}_3$ can be determined by applying (10.14) resulting in $\boldsymbol{l} \| \boldsymbol{e}_3 \| \langle\bar{1}12\rangle$; $\boldsymbol{e}_2$ is normal to $\{1\bar{1}1\}$ which has threefold

[4] The $l_4(p)$ represents an edge dislocation which will not be considered here.

symmetry, while $\boldsymbol{e}_1 \| \boldsymbol{b} \| \langle 110 \rangle$ has a twofold symmetry. These basis vectors can be used to transform the elastic constants in the slip-system for edge dislocations and will have the same structure as (10.19). Alternatively, the elastic constant matrix in this coordinate system can be determined by taking the full $[9 \times 9]$ form of (10.19) and rotating it by $\pi/2$ (in the clockwise direction) about the $\boldsymbol{e}_2$ axis. The dislocation line, $\boldsymbol{l}$, will then be perpendicular to an even-fold symmetry axis. This also means that the sextic polynomial, (10.45e), cannot be separated into a "screw" and "edge" component [46], as was done for the screw dislocations. Consequently, the characteristic roots must be determined numerically.

The characteristic roots for edge dislocations will have the form, $p_1 = \imath c_1$ and $p_{2,3} = \pm a_2 + \imath c_2$, where $\{c_1, a_2, c_2\} \in \mathbb{R}$ and are all positive. The **A**, **L** terms in (10.46) and (10.47a), (10.47b) respectively, will generally be full matrices. In this coordinate system[5], the slip direction for a pure edge dislocation is restricted to $\boldsymbol{b} = [b, 0, 0]$. The displacement vector for edge dislocations will have the form given in (10.18b). However, a surprising result for the edge displacement vectors is that the u_3 component is nonzero [43,44], resulting in the deformation being out of the slip plane. This can be compared with the isotropic case where the u_3 component is zero and the deformation is restricted to being in the slip plane (see Sect. 3.2 in [42]). For edge dislocations in elastically isotropic fcc materials, the characteristic roots will be $p_\alpha = \imath, \forall \alpha$ and the usual displacement fields will result (see Sect. 3.4 in [42]). However, care must be taken to avoid infinities (see Sect. 3.1 in [43]).

Dislocations in Elastically Anisotropic bcc Materials

As pointed out in Sect. 10.3.1 and shown in Fig. 10.3, dislocations in bcc lattices can form on several slip-systems. The most common slip-system observed in bcc lattices is $\{110\}\langle\bar{1}11\rangle$ [43]. Other slip-systems for bcc lattice can be observed, such as the $\{110\}\langle 112\rangle$, but will not be considered here. For the $\{110\}\langle\bar{1}11\rangle$ case there are a total of 12 systems (6 slip-planes $\times$ 2 slip-directions).

Essentially, the calculations of the displacement fields, $\boldsymbol{u}(x_1, x_2)$ for dislocations in bcc materials involve the same steps as in the fcc case. The main differences between the fcc and bcc case involve the definition of the slip-system and the ordering of the characteristic roots.

For screw dislocations in the $\{110\}\langle\bar{1}11\rangle$ slip system, the line vector, $\boldsymbol{l}$, is given by $\langle\bar{1}11\rangle$ and will be parallel to the $\boldsymbol{e}_3$ (see Fig. 10.1 and (10.13)), while $\boldsymbol{e}_2 \| \langle 110 \rangle$ and $\boldsymbol{e}_1 \| \langle \bar{1}1\bar{2} \rangle$. The $\boldsymbol{e}_3$ axis has a threefold system, and the transformed $[6 \times 6]$ elastic constants have the following structure (also see Tables 1.1 and 1.2, p. 7 of [43])

[5] This also includes a 30° mixed case which is not considered here.

$$
c' = \begin{pmatrix} c'_{11} & c'_{12} & c'_{13} & c'_{14} & c'_{15} & 0 \\ & c'_{11} & c'_{13} & -c'_{14} & -c'_{15} & 0 \\ & & c'_{33} & 0 & 0 & 0 \\ & & & c'_{44} & 0 & -c'_{15} \\ & & & & c'_{44} & c'_{14} \\ & & & & & \frac{1}{2}(c'_{11} - c'_{12}) \end{pmatrix} . \qquad (10.21)
$$

The characteristics roots, p_α, from the sextic equation, (see (10.45a–c)–(B.3)), for screw dislocations in the bcc $\{110\}\langle\bar{1}11\rangle$-system are chosen such that $\mathrm{Im}\{p_\alpha\} > 0$, for $\alpha = 1, 2, 3$. These roots must be determined numerically. The p_α-roots are arranged such that $p_{1,2} = \mp a_1 + \imath c_1$ and $p_3 = \imath a_2$, where $\{a_1, c_1, a_2\} \in \mathbb{R}$ and are all positive. For screw dislocations the slip direction is restricted to $\boldsymbol{b} = [0, 0, b]$, and the **A**, **L** and **D** matrices in (10.47a), (10.47b), (10.46) and (10.48) respectively, are generally full matrices (i.e. few zero elements). All the components of the corresponding displacement fields, $\boldsymbol{u}(x_1, x_2)$ for the bcc case are generally nonzero components. This is in contrast to screw dislocations in fcc materials, where $u_1 = u_2 = 0$, see (10.20). In the elastically isotropic case, $p_\alpha = \imath$, for all α, and the usual displacement fields for screw dislocations are produced.

For edge dislocations in a bcc system, the Burgers vector, $\boldsymbol{b} \| \langle\bar{1}11\rangle$, which sets the direction of $\boldsymbol{e}_1$; the line vector, $\boldsymbol{l}$, has the direction $\boldsymbol{e}_3 \| \langle 1\bar{1}2\rangle$. In this coordinate system, the $\boldsymbol{e}_1$-axis has a threefold symmetry; also the transformed elastic constants are given by (see Tables 1.1 and 1.2, p. 7 of [43])

$$
c' = \begin{pmatrix} c'_{11} & c'_{12} & c'_{12} & 0 & 0 & 0 \\ & c'_{22} & c'_{23} & 0 & c'_{25} & c'_{26} \\ & & c'_{22} & 0 & -c'_{25} & -c'_{26} \\ & & & \frac{1}{2}(c'_{22} - c'_{23}) & -c'_{26} & c'_{25} \\ & & & & c'_{55} & 0 \\ & & & & & c'_{55} \end{pmatrix} . \qquad (10.22)
$$

The slip-directions for an edge dislocation in this system are restricted to $\boldsymbol{b} = [b, 0, 0]$ and the roots from the sextic equation, (10.45a–c)–(B.3), are arranged such that, $p_1 = a_1\imath$ and $p_{2,3} = \mp a_2 + \imath c_2$, where $\{a_1, a_2, c_2\} \in \mathbb{R}$ and are positive. The **A**, **L** and **D** matrices can be determined and used to evaluate $\boldsymbol{u}(x_1, x_2)$ in a similar way to the previous cases. The displacement fields for edge dislocations in the $\{110\}\langle\bar{1}11\rangle$ system generally have nonzero components.

10.4 Computing the Contrast Factor, C_{hkl}

In this section the theoretical discussion that was presented in Sect. 10.3 is illustrated by examples of calculating the contrast factors in a single-slip system for fcc and bcc materials. This calculation can be generalised and averaged over all systems in each lattice.

10.4.1 C_{hkl} Values for fcc Materials

This example concentrates on evaluating the factors for gold (Au) on the $(\bar{1}\bar{1}\bar{1})$ slip-plane in the $[\bar{1}10]$ slip-direction. The evaluation for the contrast factor for the remaining $[0\bar{1}1]$ and $[10\bar{1}]$ slip-directions on the $(\bar{1}\bar{1}\bar{1})$ slip-plane follows the same procedure. Moreover, these calculations can be further generalised to include all 12 slip-systems in a fcc lattice.

The elastic constants for Au used here are $c_{11} = 192.9\,\mathrm{GPa}$, $c_{12} = 163.8\,\mathrm{GPa}$ and $c_{44} = 41.5\,\mathrm{GPa}$ (see Table 5.3, p. 68 of [44]).

Gold (Au) C_{hkl} Values for Screw Dislocations on $(\bar{1}\bar{1}\bar{1})\,[\bar{1}10]$

Screw dislocations in the $(\bar{1}\bar{1}\bar{1})\,[\bar{1}10]$ slip-system require the basis vectors, $\{\boldsymbol{e}_1, \boldsymbol{e}_2, \boldsymbol{e}_3\}$. On applying (10.13) and Figs. 10.1 and 10.2, the basis vectors are, $\boldsymbol{e}_1 = \frac{1}{\sqrt{6}}[1, 1, -2]$, $\boldsymbol{e}_2 = \frac{1}{\sqrt{3}}[-1, -1, -1]$ and $\boldsymbol{e}_3 = \frac{1}{\sqrt{2}}[-1, 1, 0]$. The reduced $[6 \times 6]$ transformed elastic constants for Au are (in GPa)

$$c' = \begin{pmatrix} 219.9 & 145.8 & 154.8 & 0 & 0 & -12.7 \\ & 228.8 & 145.8 & 0 & 0 & 0 \\ & & 219.9 & 0 & 0 & 12.7 \\ & & & 23.5 & 12.7 & 0 \\ & & & & 32.5 & 0 \\ & & & & & 23.5 \end{pmatrix}, \tag{10.23}$$

while the reduced compliances $\boldsymbol{S}$ (in units $\times 10^{-9}\,\mathrm{Pa}^{-1}$) have values (see Appendix B.2),

$$\mathbf{S} = \begin{pmatrix} 0.0083 & -0.0053 & 0 & 0 & 0 & 0.0045 \\ & 0.0078 & 0 & 0 & 0 & -0.0029 \\ & & 0 & 0 & 0 & 0 \\ & & & 0.5385 & -0.0210 & 0 \\ & & & & 0.0390 & 0 \\ & & & & & 0.04492 \end{pmatrix}. \tag{10.24}$$

The reduced compliances for Au in (10.24), can be used to explicitly define the sextic polynomial, (10.45a–c)–(B.3). Once the sextic polynomial is known, its characteristic roots for screw dislocations can be determined.

The characteristics roots which obey the ordering specified in Sect. 10.3.2 for screw dislocation, have the dimensionless values, $p_1 = -0.0767 + 0.4715\imath$, $p_2 = 0.6165 + 1.9233\imath$ and $p_3 = -0.5398 + 1.0442\imath$. As previously stated, the p_3 root is of importance since it is the only root that appears in the displacement field for screw dislocations. The corresponding terms given in (10.20a), A_{33} and D_3, were found to be $0.0407\imath \times 10^{-9}\,\mathrm{Pa}^{-1}$ and $12.2864b \times 10^{-9}\,\mathrm{N\,m}^{-1}$, respectively. By restricting the slip to $\boldsymbol{b} = [0, 0, b]$ and using (10.20b), the displacement field for screw dislocations in Au is, $\boldsymbol{u} = [0, 0, (-b)/(2\pi) \arctan((1.044x_2)/(x_1 - 0.5398x_2))]$.

Following the steps outlined in Sect. 10.3.1, by using (10.16) and (10.17), the elastic component of the contrast factor for screw dislocations in Au were found to be

$$\mathbf{E} = \begin{pmatrix} 0 & 0 & 0 & 0 & 0 & 0 \\ & 0 & 0 & 0 & 0 & 0 \\ & & 0.9577 & 0 & 0 & -0.5170 \\ & & & 0 & 0 & 0 \\ & & & & 0 & 0 \\ & & & & & 1.3233 \end{pmatrix} . \qquad (10.25)$$

The second component of the contrast factor is the geometric term, (10.15), and it defines the cosine direction for the scattering vector, $\boldsymbol{g}$, in the slip coordinate system, $\{\boldsymbol{e}_1, \boldsymbol{e}_2, \boldsymbol{e}_3\}$. In this example of a screw dislocation, the $\{\gamma_i; i = 1, 2, 3\}$ terms are

$$\left\{ \frac{\mathrm{h}+\mathrm{k}-2\mathrm{l}}{\sqrt{6}\sqrt{\mathrm{h}^2+\mathrm{k}^2+\mathrm{l}^2}}, -\frac{\mathrm{h}+\mathrm{k}+\mathrm{l}}{\sqrt{3}\sqrt{\mathrm{h}^2+\mathrm{k}^2+\mathrm{l}^2}}, \frac{-\mathrm{h}+\mathrm{k}}{\sqrt{2}\sqrt{\mathrm{h}^2+\mathrm{k}^2+\mathrm{l}^2}} \right\} .$$

The γ_i elements define the 4-rank tensor, (10.15), and can be expressed as the $[6 \times 6]$ $\mathbf{G}$ matrix in (10.7).

Using this approach, the Au contrast factor values for the three slip-directions on the $(\bar{1}1\bar{1})$ follow. Further, the average contrast factor (over all 12 slip-systems) for screw dislocations in Au can be determined and expressed as a linear equation, $\langle C^{\mathrm{screw}}_{\mathrm{hkl}} \rangle = 0.2815 - 0.6471\Gamma(\mathrm{h,k,l})$.

The Au contrast factor values for screw dislocations are given in Fig. 10.4a and Table 10.1. The values in Fig. 10.4a compare the $C^{\mathrm{screw}}_{\mathrm{hkl}}$ on the $(\bar{1}1\bar{1})$ slip-plane for the $[\bar{1}10]$, $[0\bar{1}1]$ and $[10\bar{1}]$ slip-directions with the average screw contrast factors, $\langle C^{\mathrm{screw}}_{\mathrm{hkl}} \rangle$ in real (anisotropic) and "isotropic" Au. From Fig. 10.4a, we notice that the difference between the anisotropic and isotropic values for $\langle C^{\mathrm{screw}}_{\mathrm{hkl}} \rangle$ can be seen in the slope and y-intercept of the linear equations. That is, the elastic anisotropy in Au for $\langle C^{\mathrm{screw}}_{\mathrm{hkl}} \rangle$ can be seen from the increased y-intercept value or $\langle C^{\mathrm{screw}}_{200} \rangle^{aniso} = 0.2815$, compared with $\langle C^{\mathrm{screw}}_{200} \rangle^{\mathrm{iso}} \sim 0.175$. Also from Fig. 10.4a, the slope of $\langle C^{\mathrm{screw}}_{\mathrm{hkl}} \rangle^{\mathrm{aniso}}$ has an increased negativity compared to the slope for $\langle C^{\mathrm{screw}}_{\mathrm{hkl}} \rangle^{\mathrm{iso}}$.

Shown in Fig. 10.4a and Table 10.1 is the variation of $C^{\mathrm{screw}}_{\mathrm{hkl}}$ for different hkl with the same d-spacing. The values of C_{hkl} for individual hkl are important when dislocations populate only a few slip-planes, such as in single-crystal and highly-preferred orientation specimens [29, 36]. The comparison between Au $\langle C^{\mathrm{screw}}_{\mathrm{hkl}} \rangle$ and $C^{\mathrm{screw}}_{\mathrm{hkl}}$ values for different hkl demonstrate how the variation in the $C^{\mathrm{screw}}_{\mathrm{hkl}}$ is "lost" when averaged over the 12 slip-systems. The

Table 10.1. The contrast values, C_{hkl}, for screw dislocations on the $(\bar{1}\,\bar{1}\,\bar{1})$ slip-plane in three different slip-directions (see Figure 10.2) in anisotropic (real) Au, Ag, and Pt. The anisotropy factor is $A = 2c_{44}/(c_{11} - c_{12})$, where c_{11}, c_{12} and c_{44} are the elastic constants for the materials (for isotropic materials A=1). The Poisson ratio, ν, for each material are also given

Material	h k l	$[\bar{1}10]$	$[0\bar{1}1]$	$[10\bar{1}]$	C_{hkl}^{screw}	$\langle C_{\mathrm{hkl}}^{screw}\rangle$
Au	111	0	0	0	0	0.0658
A=2.850	$11\bar{1}$	0	0.1316	0.1316	0.0877	–
$\nu = 0.412$	200	0.4222	0	0.4222	0.2815	0.2815
$c_{11} = 192.9$ GPa	002	0	0.4222	0.4222	0.2815	–
$c_{12} = 163.8$ GPa	220	0	0.1796	0.1796	0.1197	0.1197
$c_{44} = 41.5$ GPa	$2\bar{2}0$	0	0.1796	0.1796	0.1197	–
	311	0.2331	0	0.2331	0.1554	0.1799
	$31\bar{1}$	0.2331	0.0881	0.2625	0.1946	–
	331	0	0.1044	0.1044	0.0696	0.1040
	420	0.1520	0.0643	0.3176	0.1779	0.1779
Ag	111	0	0	0	0	0.0654
A=2.882	$11\bar{1}$	0	0.1309	0.1309	0.08726	–
$\nu = 0.354$	200	0.4244	0	0.4244	0.2830	0.2830
$c_{11} = 122.2$ GPa	002	0	0.4244	0.4244	0.2830	–
$c_{12} = 90.7$ GPa	220	0	0.1797	0.1797	0.1198	0.1198
$c_{44} = 45.4$ GPa	$2\bar{2}0$	0	0.1797	0.1797	0.1198	–
	311	0.2342	0	0.2342	0.1562	0.1805
	$31\bar{1}$	0.2342	0.0876	0.2634	0.1951	–
	331	0	0.1046	0.1046	0.0697	0.1040
	420	0.1528	0.0641	0.3188	0.1786	0.1786
Pt	111	0	0	0	0	0.0880
A=1.593	$11\bar{1}$	0	0.1760	0.1760	0	–
$\nu = 0.393$	200	0.3156	0	0.3156	0.2104	0.2104
$c_{11} = 347.0$ GPa	002	0	0.3156	0.3156	0.2105	–
$c_{12} = 251.0$ GPa	220	0	0.1779	0.1779	0.1186	0.1186
$c_{44} = 76.5$ GPa	$2\bar{2}0$	0	0.1779	0.1779	0.1186	–
	311	0.1800	0	0.1800	0.1210	0.1528
	$31\bar{1}$	0.1800	0.1178	0.2193	0.1724	–
	331	0	0.0954	0.0954	0.0636	0.1097
	420	0.1136	0.0760	0.2654	0.1517	0.1517

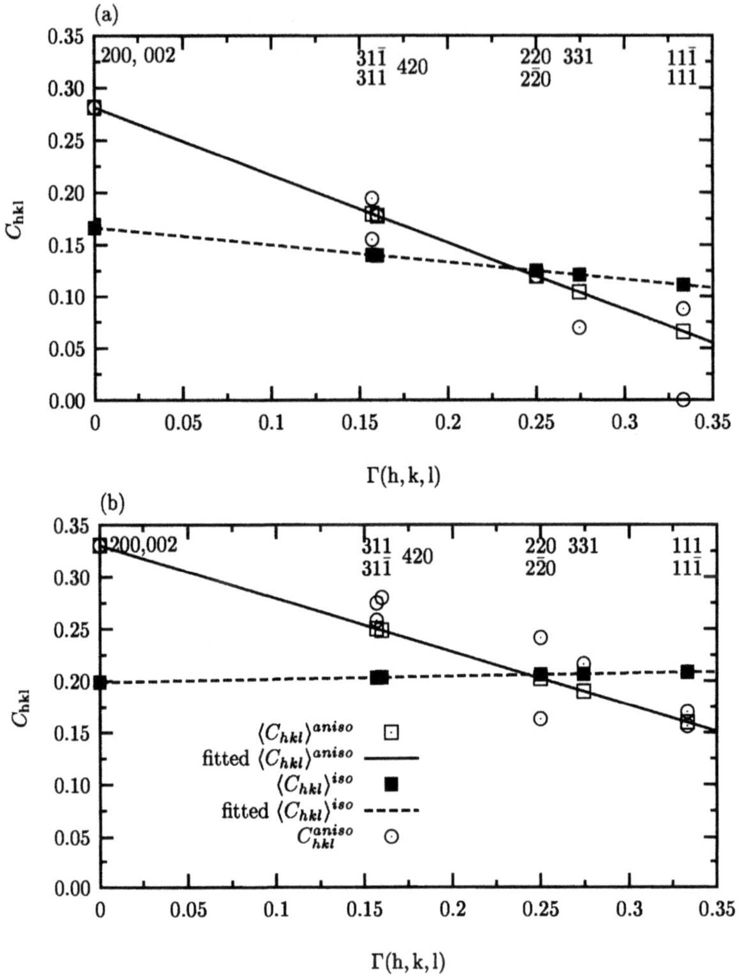

Fig. 10.4. Contrast factor values for Au. (**a**) Screw dislocations for the $(\bar{1}\bar{1}\bar{1})$ slip-plane, average values (for all slip-systems) in "isotropic" Au and average values in real Au (see Table 10.1). (**b**) Similar to (**a**), but with edge dislocations in Au (see Table 10.2)

variation in the $C_{\mathrm{hkl}}^{\mathrm{screw}}$ values for different hkl is due to the geometric component of the contrast factor. That is, the elastic component is invariant of the slip-system and is not explicitly dependent on hkl, while the geometric component is dependent on hkl, since it defines the direction cosines of the scattering vector relative to axes of the slip-system.

Also shown in Table 10.1 are the contrast values for silver (Ag) and platinum (Pt). From these values the average screw contrast factors for Ag was found to be $\langle C_{\mathrm{hkl}}^{\mathrm{screw}}\rangle = 0.2830 - 0.6526\Gamma(\mathrm{h,k,l})$, while for Pt it was found to be $\langle C_{\mathrm{hkl}}^{\mathrm{screw}}\rangle = 0.2104 - 0.3672\Gamma(\mathrm{h,k,l})$.

A similarity in the $C_{\mathrm{hkl}}^{\mathrm{screw}}$ and $\langle C_{\mathrm{hkl}}^{\mathrm{screw}}\rangle$ values for Au and Ag is noticeable. The anisotropy factor for both Au and Ag is ~ 2.9, while the c_{12}/c_{44}-ratio is 3.95 and 2.00, respectively. However, the $C_{\mathrm{hkl}}^{\mathrm{screw}}$ and $\langle C_{\mathrm{hkl}}^{\mathrm{screw}}\rangle$ for Pt differ considerably to Au and Ag. The anisotropy factor for Pt is 1.59, while the c_{12}/c_{44}-ratio is 3.28.

From these three sets of results some observations can be made concerning the (average) screw contrast values in fcc materials. The anisotropic ratio for Au and Ag are very similar, while the c_{12}/c_{44}-ratios are significantly different. For both cases, the $\langle C_{\mathrm{hkl}}^{\mathrm{screw}}\rangle$-values are also very similar. However in the case of Pt, its anisotropy factor is significantly less than that of Au, while the c_{12}/c_{44}-ratio is approximately in the mid-range for Ag and Au. These results suggest that the $\langle C_{\mathrm{hkl}}^{\mathrm{screw}}\rangle$ values in fcc materials are dependent on the anisotropy factor of the fcc materials, but weakly dependent on the c_{12}/c_{44}-ratio [28].

Gold (Au) C_{hkl} Value for Edge Dislocations on $(\bar{1}\bar{1}\bar{1})\,[\bar{1}10]$

The calculation of the contrast factors for edge dislocations in Au require care in determining the characteristic roots, $\{p_\alpha, \forall\alpha\}$ and the displacement fields, $u(x_1, x_2)$.

Using (10.14) and Figs. 10.1 and 10.2, the slip coordinate system for edge dislocations becomes $\boldsymbol{e}_1 = \frac{1}{\sqrt{2}}[-1,1,0]$, $\boldsymbol{e}_2 = \frac{1}{\sqrt{3}}[-1,-1,-1]$ and $\boldsymbol{e}_3 = \frac{1}{\sqrt{6}}[-1,-1,2]$. These basis vectors transform the elastic constants from their crystallographic coordinates into a slip coordinate system for an edge dislocation (in GPa),

$$\boldsymbol{c}' = \begin{pmatrix} 219.9 & 145.8 & 154.8 & -12.7 & 0 & 0 \\ & 228.8 & 145.8 & 0 & 0 & 0 \\ & & 219.9 & 12.7 & 0 & 0 \\ & & & 23.5 & 0 & 0 \\ & & & & 32.5 & -12.7 \\ & & & & & 23.5 \end{pmatrix}, \tag{10.26}$$

while using (10.44), the reduced compliances $\boldsymbol{S}$ (in units $\times 10^{-9}\,\mathrm{Pa}^{-1}$) have the form,

$$\mathbf{S} = \begin{pmatrix} 0.0083 & -0.0053 & 0 & 0.0045 & 0 & 0 \\ & 0.0078 & 0 & -0.0029 & 0 & 0 \\ & & 0 & 0 & 0 & 0 \\ & & & 0.0449 & 0 & 0 \\ & & & & 0.0390 & 0.0210 \\ & & & & & 0.0539 \end{pmatrix}. \tag{10.27}$$

As was carried out for the screw dislocations, the characteristic roots for edge dislocations were determined by using (10.27) in (10.45a–c)–(B.3). They were found to be $p_1 = 0.3951\imath$ and $p_{2,3} = \pm 0.5008 + 1.5300\imath$. In this case, the separation of the screw and edge components is not straight forward and characteristic roots from the sextic equation, (10.45a–c)–(B.3), must be determined numerically (see Sect. 10.3.2). The displacement field equations for edge dislocations cannot be expressed succinctly here. However, we present the intermediate results needed to determine $\boldsymbol{u}(x_1, x_2)$.

As expressed in (10.18–c), $\boldsymbol{u}(x_1, x_2)$ is given in terms of a complex potential function $f_\alpha(z_\alpha)$, which is decomposed into, **L**, **A** and **D** matrices. Since the p_α are known and following (10.48), (10.46), (10.47a) and (10.47b), the above matrices for edge dislocations in (real) Au can be evaluated directly as

$$\mathbf{L} = \begin{pmatrix} -0.3951\imath & -0.5008 - 1.5297\imath & 1.0975 - 1.2251\imath \\ 1 & 1 & 0.9355 + 0.4112\imath \\ 0.1764 & -0.8959 - 0.3938\imath & -1 \end{pmatrix}, \tag{10.28a}$$

and

$$\mathbf{A} = \begin{pmatrix} -0.0058 & -0.0267 + 0.0110\imath & -0.0205 - 0.0213\imath \\ -0.0204\imath & 2.7650 \times 10^{-6} - 0.0140\imath & 0.0058 - 0.0131\imath \\ -0.0110\imath & -0.0165 + 0.0289\imath & 0.0035 + 0.0338\imath \end{pmatrix} \times 10^{-9}\,\mathrm{Pa}^{-1} \tag{10.28b}$$

and

$$\mathbf{D} = \begin{pmatrix} -9.8316 \\ 4.9158 + 13.3867\imath \\ -0.8671 - 13.9289\imath \end{pmatrix} b \times 10^{-9}\,\mathrm{N\,m}^{-1}. \tag{10.28c}$$

Generally, the resulting displacement fields for edge dislocations in anisotropic fcc materials, such as Au, are both qualitatively and quantitatively different compared to the isotropic case [43]. For example, the $u_3(x_1, x_2)$ component in the anisotropic case is nonzero, while for the isotopic case it is zero. This feature was shown to be the case for all the fcc materials presented here (see Table 10.2). The displacement fields were used in evaluating the elastic component of the contrast factor for Au, by using (10.16) and (10.17). In general, the elastic component for edge dislocations in anisotropic materials is computationally more demanding, compared to screw dislocations, since there are fewer zero elements in the matrix and the nonzero elements require the numerical integration of (10.17) (compare (10.25) with (10.29)). Performing these operations, the symmetrical elastic components matrix for edge dislocations in Au, has the form

$$\mathbf{E} = \begin{pmatrix} 0.4129 & 0 & 0 & 0 & -0.2517 & 0.1309 \\ & 0.3714 & -0.0741 & -0.5056 & 0 & 0 \\ & & 0.0676 & 0.3299 & 0 & 0 \\ & & & 2.6324 & 0 & 0 \\ & & & & 0.1701 & -0.1060 \\ & & & & & 0.1153 \end{pmatrix} . \tag{10.29}$$

The γ_i elements of the geometric component were found to be

$$\left\{ \frac{-\mathrm{h}+\mathrm{k}}{\sqrt{2}\sqrt{\mathrm{h}^2+\mathrm{k}^2+\mathrm{l}^2}}, -\frac{\mathrm{h}+\mathrm{k}+\mathrm{l}}{\sqrt{3}\sqrt{\mathrm{h}^2+\mathrm{k}^2+\mathrm{l}^2}}, -\frac{\mathrm{h}+\mathrm{k}-2\mathrm{l}}{\sqrt{6}\sqrt{\mathrm{h}^2+\mathrm{k}^2+\mathrm{l}^2}} \right\}$$

and can be used to determine the full $[6 \times 6]$ **G** matrix in (10.7). Comparing this result with the screw case, there is no significant difference, in that the ordering of the γ_i terms reflects the change in the ordering of the basis vector from screw to edge slip-system.

The edge contrast factors for Au can be calculated for all three slip-directions on the $(\bar{1}1\bar{1})$ plane. Furthermore, the average contrast factor for edge dislocations (over all 12 slip systems) in Au can be determined and expressed as $\langle C^{\mathrm{edge}}_{\mathrm{hkl}} \rangle = 0.3305 - 0.5121\Gamma(\mathrm{h,k,l})$.

The $C^{\mathrm{edge}}_{\mathrm{hkl}}$ values for Au in Fig. 10.4b on the $(\bar{1}1\bar{1})$ slip-plane for the $[\bar{1}10]$, $[0\bar{1}1]$ and $[10\bar{1}]$ slip-directions can be compared with the average edge contrast factors, $\langle C^{\mathrm{edge}}_{\mathrm{hkl}} \rangle$ in real (anisotropic) and "isotropic" Au. In comparing the $\langle C^{\mathrm{edge}}_{\mathrm{hkl}} \rangle^{\mathrm{iso}}$ and $\langle C^{\mathrm{edge}}_{\mathrm{hkl}} \rangle^{\mathrm{aniso}}$ linear equations in Fig. 10.4b, we notice that the isotropic case has a weak positive slope over the $\Gamma(\mathrm{h,k,l})$ range and $\langle C^{\mathrm{edge}}_{200} \rangle^{\mathrm{iso}} \sim 0.20$, while the anisotropic case $\langle C^{\mathrm{edge}}_{\mathrm{hkl}} \rangle^{\mathrm{aniso}}$ has a strong negative slope $(= -0.5121)$ and $\langle C^{\mathrm{edge}}_{200} \rangle^{\mathrm{aniso}} = 0.3305$. In this case, the geometric components are the same and it clearly demonstrates the strong effects the elastic anisotropy has in determining the $\langle C^{\mathrm{edge}}_{\mathrm{hkl}} \rangle^{\mathrm{aniso}}$. Also shown in Fig. 10.4b and Table 10.2 is the variation of $C^{\mathrm{edge}}_{\mathrm{hkl}}$ for different hkl with the same d-spacing.

In addition, Table 10.2 shows the contrast factors for silver (Ag) and platinum (Pt). The edge average contrast factor for Ag and Pt was found to be, $\langle C^{\mathrm{edge}}_{\mathrm{hkl}} \rangle = 0.2995 - 0.4681\Gamma(\mathrm{h,k,l})$ and $\langle C^{\mathrm{edge}}_{\mathrm{hkl}} \rangle = 0.2416 - 0.2010\Gamma(\mathrm{h,k,l})$, respectively. From the screw case, we observed an anisotropy factor for Au and Ag ~ 2.9, while the c_{12}/c_{44}-ratio were significantly different, 3.95 and 2.00, respectively. We also observed that the $\langle C^{\mathrm{screw}}_{\mathrm{hkl}} \rangle$-values for Au and Ag were similar. However, in the case of edge dislocations, the $\langle C^{\mathrm{edge}}_{\mathrm{hkl}} \rangle$-values for Au and Ag (in Table 10.2) are different. As was pointed out above for the screw case, Pt has an anisotropy factor which is less than Au and Ag, but a c_{12}/c_{44}-ratio in the mid-range of Au and Ag.

Therefore, it can be inferred from these results that the anisotropy factor and the c_{12}/c_{44}-ratio influence the (average) contrast values for edge dislocations in fcc materials, while for screw dislocations in fcc materials, the

Table 10.2. The contrast values, C_{hkl}, for edge dislocations on the $(\bar{1}\bar{1}\bar{1})$ slip-plane in three different slip-directions (see Fig. 10.2) in anisotropic (real) Au, Ag, and Pt

Material	hkl	$[\bar{1}10]$	$[0\bar{1}1]$	$[10\bar{1}]$	$C_{\mathrm{hkl}}^{\mathrm{edge}}$	$\langle C_{hkl}^{\mathrm{edge}} \rangle$
Au	111	0.1701	0.1701	0.1701	0.1701	0.1598
A=2.850	$11\bar{1}$	0.0061	0.2315	0.2314	0.1563	–
$\nu = 0.412$	200	0.4568	0.0778	0.4568	0.3305	0.3305
$c_{11} = 192.9\,\mathrm{GPa}$	002	0.0778	0.4568	0.4568	0.3305	–
$c_{12} = 163.8\,\mathrm{GPa}$	220	0.0346	0.3452	0.3452	0.2416	0.2025
$c_{44} = 41.5\,\mathrm{GPa}$	$2\bar{2}0$	0.4129	0.0385	0.0385	0.1633	–
	311	0.3182	0.1876	0.3182	0.2746	0.2501
	$31\bar{1}$	0.1559	0.0921	0.5264	0.2581	–
	331	0.0763	0.2863	0.2863	0.2163	0.1900
	420	0.1563	0.1986	0.4847	0.2799	0.2486
Ag	111	0.0963	0.0963	0.0963	0.0963	0.1434
A=2.882	$11\bar{1}$	0.0059	0.2358	0.2358	0.1591	–
$\nu = 0.354$	200	0.4178	0.0629	0.4178	0.2994	0.2995
$c_{11} = 122.2\,\mathrm{GPa}$	002	0.0629	0.4178	0.4178	0.2995	–
$c_{12} = 90.7\,\mathrm{GPa}$	220	0.0117	0.2852	0.2852	0.1940	0.1824
$c_{44} = 45.5\,\mathrm{GPa}$	$2\bar{2}0$	0.4372	0.0377	0.0377	0.1709	–
	311	0.2554	0.1328	0.2554	0.2146	0.2260
	$31\bar{1}$	0.1384	0.0796	0.5007	0.2395	–
	331	0.0317	0.2130	0.2130	0.1526	0.1711
	420	0.1239	0.1557	0.4219	0.4219	0.2246
Pt	111	0.1777	0.1777	0.1777	0.1777	0.1746
A=1.593	$11\bar{1}$	0.0007	0.2600	0.2600	0.1736	–
$\nu = 0.393$	200	0.3421	0.0406	0.3421	0.2416	0.2416
$c_{11} = 347.0\,\mathrm{GPa}$	002	0.0406	0.3421	0.3421	0.2416	–
$c_{12} = 251.0\,\mathrm{GPa}$	220	0.0517	0.2831	0.2831	0.2060	0.1914
$c_{44} = 76.5\,\mathrm{GPa}$	$2\bar{2}0$	0.4673	0.03144	0.0314	0.1767	–
	311	0.2588	0.1391	0.2588	0.2189	0.2100
	$31\bar{1}$	0.0959	0.0746	0.4691	0.2132	–
	331	0.1034	0.2421	0.2421	0.2421	0.1865
	420	0.1230	0.1491	0.3913	0.2212	0.2094

(average) contrast value is strongly dependent on the anisotropy factor and weakly dependent on the c_{12}/c_{44}-ratio [28].

10.4.2 C_{hkl} Values for bcc Materials

In this example we present calculations of the displacement fields and contrast factor for screw and edge dislocations in α-Fe on the (110) slip-plane in the $[\bar{1}11]$ direction. These calculations can be extended to determine C_{hkl} and the average contrast factor, $\langle C_{\mathrm{hkl}}\rangle$ over all 12 slip-systems for the $\{110\}\langle\bar{1}11\rangle$ system.

Essentially the same procedure as for determining the C_{hkl}-values for fcc lattice (see Sect. 10.4.1) was followed for edge and screw dislocations in the bcc $\{110\}\langle\bar{1}11\rangle$ slip-system. Hence, only the main results, such as the p_α-roots, γ_i elements, and the calculated C_{hkl} and $\langle C_{\mathrm{hkl}}\rangle$ values are presented.

The elastic constants for α-Fe used in these calculations were $c_{11} = 230.1$ GPa, $c_{12} = 134.6$ GPa and $c_{44} = 116.6$ GPa (see Table 5.3, p. 68 of [44]).

Iron (α-Fe) C_{hkl}-Values for Screw Dislocations on $\{110\}\langle\bar{1}11\rangle$

The basis vectors for screw dislocations in this system, can be determined from (10.13) and Figs. 10.1 and 10.3, and result in $\boldsymbol{e}_1 \| \frac{1}{\sqrt{6}}[-1,1,-2]$, $\boldsymbol{e}_2 \| \frac{1}{\sqrt{2}}[1,1,0]$ and $\boldsymbol{e}_3 \| \frac{1}{\sqrt{3}}[-1,1,1]$. The transformed elastic constants for α-Fe for screw dislocations have a matrix structure given by (10.21). The corresponding reduced compliances can be determined from (10.44) and substituting into (10.45e), the characteristic roots for screw dislocations in α-Fe were found to be $p_{1,2} = \mp 0.3839 + 0.6398\imath$ and $p_3 = 1.9093\imath$. This enables the $\mathbf{A}$, $\mathbf{L}$ and $\mathbf{D}$ matrices for the displacement field, $\boldsymbol{u}(x_1, x_2)$ and the elastic component of the contrast factors to be determined. In addition, the γ_i elements needed in the geometric components of this slip-system were found to be

$$\left\{\frac{-\mathrm{h}+\mathrm{k}-2\mathrm{l}}{\sqrt{6}\sqrt{\mathrm{h}^2+\mathrm{k}^2+\mathrm{l}^2}}, \frac{\mathrm{h}+\mathrm{k}}{\sqrt{2}\sqrt{\mathrm{h}^2+\mathrm{k}^2+\mathrm{l}^2}}, \frac{-\mathrm{h}+\mathrm{k}+\mathrm{l}}{\sqrt{3}\sqrt{\mathrm{h}^2+\mathrm{k}^2+\mathrm{l}^2}}\right\}.$$

Following the steps discussed above, the contrast factor for screw dislocations on the (1 1 0) slip-plane for the $[\bar{1}11]$ and $[1\bar{1}1]$ slip-directions in α-Fe are given in Fig. 10.5 and Table 10.3. These results also show the variation of the C_{hkl} for individual slip-systems, and for different hkl in the {hkl} that have the same d-spacing. The average contrast factor for α-Fe for screw dislocations was found to be $\langle C_{\mathrm{hkl}}^{\mathrm{screw}}\rangle = 0.3055 - 0.8060\Gamma(\mathrm{h,k,l})$.

Also shown in Table 10.3 are the $C_{\mathrm{hkl}}^{\mathrm{screw}}$ and $\langle C_{\mathrm{hkl}}^{\mathrm{screw}}\rangle$ values for molybdenum (Mo) and niobium (Nb). The anisotropy factors in this case ranges from 0.51 for Nb, 0.72 for Mo and 2.44 for α-Fe, while the c_{12}/c_{44}-ratio is 1.15 for α-Fe, 1.50 for Mo and 4.69 for Nb. The linear fits for Mo were found to be $\langle C_{\mathrm{hkl}}^{\mathrm{screw}}\rangle = 0.1912 - 0.2786\Gamma(\mathrm{h,k,l})$, and the linear equation for Nb was found to be $\langle C_{\mathrm{hkl}}^{\mathrm{screw}}\rangle = 0.1651 - 0.1146\Gamma(\mathrm{h,k,l})$. From these results, it is clear

Table 10.3. The contrast values, C_{hkl}, for screw dislocations on the (110) slip-plane in two different slip-directions (see Fig. 10.2) in anisotropic (real) $\alpha-$Fe, Mo and Nb. The anisotropy factor, $A = 2c_{44}/(c_{11} - c_{12})$ (for isotropic materials $A = 1$) and Poisson ratio, ν, for each material is also given

Material	hkl	$[\bar{1}11]$	$[1\bar{1}1]$	$C_{\text{hkl}}^{\text{screw}}$	$\langle C_{\text{hkl}}^{\text{screw}} \rangle$
$\alpha-$Fe	1 1 0	0.0201	0.0201	0.0201	0.1040
A=2.442	1 0 $\bar{1}$	0.1879	0.0201	0.1040	–
$\nu = 0.273$	2 0 0	0.3055	0.3055	0.3055	0.3055
$c_{11} = 230.1$ GPa	0 0 2	0.3055	0.3055	0.3055	–
$c_{12} = 134.6$ GPa	2 2 0	0.0201	0.0201	0.0201	0.1040
$c_{44} = 116.6$ GPa	2 0 2	0.0201	0.1879	0.1040	–
	2 1 1	0.0201	0.1443	0.0822	0.1040
	3 1 0	0.1826	0.1826	0.1826	0.2330
	2 2 2	0.0491	0.0491	0.0491	0.03684
	2 2 $\bar{2}$	0.0491	0.0491	0.0491	–
Mo	1 1 0	0.0041	0.0041	0.0041	0.1216
A=0.722	1 0 $\bar{1}$	0.2391	0.0041	0.1216	–
$\nu = 0.303$	2 0 0	0.1912	0.1912	0.1912	0.1912
$c_{11} = 463.0$ GPa	0 0 2	0.1912	0.1912	0.1912	–
$c_{12} = 161.0$ GPa	2 2 0	0.0041	0.0041	0.0041	0.1216
$c_{44} = 109.0$ GPa	2 0 2	0.0041	0.2391	0.1216	–
	2 1 1	0.0041	0.1936	0.0988	0.1216
	3 1 0	0.0956	0.0956	0.0956	0.1662
	2 2 2	0.1312	0.1312	0.1312	0.0984
	2 2 $\bar{2}$	0.1312	0.1312	0.1312	–
Nb	1 1 0	0.0163	0.0163	0.0163	0.1364
A=0.513	1 0 $\bar{1}$	0.2565	0.0163	0.1364	–
$\nu = 0.392$	2 0 0	0.1651	0.1651	0.1651	0.1651
$c_{11} = 246.5$ GPa	0 0 2	0.1651	0.1651	0.1651	–
$c_{12} = 134.5$ GPa	2 2 0	0.0163	0.0163	0.0163	0.1364
$c_{44} = 28.7$ GPa	2 0 2	0.0163	0.2565	0.1364	–
	2 1 1	0.0163	0.2187	0.1175	0.1364
	3 1 0	0.0827	0.0827	0.0827	0.1548
	2 2 2	0.1692	0.1692	0.1692	0.1269
	2 2 $\bar{2}$	0.1692	0.1692	0.1692	–

that the anisotropy factor has a strong influence on the slope of $\langle C_{\text{hkl}}^{\text{screw}}\rangle$ for bcc materials. That is, as the anisotropy factor increases from Nb to α-Fe, the slope in $\langle C_{\text{hkl}}^{\text{screw}}\rangle$ becomes increasingly negative. On the other hand, for these results no such clear trend in comparing c_{12}/c_{44}-ratio with the linear fits of $\langle C_{\text{hkl}}^{\text{screw}}\rangle$ can be seen.

Iron (α-Fe) C_{hkl}-Values for Edge Dislocations on $\{110\}\,\langle\bar{1}11\rangle$

The case of edge dislocations in α-Fe on the $(110)\,[\bar{1}11]$ slip-system follows similar steps as for the screw case. The basis vectors that define this system (see (10.14) and Figs. 10.1 and 10.3) were found to be $\boldsymbol{e}_1 \| \frac{1}{\sqrt{3}}[-1,1,1]$, $\boldsymbol{e}_2 \| \frac{1}{\sqrt{2}}[1,1,0]$ and $\boldsymbol{e}_3 \| \frac{1}{\sqrt{6}}[1,-1,2]$. The transformed elastic constants for α-Fe for edge dislocations have the matrix structure given in (10.22). The corresponding reduced compliances can be determined from (10.44) and substituting into (10.45e); the characteristic roots for edge dislocations in α-Fe were found to be $p_1 = 2.1415\imath$ and $p_{2,3} = \mp 0.2002 + 0.6473\imath$. This enables the **A**, **L** and **D** quantities for the displacement field, $\boldsymbol{u}(x_1, x_2)$ to be determined. The elastic component can then be evaluated, while the γ_i elements for this slip-system were found to be

$$\left\{\frac{-\text{h}+\text{k}+\text{l}}{\sqrt{3}\sqrt{\text{h}^2+\text{k}^2+\text{l}^2}}, \frac{\text{h}+\text{k}}{\sqrt{2}\sqrt{\text{h}^2+\text{k}^2+\text{l}^2}}, \frac{\text{h}-\text{k}+2\text{l}}{\sqrt{6}\sqrt{\text{h}^2+\text{k}^2+\text{l}^2}}\right\}$$

from which the geometric component for edge dislocations can be determined.

The $C_{\text{hkl}}^{\text{edge}}$ values for α-Fe on the (111) slip-plane in the $[\bar{1}11]$ and $[1\bar{1}1]$ slip-directions are given in Fig. 10.5b and Table 10.4. The average contrast factor for edge dislocations was determined to be $\langle C_{\text{hkl}}^{\text{edge}}\rangle = 0.2648 - 0.3468\Gamma(\text{h},\text{k},\text{l})$.

The edge C_{hkl} and $\langle C_{hkl}^{\text{edge}}\rangle$ values for Mo and Nb are also given in Table 10.4. The fitted equation for Mo was found to be $\langle C_{\text{hkl}}^{\text{edge}}\rangle = 0.1326 + 0.2412\Gamma(\text{h},\text{k},\text{l})$, while for Nb was $\langle C_{\text{hkl}}^{\text{edge}}\rangle = 0.1188 + 0.4475\Gamma(\text{h},\text{k},\text{l})$. Here we notice that the slopes in $\langle C_{\text{hkl}}^{\text{edge}}\rangle$ are positive. That is, going from screw to edge dislocations, the slope in $\langle C_{\text{hkl}}\rangle$ has changed signs from negative to positive. From these results we observe that as the c_{12}/c_{44} ratio increases from α-Fe to Nb, the y-intercept in $\langle C_{\text{hkl}}^{\text{edge}}\rangle$ decreases, while as the anisotropy factor increases from Nb to α-Fe, the sign in the slope for $\langle C_{\text{hkl}}^{\text{edge}}\rangle$ changes from positive to negative. This observation suggests that the anisotropy factor and the c_{12}/c_{44} ratio both affect the (average) contrast factor for edge dislocations in bcc materials [28].

10.4.3 An Unusual Case

Recently Scardi [37] suggested the problem of determining the contrast factor for Cu_2O (elastic constants, $c_{11} = 121\,\text{GPa}$, $c_{12} = 105\,\text{GPa}$ and $c_{44} = 12.1\,\text{GPa}$). This case is unusual from both the crystallographic and physical points of view.

Table 10.4. The contrast values, C_{hkl}, for edge dislocations on the (110) slip-plane in two different slip-directions (see Fig. 10.3) in anisotropic (real) α–Fe, Mo and Nb

Material	hkl	$[\bar{1}11]$	$[1\bar{1}1]$	$C^{\mathrm{edge}}_{\mathrm{hkl}}$	$\langle C^{\mathrm{edge}}_{\mathrm{hkl}}\rangle$
α–Fe	1 1 0	0.0752	0.0752	0.0752	0.1781
A=2.442	1 0 $\bar{1}$	0.3947	0.0225	0.2086	–
$\nu = 0.273$	2 0 0	0.3596	0.3596	0.3596	0.2648
$c_{11} = 230.1$ GPa	0 0 2	0.0751	0.0751	0.0751	–
$c_{12} = 134.6$ GPa	2 2 0	0.0752	0.0752	0.0752	0.1781
$c_{44} = 116.6$ GPa	2 0 2	0.0225	0.3947	0.2086	–
	2 1 1	0.0601	0.2415	0.1508	0.1781
	3 1 0	0.2472	0.2472	0.2472	0.2335
	2 2 2	0.0797	0.0797	0.0797	0.1492
	2 2 $\bar{2}$	0.0797	0.0797	0.0797	–
Mo	1 1 0	0.1531	0.1531	0.1531	0.1929
A=0.722	1 0 $\bar{1}$	0.3519	0.0114	0.1817	–
$\nu = 0.303$	2 0 0	0.1705	0.1705	0.1705	0.1326
$c_{11} = 463.0$ GPa	0 0 2	0.0567	0.0567	0.0567	–
$c_{12} = 161.0$ GPa	2 2 0	0.1531	0.1531	0.1531	0.1929
$c_{44} = 109.0$ GPa	2 0 2	0.0114	0.3519	0.1817	–
	2 1 1	0.0879	0.2058	0.1469	0.1929
	3 1 0	0.1493	0.1493	0.1543	0.1543
	2 2 2	0.1297	0.1297	0.1297	0.2130
	2 2 $\bar{2}$	0.1297	0.1297	0.1297	–
Nb	1 1 0	0.3054	0.3054	0.3054	0.2307
A=0.513	1 0 $\bar{1}$	0.3509	0.0305	0.1907	–
$\nu = 0.392$	2 0 0	0.1532	0.1532	0.1532	0.1188
$c_{11} = 246.5$ GPa	0 0 2	0.0500	0.0500	0.0500	–
$c_{12} = 134.5$ GPa	2 2 0	0.3054	0.3054	0.3054	0.2307
$c_{44} = 28.7$ GPa	2 0 2	0.0305	0.3509	0.1907	–
	2 1 1	0.1832	0.2838	0.2335	0.2307
	3 1 0	0.2067	0.2067	0.2067	0.1591
	2 2 2	0.2275	0.2275	0.2275	0.2680
	2 2 $\bar{2}$	0.2275	0.2275	0.2275	–

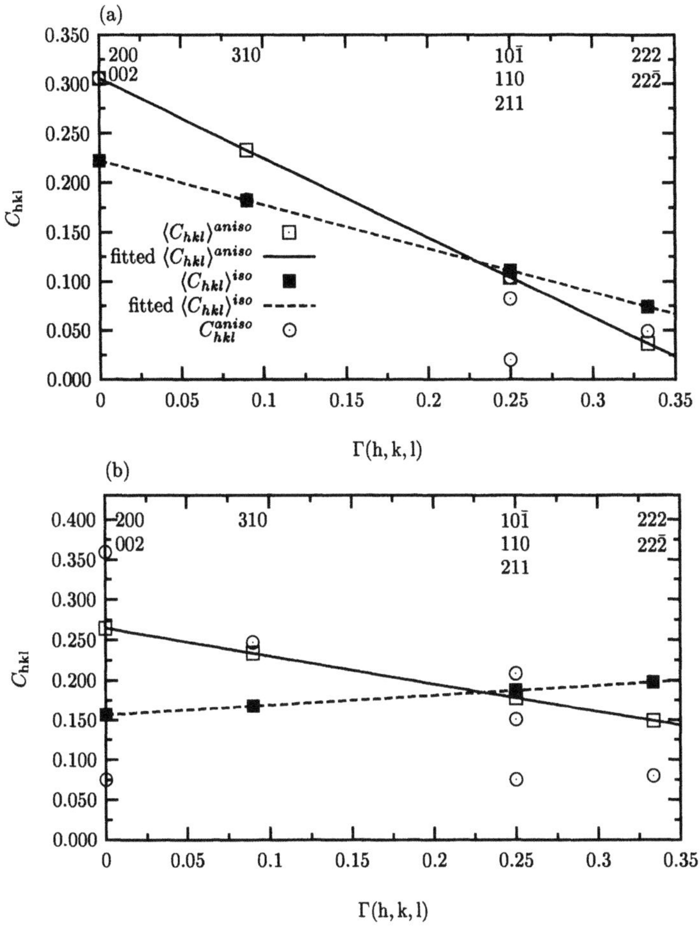

Fig. 10.5. Contrast factor values for α-Fe: (**a**) Screw dislocations for the (1 1 0) slip-plane, average values (for all slip-systems) in "isotropic" α-Fe and average values in real α-Fe (see Table 10.3). (**b**) Similar to (**a**), but with edge dislocations in α-Fe (see Table 10.4)

Cu_2O Slip-System

Referring to Fig. 10.6a it can be seen that there exists an fcc sublattice of Cu atoms (small spheres), with oxygen atoms (large spheres) on the tetrahedron face. Since Cu is the stronger scatterer, we expect the X-ray diffraction pattern to closely resemble that of a typical fcc lattice [37]. However, elastic properties of Cu_2O do not have a typical fcc slip-system. This can be explained by the presence of the O atoms on the tetrahedron face, which prevents slipping in the ⟨110⟩-directions. Consequently, the slip-system for Cu_2O resembles a simple cubic system. That is, slip occurs along the edges

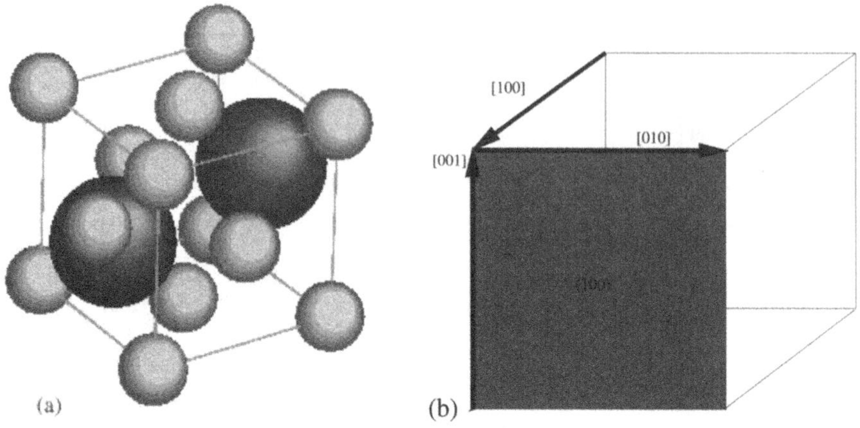

Fig. 10.6. The unusual case of Cu_2O: (**a**) A schematic of the Cu_2O lattice. The small spheres are the Cu and large spheres are the O. (**b**) A schematic representation of the $\{100\}\langle 001\rangle$ slip-system for the Cu_2O lattice

of the lattice, $\{100\}\,\langle 001\rangle$, schematically given in Fig. 10.6b [37][6]. The total number of slip-systems for Cu_2O is 6 slip-planes $\times$ 4 slip-directions per plane $=$ 24 systems. However, on appealing to symmetry only 6 slip-systems (3 slip-planes with 2 slip-directions perpendicular to each other – see Fig. 10.6b) are required to completely describe it.

Not only does Cu_2O have an unusual slip-system for a real material, it also has an anisotropy factor, $A = 1.51$ and a large $c_{12}/c_{44} = 8.68$. These two facts put Cu_2O outside the range of applicability for existing methods of determining the contrast factor to be used in X-ray line-profile analysis [28]. Consequently, this example highlights the need to be able to determine the contrast factor from a first-principles approach.

$C_{\mathbf{hkl}}$ Values for Screw Dislocations in Cu_2O

To illustrate the calculation of the contrast factor for this example we essentially concentrate on the (100) [001] slip-system.

Following the same steps as in the fcc and bcc examples (10.13) and Figs. 10.1 and 10.2, the slip-system for screw dislocations is defined by the basis vectors $\boldsymbol{e}_1 = [0, -1, 0]$, $\boldsymbol{e}_2 = [1, 0, 0]$ and $\boldsymbol{e}_3 = [0, 0, 1]$. The transformed $[6 \times 6]$ $\mathbf{c}'$ matrix has the following form (in GPa),

[6] Also see [43], p. 147, Sect. 5.1, in which observations of $\{100\}\,\langle 001\rangle$ have been made.

$$\mathbf{c}' = \begin{pmatrix} 121.0 & 105.0 & 105.0 & 0 & 0 & 0 \\ & 121.0 & 105.0 & 0 & 0 & 0 \\ & & 121.0 & 0 & 0 & 0 \\ & & & 12.1 & 0 & 0 \\ & & & & 12.1 & 0 \\ & & & & & 12.1 \end{pmatrix}, \tag{10.30}$$

while, the reduced compliances $\mathbf{S}$ (in units $\times 10^{-9}\,\mathrm{Pa}^{-1}$) have the form,

$$\mathbf{S} = \begin{pmatrix} 0.0335 & -0.0290 & 0 & 0 & 0 & 0 \\ & 0.0335 & 0 & 0 & 0 & 0 \\ & & 0 & 0 & 0 & 0 \\ & & & 0.0826 & 0 & 0 \\ & & & & 0.0826 & 0 \\ & & & & & 0.0826 \end{pmatrix}. \tag{10.31}$$

Using (10.31) in (10.45a–c)–(B.3), the resulting p_α-roots for screw (and edge) dislocations are $p_{1,2} = \pm 0.5625 + 0.8268\imath$ and $p_3 = \imath$. From the values of the characteristic roots, p_3 will produce $\boldsymbol{u}(x_1, x_2)$ that correspond to screw dislocations in an isotropic material, while $p_{1,2}$ will result in an edge dislocations in an elastically anisotropic material.

The $\boldsymbol{u}(x_1, x_2)$ for screw dislocations reduces to $\boldsymbol{u} = [0, 0, \frac{-b}{2\pi}\arctan(\frac{x_2}{x_1})]$. This results in an elastic component of the contrast factor whose only nonzero elements are $E_{33} = E_{66} = 1$. The γ_i elements for the geometric term have components

$$\left\{ -\frac{\mathrm{k}}{\sqrt{\mathrm{h}^2+\mathrm{k}^2+\mathrm{l}^2}}, \frac{\mathrm{h}}{\sqrt{\mathrm{h}^2+\mathrm{k}^2+\mathrm{l}^2}}, \frac{\mathrm{l}}{\sqrt{\mathrm{h}^2+\mathrm{k}^2+\mathrm{l}^2}} \right\}.$$

Averaging over all 6 of the slip-systems for $\{100\}\langle 001\rangle$, results in $\langle C_{\mathrm{hkl}}\rangle = \frac{4}{6}\Gamma(\mathrm{h,k,l})$. That is, in this case the $\langle C_{\mathrm{h00}}\rangle = 0$, clearly demonstrating that the geometric components will be the dominant term. It also highlights the "invisibility criterion", where in this particular case $\boldsymbol{g}_{\mathrm{h00}} \cdot \boldsymbol{b} = 0$ [41,43].

C_{hkl} Values for Edge Dislocations in Cu_2O

The slip-system for edge dislocations in Cu_2O is given by the following basis vectors, $\boldsymbol{e}_1 = [0,0,1]$, $\boldsymbol{e}_2 = [1,0,0]$ and $\boldsymbol{e}_3 = [0,1,0]$. However, transformed elastic constants and the reduced compliances do not change from (10.30) and (10.31), respectively.

The $p_{1,2}$ characteristic roots are the only terms determining the edge dislocation contrast factor values. As demonstrated above the $\mathbf{A}$, $\mathbf{L}$ and $\mathbf{D}$ matrices can be evaluated using $p_{1,2}$. The general form of the displacement fields for edge dislocations in Cu_2O is also different to the previously encountered fcc and bcc cases. That is, for real fcc and bcc materials there has been

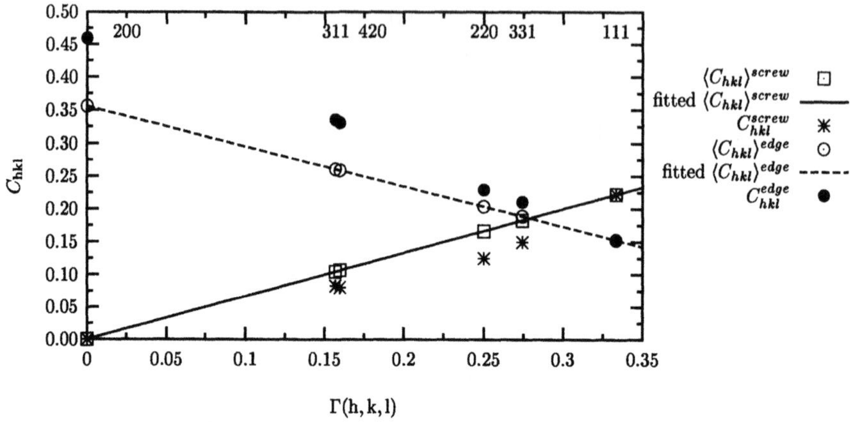

Fig. 10.7. The C_{hkl} and $\langle C_{\mathrm{hkl}} \rangle$ values for edge and screw dislocations in Cu_2O (also see Table 10.5)

a nonzero u_3 component; however, in this case the anisotropy is still seen in the u_1 and u_2 components and not in the u_3 component with $u_3 = 0$.

The elastic component matrix for edge dislocations in Cu_2O is given by

$$\mathbf{E} = \begin{pmatrix} 0.6086 & 0 & 0 & 0 & -0.5209 & 0 \\ & 0.6086 & 0 & -0.6809 & 0 & 0 \\ & & 0 & 0 & 0 & 0 \\ & & & 2.1022 & 0 & 0 \\ & & & & 0.4592 & 0 \\ & & & & & 0 \end{pmatrix} . \tag{10.32}$$

The geometric element has components

$$\left\{ \frac{\mathrm{l}}{\sqrt{\mathrm{h}^2+\mathrm{k}^2+\mathrm{l}^2}}, \frac{\mathrm{h}}{\sqrt{\mathrm{h}^2+\mathrm{k}^2+\mathrm{l}^2}}, \frac{\mathrm{k}}{\sqrt{\mathrm{h}^2+\mathrm{k}^2+\mathrm{l}^2}} \right\} .$$

Combining these terms in (10.8b), the contrast factors for edge dislocations can be determined for each slip system and averaged over the $\{100\}\langle 001\rangle$ system to produce the linear relationship $\langle C^{\mathrm{edge}}_{\mathrm{hkl}} \rangle = 0.3560 - 0.6094\Gamma(\mathrm{h,k,l})$.

In comparing $\langle C^{\mathrm{screw}}_{\mathrm{hkl}} \rangle$ and $\langle C^{\mathrm{edge}}_{\mathrm{hkl}} \rangle$ for Cu_2O in Table 10.5, we notice that screw dislocations produce a positive slope and zero intercept. However, in the case of edge dislocations, the slope is negative and the y-intercept is non-zero, which reflects the elastic anisotropy.

10.5 Conclusion

This article has presented a first-principles evaluation of the contrast factors needed to quantify the dislocation broadening from X-ray line profiles. In

Table 10.5. Showing the contrast values, C_{hkl} and $\langle C_{\mathrm{hkl}} \rangle$, for screw and edge dislocations on the (100) slip-plane in two different slip-directions (see Fig. 10.6b) in Cu_2O

Cu_2O	hkl	[001]	[010]	C_{hkl}	$\langle C_{\mathrm{hkl}} \rangle$
Screw	1 1 1	0.2222	0.2222	0.2222	0.2222
A=8.68	2 0 0	0	0	0	0
$c_{11} = 121\,\mathrm{GPa}$	2 2 0	0	0.25	0.1250	0.1667
$c_{12} = 105\,\mathrm{GPa}$	3 1 1	0.0826	0.0826	0.0826	0.1047
$c_{44} = 12.1\,\mathrm{GPa}$	3 3 1	0.0499	0.2493	0.1496	0.1828
	4 2 0	0	0.1600	0.0800	0.1067
Edge	1 1 1	0.1528	0.1528	0.1528	0.1528
	2 0 0	0.4593	0.4593	0.4593	0.3560
	2 2 0	0.1148	0.3438	0.2293	0.2036
	3 1 1	0.3353	0.3353	0.3353	0.2603
	3 3 1	0.1124	0.3086	0.2105	0.1888
	4 2 0	0.2939	0.3674	0.3307	0.2584

addition, a detailed discussion of dislocations in elastically anisotropic cubic materials has been presented. This has involved developing an understanding of the symmetry properties of the slip-system, which influences the separation of the "screw" and "edge" components of the compliance polynomial. It has been shown that deriving the fundamental physical properties of dislocations in real cubic materials is paramount in determining the dislocation contrast factor.

An overview of the broadening mechanisms of line profiles was presented, where it was pointed out that the lattice faulting/twinning, size and dislocation components of the broadening are generally convoluted together and it is necessary to separate these contributions (see Sect. 10.2). From the equations describing dislocation broadening, (10.5) and (10.6), it is evident that the dislocation contrast factor is critical to quantify the dislocation broadening.

In determining the dislocation contrast factor for cubic lattice materials, we have drawn on the approach of [31] and [32], who carried out the analysis for hexagonal lattice materials. This approach separates the contrast factor calculations into a geometric component and an elastic component. The geometric component describes the orientation of the diffraction vector, $\boldsymbol{g}$, in the slip-coordinate system of the dislocation, while the elastic component takes into account elastic properties of the material and the displacement fields of the dislocation in the lattice (see Sect. 10.3.1). Using a combination of the established physical theory [43] and mathematical physics [44] of dislocations in elastically anisotropic cubic materials, important properties of dislocations were discussed (see Sect. 10.3.2). It was shown that the characteristic roots, $\{p_\alpha, \forall\alpha\}$, from the compliance polynomial are dependent on the symmetry of the slip-system and that each root can be related to an edge and screw component of the displacement fields. This discussion has also shown how the p_α-roots are important in determining the **A**, **D** and **L** matrices which are used to find the displacement fields of dislocations (10.18–c). This discussion also shows that the properties of dislocations in anisotropic materials are both quantitatively and qualitatively different from dislocations in isotropic materials [43].

The numerical evaluations of the dislocation contrast factor were carried out for a range of fcc and bcc materials (see Sect. 10.4). These numerical examples highlight how a first-principles evaluation of the contrast factors can be applied to single slip-systems and can be extended to determine the average contrast factor over all slip-systems in the lattice [31,32]. The advantage of this approach is that a physical basis for the contrast factors is developed, and in turn it allows models to be developed for dislocations populating selected slip-planes for single crystal and textured materials; it also enables untextured specimens to be analysed. Moreover by developing a first-principles approach and carrying out numerical evaluations for a range of materials, empirical observations concerning the elastic properties of the materials, the type of dislocation and their affects on the contrast factors were made (see Sects. 10.4.1 and 10.4.2). Although there exist parameterised methods for determining the average contrast factor [28,34,35], such approaches do not give an insight into the displacement fields of the dislocations and do not enable the contrast factor for individual slip-system to be determined. Furthermore, the calculations presented here show how a first-principles approach is necessary in order to make the essential empirical observations needed to develop a parameterized method. To further supplement the advantages of a first-principle approach, the atypical material Cu_2O was suggested for analysis by [37] (see Sect. 10.4.3). This particular case is outside the range of applicability of the [28] method, with an anisotropy factor of 1.51 and $c_{12}/c_{44} = 8.68$. Further, the slip-system for Cu_2O was shown to be unusual, in that it followed a simple cubic slip-system, where the slip-planes are on the face of the cubes and the slip-directions along its edges.

In developing a first-principles approach for determining the contrast factors for real cubic materials, we have shown that an understanding of the elastic properties and dislocations in anisotropic materials, as well as the geometrical properties of the slip-system, is critical. This approach provides a greater flexibility in developing a particular problem and models for dislocations in various slip-systems, and how they may affect the line profile broadening. Moreover, the application of contrast factors in quantifying the dislocation broadening from X-ray line profiles places it on a level footing with other methods that are used to quantify the dislocation content in materials.

Acknowledgments

The first author would like to acknowledge that the basic research on contrast factor was carried out in March–October 1999, while he was a postdoctoral fellow at National Institute of Standards and Technology (NIST), USA; he would like to thank Dr Mark Vaudin for discussions concerning dislocations in real materials, and Dr. James P. Cline for encouraging this research work. We would also like to thank Dr. Walter Kalceff, Department of Applied Physics, UTS, for providing useful comments concerning this paper.

A Transforming the Elastic Constants and Diffraction Vector

The purpose of this appendix is to provide a self-contained outline of the mathematical tools needed to transform the elastic constants from their crystallographic coordinates to the coordinate slip-system of the dislocation. It essentially draws on the material presented in [42]. In turn, the transformation matrix that defines this mapping also defines the transformation of the diffraction vector in that coordinate slip-system.

A.1 Transforming the Elastic Constants

The elastic constants are defined as a 4-rank tensor, $\{c_{ijkl}; \forall\, i, j, k, l = 1, 2, 3\}$. This can be expressed as a matrix, $\boldsymbol{c} = \{c_{mn};\ m, n = 1, 2, 3 \ldots 9\}$ by pairing the indices ij and kl, such that $ij(kl) \to m(n)$, where $11 \to 1$, $22 \to 2$, $33 \to 3$, $23 \to 4$, $31 \to 5$, $12 \to 6$, $32 \to$, $13 \to 8$ and $21 \to 9$ [42].

The elastic constants, $\boldsymbol{c}$, are defined in the crystallographic coordinate system of the lattice and must be transformed into the coordinate system of the dislocation. This can be carried out by defining a set of basis vectors, $\{\boldsymbol{e}_i, i = 1, 2, 3\}$. The transformation matrix, $\mathbf{P}$ can be defined using these sets of vectors,

$$\mathbf{P} = \begin{bmatrix} \boldsymbol{e}_1 \\ \boldsymbol{e}_2 \\ \boldsymbol{e}_3 \end{bmatrix}, \qquad (10.33)$$

where $\boldsymbol{e}_i$ are expressed as row-vectors. This projection matrix enables both the $\boldsymbol{c}$ and $\boldsymbol{g}$ to be transformed into the coordinate system given by the basis vector, $\{\boldsymbol{e}_i, i = 1, 2, 3\}$.

In order to transform the $\boldsymbol{c}$ we define the 4-rank tensor, Q_{mnkl}, such that

$$Q_{mnkl} = P_{km} P_{ln} \qquad (10.34)$$

and on using the above algorithm Q_{mnkl} can be expressed as a $[9 \times 9]$ matrix, $\mathbf{Q}$. In transforming the elastic constants, such that $\boldsymbol{c} \to \boldsymbol{c}'$, we have

$$\begin{aligned} c'_{ijkl} &= Q_{ghij}\, c_{ghmn}\, Q_{mnkl} \\ &= Q^T_{ijgh}\, c_{ghmn}\, Q_{mnkl}\,, \end{aligned} \tag{10.35}$$

where T represents the transpose and the indices in (10.35) correspond to typical matrix multiplication. This enable us to express (10.35) as

$$\mathbf{c}' = \mathbf{Q}^T\, \mathbf{c}\, \mathbf{Q}\,. \tag{10.36}$$

Often the transformed matrix can be expressed as a reduced $[6 \times 6]$ matrix for a given slip-system. The $\boldsymbol{c}'$ can be used to evaluate the compliances, $\boldsymbol{s}$ and reduced compliances, $\boldsymbol{S}$, (10.44) (see Appendix B.2). These terms are essential in defining the sextic polynomial and characteristic roots, (10.45a–c)–(B.3), of the dislocation. This is the first step in determining the displacement fields of the dislocations in a given cubic lattice.

A.2 Transforming the Diffraction Vector

The transformation of the diffraction vector can be expressed as

$$\boldsymbol{g}' = \mathbf{P}\,\boldsymbol{g}\,, \tag{10.37}$$

where, $\boldsymbol{g}$ is a unit-vector and elements of $\boldsymbol{g}'$ define the direction cosines between the scattering vector and the ith axis. This in turn defines the, $\{\gamma_i; i = 1, 2, 3\}$ elements which are needed to determine the 4-rank geometric tensor of the contrast factor. It is also worth noting that $\mathbf{P}$ is an orthonormal matrix, such that $\mathbf{P}^T\mathbf{P} = \mathbf{I}$, where $\mathbf{I}$ is an identity matrix.

B Determining A, L and D for $\boldsymbol{u}(x_1, x_2)$

This appendix provides a brief outline of the algorithm for determining the **A**, **L** and **D** matrices needed to evaluate $\boldsymbol{u}(x_1, x_2)$. For a detailed discussion related to the mathematical physics of dislocations see [44] (Chaps. 6, 7 and 10). In addition, [42] and [43] present discussions that concentrate on the physics and observations of dislocations in real materials.

B.1 General Field Equations

The general field equations for plane strain in an anisotropic elastic body are

$$\sigma_{ij} = c_{ijkl}\, \epsilon_{kl}\,, \tag{10.38}$$

where σ_{ij} is the stress tensor, ϵ_{kl} the strain tensor and c_{ijkl} are the 4-rank tensor containing the elastic constants. Using the transformation laws for tensors, (10.38) can be expressed as

$$\epsilon_{ij} = s_{ijkl}\, \sigma_{kl}\,, \tag{10.39}$$

where we define the compliance tensor, s_{ijkl} as

$$c_{ijkl}\, s_{klmn} = \frac{1}{2}\left(\delta_{im}\delta_{jn} + \delta_{in}\delta_{jm}\right)\,. \tag{10.40}$$

Expressing the solution in the form $u_k \equiv u_k(x_1, x_2, x_3)$ the strain terms in (10.38) can be defined as

$$\epsilon_{kl} = \frac{1}{2}\left(\partial_m\, u_k + \partial_l\, u_m\right) \quad \text{for} \quad k, l, m = 1, 2, 3\,. \tag{10.41}$$

The equilibrium conditions demand that

$$\partial_j \sigma_{ij} = 0\,, \tag{10.42}$$

where $\partial_i \equiv \partial\,(\cdot)\,/\partial\, x_i$ for convenience. The solution of the field equations must obey the boundary conditions of dislocations (see [44], Sect. 10.2). The problem can be simplified if a coordinate system is orientated such that the x_3 axis is parallel to the dislocation line vector, $\boldsymbol{l}$. The displacement fields, (10.18–c), become independent of x_3 as given in (10.18–c). The **A**, **L**, and **D** matrices in (10.18–c) need to be evaluated; this is outlined below.

B.2 Determining the Reduced Elastic Compliances

The transformation of the elastic constants into the slip-coordinate system enables the displacement fields of the dislocation (10.18a) to be evaluated. The elastic compliances are essentially the inverse of the [6 × 6] elastic constants[7] and arise after expressing the strain in terms of stress, with compliance becoming the constant of proportionality (see [44], (10.4) p. 135),

$$\boldsymbol{c}'\boldsymbol{s}' = \mathbf{I}\,, \tag{10.43}$$

where the compliance has units of $10^{-9}\,\text{Pa}^{-1}$ if the elastic constants have units of GPa. The reduced compliances arise from a further simplification of the field equations. That is, if the coordinate system for the displacement equations is only in terms of x_1 and x_2, (see Sect. 10.3.2), they reduce to five simultaneous equations to be solved,

$$S_{MN} = s'_{MN} - s'_{M3}\, s'_{N3}/s'_{33}\,, \tag{10.44}$$

where $S_{M3} = S_{3N} = 0$ for $M, N = 1, 2, 4, 5, 6$, can be thought of as a [5 × 5] matrix[8], such that $\boldsymbol{S} \equiv [S_{MN}]$ and will also have physical units of $10^{-9}\,\text{Pa}^{-1}$.

[7] The [9 × 9] matrix representation of the elastic constants is singular since the 7th-9th column vectors are linearly dependent.

[8] An alternative method for determining $\boldsymbol{S}$ is to drop the third column and row of $\boldsymbol{c}$ and take its inverse.

B.3 Determining the Compliance Polynomial

Using the reduced compliances, the polynomial can be determined using the following

$$l_2(p) = S_{55}\,p^2 - 2S_{45}\,p + S_{44} \tag{10.45a}$$

$$l_3(p) = S_{15}\,p^3 - (S_{14} + S_{56})\,p^2 + (S_{25} + S_{46})\,p - S_{24} \tag{10.45b}$$

$$l_4(p) = S_{11}\,p^4 - 2S_{16}\,p^3 + (2S_{12} + S_{66})\,p^2 - S_{26}\,p + S_{22}\,, \tag{10.45c}$$

where the indices of S_{MN} are taken from (10.44). These terms are used to form the sextic equations,

$$\begin{aligned} l(p) &= l_2(p)\,l_4(p) - l_3^2(p) \\ &= 0\,. \end{aligned} \tag{10.45d}$$

Using roots which have positive imaginary parts, the ordering of the roots become important in determining the coefficients for (10.18b).

B.4 Determining the L Matrix

Using the selected roots from the sextic equation, the **L** matrix becomes

$$\mathbf{L} = \begin{bmatrix} -p_1 & -p_2 & -p_3\lambda_3 \\ 1 & 1 & \lambda_3 \\ -\lambda_1 & -\lambda_2 & -1 \end{bmatrix}, \tag{10.46}$$

where $\lambda_1 = -\frac{l_3(p_1)}{l_2(p_1)}$, $\lambda_2 = -\frac{l_3(p_2)}{l_2(p_2)}$ and $\lambda_3 = -\frac{l_3(p_3)}{l_4(p_3)}$ and $\mathbf{L} \equiv [L_{k\,\alpha}]$ for k, $\alpha = 1, 2, 3$.

B.5 Determining the $A_{k\,\alpha}$ Coefficients

Calculating the coefficients $A_{k\alpha}$ for $\alpha = 1, 2$, we have

$$\begin{aligned} A_{1\alpha} &= S_{11}\,p_\alpha^2 - S_{16}\,p_\alpha + S_{12} + \lambda_\alpha(S_{15}\,p_\alpha - S_{14}) \\ A_{2\alpha} &= \frac{1}{p_\alpha}\left[S_{12}\,p_\alpha^2 - S_{26}\,p_\alpha + S_{22} + \lambda_\alpha(S_{25}\,p_\alpha - S_{24})\right] \\ A_{3\alpha} &= \frac{1}{p_\alpha}\left[S_{14}\,p_\alpha^2 - S_{46}\,p_\alpha + S_{24} + \lambda_\alpha(S_{45}\,p_\alpha - S_{44})\right] \end{aligned} \tag{10.47a}$$

and for $\alpha = 3$, we have

$$\begin{aligned} A_{13} &= \lambda_3(S_{11}\,p_3^2 - S_{16}\,p_3 + S_{12}) + S_{15}\,p_3 - S_{14} \\ A_{23} &= \frac{1}{p_3}\left[\lambda_3(S_{12}\,p_3^2 - S_{26}\,p_3 + S_{22}) + S_{25}\,p_3 - S_{24}\right] \\ A_{33} &= \frac{1}{p_3}\left[\lambda_3(S_{14}\,p_3^2 - S_{46}\,p_3 + S_{24}) + S_{45}\,p_3 - S_{44}\right]. \end{aligned} \tag{10.47b}$$

B.6 Determining the D_α Coefficients

The D_α coefficients that are needed in the evaluation of the potential function can be simply expressed as

$$D_\alpha = -\frac{\sum_{k=1}^{3} L_{k\,\alpha}\, b_k}{2\sum_{m=1}^{3} A_{m\,\alpha}\, L_{m\,\alpha}} \quad \text{for} \quad \alpha = 1, 2, 3\,, \tag{10.48}$$

where the $A_{m\alpha}$ terms are the same as those in (10.18b); b_k are the components of the Burgers vector, $\boldsymbol{b}$; and the coefficients require the characteristics roots, p_α.

The p_α and $\boldsymbol{L}$ terms have no physical units. If the elastic constants have units of GPa ($\times 10^9$ Pa), then $\boldsymbol{A}$ has units of $10^{-9}\,\mathrm{Pa}^{-1} (\equiv 10^{-9}\,\mathrm{m}^2/\mathrm{N})$ and $\boldsymbol{D}$ has units $10^{-9}\,\mathrm{N\,m}^{-1}$ for (10.18–c) to have units of meters, m.

References

1. B.E. Warren: *X-ray Diffraction.* (Addison-Wesley, Massachusetts 1969)
2. A.R. Stokes and A.J.C. Wilson: Proc. Camb. Phil. Soc. **38**, 313 (1942)
3. A.R. Stokes and A.J.C. Wilson: Proc. Phys. Soc. Lond. **56**, 174 (1944)
4. B.E. Warren and B.L. Averbach: J. Appl. Phys. **21**, 595 (1950)
5. B.E. Warren and B.L. Averbach: J. Appl. Phys. **23**(4), 497 (1952)
6. G.K. Williamson and W.H. Hall: Acta Met. **1**, 22 (1953)
7. A. Krivoglaz and K.P. Ryaboshapka: Fiz. metal. metalloved. **15**(1), 18 (1963)
8. M. Wilkens: NBS Spec. Publ. 317 **2**, 1195 (1970); Proceedings of *Fundamental aspects of dislocation theory.* J.A. Simmons and R. de Witt and R. Bullough (Eds.)
9. M. Wilkens: Phys. Stat. Sol. A **2**, 359 (1970)
10. M.A. Krivoglaz, O.V. Martynenko and K.P. Ryaboshapka: Phys. Met. Metall. **55**(1), 1 (1983)
11. I. Groma, T. Ungár and M. Wilkens: J. Appl. Cryst. **21**, 47 (1988)
12. J.G. van Berkum, R. Delhez, T. de Keijser and E.J. Mittemeijer: Acta. Cryst. **A52**, 730 (1996)
13. I. Groma: Phys. Rev. B **57**(13), 7535 (1998)
14. I. Groma and G. Monnet: J. Appl. Cryst. **35**, 589 (2002)
15. J.I. Langford and D. Louër: J. Appl. Cryst. **15**, 20 (1982)
16. D. Louër, J.P. Auffrédic and J.I. Langford: J. Appl. Cryst. **16**, 183 (1983)
17. J.I. Langford: NIST Spec. Pub. **846**, 110 (1992)
18. C.E. Krill and R. Birringer: Phil. Mag. **A77**(3), 621 (1998)
19. J.I. Langford, D. Louër and P. Scardi: J. Appl. Cryst. **33**, 964 (2000)
20. P. Scardi, M. Leoni and Y.H. Dong: Eur. Phys. J. B **18**, 23 (2000)
21. G. Ribárik ,T. Ungár and J. Gubicza: Appl. Cryst. **34**, 669 (2001)
22. N. Armstrong, W. Kalceff, J.P. Cline and J. Bonevich: J. Res. Nat. Inst. Stand. Techn. (2001) Submitted. *Proceedings of Accuracy in Powder Diffraction III, 22–25 April 2001, NIST* (Gaithersburg, USA 2001)
23. P. Scardi and M. Leoni: Acta Cryst. **A57**, 604 (2001)
24. T. Ungár and A. Borbély: J. Appl. Lett. **69**(21), 3173 (1996)

25. E. Wu, E.M.A. Gray and E.H. Kisi: J. Appl. Cryst. **31**, 356 (1998)
26. E. Wu, E.H. Kisi and E.M.A. Gray: J. Appl. Cryst. **31**, 363 (1998)
27. T. Ungár, M. Leoni and P. Scardi: J. Appl. Cryst. **32**, 290 (1999)
28. T. Ungár, I. Dragomir, A. Révész and A. Borbély: J. Appl. Cryst. **32**, 992 (1999)
29. R.W.Cheary, E. Dooryhee, P. Lynch, N. Armstrong and S. Dligatch: J. Appl. Cryst. **33**, 1271 (2000)
30. M. Wilkens: Phys. Stat. Sol. **104**, K1 (1987)
31. P. Klimanek and R. Kužel: J. Appl. Cryst. **21**, 59 (1988)
32. R. Kužel and P. Klimanek: J. Appl. Cryst. **21**, 363 (1988)
33. T. Ungár and G. Tichy: Phys. Stat. Sol. (a) **171**, 425 (1999)
34. I.C. Dragomir and T. Ungár: J. Appl. Cryst. **35**, 556 (2002)
35. I.C. Dragomir and T. Ungár: Powd. Diffract. **17**(2), 104 (2002)
36. P. Lynch: *X-ray diffraction line broadening from gold thin film structures.* PhD thesis, Department of Applied Physics, University of Technology, Sydney, Australia 2002
37. P. Scardi. Private communication (25 March – 03 April 2002)
38. M. Wilkens: *X-ray line broadening of plastically deformed crystals.* In: "Proc. 5th Riso Int. Symp. Mater. Sci. 1984". N.H. Andersen (Ed.) pp. 153–168
39. A. Guinier: *X-ray diffraction in crystals, imperfect crystals and amorphous bodies.* (Freeman, San Francisco 1963). Reprinted by Dover (New York) 1994
40. T. Ungár, J. Gubicza, G. Ribárik and A. Borbély: J. Appl. Cryst. **34**, 298 (2001)
41. D. Hull and D.J. Bacon: *Introduction to dislocations.* 4th edn. (Butterworthh-Heinemann, Oxford 2001)
42. J.P. Hirth and J. Lothe: *Theory of dislocations.* (Wiley, New York 1982)
43. J.W. Steeds: *Introduction to anisotropic elasticity theory of dislocations.* (Clarendon, Oxford 1973)
44. C. Teodosiu: *Elastic models of crystal defects*: (Springer, Berlin Heidelberg New York 1982)
45. J.D. Eshelby, W.T. Read and W. Shockley: Acta Metall. **1**, 251 (1953)
46. L.J. Teotonico: *Dislocations in anisotropic elastic media.* In: "Mathematical theory of dislocations". T. Mura (Ed.) (Amer. Soc. Mech. Eng., New York 1969) pp. 49–69

11 X-ray Peak Broadening Due to Inhomogeneous Dislocation Distributions

I. Groma and A. Borbély

An overview of the theory of X-ray line broadening caused by dislocation is presented. It is shown that the properties of the tails of the profile are determined by the average dislocation density, the dislocation density fluctuation, and the dislocation–dislocation correlation length. The obtained asymptotic behaviour is compared with the predictions of earlier models. The influence of the finite coherent domain size is also analysed. For the determination of the statistical parameters characterising the dislocation ensemble investigated a generalised form of Wilson's variance method is outlined. The proposed evaluation method is demonstrated on line profiles measured on plastically deformed Cu single and AlMg polycrystals.

11.1 Introduction

X-ray line broadening analysis is proved to be an extremely powerful method to study the defect properties of crystalline materials (see e.g. [8, 12, 14, 15, 20–27, 29–34]). Since different type of defects create strongly different profile shapes, in principle the defect content of the sample investigated can be determined from a detailed analysis of the measured intensity distribution. However, this is not an easy task. There are several open theoretical and experimental problems which need to be solved for achieving a really satisfactory profile analysis method.

It is well known, the strongest line broadening (i.e. the slowest decay rate of the diffracted intensity at the tail of the intensity distribution) is caused by finite grain size and 3 dimensional defects (like sub-grains), resulting that coherent X-ray diffraction takes place only in domains with finite sizes. The theoretical description of the profile shape caused by finite domain size is relatively straightforward. The Fourier transform of the intensity distribution is proportional to the auto-correlation function of the domain shape, i.e. the overlapping volume of the domain itself and the domain shifted by the Fourier parameter. If the domain size has a distribution the scattered intensity is the weighted average of the intensity profiles created by the different size domains. Although, the mathematical formulation of the problem is relatively simple, the determination of the grain size distribution from a measured X-ray profile is still a challenge. This problem is beyond the scope of the present chapter, we refer the reader to Ungár et al. [25–27], Scardi et al. [29–31], and Amstrong and Kalceff [1].

For crystalline materials another possible source of line broadening is the displacement of the atoms from the periodic lattice positions. The first theory describing this internal strain induced Bragg peak broadening was proposed by Warren and Averbach [35, 36]. By the Taylor expansion of the Fourier transform of the peak they found that the strain induced broadening scales with the spatial average of the square of the strain $\langle \varepsilon^2 \rangle$. However, if the sources of the strain are dislocations $\langle \varepsilon^2 \rangle$ diverges logarithmically with the specimen size [39]. This is a consequence of the 1/r type decay of the strain field created by a dislocation. This means that in its original form the Warren–Averbach method cannot be applied for dislocations. (Other defects like point defects, stacking faults, APBs, or disclinations create weaker broadening. They are not treated in this chapter.)

For a completely random dislocation distribution Krivoglaz and co-workers [17–19] worked out an analytical expression. They found that the Fourier transform of the intensity distribution has the form

$$A(L) = \exp\left\{ \rho L^2 \ln\left(\frac{R_{\mathrm{cryst}}}{L} \right) \right\} , \tag{11.1}$$

where L is the Fourier parameter, ρ is the dislocation density, and R_{cryst} is a parameter proportional to the crystal dimension. A serious problem with this result is that the $A(L)$ diverge with the crystal size which is not observed experimentally. On an indirect way this indicates, that a real dislocation distribution cannot be completely random.

The problem was first resolved by Wilkens [37–42] by calculating the line profile for a so-called restricted random dislocation distribution. In his model the restriction of the randomness means that there are areas with cross-section A for which the number of intersecting dislocations are precisely $N_A = \rho A$. For a completely random distribution only the expected value of the intersecting dislocations would be N_A. With this a characteristic length scale the dislocation–dislocation correlation length $R_c = \sqrt{A}/\pi$ is introduced. This formally eliminate the crystal size dependence obtained by Krivoglaz and Pyaboshapka [17] because according to Wilkens results [39] for small enough Fourier parameters the Fourier coefficients follow expression (11.1) but the crystal size is replaced by R_c. A considerable achievement of Wilkens theory is that for this special dislocation–dislocation correlation he also derived an approximate formula for larger Fourier parameters (see [40]) i.e. he gave the shape of the entire profile. Nevertheless, the Wilkens theory is not satisfactory for many reasons. First of all, the concept of restricted random dislocation distribution is quite artificial, not being proved by any dislocation dynamic model. It is a rather formal solution of the divergence problem. As it will be shown later the crystal size dependence can be eliminated under much more general conditions. Beside these, a significant shortcoming of the Wilkens model is that it predicts symmetric profile shape in contrast with common experimental findings. Although, asymmetric broadening can be understood by combining Wilkens theory with the concept of the quasi-composite model [22] it is based on a particular dislocation model that we need to assume in order to be able to evaluate the measured line profiles. Furthermore, according to Gaál's investigations [9] a polarised dislocation dipole ensemble can also produce an asymmetric profile. So, asymmetric broadening can appear under much more general conditions than those considered in the quasi-composite model.

The purpose of the present chapter is to demonstrate that the tails of the intensity distributions can be described by an analytical expression, the derivation of which does not require any assumption about the dislocation distribution. The tails of the Bragg peak depend on some fundamental statistical parameters of the dislocation network, like the average dislocation density and the average dislocation density fluctuation. It is shown that by a generalised form of the variance method originally proposed by Wilson [43,44] these parameters can be determined with high accuracy from the measured profile. The influence of finite coherent domain size is also analysed.

11.2 Properties of the Fourier Transform of Intensity Distribution Induced by Dislocations

Within the framework of the kinematic scattering theory the scattered intensity on uniform atoms positioned at the points $\boldsymbol{r}_i$ $i = 1, \ldots, N$ has the form

$$I(\boldsymbol{\kappa}) = S^2 \sum_{j,l}^{N} \exp\left[i\boldsymbol{\kappa}(\boldsymbol{R}_j - \boldsymbol{R}_l)\right] , \tag{11.2}$$

where $\boldsymbol{\kappa}$ is the scattering vector, and S is the atomic scattering factor (a constant in the problem considered). For a crystalline material in which the atoms are displaced from their perfect periodic positions by a displacement field $\boldsymbol{u}(\boldsymbol{r})$ (11.2) simplifies to (keeping only the first order term)

$$\begin{aligned} I(q) = \frac{S^2}{V} \int dL \int dr^3 \exp\left\{2\pi i \boldsymbol{g}\left[\boldsymbol{u}\left(\boldsymbol{r} + \frac{L}{2}\boldsymbol{t}\right) - \boldsymbol{u}\left(\boldsymbol{r} - \frac{L}{2}\boldsymbol{t}\right)\right]\right\} \\ \times \exp(2\pi i q L) \,, \end{aligned} \tag{11.3}$$

where V is the crystal volume, $\boldsymbol{g}$ is the reciprocal lattice vector of the Bragg peak considered, $\boldsymbol{t}$ is a unit vector parallel to $\boldsymbol{g}$, and $q = 2(\sin(\Theta) - \sin(\Theta_0))/\lambda$ in which λ is the wavelength, and Θ_0 is the exact Bragg angle. It is important to note that the above given expression corresponds to the so called deformation broadening where the originally 3 dimensional intensity distribution is integrated for the plane perpendicular to the diffraction vector. (It can be measured by the appropriate swinging of the sample.) As it can be seen, the intensity distribution given by (11.3) is the Fourier transform of

$$A(L) = \frac{1}{V}\int dr^3 \exp\left\{2\pi i \boldsymbol{g}\left[\boldsymbol{u}\left(\boldsymbol{r}+\frac{L}{2}\boldsymbol{t}\right)-\boldsymbol{u}\left(\boldsymbol{r}-\frac{L}{2}\boldsymbol{t}\right)\right]\right\} . \quad (11.4)$$

First its properties are investigated.

To derive the form of $A(L)$ for dislocated crystals let us consider N straight dislocations with line vector parallel to the z-axis, with Burgers vectors $\pm\boldsymbol{b}$, and positioned at the points $\boldsymbol{r}_i$, $i = 1,\ldots,N$ in the xy-plane. In this case, the total displacement field appearing in (11.4) is

$$\boldsymbol{u}(\boldsymbol{r}) = \sum_{j=1}^{N} \boldsymbol{u}_{\text{sing}}(\boldsymbol{r}-\boldsymbol{r}_j) , \quad (11.5)$$

in which $\boldsymbol{u}_{\text{sing}}(\boldsymbol{r})$ is the displacement field of a single dislocation. For the further analysis it is useful to introduce the quantities

$$E(L) = \ln\{A(L)\} \quad (11.6)$$

and

$$B(\boldsymbol{r},L) = 1-\exp\left\{2\pi i \boldsymbol{g}\left[\boldsymbol{u}_{\text{sing}}\left(\boldsymbol{r}+\frac{L}{2}\boldsymbol{e}\right)-\boldsymbol{u}_{\text{sing}}\left(\boldsymbol{r}-\frac{L}{2}\boldsymbol{e}\right)\right]\right\} , \quad (11.7)$$

where $\boldsymbol{e}$ is the projection of $\boldsymbol{t}$ to the xy-plane. After substituting (11.5) into (11.4, 11.6) one can find that $E(L)$ can be transformed into a power series of B

$$\begin{aligned} E(L) = -\frac{1}{F}\int dr^2 \Big\{ &\int dr^2 \left[\rho_+(\boldsymbol{r}_1)B(\boldsymbol{r}-\boldsymbol{r}_1,L)+\rho_-(\boldsymbol{r}_1)B^*(\boldsymbol{r}-\boldsymbol{r}_1,L)\right] \\ &+\int dr_1^2 \int dr_2^2 \left[d_{++}(\boldsymbol{r}_1,\boldsymbol{r}_2)B(\boldsymbol{r}-\boldsymbol{r}_1,L)B(\boldsymbol{r}-\boldsymbol{r}_2,L)\right. \\ &+d_{+-}(\boldsymbol{r}_1,\boldsymbol{r}_2)B(\boldsymbol{r}-\boldsymbol{r}_1,L)B^*(\boldsymbol{r}-\boldsymbol{r}_2,L) \\ &+d_{-+}(\boldsymbol{r}_1\boldsymbol{r}_2)B^*(\boldsymbol{r}-\boldsymbol{r}_1,L)B(\boldsymbol{r}-\boldsymbol{r}_2,L) \\ &\left.+d_{--}(\boldsymbol{r}_1,\boldsymbol{r}_2)B^*(\boldsymbol{r}-\boldsymbol{r}_1,L)B^*(\boldsymbol{r}-\boldsymbol{r}_2,L)\right]+\ldots\Big\} , \end{aligned} \quad (11.8)$$

where $\rho_+(\boldsymbol{r})$ and $\rho_-(\boldsymbol{r})$ are the density of positive and negative sigh dislocations, $d_{i,j}(\boldsymbol{r}_1,\boldsymbol{r}_2)$ is the correlation function of dislocations with corresponding signs, F is the cross-section of the sample, and * indicate complex conjugation. (In order to obtain (11.8) the summations were replaced by weighted integrals. This involves the assumption that the discrete dislocation density functions can be replaced by their statistical averages.)

In the first step the leading term of (11.8)

$$f^0(\boldsymbol{r}, L) = \int dr_1^2 \rho_+(\boldsymbol{r} - \boldsymbol{r}_1) B(\boldsymbol{r}_1, L) \tag{11.9}$$

is investigated. One can try to approximate $f^0(\boldsymbol{r}, L)$ by a power series of the Fourier parameter L. However, as it was first pointed out by Krivoglaz and Ryaboshapka [17] at $L = 0$ the second (and higher) derivatives of $f^0(\boldsymbol{r}, L)$ are infinite. The divergence is due to the fact that the distortion created by a dislocation is inversely proportional to the distance from the dislocation line, $\beta_{i,j}(\boldsymbol{r}) = \partial(\boldsymbol{u}_{\text{sing}})_i(\boldsymbol{r})/\partial r_j \approx 1/|\boldsymbol{r}|$. The calculation method explained below is based on the fact that, if the coefficient before $B(\boldsymbol{r}_1, L)$ in the integrand of expression (11.9) is zero at $\boldsymbol{r}_1 = 0$, then the second derivative of $f^0(\boldsymbol{r}, L)$ would be finite at $L = 0$. So, instead of $f^0(\boldsymbol{r}, L)$ the expression

$$f^1(\boldsymbol{r}, L) = \int dr_1^2 \left\{\rho_+(\boldsymbol{r} - \boldsymbol{r}_1) - \rho_+(\boldsymbol{r} - 2\boldsymbol{r}_1)\right\} B(\boldsymbol{r}_1, L) \tag{11.10}$$

is considered, where the above singularity in the second derivative of $f^1(\boldsymbol{r}, L)$ does not appear. The newly defined $f^1(\boldsymbol{r}, L)$ and $f^0(\boldsymbol{r}, L)$ are not independent from each other. One can find (see [10, 11]) that

$$f^1(\boldsymbol{r}, L) = f^0(\boldsymbol{r}, L) - \frac{1}{4} f^0(\boldsymbol{r}, 2) \,. \tag{11.11}$$

This is strictly related to the fact that in any elastic medium for any type of straight dislocation $\boldsymbol{u}_{\text{sing}}(\boldsymbol{r})$ fulfils the "expansion" relation

$$\boldsymbol{u}_{\text{sing}}(\boldsymbol{r}_1) - \boldsymbol{u}_{\text{sing}}(\boldsymbol{r}_2) = \boldsymbol{u}_{\text{sing}}(k\boldsymbol{r}_1) - \boldsymbol{u}_{\text{sing}}(\boldsymbol{r}_2) \,, \tag{11.12}$$

in which k is an arbitrary constant. Equation (11.12) is equivalent with the $1/r$ type decay of the stress field of a straight dislocation.

Since the first two derivatives of $f^1(\boldsymbol{r}, L)$ are finite, for small enough L values $f^1(\boldsymbol{r}, L)$ can be approximated by a second order polynomial leading to the function equation

$$f^0(\boldsymbol{r}, L) - \frac{1}{4} f^0(\boldsymbol{r}, 2L) = a_1(\boldsymbol{r}) L + a_2(\boldsymbol{r}) L^2 \,. \tag{11.13}$$

Its general solution is

$$f^0(\boldsymbol{r}, L) = 2a_1(\boldsymbol{r}) L - \frac{1}{\ln(2)} a_2(\boldsymbol{r}) L^2 \ln(L) + \frac{1}{2} W L^2 \,, \tag{11.14}$$

where the term $WL^2/2$ is the solution of the homogeneous equation

$$f^0(\boldsymbol{r}, L) - \frac{1}{4} f^0(\boldsymbol{r}, 2L) = 0 \,. \tag{11.15}$$

So, the coefficient W cannot be determined from (11.13). It is the second derivative of

$$k(\boldsymbol{r},L)=f^0(\boldsymbol{r},L)+\frac{1}{\ln(2)}a_2(\boldsymbol{r})L^2\ln(L) \tag{11.16}$$

at $L=0$ which is already finite. It is important to note that in agreement with the result of Krivoglaz and Ryaboshapka [17] obtained for a completely random dislocation distribution, W diverges logarithmically with the specimen size. However, the higher order terms in (11.8) depending on the dislocation-dislocation correlation can cancel this divergence, since it can be found that the leading terms in their expansion are proportional to L^2. For details see [11]. According to this, it can be stated that by taking into account dislocation–dislocation correlation the unphysical crystal size dependence of the line profile can be eliminated under conditions much more general than those considered in the theory of Wilkens.

The actual values of $a_1(\boldsymbol{r})$ and $a_2(\boldsymbol{r})$ can be obtained from the derivatives of $f^1(\boldsymbol{r},L)$. Without going into the details (see [11]) we give that

$$a_2(\boldsymbol{r})=-\ln(2)\Delta\rho_+(\boldsymbol{r})\,, \tag{11.17}$$

where

$$\Delta=|\boldsymbol{g}||\boldsymbol{b}|C=\oint\limits_{|\boldsymbol{r}|=1} g_i\beta_{ij}(\boldsymbol{r})\beta_{ij}(\boldsymbol{r})g_i ds\,. \tag{11.18}$$

(The function $a_1(\boldsymbol{r})$ does not have contribution to the final result, so it is not given.)

The parameter C is often referred to contrast factor. Its value was determined by Wilkens for isotropic media [40], and by Klimanek and Kužel for hexagonal materials [16]. The general case was considered by Amstrong [2], and Borbély et al. [6]. For obtaining numerical values of C an interactive code is available on the web site `http://metal.elte.hu/anizc`.

The method outlined above can be generalised for the investigation of the properties of the higher order terms in (11.8). It can be found [11] that up to forth order in L

$$\begin{aligned}\ln(A(L))=&\Lambda\langle\rho\rangle L^2\ln\left(\frac{L}{R_1}\right)+i\langle s^{(2)}\rangle L^3\ln\left(\frac{L}{R_2}\right)\\&+\frac{1}{2}\Lambda^2\left[\langle\rho^2\rangle-\langle\rho\rangle^2\right]L^4\ln\left(\frac{L}{R_3}\right)\ln\left(\frac{L}{R_4}\right)\,,\end{aligned} \tag{11.19}$$

where $\langle\rho\rangle$ is the average dislocation density, $\langle\rho^2\rangle-\langle\rho\rangle^2$ is the average of the dislocation density fluctuation, R_1, R_2, R_3, and R_4 are parameters with length dimension determined by the different order dislocation-dislocation correlation functions, and

$$\langle s^{(2)}\rangle=\pi\frac{g_lg_m}{|\boldsymbol{g}|}\Lambda\int\int T(\boldsymbol{r},\boldsymbol{r}-\boldsymbol{r}_1)\beta_{lm}(\boldsymbol{r}_1)dr_1^2dr^2\,. \tag{11.20}$$

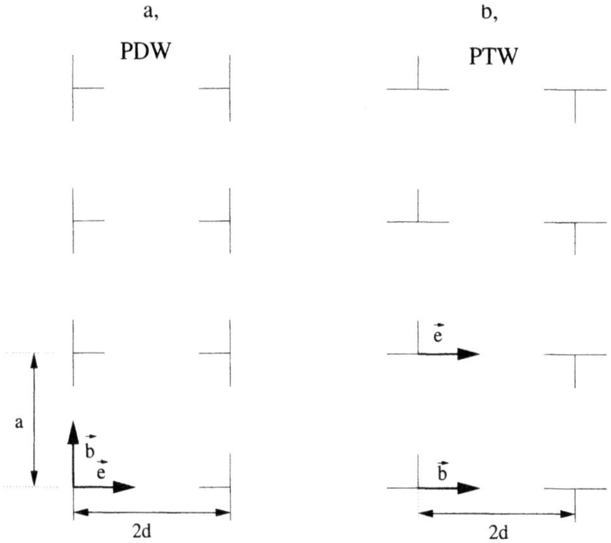

Fig. 11.1. The two types of dislocation walls investigated. (**a**) Polarised dipolar wall where the Burgers vector is perpendicular to the dipole direction. (**b**) Polarised tilt wall, where the Burgers vector is parallel to the dipole direction [13]

In (11.20) $\beta_{lm}(\boldsymbol{r})$ is the distorsion created by a single dislocation, and

$$\begin{aligned} T(\boldsymbol{r}_1, \boldsymbol{r}_2) = & [\rho_+(\boldsymbol{r}_1) + \rho_-(\boldsymbol{r}_1)] [\rho_+(\boldsymbol{r}_1) - \rho_-(\boldsymbol{r}_1)] \\ & + d_{--}(\boldsymbol{r}_1, \boldsymbol{r}_2) + d_{++}(\boldsymbol{r}_2, \boldsymbol{r}_1) - d_{--}(\boldsymbol{r}_1, \boldsymbol{r}_2) - d_{--}(\boldsymbol{r}_2, \boldsymbol{r}_1) \\ & \times d_{-+}(\boldsymbol{r}_1, \boldsymbol{r}_2) + d_{-+}(\boldsymbol{r}_2, \boldsymbol{r}_1) - d_{+-}(\boldsymbol{r}_1, \boldsymbol{r}_2) - d_{+-}(\boldsymbol{r}_2, \boldsymbol{r}_1) \,. \end{aligned} \tag{11.21}$$

According to the well known properties of the Fourier transformation the real part of $\ln(A(L))$ given by (11.19) determines the symmetric, while the imaginary one the antisymmetric part of the intensity distribution. It can be seen, that the symmetric part of the X-ray peak depends on the average dislocation density and the dislocation density fluctuation. The quantities $\langle \rho \rangle$ and $\langle \rho^2 \rangle$ have well defined physical meanings, resulting that the interpretation of the parameters one can obtain from the symmetrical part of the intensity distribution does not require any a priori assumption about the dislocation distribution. However, this is not the case for the antisymmetrical part. In the imaginary part of $\ln(A(L))$ the coefficient $\langle s^{(2)} \rangle$ is a complicated function of the dislocation–dislocation correlation function. For two experimentally interesting dislocation configurations sketched in Fig. 11.1 $\langle s^{(2)} \rangle$ was determined by Groma and Monnet [13]. The dislocation mechanism they form by, and their possible role in the plastic deformation of crystalline material are explained in details in [24]. For the one called polarised dipolar wall (Fig. 11.1a)

$$\langle s^{(2)} \rangle = \pi \frac{\langle \eta \rangle |\boldsymbol{g}||\boldsymbol{b}|}{a^2} \Lambda \sin^2(\Psi) \left[\cos^2(\gamma) - \frac{\nu}{1-\nu} \sin^2(\gamma) \right] , \tag{11.22}$$

while for the other one often referred to as polarised tilt wall (Fig. 11.1b)

$$\langle s^{(2)} \rangle = \pi^2 \frac{\langle \eta \rangle |\boldsymbol{g}||\boldsymbol{b}|}{a^2} \frac{\Lambda}{1-\nu} \sin^2(\Psi) \sin(2\gamma) \frac{d}{a} \exp\left\{ -\frac{2\pi d}{a} \right\} , \qquad (11.23)$$

where $\boldsymbol{b}$ is the Burgers vector, $\langle \eta \rangle$ is the number of walls per unit length, ν is the Poisson ratio, Ψ is the angle between the diffraction vector $\boldsymbol{g}$ and the dislocation line vector $\boldsymbol{l}$, γ is the angle between the vectors $\boldsymbol{b} - \boldsymbol{l}(\boldsymbol{lb})$ and $\boldsymbol{g} - \boldsymbol{l}(\boldsymbol{lg})$, and a and d are defined according to Fig. 11.1.

11.3 Peak Broadening Due to Narrow Dislocation Dipoles

Most dislocation ensembles produced by plastic deformation consist of relatively narrow dislocation dipoles. So, it is interesting to investigate the X-ray peak broadening created by idealised point-like dipoles [18]. A dipole is point-like if at any distance from it its displacement field has the form

$$\boldsymbol{u}_d = \boldsymbol{p} \frac{d\boldsymbol{u}_{\text{sing}}}{d\boldsymbol{r}} , \qquad (11.24)$$

where $\boldsymbol{p}$ is the dipole moment. With other words, condition (11.24) means that the dislocation dipole cannot be resolved into two dislocations at any length scale. This is certainly a strongly idealised dislocation configuration. The limits of the obtained results arising from this will be discussed at the end of the section.

By a procedure similar to the one applied for individual dislocations we can found that in this case

$$E(L) = \frac{1}{F} \int dr^2 \int dr_1^2 P(\boldsymbol{r}_1) \qquad (11.25)$$
$$\times \left[1 - \exp\left\{ 2\pi i \boldsymbol{g} \left[\boldsymbol{u}_d \left(\boldsymbol{r} + \frac{L}{2}\boldsymbol{e} \right) - \boldsymbol{u}_d \left(\boldsymbol{r} - \frac{L}{2}\boldsymbol{e} \right) \right] \right\} \right] ,$$

where $P(\boldsymbol{r})$ is the dipole density. If the system is homogeneous and L is small enough (11.26) simplifies to

$$E(L) = P \int dr^2 \left[1 - \exp\left\{ 2\pi i \boldsymbol{g}\boldsymbol{e} \left[\frac{d\boldsymbol{u}_d(\boldsymbol{r})}{d\boldsymbol{r}} \right] L \right\} \right] . \qquad (11.26)$$

Since the derivative of the displacement field $\boldsymbol{u}_d(\boldsymbol{r})$ has the form

$$\frac{d\boldsymbol{u}_d(\boldsymbol{r})}{d\boldsymbol{r}} = \frac{K(x/y)}{x^2 + y^2} , \qquad (11.27)$$

where $K(x/y)$ is a trigonometric polynomial, with the variable substitution $\boldsymbol{r} \to \sqrt{|L|}/\boldsymbol{\xi}$ in the integrand of (11.26) we get

$$E(L) = P \int \left[1 - \cos \left\{ 2\pi \boldsymbol{g e} \left[\frac{d\boldsymbol{u}_d(\boldsymbol{\xi})}{d\boldsymbol{\xi}} \right] \right\} \right] d\boldsymbol{\xi}^2 |L| = -G|L| \,, \tag{11.28}$$

(the imaginary part of $E(L)$ vanishes). The above result means that for a system of point-like dislocation dipoles the Fourier transform of the intensity distribution can be approximated by

$$A(L) = \exp\{-G|L|\} \,, \tag{11.29}$$

i.e. the X-ray peak has Lorentzian shape. This result was first derived by Krivoglaz [18]. However, it is important to note that for a real dislocation configuration in which the dipoles have a finite dipole width, (11.28) is valid only for L values larger than the dipole width. For small enough Fourier parameter L, $A(L)$ always follows the form given by (11.19). As it is explained in the next section this results that the tail of the intensity distribution always deviates from the Lorentz function. It decays faster than $1/q^2$, that the Lorentzian would follow.

11.4 Asymptotic Properties of the Intensity Distribution

In the previous section the analytical behaviour of the Fourier transform of the X-ray peak was analysed. In this section the connection between (11.19) and the different order moments of the scattered intensity are established. From the integration limit dependence of the moments the properties of the tails of the peak profiles are derived.

As it is well known the kth order moments of $I(q)$ defined as

$$m_k = \frac{\int_{-\infty}^{\infty} q^k I(q) dq}{\int_{-\infty}^{\infty} I(q) dq} \tag{11.30}$$

can be determined from the Fourier transform of $I(q)$. Namely

$$m_k = \frac{(i)^k}{(2\pi)^k} \frac{1}{A(0)} \frac{d^k}{dL^2} A(L) \bigg|_{L=0} \,. \tag{11.31}$$

By substituting (11.28) into (11.31) it can be seen, that for peak broadening created by dislocations the second and the higher order moments are infinite. As a consequence of these, the moments themselves cannot be used for the evaluation of measured peak profiles.

In order to obtain applicable results let us analyse the integration limit dependence of the different order moments, i.e. the properties of the function

$$v_k(q) = \frac{\int_{-q}^{q} q'^k I(q') dq'}{\int_{-\infty}^{\infty} I(q') dq'} \,, \tag{11.32}$$

in which q' is measured from the centre of gravity of the intensity distribution. In the following the function $\nu_k(q)$ will be referred to as kth order restricted moment. As a first step the behaviour of the second order restricted moment $\nu_2(q)$ is analysed. For overcoming the difficulties arising from the singularity mentioned above let us introduce the function

$$A_1(L) = A(L) - \frac{1}{4}A(2L) \quad (11.33)$$

and its Fourier transform

$$I_1(q) = \int_{-\infty}^{\infty} A_1(L)\exp(i2\pi Lq)dL\,. \quad (11.34)$$

The second derivative of $A_1(L)$ at $L = 0$ is already finite. So, from (11.19, 11.31, 11.33) one can find that

$$\lim_{q\to\infty}\int_{-q}^{q} q'^2 I_1(q')dq' = -\frac{1}{4\pi^2}\frac{d^2}{dL^2}A_1(L)\bigg|_{L=0} = \frac{\Lambda}{2\pi^2}\langle\rho\rangle\ln(2)\,. \quad (11.35)$$

On the other hand, it can be obtained from the (11.34) definition of $I_1(q)$ that

$$\int_{-q}^{q} q'^2 I_1(q')dq' = [v_2(q) - v_2(2q)]\int_{-\infty}^{\infty} I(q')dq'\,. \quad (11.36)$$

By combining equations (11.35, 11.36) one concludes that for large enough q values $\nu_2(q)$ fulfil the function equation

$$v_2(q) - v_2(2q) = \frac{\Lambda}{2\pi^2}\langle\rho\rangle\ln(2)\,. \quad (11.37)$$

Its general solution is

$$v_2(q) = \frac{\Lambda}{2\pi^2}\langle\rho\rangle\ln\left(\frac{q}{q_0}\right)\,, \quad (11.38)$$

where q_0 is a constant the value of which is not determined by (11.37).

With a similar procedure outlined above it can be found (see [9]) that

$$v_3(q) = \frac{3}{4\pi^3}\langle s^{(2)}\rangle\ln\left(\frac{q}{q_1}\right) \quad (11.39)$$

and

$$v_4(q) = \frac{\Lambda}{4\pi^2}\langle\rho\rangle q^2 + \frac{3\Lambda^2}{4\pi^4}\langle\rho^2\rangle\ln^2\left(\frac{q}{q_2}\right)\,, \quad (11.40)$$

where q_1 and q_2 are constants. It follows from equations (11.38–11.40) that the tail of the diffracted intensity is described by the expression:

$$I(q) = \frac{\Lambda}{4\pi^2}\langle\rho\rangle\frac{1}{|q|^3} + \frac{3}{8\pi^3}\langle s^{(2)}\rangle\frac{q}{|q|^5} + \frac{3\Lambda}{8\pi^4}\langle\rho^2\rangle\frac{\ln(|q|/q_2)}{|q|^5}\,. \tag{11.41}$$

It is important to note that (11.41) corresponds to the normalisation

$$\int_{-\infty}^{\infty} I(q)dq = 1\,.$$

As it can be seen the leading term in the tail of $I(q)$ is inversely proportional to the cube of the diffraction parameter q, and it is independent from the actual form of the dislocation arrangement. It is determined only by the length of dislocation lines per unit volume. In other words, if the dislocations are rearranged (keeping the total length of dislocation line constant) the far enough tails of the peak profile remain unchanged.

This asymptotic behaviour is certainly also valid for the narrow dipole system considered in the previous section. In this case the central part of the peak is Lorentzian but the tail part decays with $1/q^3$. The point up to which the peak can be well approximated with a Lorentz curve depends on the average dipole distance. For narrower dipoles the interval is larger.

11.5 Evaluation Procedure

A commonly applied method for the evaluation of the measured X-ray Bragg peak profiles is that the function given by the expression (11.19) is fitted to the Fourier transform of the measured intensity distribution for small enough Fourier parameters [8]. For the determination of the dislocation density and the correlation length R_1 a frequently used simple method is to fit a straight line to the initial part of the $\ln(|A(L)|)/L^2$ versus $\ln(L)$ plot. According to (11.19) the slope of the straight line is proportional to the dislocation density, and R_1 can be determined from the intercept [8,10]. This is demonstrated on a profile obtained on a fatigued austenitic steel (Fig. 11.2). (We mention here that the profile analysed was measured on a double crystal diffractometer with negligible instrumental broadening.) However, for strongly inhomogeneous dislocation systems often developing during plastic deformation, the third term in (11.19) proportional to the dislocation density fluctuation cannot be neglected. In this case, according to (11.19) there are five free parameters we have to determine from the Fourier coefficients. A further complication is that expression (11.19) is valid only for small Fourier parameters. The range in which it describes well the Fourier coefficients depends on the length parameters R_1, R_2, and R_3 that are not known in advance. This makes difficult to choose the correct fitting interval. In most cases this results in a

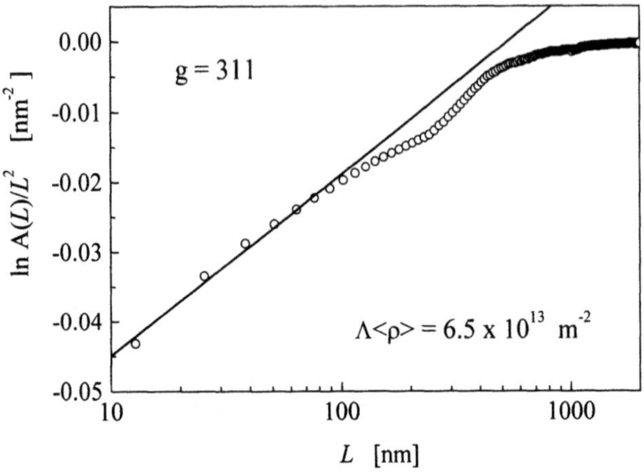

Fig. 11.2. The $\ln(|a(L)|)/L^2$ versus $\ln(L)$ plot with the straight line fitted at the initial part of the curve. The peak profile measurement was performed on a fatigued austenitic steel

quite large scatter of the parameters determined. As a consequence of this the method explained above can be applied only under restricted circumstances.

An alternative approach for the evaluation of the X-ray peaks first proposed by Groma et al. [11, 12] is to use the second, third, and fourth order restricted moments. The method is a generalisation of Wilson's variance method [20, 43, 44] commonly applied for the determination of the coherent domain size. In the original form of the variance method only the second order restricted moment (the variance) is analysed, and the dislocation induced profile broadening was not taken into account. In the following first it is outlined how the dislocation density, the fluctuation of the dislocation density, and the parameter $\langle s^{(2)} \rangle$ can be determined if the coherent domain size is so large that the broadening caused by the finite size can be neglected beside the strain induced one. After this the influence of the finite domain size is investigated.

According to (11.38) the second order restricted moment increases logarithmically with the scattering parameter q if q is large enough. So, by fitting a straight line to the asymptotic part of the $\nu_2(q) \rightarrow \ln(q)$ plot the average dislocation density can be determined from the slope of the fitted straight line. This is demonstrated on a measurement obtained on a Cu single crystal deformed by compression up to 40 MPa resolved shear stress. Figure 11.3a shows the measured intensity distribution. The corresponding $\nu_2 \rightarrow \ln(q)$ plot and the fitted straight line are drawn in Fig. 11.3b. As it can be seen, there is a well defined regime where the linear relationship between $\nu_2(q)$ and $\ln(q)$ is satisfied. There are about 100 data points in this regime, so the dislocation density can be obtained with high accuracy from the slope of the fitted line. It is important to stress however, that in the regime where $\nu_2(q)$ increases linearly with $\ln(q)$ the scattered intensity relative to the peak maximum is

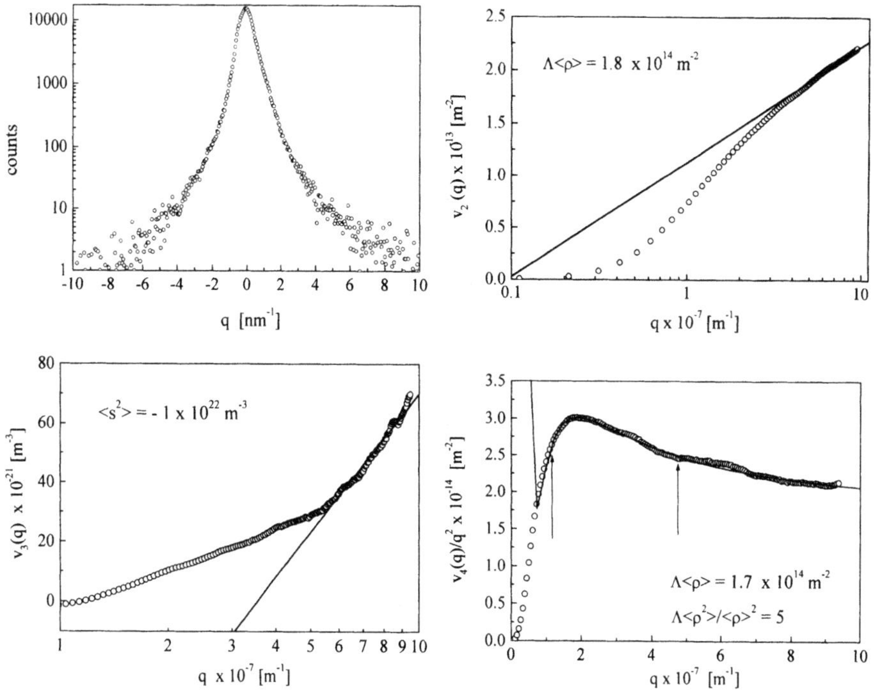

Fig. 11.3. (**a**) 200 reflection of a Cu single crystal deformed in compression. (**b**) The second order restricted moment together with the fitted straight line (according to (11.38)). (**c**) The third order restricted moment together with the fitted straight line (according to (11.39)). (**d**) The function $f(q)$ defined by (11.42) together with the fitted theoretical curve given by (11.43). The arrows indicate the fitting interval

smaller than about 2×10^{-3}. It means, the applicability of the above analysis requires that the tail of the Bragg peak can be measured up to about 10^{-4} relative intensity with appropriate precision. Certainly, the same holds for the Fourier analysis explained above.

A similar procedure can be used for the determination of $\langle s^{(2)} \rangle$. Namely, according to (11.39) the asymptotic part of $\nu_3(q)$ also varies linearly with $\ln(q)$. So, by fitting a straight line to the asymptotic part of $\nu_3(q) \rightarrow \ln(q)$ plot the parameter $\langle s^{(2)} \rangle$ can be obtained from the slope of the straight line. This is demonstrated in Fig. 11.3c corresponding to the same measurement explained above. Since the third order restricted moment measures the anti-symmetry of the Bragg peak which is relatively small in most cases, the scatter of the data points in this plot are larger than in the $\nu_2(q) \rightarrow \ln(q)$ plot (Fig. 11.3b). In spite of this the slope of the fitted strait line can be determined within about 20% precision.

Both the dislocation density and its fluctuation can be obtained from the fourth order restricted moment, the asymptotic behaviour of which is given by (11.40). For further considerations it is useful to introduce the quantity

$$f(q) = v_4 \frac{(q)}{q^2} . \tag{11.42}$$

For large enough q values it varies as

$$f(q) = \frac{\Lambda}{4\pi^2} \langle \rho \rangle + \frac{3\Lambda^2}{4\pi^4} \langle \rho^2 \rangle \frac{\ln^2(q/q_2)}{q^2} . \tag{11.43}$$

Two important properties of this function are that its asymptotic value is

$$f_{as} = \frac{\Lambda}{4\pi^2} \langle \rho \rangle , \tag{11.44}$$

and it has a maximum at $q_m = e \cdot q_2$ with a value of

$$f_{\max} = \frac{\Lambda}{4\pi^2} \langle \rho \rangle + \frac{3\Lambda^2}{4\pi^4} \langle \rho^2 \rangle \frac{1}{q_m^2} . \tag{11.45}$$

The difference between the maximum and the asymptotic value can be related to the relative dislocation density fluctuation as follows:

$$\frac{f_{\max} - f_{as}}{f_{as}^2} = 12 \frac{\langle \rho^2 \rangle}{\langle \rho \rangle^2} \frac{1}{q_m^2} . \tag{11.46}$$

In Fig. 11.3d the quantity $f(q)$ is plotted as a function of q together with the fitted theoretical function (11.43). The fit involving three free parameters was performed between the limits indicated by the two arrows in Fig. 11.3d. It is remarkable that the fitted curve follows the measured data points extremely well even for q values considerable larger than the upper fitting limit. This indicates that the function (11.43) obtained theoretically describes the measured curve with high accuracy [32–34]. It is important to note that the determination of the parameters requires a nonlinear fitting procedure, which demands an initial guess for the parameter values. For this the dislocation density and $q_2 = q_0$ obtained from the second order restricted moment can be applied. For the relative fluctuation $\langle \rho^2 \rangle / \langle \rho \rangle^2 = 1$ is usually a good initial guess.

In order to determine the regime where the asymptotic expression given above describes the tails of the profile, the measured profile and the calculated $f(q)$ quantity are plotted in Fig. 11.4. It can be seen that at the maximum of $f(q)$ the relative intensity is approximately 10^{-2} and the theoretical expression given by (11.43) describes well the tail part of the intensity distribution from about 10^{-1} relative intensity.

The advantages of the method can be summarised as follows. The regime from which the information is obtained consists of approximately 100 data points. So, one can easily judge the reliability of the results. Furthermore, since the average dislocation density $\langle \rho \rangle$ can be determined from both the second and fourth order restricted moments, the evaluation procedure offers the possibility an internal checking. Finally, it can be directly checked whether

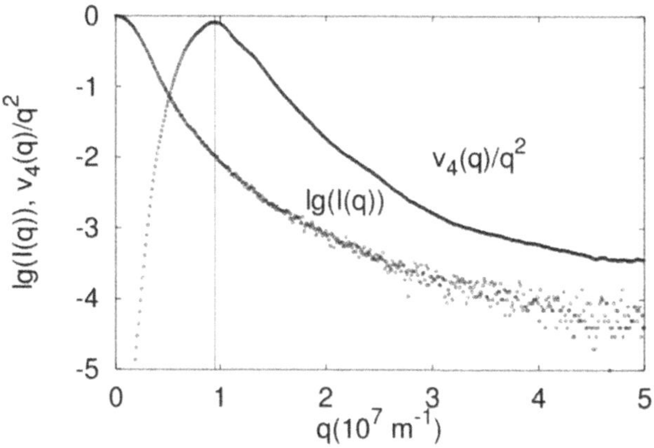

Fig. 11.4. High resolution X-ray peak which show pure strain (AlMg at Rt), pure size (nano Fe), and combined (AlMg at 400°C) type broadening. Maximum peak intensities are 35 000 for (RT), 31 000 for 400°C AlMg and 2000 counts for nano Fe

the profile is measured in a sufficiently wide q range, because in order to obtain proper values of the fitted parameters, the downhill part of $f(q)$ has to be reached. This cannot be directly seen if the Fourier coefficients are analysed. This visual checking can indicate the presence of peak overlapping too.

11.6 Influence of the Finite Coherent Domain Size

According to Wilson [43] if the X-ray diffraction occurs on a crystalline having small ($< 1\,\mu$m) coherent domain size the tail of the intensity distribution decays as

$$I(q) = \frac{1}{\pi^2 \varepsilon_f} \frac{1}{q^2}, \tag{11.47}$$

where ε_f is the average column length [7] or area weighted particle size [43] measured in the direction of the diffraction vector $\boldsymbol{g}$ (it will be simply referred as particle size in the following). An important feature of the above result is that the intensity distribution is independent from $|\boldsymbol{g}|$.

On the basis of this for isotropic particle shape Ungár and Borbély [25] have proposed a method referred to as modified Warren–Averbach (WA) [35,36] analysis for the determination of the particle size and the dislocation density by the simultaneous evaluation of different Bragg reflections. A fundamental assumption of both the classical and the modified (WA) analysis is that the total broadening created by the size and the strain is the convolution of the two separated broadening terms. If the strain is due to dislocations, the Fourier transform of the intensity distribution fulfils the relation

$$\ln(|A(L)| = \ln(|A_s(L)| + |E(L)|, \tag{11.48}$$

where $A_s(L)$ corresponds to the size induced broadening and $E(L)$ is given by (11.19). Since $E(L)$ depends on $|\boldsymbol{q}|$ only through $\Lambda(\lambda \propto |\boldsymbol{g}|^2)$ and $A_s(L)$ is independent from the diffraction vector, by plotting the Fourier coefficients measured at different $\boldsymbol{g}$ versus $\Lambda|\boldsymbol{q}|^2$ the size and strain broadening can be separated. Certainly, the analysis requires the knowledge of $\Lambda(\boldsymbol{g})$ in advance, which is often nontrivial to determine, because the $\Lambda(\boldsymbol{g})$ we need is an appropriate weighted average of the contrast factors of different types of dislocations distributed on different slip systems. It has to be mention that the applicability of the method relays on two conditions. It need to be assumed that the diffraction vector dependence of the parameters R_1, R_3 and R_4 can be neglected. Beside this, one has to assume that the mean value of the square of the contrast factor (appearing in the third term in (11.19)) can be approximated by the square of the mean value of the contrast factor, i.e. $\langle \Lambda^2 \rangle \approx \langle \Lambda \rangle^2$. For details of the evaluation method the reader is referred to [3, 5].

An alternative approach which does not require the simultaneous evaluation of different Bragg peaks is to use the second and fourth order restricted moments (see [4]). If the size broadening has measurable contribution it can be found from (11.38, 11.45, 11.47) that for large enough q values

$$v_2(q) = \frac{2}{\pi^2 \varepsilon_f} q + \frac{\Lambda}{2\pi^2} \langle \rho \rangle \ln\left(\frac{q}{q_0}\right) \tag{11.49}$$

and

$$f(q) = \frac{2}{3\pi^2 \varepsilon_f} q + \frac{\Lambda}{4\pi^2} \langle \rho \rangle . \tag{11.50}$$

(In both cases only the two leading terms are kept.) As can be seen, in $\nu_2(q)$ the leading term related to finite size is linear in q, while dislocations result in a logarithmic q dependence. The difference between the size and strain contribution is more pronounced on $f(q)$. In the asymptotic regime the term due to finite size is again proportional to q, but the dislocation created internal stain results a positive constant shift of the corresponding straight line. These characteristic behaviours can be easily realised by eye and one can determine whether the broadening is influenced by size effect. In order to show how the size and dislocation structure parameters can be obtained from the quantitative analysis of in $\nu_2(q)$ and $f(q)$, line profiles measured on three different materials corresponding to pure strain, pure size, and mixed type broadening are investigated.

Figure 11.5 shows high resolution X-ray peaks obtained on an AlMg sample deformed in compression at room temperature (RT), an AlMg sample hot rolled at 400°C, and a ball milled then annealed iron powder. The measurements were performed on a double-crystal diffractometer with negligible instrumental broadening. The obtained $f(q)$ functions are presented

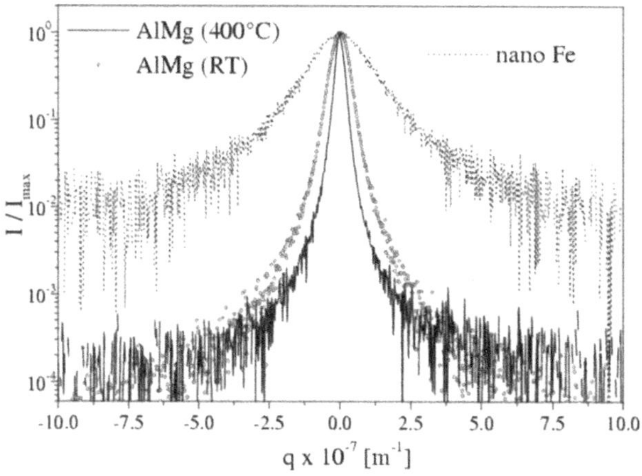

Fig. 11.5. Three different characteristic behaviours of the forth order restricted moment divided by q^2

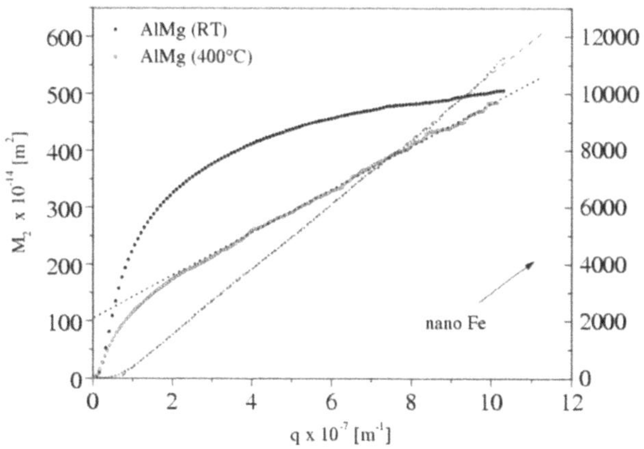

Fig. 11.6. Three different characteristic behaviours of the second order restricted moment

in Fig. 11.6 for the three different samples. The characteristic feature of the curve of the AlMg alloy deformed in compression ($\varepsilon = 5\%$) is the same as the one obtained on the deformed Cu single crystal (Fig. 11.3d). It tends to a constant value, i.e. the coherent domain size is so big that it does not have measurable influence. The broadening is induced only by the dislocations. The behaviour of $f(q)$ is much simpler for the iron powder. Apart from the vicinity of the origin, it can be fitted by a straight line with an intercept close to zero. According to (11.50) this indicates that peak broadening is only due to size effect. After annealing dislocations did not remain in the powder. The curve obtained on the hot rolled AlMg sample is a combination of the two

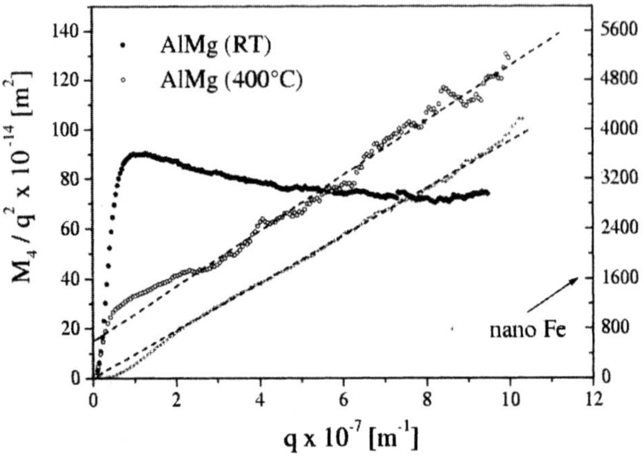

Fig. 11.7.

previous ones. The asymptotic part is linear indicating a finite size effect, but the fitted straight line has an intercept at a well defined positive value which is proportional to the dislocation density stored in the sample.

The second order restricted moments plotted in Fig. 11.7 also have very typical behaviour. The curves corresponding to the iron powder and the hot rolled AlMg sample show the expected linear behaviour. In contrast, the compressed AlMg sample exhibits logarithmic q dependence in complete agreement with the results obtained above. Although, the second order restricted moments could be used for the determination of both the domain size and the dislocation density, according to our experience, if the broadening is influenced by the size, the dislocation density cannot be determined with satisfactory accuracy from the second order restricted moment (it gives only the size with high precision). For the determination of the dislocation density the analysis of $f(q)$ is recommended. However, the simultaneous evaluation of $f(q)$ and $\nu_2(q)$ offers a kind of internal consistency checking. Namely, as it is well known, the Bragg peaks measured on samples with small particle sizes often have relatively high (about 10^{-2}) background to peak maximum ratio (for dislocated single crystals it is less than 10^{-4}), which are the main obstacles in the straightforward and unique evaluation. From the method presented here the background level can be determined from the condition that both $f(q)$ and $\nu_2(q)$ have to give the same particle size.

11.7 Conclusions

A general theory of X-ray Bragg peak broadening created by dislocations is outlined. In contrast with the models proposed earlier [17,39] the derivation of the results do not require any assumption about the actual dislocation arrangement. Although, in the formal derivation only straight dislocations are considered, in most cases this does not restrict the applicability of the results obtained. On the 10–100 nm length scale, the system is "seen" in a common X-ray scattering measurement, the dislocations are more or less straight. In other words, the average radius of curvature of the dislocation loops is larger than the smallest Fourier parameter reached in the profile measurement. Certainly, if the dislocation system consists of small separate loops the above results cannot be applied.

The general theoretical approach applied reveals an important nontrivial consequence. Namely, in order to obtain the Fourier transform of the intensity distribution (11.19) only the "expansion" relation given by (11.12) was used which is equivalent with the $1/r$ type decay of the stress field of a straight dislocation. However, a point like force acting on the edge of a wedge results also a stress field proportional to $1/r$. So, e.g. a triple junction creates X-ray broadening which cannot be distinguished from peak broadening caused by dislocations.

It is also shown that the non-analytical character of the Fourier coefficients results in a singular behaviour of the different order moments of the scattered intensity. It was found from the detailed analysis of the integration limit dependence of the moments that for dislocation created broadening the tails of the intensity distribution decay as $\Lambda\langle\rho\rangle/(4\pi^2 q^3)$. It is important to stress that this tail behaviour is mathematically equivalent with the $\exp\{\Lambda\langle\rho\rangle L^2 \ln(L/R_1)\}$ variation of the Fourier coefficients at the vicinity of the origin. Nevertheless, according to our experience more accurate result can be obtained from the evaluation of the different order restricted moments (calculated from the measured intensity distribution) than from the analysis of the Fourier coefficients.

It is shown that using the second and fourth order restricted moments the size and strain created broadening can be separated on a single profile. (The other types of crystal defects create faster tail decay, so they do not influence the analysis.) The determination of the coherent domain size and the dislocation density do not require the simultaneous evaluation of Bragg peaks measured at different reflections. This certainly does not mean that measuring as many as possible reflections would not be useful. The particle shape and size distribution and the population of the different slip systems cannot be determined from a single profile. However, a strong advantage of the single profile analysis with the momentum technique proposed is that it does not require any a priori assumption. It provides data from the profiles with a quite simple procedure even if we do not know anything about the sample. Then the data can be analysed and compared with the prediction of different models. However, as a final note we need to stress that the momentum technique suggested can only be used if the different Bragg peaks are well separated and their tails can be measured at least up to 10^{-3} relative intensity. If these cannot be achieved the profile analysis requires more specific model of the coherent domain geometry and the dislocation network. There are several methods proposed for this case [27–31] but they are beyond the scope of this chapter.

Acknowledgements

The authors are indebted to Professor J. Driver and Dr. A. Révész for providing the AlMg and the iron samples, respectively. We also grateful to Dr. F. Székely for performing and analysing the measurements on the Cu single crystal. The financial support of Hungarian Scientific Research Fund (OTKA) under Contract Nos. T-034999 and T-030791 is also acknowledged.

References

1. N. Amstrong, W. Kalceff: J. Appl. Cryst. **32**, 600 (1999)
2. N. Amstrong: this issue (2002) Chap. ??
3. A. Borbély, J.H. Driver, T. Ungár: Acta Mater. **48**, 2005 (2000)
4. A. Borbély, I. Groma: Appl. Phys. Letter **79**, 1772 (2001)
5. A. Borbély, G. Guinglianda, J.H. Driver: Z. Metallkunde **7**, 689 (2002)
6. A. Borbély, G. Ribárik, J. Dragomir, T. Ungár; J. Appl. Cryst. (2002) in preparation
7. E.F. Bertaut: Acta Cryst. **3**, 14 (1950)
8. D. Breuer, P. Klimanek, W. Pantleon: J. Appl. Cryst. **33**, 1284 (2000)
9. I. Gaál, in *Proceedings of 5th Int. Symp. Metall. Mater. Sci.*. Ed. by N. Andersen, M. Eldrup, N. Hansen, D.J. Jensen, T. Leffer, H. Lilholt, O. Pedersen, B. Singer (Riso Nat. Lab, Roskilde, Denmark 1984) p. 245
10. I. Groma, T. Ungár, M. Wilkens: J. Appl. Cryst. **21**, 47 (1988)
11. I. Groma: Phys. Rev. B **57**, 7535 (1998)
12. I. Groma, F. Székely: J. Appl. Cryst. **33**, 1328 (2000)
13. I. Groma, G. Monnet: J. Appl. Cryst. **35**, 589 (2002)
14. M. Hacker, E. Thiele, C. Holste: Z. Metallkunde **88**, 321 (1997)
15. V. Kaganer, R. Köhler, M. Schmidbauer, R. Opitz: Phys. Rev. B **55**, 1793 (1997)
16. P. Klimanek, J. Kužel: J. Appl. Cryst. **21**, 59 (1988)
17. M. Krivoglaz, K. Ryaboshapka: Phys. Met. Metall. **15**, 18 (1963)
18. M. Krivoglaz, in *Theory of X-ray and thermal neutron scattering by real crystals* (Plenum, New York 1969) p. 258
19. M. Krivoglaz, O. Martynenko, K. Ryaboshapka: Fiz. Met. Metall. **55**, 1 (1983)
20. J. Langford: J. Appl. Cryst. **1**, 48 (1968)
21. J.I. Langford, D. Louer, P. Scardi: J. Appl. Cryst. **33**, 964 (2000)
22. T. Ungár, H. Mughrabi, D. Ronnpagel, M. Wilkens: Acta Met. **32**, 333 (1984)
23. T. Ungár, I. Groma, M. Wilkens: J. Appl. Cryst. **22**, 26 (1989)
24. T. Ungár, M. Zehetbauer: Scripta Mater. **35**, 1467 (1996)
25. T. Ungár, A. Borbély: Appl. Phys. Letter **69**, 3173 (1996)
26. T. Ungár, in *Defect and microstructure analysis by diffraction*. Ed. by R. Snyder, J. Fiala, H. Bunge (Oxford University Press, Place?? 1999) pp. ??
27. T. Ungár, J. Gubicza, P. Hanák, I: Alexandrov: Mat. Sci. Eng. A **319**, 274 (2001)

28. G. Ribarik, T. Ungár, J. Gubicza: J. Appl. Crys. **34**, 669 (2001)
29. P. Scardi, M. Leoni: J. Appl. Cryst. **32**, 671 (1999)
30. P. Scardi, M. Leoni, Y.H. Dong: European Phys. J. B **18**, 23 (2000)
31. P. Scardi, Y.H. Dong, M. Leoni: Materials Sci. Forum **378**, 132 (2001)
32. F. Székely, I. Groma, J. Lendvai: Phys. Rev. B **62**, 3093 (2000)
33. F. Székely, I. Groma, J. Lendvai: Mat. Sci. Eng. A **309**, 352 (2001)
34. F. Székely, I. Groma, J. Lendvai: Scripta Mater. **45**, 55 (2001)
35. B. Warren, B. Averbach: J. Appl. Phys. **21**, 595 (1950)
36. B. Warren, B. Averbach: J. Appl. Phys. **28**, 497 (1952)
37. M. Wilkens: Phys. Stat. Sol. **2**, 807 (1962)
38. M. Wilkens: Acta. Met. **17**, 1155 (1969)
39. M. Wilkens, in *Fundamental Aspects of Dislocation Theory*. Ed. by J. Simmons, R. de Witt, R. Bullough. (Nat. Bur. Stand. No. II 317 (U.S.) Spec. Publ. 1969) p. 1105
40. M. Wilkens: Phys. Stat. Sol. a **2**, 359 (1970)
41. M. Wilkens: Kristall und Technik **11**, 1159 (1976)
42. M. Wilkens: Phys. Stat. Sol. a **104**, 344 (1987)
43. A. Wilson: Il Nuovo Cimento **1**, 277 (1955)
44. A. Wilson: *Mathematical Theory of X-ray Powder Diffractometry*. (Centrex Publishing, Eindhoven 1963)

12 Determination of Non-uniform Dislocation Distributions in Polycrystalline Materials

J.-D. Kamminga, L.J. Seijbel, and R. Delhez

Diffraction line broadening caused by dislocations is discussed for polycrystalline materials that consist of crystallites in which non-uniform dislocation distributions occur; i.e. not all glide systems contain the same density and/or type of dislocations. The discussion is illustrated by two practical examples. It is shown that for 111 textured Al layers sputter deposited on silicon, during deformation due to a biaxial stress, predominantly screw dislocations with Burgers vectors inclined to the specimen surface are formed. Whether one or more slip systems are active in each grain cannot be determined with the present experiments. It is also shown that in initially stressed Ni layers electrodeposited on steel the internal stress favours the annihilation of dislocations with Burgers vectors close to the maximum resolved shear stress.

12.1 Introduction

Driven by the renewed interest in dislocations and recent developments in the field of diffraction line calculation, diffraction line shape analysis of dislocation distributions has become more and more a subject of investigation in recent years. One of the important developments in the field of the characterisation of diffraction line broadening by dislocations is the development of the so-called whole diffraction pattern-fitting programs, in which the reflection dependence of diffraction line broadening due to dislocations in cubic and hexagonal materials has been fully exploited [1,2]. The fitting procedures have made the dislocation distributions experimentally more accessible; fitting routines can now deliver dislocation distribution parameters on a routine basis.

One underlying assumption in most of the work concerning diffraction line broadening due to dislocations is the presence of a uniform dislocation distribution. A uniform dislocation distribution in the present context means that dislocations are distributed similarly at all crystallographically equivalent slip systems. In practice it often happens that dislocation distributions are non-uniform. For example, during deformation dislocations will form predominantly with a glide direction close to the maximum resolved shear stress. Consequently, within the crystallites different dislocation densities will occur on glide systems oriented differently with respect to the stress axes, and crystallites with different crystallographic orientations with respect to the stress axes will contain different dislocation distributions. These two related aspects of diffraction line analysis of non-uniform dislocation distributions are relatively unexplored [3–6].

The present chapter considers diffraction line shape analysis of polycrystalline materials with non-uniform dislocation distributions. The analysis for polycrystalline materials is confusing because the orientation distribution of the dislocations within the individual crystallites mixes up with the orientation distribution of the crystallites themselves. Therefore, in a theoretical introduction the effect of (non-)uniform dislocation distributions on the diffraction line broadening will be discussed. Then, two cases will be treated. First, the dislocation distribution in a heavily textured Al layer on silicon due to biaxial deformation is determined. It is shown that the pronounced texture can simplify the analysis considerably. Second the much more complex problem will be treated of a steel sheet covered with a Ni layer, virtually without texture. Surprisingly a rough model delivers a rather detailed picture of the dislocation distribution evolution in the Ni layer during annealing.

In the present chapter the Wilkens model for symmetric diffraction line broadening caused by dislocations is used in its simplest form [7]. It is realised that more advanced models for asymmetric line broadening exist (see e.g. [8]). However, the analysis using the simple form is already complicated and the use of more advanced models would complicate the analysis further. Fortunately, the diffraction line profiles obtained in the current work turned out to be more or less symmetric.

In the Wilkens model, the so-called contrast factors take into account the types of dislocations present, the orientations of the dislocations with respect to the diffracting lattice plane (diffraction vector), and the elastic constants of the material investigated. In this chapter the contrast factors are calculated for the elastically isotropic case (see e.g. [4]). Clearly, for the almost elastically isotropic Al layers this leads to small errors. However, for the Ni layers this could induce errors in the dislocation distribution parameters obtained, and therefore the values obtained for these parameters have to be considered with some suspicion.

Throughout this chapter, the reader is expected to be familiar with the representation of diffraction line profiles in terms of Fourier coefficients (see e.g. [9]). In practice, possibly present instrumental broadening and size broadening must be dealt with, before the strain broadening can be investigated. The removal of instrumental broadening and size broadening has been discussed elsewhere (see e.g. [9, 10], respectively). The present chapter focuses on the interpretation of diffraction lines that are exclusively broadened due to the presence of dislocations. Therefore, in the theory section it is always assumed that the diffraction line shapes are exclusively broadened due to dislocations.

Table 12.1. Slip systems in FCC metals. There are four different slip planes. In each slip plane there are three possible slip directions. Consequently, 12 equivalent slip systems exist

Slip plane	[1 1 1]	[1 1 $\bar{1}$]	[1 $\bar{1}$ 1]	[$\bar{1}$ 1 1]
	[$\bar{1}$ 1 0]	[$\bar{1}$ 1 0]	[1 1 0]	[1 1 0]
Slip directions	[0 $\bar{1}$ 1]	[0 1 1]	[0 1 1]	[0 $\bar{1}$ 1]
	[$\bar{1}$ 0 1]	[1 0 1]	[$\bar{1}$ 0 1]	[1 0 1]

12.2 Theory

12.2.1 Background

Diffraction Line Broadening Due to Dislocations in Single Crystals: Wilkens Model

A model proposed by Wilkens will be used to interpret the diffraction line broadening due to dislocations. The model presupposes that in the diffracting crystallites so-called restrictedly random dislocation distributions are present (see [7,11]). This distribution is characterised by the dislocation density ρ and the extent of the strain field of the individual dislocations that is characterised by the so-called outer cut-off radius R_e. In the simplest form, the diffraction line broadening due to a single set of dislocations is derived [11]. In a set of dislocations, all dislocations belong to the same slip system and are of the same character. A slip system consists of a slip plane and a slip direction (see Table 12.1). The character of the dislocation is identified by the angle between the line vector (which always lies in the slip plane) and the slip direction; for screw dislocations this angle is zero, for edge dislocations the line vector is normal to the slip direction. In the model, diffraction line profiles are given in terms of Fourier coefficients. It holds for the strain cosine Fourier coefficients $A(L)$ of a diffraction line profile broadened due to a restrictedly random distribution of dislocations, where all dislocations belong to a single set [11]:

$$A(L) = e^{-PL^2(Q-\ln L)} \tag{12.1}$$

$$P = \tfrac{\pi}{2} g^2 b^2 C \rho \tag{12.2}$$

$$Q = \ln R_e + 2\ln 2 - \tfrac{1}{3} - \ln \sigma|\mu| \,, \tag{12.3}$$

where L is the correlation length, b is the length of the Burgers vector $\boldsymbol{b}$, σ equals $|\sin\psi|$ where ψ is the angle between the line vector $\boldsymbol{l}$ of the dislocation and the vector $\boldsymbol{g}$, where $\boldsymbol{g}$ is the diffraction vector at which Bragg's law holds exactly for the reflection considered (length g; for cubic material it holds: $g = (h^2 + k^2 + l^2)^{1/2}/a_0$, where a_0 is the lattice parameter of the material considered). The scalar product of $\boldsymbol{g}$ and $\boldsymbol{b}$ is denoted by μ. C is the so-called contrast factor for the particular dislocation type, slip system and reflection (see e.g. [4]). The above expressions yield a proper description of the line profile if $R_e\sqrt{\rho} \geq 1$ and $L \leq 0.5R_e/\sigma|\mu|$ [7]. Further, the Wilkens model assumes symmetric line profiles (which is correct; see [12]) and therefore the sine Fourier coefficients of the line profile are zero.

In practice, more than one set of dislocations will be present in a crystal. The Wilkens model has been extended to crystals with more dislocation sets [7]. It was proposed that the diffraction line profile from a crystal with various sets of dislocations is the convolution of the line profiles from the single sets. Then, if more than one set i of dislocations is present (12.1) can still be used if P and Q are replaced by $\bar{P}$ and $\bar{Q}$ where (see [4]):

$$\bar{P} = \sum_i P_i = \frac{\pi}{2} g^2 b^2 \sum_i C_i \rho_i \,, \tag{12.4}$$

$$\bar{Q} = \frac{\sum_i P_i Q_i}{\sum_i P_i} = 2\ln 2 - \frac{1}{3} - \frac{\sum_i C_i \rho_i \ln(\sigma_i |\mu_i| / R_{e,i})}{\sum_i C_i \rho_i} \,, \tag{12.5}$$

(a bar indicates an average over all dislocation sets present in the crystal).

By assuming that the diffraction line profile from a crystal with various sets of dislocations is the convolution of the line profiles from the single sets, the elastic interaction of the various sets is neglected. It has been obtained in [12] that this leads to minor errors only.

Diffraction Line Broadening Due to Dislocations in Polycrystalline Materials

Although the above expressions have proved to be useful in a lot of investigations, they are not sufficiently general for the present purpose. The present investigations concern polycrystalline materials in which the dislocation distributions in the individual crystallites may be different. For a diffraction line to which crystallites α with different values $\bar{P}_\alpha$ and $\bar{Q}_\alpha$ contribute, the observed Fourier coefficients are the volume weighted averages of the Fourier coefficients for the diffracting crystallites:

$$\langle \bar{A}(L) \rangle = \sum_\alpha f_\alpha e^{-\bar{P}_\alpha L^2 (\bar{Q}_\alpha - \ln L)} \,, \tag{12.6}$$

where f_α denotes the volume of crystallite α divided by the volume of all crystallites that contribute to the diffraction line. The brackets $\langle\rangle$ denote an average over the diffracting crystallites.

If in a polycrystalline material all diffracting crystallites have identical values for $\bar{P}_\alpha$ and $\bar{Q}_\alpha$ (12.6) reduces to (12.7):

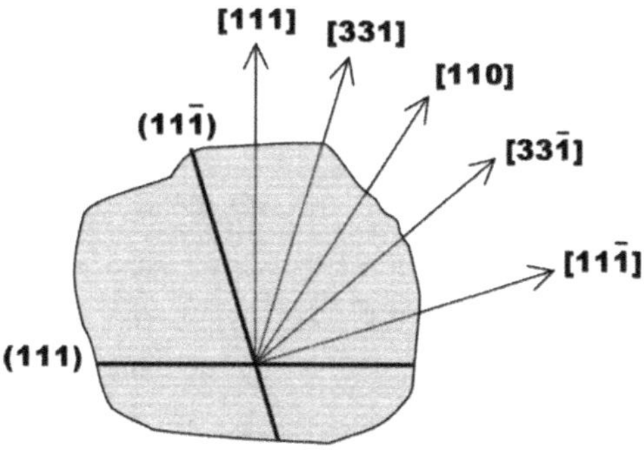

Fig. 12.1. The $(1\,\bar{1}\,0)$ plane of a single crystal of cubic material. All directions indicated are in the $(1\,\bar{1}\,0)$ plane and planes indicated are perpendicular to the $(1\,\bar{1}\,0)$ plane. Glide planes, type $\{1\,1\,1\}$, can be either normal or inclined to the $[1\,1\,1]$ direction (the $(1\,1\,1)$ plane and e.g. the $(1\,1\,\bar{1})$ plane, respectively). If, for some reason, only dislocations exist in the glide planes normal to the $[1\,1\,1]$ direction, the 111 diffraction profile measured with a diffraction vector parallel to the $[1\,1\,1]$ direction is not broadened ($C = 0$ since all Burgers vectors are normal to the diffraction vector). However, for the $[1\,1\,\bar{1}]$ direction $C \neq 0$ and broadening is present. Similarly, for this strongly non-uniform dislocation distribution 331 diffraction lines measured along the $[3\,3\,1]$ and $[3\,3\,\bar{1}]$ directions, will be broadened differently. If a powder is created in which such single crystallites are randomly oriented, diffraction line broadenings corresponding to all $\langle 3\,3\,1\rangle$ directions will contribute to the 331 diffraction line. However, in that case the analysis becomes complicated, since the values of $\bar{P}$ and $\bar{Q}$ are different for the identical, but differently oriented crystallites

$$\langle \bar{A}(L)\rangle = e^{-\bar{P}L^2(\bar{Q}-\ln L)}, \tag{12.7}$$

which has the same form as (12.1). Care should be taken in the use of this equation. Even if all diffracting crystallites are identical, the values of $\bar{P}$ and $\bar{Q}$ may be different for crystallites oriented differently in the specimen (see Fig. 12.1). Then, (12.7) cannot be used.

If the parameters $\bar{P}_\alpha$ and $\bar{Q}_\alpha$ given by (12.4) and (12.5) are different for each crystallite α, the different values of $\bar{P}_\alpha$ and $\bar{Q}_\alpha$ can in principle be obtained by fitting to (12.6). However, in practice, the different values of $\bar{P}_\alpha$ and $\bar{Q}_\alpha$ will be difficult to resolve.

12.2.2 Analysis for Different Kinds of Polycrystalline Specimens

In polycrystalline materials, the determination of dislocation distribution parameters from diffraction line profiles is no sinecure. The discussion is complicated because in the analysis for polycrystalline materials the orientation distribution of the dislocations within the individual crystallites mixes up with the orientation distribution of the crystallites in the specimen. To clarify the analysis for a polycrystalline material, the relatively simple case of a single crystal of FCC metal is considered first.

Table 12.2. Definition of the pure screw (type 1 to 6) and pure edge (type 7 to 18) dislocation types for FCC metals

Type	screw						edge											
Type no.	1	2	3	4	5	6	7	8	9	10	11	12	13	14	15	16	17	18
	1	−1	−1	1	0	0	1	1	−1	−1	−1	−1	1	1	0	0	0	0
± Slip dir.	1	1	0	0	1	−1	1	1	1	1	0	0	0	0	1	1	−1	−1
	0	0	1	1	1	1	0	0	0	0	1	1	1	1	1	1	1	1
	1	−1	−1	1	0	0	1	−1	1	1	1	1	−1	1	2	2	−2	2
l	1	1	0	0	1	−1	−1	1	1	1	2	−2	2	2	1	−1	1	1
	0	0	1	1	1	1	2	2	−2	2	1	1	1	−1	−1	1	1	1

All dislocations in FCC metals belong to one of the 12 slip systems shown in Table 12.1 [13] (the possible presence of partial dislocations is ignored). For pure screw dislocations, the line vector is parallel to the slip direction, and a slip plane cannot be indicated. Thus 6 possible sets of screw dislocations exist with the 6 equivalent $\langle 1\,1\,0 \rangle$ slip directions (within a set opposite slip directions are allowed). For pure edge dislocations the direction of the line vector is given by the cross product of the slip plane and the slip direction; 12 sets of pure edge dislocations exist. Consequently, if we restrict ourselves to only pure screw and pure edge dislocations, there are already 18 different sets i characterised by ρ_i and $R_{e,i}$ (see Table 12.2). Since from each diffraction line profile only two parameters, $\bar{P}_\alpha$ and $\bar{Q}_\alpha$, can be obtained, at least 18 reflections from the single crystal must be measured to determine the dislocation distribution completely. Even if this is possible, it should be recognised that dislocations are in general not of pure edge or pure screw character.

For a polycrystalline material the analysis is further complicated because the orientation distribution of the dislocations within the individual crystallites mixes up with the orientation distribution of the crystallites in the specimen and, in general, the analysis must be performed on the basis of (12.6), because the values for $\bar{P}_\alpha$ and $\bar{Q}_\alpha$ may be different for different crystallites α.

Different levels of complexity can be distinguished for different kinds of dislocation distributions. Next, diffraction line broadening of polycrystalline materials consisting of crystallites with uniform and non-uniform dislocation distributions will be discussed separately.

Uniform Dislocation Distributions

First the case of uniformly distributed dislocations is considered. Throughout the chapter, a uniform dislocation distribution in a crystal implies, that for each dislocation character present in that crystallite, dislocations are distributed over all equivalent slip systems uniformly with respect to both density and arrangement (i.e. outer cut-off radius).[1] For a crystallite with a uniform dislocation distribution, the values of $\bar{P}_\alpha$ and $\bar{Q}_\alpha$ are equal for diffraction along equivalent directions. For example, in a single crystal of cubic material with a uniform dislocation distribution, 111 diffraction lines measured along the 4 possible $\langle 1\,1\,1\rangle$ directions will be identical in shape.

Identical Crystallites

Obviously, if a polycrystalline material consists of differently oriented, but identical crystallites with such uniform dislocation distributions, (12.7) applies. Furthermore, the diffraction line shape of a given reflection hkl is then independent of the orientation of the diffraction vector in the reference frame of the polycrystalline specimen. Since the diffraction line profiles from each diffracting crystallite are identical in shape, possibly present crystallographic texture does not effect this result. However, reflections different in hkl yield different line broadening (cf. (12.4) and (12.5)). From the hkl dependence of the diffraction line shape, the character of the dislocations can be obtained. Methods for the characterisation of a dislocation distribution from diffraction line shape presented in a lot of recent investigations are intended for the specimens with these uniform dislocation distributions. This also holds for the recently developed Whole Powder Pattern Fitting procedures [1, 2].

Different Crystallites

If a polycrystalline material consists of crystallites with different, but uniform, dislocation distributions, diffraction line analysis must be performed on the basis of (12.6). An example of such an analysis was presented in a recent investigation to the recrystallisation of nickel layers [10]. Nickel layers were characterised as a collection of not yet recrystallised grains with a high dislocation density, and recrystallised grains with low dislocation density. For such specimens one might be able to acquire the different values for $\bar{P}_\alpha$ and $\bar{Q}_\alpha$ for the two types of crystallites, however if more types of crystallites are present it will be very difficult to unravel all different values $\bar{P}_\alpha$ and $\bar{Q}_\alpha$ from the measured diffraction line profiles. Note that in this case the diffraction line shape of a given reflection hkl is independent of the orientation of the diffraction vector in the reference frame of the polycrystalline specimen, as long as crystallographic texture is absent (or if the same crystallographic texture applies to all types of crystallites). If different crystallographic textures exist in the different types of crystallites, diffraction line shape of a reflection hkl will depend on the orientation of the diffraction vector.

---

[1] If a polycrystalline material consists of crystallites with different dislocation distribution, the dislocation distribution within each individual crystallite may still be uniform.

Non-uniform Dislocation Distributions

If a polycrystalline specimen consists of crystallites with non-uniform dislocation distributions, $\bar{P}_\alpha$ and $\bar{Q}_\alpha$ depend on crystallite orientation (see Fig. 12.1). In practice non-uniform dislocation distributions often occur. If for example dislocations in FCC metals are generated by plastic deformation, in each crystal usually one of the slip systems lies close to a plane and direction of maximum resolved shear stress, because there are many slip systems (see Table 12.1). The applied stress will be relieved by slip on this particular system and consequently, predominantly dislocations that belong to this slip system are formed. After deformation the dislocation distribution in each crystallite will exhibit pronounced orientation dependence with respect to the direction of the applied stress.

In practice, non-uniform dislocation distributions are recognised by a dependence of the diffraction line shape on the orientation of the diffraction vector of a given hkl.[2] If orientation dependence of diffraction line shape is absent, this should be considered as a strong indication for a uniform distribution of dislocations; if orientation dependence is measured, the dislocation distributions in the crystallites are probably non-uniform. Note that these findings follow directly from symmetry reasons and therefore this reasoning also holds if the Wilkens model does not apply.

Identical Crystallites

For a polycrystalline specimen that consists of identical crystallites with a non-uniform dislocation distribution, in general (12.7) cannot be used, because $\bar{P}_\alpha$ and $\bar{Q}_\alpha$ depend on crystallite orientation. However, if the texture in the specimen and the orientation of the diffraction vector are such, that only crystallites with the same values for $\bar{P}_\alpha$ and $\bar{Q}_\alpha$ contribute to the diffraction line, the analysis can still be performed on the basis of (12.7). In the experimental illustrations section, an example is given in which this is explored.

---

[2] For crystallographically textured materials, dependence of diffraction line shape on specimen orientation cannot be considered as proof for a non-uniform dislocation distribution in the crystallites (see the treatment on uniform dislocation distributions). Furthermore, for a specimen in which identical crystallites with non-uniform dislocation distributions are randomly oriented (also with respect to the dislocation distribution!), the shape of diffraction lines of a given hkl does not depend on the orientation of the diffraction vector in the specimen reference frame. In practice, this kind of specimens will be rare. For example, in the plastically deformed FCC material described, crystallites may be randomly oriented crystallographically, but after deformation pronounced preference ("texture") will exist in the orientation of the dislocation distributions due to the specific direction of the (macroscopically) applied stress.

Different Crystallites

For a polycrystalline specimen that consists of crystallites with different non-uniform dislocation distributions, it is obvious that it will be difficult to unravel dislocation distribution parameters from the diffraction line profiles. In the section with experimental illustrations also such a specimen is analysed.

12.3 Experimental Illustrations

Two illustrations will be given for the analysis of non-uniform dislocation distributions in polycrystalline materials. First, Al layers with a 111 fibre texture will be discussed, where the Al layers can be considered as composed of identical crystallites. Second, the more difficult case of Ni layers on steel will be discussed, in which the dislocation configuration is different for differently oriented crystallites of the Ni layers.

12.3.1 Al Layers

An Al layer was sputtered on an oxidised Si(100) wafer to a thickness of 1 µm. A strong and sharp 111 fibre texture was present in the layer. After deposition, the layer has been annealed for one hour at 673 K in an Ar atmosphere. The annealing caused growth of the crystallites of the layer, such that after the annealing the grain size was approximately equal to the layer thickness (1 µm). Subsequently, the layer has been cooled down to and further kept at room temperature. Due to the differences in thermal expansion coefficient of Si and Al, a biaxial tensile stress was imposed to the layer during cooling down. The 331 diffraction lines at $\omega = 22.0°$ and $\omega = 48.5°$ (ω denotes the specimen tilt angle) have been recorded many times during 2 days with a Siemens D500 ω type diffractometer, equipped with an incident beam monochromator set to select Co$K\alpha_1$ radiation in combination with a Braun scanning position sensitive detector. Both peak positions and peak shapes changed considerably during annealing at room temperature. The time dependence of the diffraction line shape and position are discussed elsewhere [4]. In the present chapter only the $\omega = 22.0°$ and $\omega = 48.5°$ 331 diffraction lines recorded approximately 165 min. after cooling down are considered.

The contribution of the instrumental broadening to the diffraction lines of the Al layer has been removed by means of measurements on a strain free Al standard powder using the Stokes correction (e.g. see [9]). The 331 diffraction lines have been recorded from the Al standard powder using the same instrumental settings (at both $\omega = 22.0°$ and $\omega = 48.5°$). For grains with a diameter of approximately 1 µm, size broadening can be neglected safely. Furthermore, it is believed that diffraction line broadening caused by lattice defects other than dislocations can be neglected (note that e.g. stacking faults are improbable in Al). Therefore the obtained Fourier coefficients can be regarded to stem from diffraction lines broadened exclusively due to the presence of dislocations. Because the diffraction lines were approximately symmetric, the moduli of the Fourier coefficients have been used for the analysis (for symmetric diffraction lines, the moduli of the Fourier coefficients equal the cosine Fourier coefficients).

The obtained Fourier coefficients, corrected for instrumental broadening, were different for the two reflections under investigation. This suggests that the dislocation distribution in the individual crystallites is non-uniform. At the annealing temperature the Al layer recrystallises and a microstructure is formed with grains (more or less identical in size) with low dislocation density. Cooling down the specimen induces a tensile, biaxial stress equal for each Al crystallite and dislocations are produced in the crystallites during the simultaneously occurring plastic deformation. Since all crystallites have identical orientation with respect to the stress axes, the dislocation distributions produced in each crystallite are expected to be similar. In view of the microstructure and thermal history of the specimens it therefore seems justified in this case to regard all crystallites as identical; Fig. 12.1 can be considered to represent each crystallite in the layer.

Obviously, dislocations will be generated predominantly on slip directions close to the maximum resolved shear stress. In the present case, it is therefore sufficient to consider only the $\langle 110\rangle$ slip directions that have angle of 35.26° with respect to the [111] surface normal (namely the [110], [101], and [011] directions). All other slip directions are parallel to the specimen surface. Based on the symmetry of the problem, two routes can be followed for the interpretation of the Fourier coefficients of the diffraction lines.

Route 1: It may be assumed that in each crystallite, dislocations are distributed uniformly over the three inclined slip directions. Then it follows from symmetry that for each single crystallite, P and Q will be equal for diffraction along the [331], [313], and [133] directions, that all have an angle of 22.0° with the surface normal. Similarly, the values of P and Q will be equal for diffraction along the $[33\bar{1}]$, $[3\bar{1}3]$, and $[\bar{1}33]$ directions that make an angle of 48.5° with the surface normal. Consequently, under these assumptions (12.7) can be used, which simplifies the problem for reflections measured at both specimen tilt angles.

Figure 12.2 shows the experimental Fourier coefficients and the fits to (12.7) for both reflections. Indeed, (12.7) can be accurately matched to these Fourier coefficients. The results obtained for $\bar{P}$ and $\bar{Q}$ for the different reflections are different for the different reflections as a result of the marked non-uniform dislocation distribution.

For the interpretation of $\bar{P}$ and $\bar{Q}$ in terms of dislocation density and outer cut-off radius, the values of C_i and $\sigma_i|\mu_i|$ are required for each reflection and dislocation type. In Table 12.2 all pure dislocation types relevant for FCC metals have been numbered. The values of C_i and $\sigma_i|\mu_i|$ for diffraction along

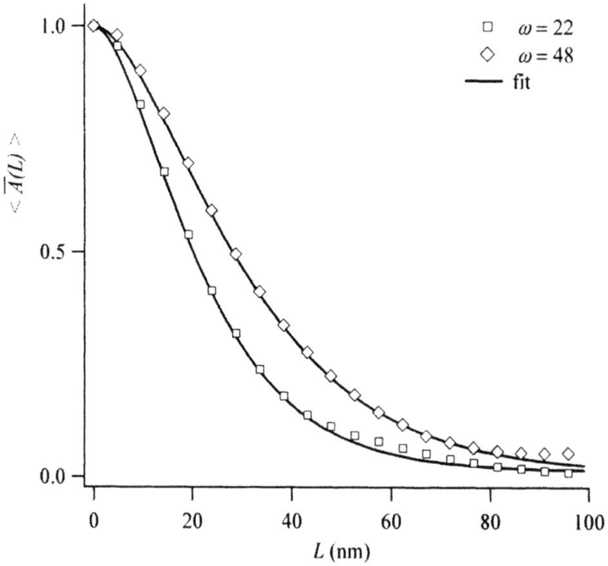

Fig. 12.2. Fourier coefficient curves of the 331 diffraction line profiles measured from the Al layer (containing a strong 111 texture) at $\omega = 22.0°$ and $\omega = 48.5°$. Fits of the data to (12.7) are also shown. It holds for the fits: $\bar{P} = 8.0 \times 10^{14}\,\mathrm{m}^{-2}$, $\bar{Q} = -15.61$, and $\bar{P} = 4.0 \times 10^{14}\,\mathrm{m}^{-2}$, $\bar{Q} = -15.21$ for $\omega = 22.0°$ and $48.5°$, respectively

the [3 3 1] and [3 3 $\bar{1}$] directions and the corresponding values for P_i/ρ_i and $Q_i - \ln R_{e,i}$ are given in Table 12.3 for all dislocation types.

As stated above, in this case it was assumed that for each dislocation character, identical dislocation distributions for the three slip directions inclined to the specimen surface exist. Hence, for a given dislocation character (edge or screw), ρ_i and $R_{e,i}$ are equal for the three directions at a given specimen tilt angle, and values $\bar{P}^{\omega=22}_{\{331\}}/\bar{P}^{\omega=48.5}_{\{33\bar{1}\}}$ and $\bar{Q}^{\omega=48.5}_{\{33\bar{1}\}} - \bar{Q}^{\omega=22}_{\{331\}}$ are easily obtained using (12.4) and (12.5). The values are shown with the experimental values in Table 12.4 together with the values of $\bar{P}^{\omega=22}_{\{331\}}/\bar{P}^{\omega=48.5}_{\{33\bar{1}\}}$ and $\bar{Q}^{\omega=48.5}_{\{33\bar{1}\}} - \bar{Q}^{\omega=22}_{\{331\}}$ obtained, assuming that the total number of screw and edge dislocations are equal and that $R_{e,i}$ is the same for both dislocation characters. Clearly, the experiments indicate that predominantly screw dislocations with inclined slip directions (or Burgers vectors) occur. Now, starting from this dislocation distribution the values of ρ and R_e are readily obtained. The value of ρ follows from Table 12.3 and $\bar{P}$ using

$$\bar{P} = \sum_i^{1,4,5} \frac{\rho_i P_i}{\rho_i} = \frac{1}{3}\rho \sum \frac{P_i}{\rho_i}, \tag{12.8}$$

Table 12.3. Values of C_i, $\sigma_i|\mu_i|$, P_i/ρ_i and $Q_i - \ln R_{e,i}$ for diffraction along the [3 3 1] and [3 3 $\bar{1}$] directions. Slip direction and line vector of each dislocation type are given in Table 12.2.

g	[3 3 1]				[3 3 $\bar{1}$]			
Type	C_i	$\sigma_i\|\mu_i\|$	P_i/ρ_i	$Q_i - \ln R_{e,i}$	C_i	$\sigma_i\|\mu_i\|$	P_i/ρ_i	$Q_i - \ln R_{e,i}$
1	0.05	0.69	0.74	1.42	0.05	0.69	0.74	1.43
2	0	0	0	0	0	0	0	0
3	0.09	0.95	1.41	1.11	0.24	1.52	3.64	0.63
4	0.24	1.52	3.64	0.63	0.09	0.95	1.41	1.11
5	0.24	1.52	3.64	0.63	0.09	0.95	1.41	1.11
6	0.09	0.95	1.41	1.11	0.24	1.52	3.64	0.63
7	0.49	2.95	7.24	−0.03	0.49	2.95	7.24	−0.03
8	0.49	2.95	7.24	−0.03	0.49	2.95	7.24	−0.03
9	0.12	0	0	0	0.03	0	0	0
10	0.03	0	0	0	0.12	0	0	0
11	0.01	0.35	0.11	2.10	0.10	1.32	1.47	0.77
12	0.19	0.98	2.88	1.07	0.25	1.85	3.79	0.44
13	0.25	1.85	3.79	0.44	0.19	0.98	2.88	1.07
14	0.10	1.32	1.47	0.77	0.01	0.35	0.11	2.10
15	0.10	1.32	1.47	0.77	0.01	0.35	0.11	2.10
16	0.25	1.85	3.79	0.44	0.19	0.98	2.88	1.07
17	0.19	0.98	2.88	1.07	0.25	1.85	3.79	0.44
18	0.01	0.35	0.11	2.10	0.10	1.32	1.47	0.77

Table 12.4. Values calculated using (12.4) and (12.5) for and for $\bar{P}^{\omega=22}_{\{331\}}/\bar{P}^{\omega=48.5}_{\{33\bar{1}\}}$ and $\bar{Q}^{\omega=48.5}_{\{33\bar{1}\}} - \bar{Q}^{\omega=22}_{\{331\}}$ distributions described in the text with only screw, only edge, and equal densities of screw and edge dislocations together with the experimental values

	Experiment	Screw	Screw + Edge	Edge
$\bar{P}^{\omega=22}_{\{331\}}/\bar{P}^{\omega=48.5}_{\{33\bar{1}\}}$	2.00	2.26	1.49	1.22
$\bar{Q}^{\omega=48.5}_{\{33\bar{1}\}} - \bar{Q}^{\omega=22}_{\{331\}}$	0.40	0.47	0.13	0.14

and the outer cut-off radius follows using:

$$\bar{Q} = \frac{\sum P_i/\rho_i(Q_i - \ln R_{e,i})}{\sum P_i/\rho_i} + \ln R_e \,. \tag{12.9}$$

It follows: $\rho = 3.0 \times 10^{14}\,\mathrm{m}^{-2}$, $R_e = 82\,\mathrm{nm}$, and $\rho = 3.4 \times 10^{14}\,\mathrm{m}^{-2}$, $R_e = 76\,\mathrm{nm}$, for the reflections measured at $\omega = 22°$ and $\omega = 48.5°$, respectively; the values obtained for both specimen tilt angles are equal within experimental precision.

Route 2: It may also be assumed that in each crystallite dislocations with only one specific slip direction, [110], [101], or [011] occur. Then, diffraction lines obtained from a single crystallite in the [331], [313], and [133] directions (all at 22.0° from the surface normal) have different values for P and Q. Consequently, (12.7) cannot be used. This also holds for the diffraction lines along $[3\,3\,\bar{1}]$, $[3\,\bar{1}\,3]$, and $[\bar{1}\,3\,3]$ directions obtained at 48.5° from the surface normal.

From the rotational symmetric state of stress causing the deformation it is expected that, although in each crystallite slip in only one direction occurs, the single crystallites will be distributed in a rotationally symmetric way. Then, the fractions f_α in (12.6) are 1/3 for diffraction along the three [331], [313], and [133] directions, (and similarly for the $[3\,3\,\bar{1}]$, $[3\,\bar{1}\,3]$, and $[\bar{1}\,3\,3]$ directions). Thus, $A(L)$ is given by the sum of three (equally weighted) different peak profiles with different values for $\bar{P}_\alpha$ and $\bar{Q}_\alpha$ (cf. (12.6)).

Unfortunately, the different values of $\bar{P}_\alpha$ and $\bar{Q}_\alpha$ cannot be reliably obtained from fitting the acquired Fourier coefficients. The following procedure has been followed to check the different possibilities of dislocation distributions.

Starting from a guessed dislocation distribution (see below) and guessed initial values for ρ_i and $R_{e,i}$ the Fourier coefficient curves corresponding to the peak profiles from a single crystal, for diffraction along the [331], [313], and [133] directions have been calculated. Fourier coefficients of the peak profiles from the polycrystalline specimen have been calculated by summing the Fourier coefficients of the three different single crystal profiles and normalising the result such that $\langle \bar{A}(0) \rangle = 1$ (i.e. (12.6) with $f_\alpha = 1/3$ for diffraction along the [331], [313], and [133] directions). The values of ρ_i and $R_{e,i}$ have been fitted such that good correspondence was obtained between the experimental values for the Fourier coefficients and the calculated profiles for $\omega = 22°$.

Three different distributions were used:

- The first distribution was assumed to contain only one type of screw dislocation with inclined Burgers vector present in each single crystal. For the calculations, type 1 has been used (cf. Table 12.3). Of course, from the symmetry of the problem it follows that using one of the other screw dislocations with inclined Burgers vector (i.e. type 4 or type 5) yields the same results.
- The second distribution consisted of only one type of edge dislocation with inclined Burgers vector. Dislocation type 16 has been used for the calculations. Again, using one of the other edge dislocations with inclined Burgers vector yields the same results. Also, using two types of edge dislocations with the same Burgers vectors (e.g. type 15 and type 16) and half the dislocation density of each set yielded equal results.
- The third distribution consisted of a mixture of equal densities of screw and edge dislocations of type 1 and type 16, respectively.

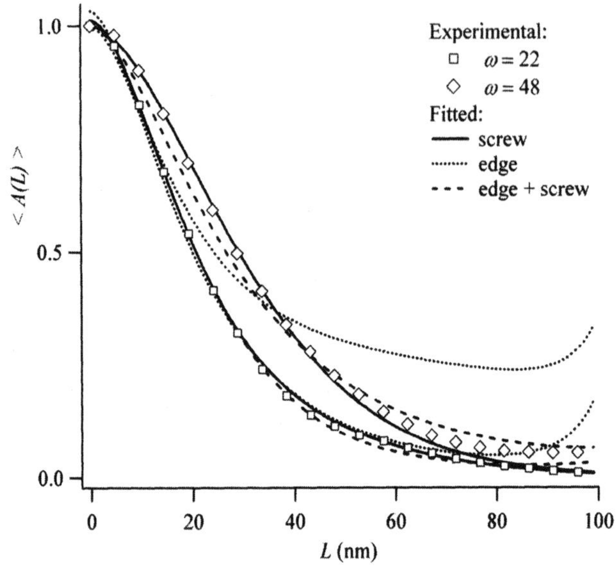

Fig. 12.3. Fourier coefficient curves of the 331 diffraction line profiles measured from the Al layer at $\omega = 22.0°$ and $\omega = 48.5°$. Fourier coefficient curves corresponding to three different dislocation distributions mentioned in the text (i.e. only screw, only edge, both screw and edge dislocations with Burgers vectors inclined to the specimen plane) have been fitted, using ρ and R_e as fitting parameters, to the data at $\omega = 22.0°$. Subsequently, using the fitted values for ρ and R_e, the Fourier coefficient curves at $\omega = 48.5°$ have been constructed for each dislocation distribution. Obviously, the distribution with only screw dislocations shows the best correspondence with the measured data

Figure 12.3 shows the results for three different dislocation distributions (lower lines). For the three distributions, dislocation densities of 2.2, 2.5, and $2.3 \times 10^{14}\,\mathrm{m}^{-2}$ were obtained and outer cut-off radii of about 101, 107, and 117 nm, respectively. Subsequently, for each proposed dislocation distribution, using the obtained values for ρ_i and $R_{e,i}$, the Fourier coefficient curves for the 331 reflections at $\omega = 48.5°$ have been calculated. Figure 12.3 shows the experimental values together with the calculated Fourier coefficient curves, for the three dislocation distributions. The data are well described, using only inclined screw dislocations in the single crystal.

Comparing Figs. 12.2 and 12.3 shows that the assumption that in each crystallite screw dislocations are distributed uniformly over the three inclined slip directions and the assumption that in each crystallite screw dislocations with only one inclined slip direction occur both yield good results. Therefore the experiments cannot discriminate between the two cases. The dislocation densities and outer cut-off radii obtained differ distinctly for the two different assumptions made. Thus a precise value of these parameters for the present specimen cannot be given. Anyhow, irrespective the assumption made it can be concluded that predominantly screw dislocations with inclined Burgers vector occur in the annealed Al layers.

12.3.2 Ni Layers

Nickel layers were electrodeposited at room temperature on a steel plate of approximately 20×30 cm to a thickness of 2 µm. Two specimens (approximately 2×3 cm) were taken from the central region of this plate. One specimen was measured in the as-prepared condition. The other specimen was annealed for three hours at 525 K in an Ar atmosphere. Diffraction lines were measured using a Bruker AXS D5005 X-ray diffractometer. The XRD measurements were made using Cu$K\alpha$ radiation applying a diffracted beam monochromator. To measure both the stress and the dislocation distribution the specimen has been tilted in the diffractometer in the ω direction. Diffraction lines were obtained at specimen tilt angles ω of 0°, 12.92°, 15.98°, 18.43°, 26.57°, and 33.21°. From the peak positions it was obtained, using the well-known $\sin^2 \Psi$ method [14], that a growth stress of 210 MPa was present in the as-deposited nickel layer. After annealing a stress of −80 MPa was measured. The latter stress is believed to be thermal stress caused by the slight difference of thermal expansion coefficients of the nickel and the steel; the growth stress in the nickel layer is completely relaxed during the annealing. The contribution of the instrumental broadening to the diffraction lines of the nickel layers has been removed by means of strain free, well-annealed nickel powder. The diffraction lines obtained from this powder showed negligible contributions of lattice deformation and finite dimensions of the crystallites to the line broadening.

Figure 12.4 shows the Fourier coefficient curves of the 311 diffraction lines as obtained from the as-prepared specimen. Fourier coefficient curves of diffraction peaks obtained at various specimen tilt angles have been plotted. Clearly, the diffraction line shape does not depend on the specimen tilt angle; a specimen orientation dependence of size and/or strain broadening does not exist. Fitting the data in the figure to (12.7) convoluted with a size profile of spherical particles with radius R_{cr} (see [15]) yields: $\bar{P} = 2.0 \times 10^{15}\,\mathrm{m}^{-2}$, $\bar{Q} = -16.27$, and $R_{cr} = 64$ nm. Figure 12.4 also shows the fitted curve, and separately the size Fourier coefficient curve.

Figure 12.5 shows the data of the annealed specimen. The Fourier coefficient curves in Fig. 12.5 are clearly wider than those in Fig. 12.4 indicating that the "overall" dislocation density decreased and/or that grain growth occurred during annealing. Figure 12.5 also shows the size Fourier coefficients obtained from the as-prepared fit. The curve intersects the measured profiles at correlation distances between 50 and 70 nm. This proves that grain growth has occurred during annealing.

A pronounced diffraction line shape dependence on the specimen tilt angle exists. The inset shows that the curves are shifted upwards for an increasing specimen tilt angle up to $\omega = 26.57°$. However, for $\omega = 33.21°$ the curve

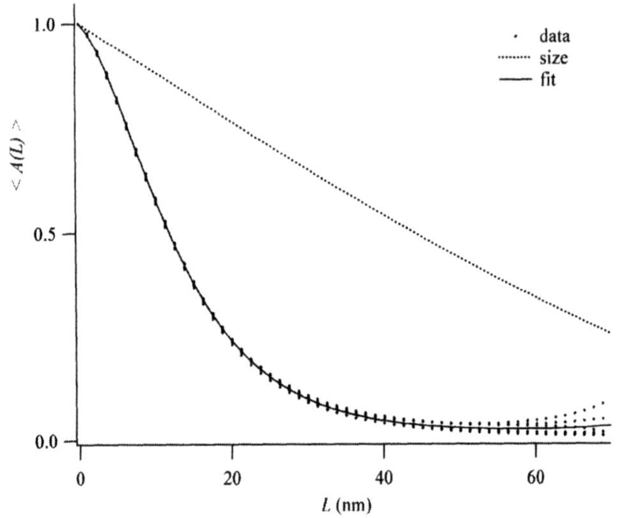

Fig. 12.4. Fourier coefficient curves of the 311 diffraction line profiles measured from the as-deposited nickel layer at specimen tilt angles ω of $0°$, $12.92°$, $15.98°$, $18.43°$, $26.57°$ and $33.21°$. Data corresponding to different specimen tilt angles are virtually indistinguishable. A fit of (12.7) convoluted with a size profile of equally sized spherical crystallites (radius R_{cr}) to all the data is also shown (*solid line*). The size Fourier coefficients are indicated by the *dotted line*

has shifted downwards. The anisotropic diffraction line broadening can be attributed to a non-uniform dislocation distribution. During annealing, the growth stress is relaxed; dislocations move under the influence of shear and are activated when the temperature rises. Obviously, since the plane stress state has maximum shear stress at an angle $\eta = 45°$ with respect to the specimen surface, dislocations with a glide direction at approximately $45°$ to the surface are in particular mobile during the annealing and therefore have the opportunity of being annihilated by other dislocations, grain boundaries or the surface. For this reason, lower dislocation densities for dislocation sets with glide directions close to $45°$ with the specimen surface are assumed than for other glide directions. The model dislocation distribution shown in Fig. 12.6 characterises the dislocation distribution in the annealed layer. If the angle η between the surface normal and the Burgers vector is close to $45°$ (or $135°$) a dislocation density ρ_1 is present for the particular dislocation type. If the angle is far from $45°$ (or $135°$) a considerably larger dislocation density ρ_2 exists.

Diffraction line profiles have been calculated according to the model in Fig. 12.6. Consider the crystallite in Fig. 12.7. The 311 diffraction line is measured along the [311] direction in the crystal. In this particular case, the [322] direction represents the specimen surface normal. Of the six possible $\langle 110 \rangle$ glide directions, the [011] direction is in the $(0\,\bar{1}\,1)$ plane and makes an angle

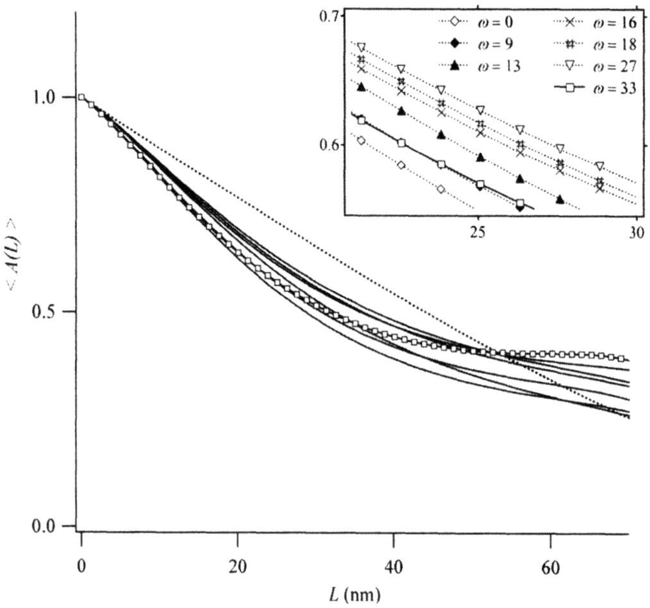

Fig. 12.5. Fourier coefficient curves of the 311 diffraction line profiles measured from the Ni layer of Fig. 12.4 after annealing 3 h at 525 K. The curves shown correspond to various specimen tilt angles ω (solid lines, the solid line with markers is for $\omega = 33.21°$, in main graph). In this case the data corresponding to different specimen tilt angles are clearly different. A magnification of a part of the data is given in the inset. Increasing ω accompanies a shift upwards of the Fourier coefficient curves until $\omega = 26.57°$. Upon increasing the specimen tilt further to $\omega = 33.21°$ the curve is shifted downwards. The dotted line in the main graph is the size Fourier coefficient curve shown in Fig. 12.4. In the legend of the inset, the values of the specimen tilt angles have been rounded

η of about 46° with the specimen surface. The other $\langle 110 \rangle$ type directions are at angles of 30, 80, or 90° with respect to the [322] direction. Consequently, if e.g. only screw dislocations are considered, dislocations of type 5 (cf. Table 12.2) have density ρ_1 and dislocations of the other types occur with a density ρ_2. From a given outer cut-off radius, the dislocation distribution now yields the 311 diffraction line shape of this particular crystal.

In a diffraction experiment all crystallites with a $\langle 311 \rangle$ direction parallel to the diffraction vector, with the particular ω with respect to the specimen surface normal, contribute to the 311 diffraction peak. Thus the diffraction line is obtained by summing the contributions of all crystallites with orientations obtained by rotation a around the [311] direction. The Fourier coefficients of the diffraction line shape are then given by (12.6). The presence of texture is neglected in this case and for each crystallite orientation, characterised by ω and α, the same fraction f_α is used. The absence of texture was indicated by a constant integral intensity of the diffraction peaks as a function of ω.

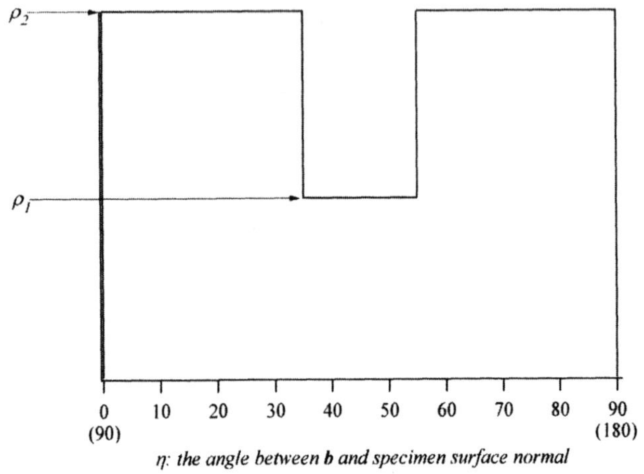

Fig. 12.6. Proposed dislocation distribution for the annealed nickel layer. Dislocations with a Burgers vector with an angle close to 45° with the specimen surface normal occur with lower density than other dislocations

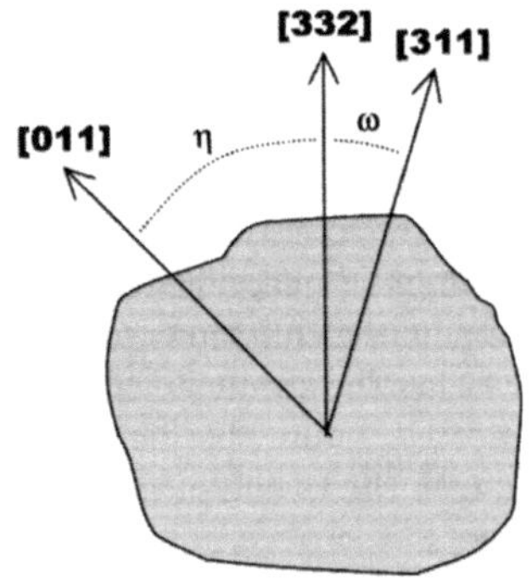

Fig. 12.7. A single crystal of cubic material in an orientation such that the [3 3 2] direction corresponds to the specimen surface normal. It contains dislocations with [0 1 1] Burgers vectors. If a 311 diffraction line is measured along the [3 1 1] direction, the present crystallite will contribute to the diffraction pattern. For a random powder, all crystallites with a ⟨3 1 1⟩ direction parallel to the diffraction vector will contribute. All possible orientations of the diffracting crystallites are obtained by rotation of the crystal in the figure around an axis parallel to the [3 1 1] direction. The ⟨1 1 0⟩ Burgers vectors will have different angles η with the surface normal for each crystallite orientation. Consequently, if the dislocation density depends on η, different dislocation distributions may be present in crystallites with different orientations that contribute to the 311 diffraction line (cf. Fig. 12.6)

Figure 12.8 shows results of the calculations, using an outer cut-off radius of 60 nm, $\rho_1 = 1.5 \times 10^{14}\,\mathrm{m}^{-2}$, $\rho_2 = 3.0 \times 10^{14}\,\mathrm{m}^{-2}$, and assuming that only edge dislocations are present. Crystallite size broadening was assumed to be absent. As for the experimental data (Fig. 12.5), the curves shift upwards for

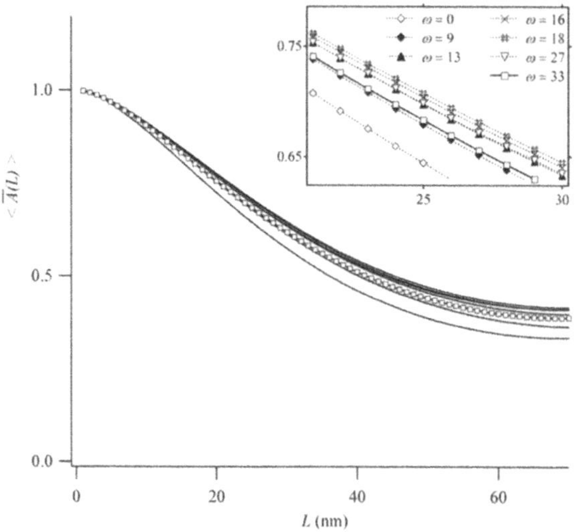

Fig. 12.8. Fourier coefficient curves of 311 diffraction line profiles for various ω tilts of the specimen, simulated from the dislocation distribution given in Fig. 12.6, assuming that only edge dislocations are present. A magnification of part of the data is given in the inset. Each increased ω shifts the Fourier coefficient curves upwards until $\omega = 18.43°$. However, increasing the specimen tilt further to $\omega = 33.21°$ the curve is shifted downwards. In the legend of the inset, the values of the specimen tilt angles have been rounded

increasing ω tilt, and then shift downwards again for the highest tilt angles. This general trend was also observed in simulations using distributions containing only screw dislocations and distributions with both screw and edge dislocations, although the curves were at slightly different vertical positions. However, in the simulations, the highest curve is obtained $\omega = 18.43°$ instead of 26.57°, as shown in Fig. 12.5. In view of the crude modelling (Fig. 12.6), the correspondence of the simulations with the data is believed to be reasonable. It should be noted that in the simulations crystallite size broadening was neglected and the outer cut-off radius was set (arbitrarily) to 60 nm. Further, contrast factors for elastically isotropic material were used, which surely influences the simulations. However the general trends indicated by the curves should not be influenced by the above approximations and it is believed that the dislocation distribution shown in Fig. 12.6 represents qualitatively the non-uniformity of the dislocation distribution in the specimen.

12.4 Discussion

The examples treated above substantiate the possibilities of diffraction line shape analysis to study dislocation distributions and their development in materials during thermal or mechanical treatments. However, also the weaknesses have become apparent. The weakness of these diffraction line shape investigations is without doubt the ambiguity of the results. For the Al layer, the experiments were well described both by a distribution with three sets of dislocations occurring with equal density in each grain and with only one set of dislocations occurring in each grain (cf. Figs. 12.2 and 12.3). However, the outer cut-off radii and dislocation densities obtained from both distributions were quite different: using results of the reflection measured at $\omega = 22°$, 82 and 101 nm and 3.0 and $2.2 \times 10^{14}\,\mathrm{m}^{-2}$ were found, respectively. Thus, from the present experiments, we cannot discriminate between one or three sets of dislocations being present in the crystallites and the values obtained for the outer cut-off radius and the dislocation density are unreliable. This weakness is not inherent to diffraction line analysis; if more (different) reflections had been measured, a more accurate description of the dislocation distribution would have been possible.

The character and type of the dislocations could be pinned down accurately. The type of the dislocation distribution assessed could be understood in view of the thermal history of the specimen; inclined Burgers vectors were to be expected. However, why the character of the dislocations is predominantly screw is not yet clear.

For the nickel layers the results were of similar nature. Whereas quantitative information was difficult to obtain, the process occurring during annealing could be pinned down accurately; in particular dislocations with glide directions close to 45° with the surface annihilated during annealing. In this case, the character of the dislocations could not be established, because similar results were obtained for different kinds of dislocation character. Apparently, it is mainly the Burgers vector that determines the diffraction line broadening in this case, whereas the character of the dislocations is of less influence.

It is the opinion of the authors that the difficulty in the quantification of the results is systematic for diffraction line shape studies to dislocation distributions. It is realised that apart from the quantification problems arising in the analyses discussed above, an additional important source of errors is present. The possible incompatibility of the model (restrictedly random) dislocation distribution that forms the basis of the Wilkens model with a real dislocation distribution can also obscure the quantification of the results obtained by diffraction line broadening. Therefore, values obtained should be considered as order of magnitude estimates. The use of a more advanced model available [8] does not solve this problem. Although this model does not necessarily describe restrictedly random dislocation distributions, dislocations are still supposed to be straight and infinitely long. At present, sufficient knowledge of the line broadening of diffraction line shapes due to curved dislocations (which are necessarily neither of pure screw nor edge character) is lacking and to improve the quantification of the results, in our opinion, theoretical work in this field is necessary.

If a series of comparably treated specimens is investigated, the changes in the values obtained for the dislocation densities and the outer cut-off radii are meaningful. The strength of diffraction line broadening analysis for the determination of dislocation distributions is that it reveals the orientation dependent dislocation dynamical processes occurring during plastic deformation.

12.5 Conclusions

In practice, for the diffraction line shape analysis of dislocation distributions uniform dislocation distributions are often assumed, while in a lot of practical specimens non-uniform dislocation distributions can be expected. Non-uniform dislocation distributions necessitate a more complicated analysis.

Non-uniform dislocation distributions can be recognised by a specimen tilt angle dependence of the diffraction line broadening.

In plastically deformed 111 textured Al layers, after deformation due to a biaxial stress, predominantly screw dislocations occur with Burgers vectors inclined to the specimen surface. Whether one or more slip systems are active in each grain could not be determined with the present experiments.

During annealing of initially stressed Ni layers the internal stress favours the annihilation of dislocations with Burgers vectors close to the maximum resolved shear stress.

Diffraction line broadening analysis can be used conveniently to gain insight in the orientation dependent dislocation dynamical processes occurring during plastic deformation of real materials by means of relatively simple experiments. The values of the dislocation distribution parameters obtained should be regarded with some suspicion.

Acknowledgement

This research was carried out under project number MS97007 in the framework of the Strategic Research program of the Netherlands Institute for Metals Research in the Netherlands (`www.nimr.nl`).

Note Added in Proof

In this chapter, diffraction line broadenings of the 311 diffraction lines from a nickel specimen, recorded at different specimen tilt angles are compared (Fig. 12.5). The experiments are compared to line profiles from simulations based on contrast factors calculated for elastically isotropic material (Fig. 12.8). We did the same calculations using the contrast factors for the elastically anisotropic case. The values for the contrast factors of the 311 reflection for dislocation types 1 to 16 (see Table 12.2) for nickel are: 0.249228,

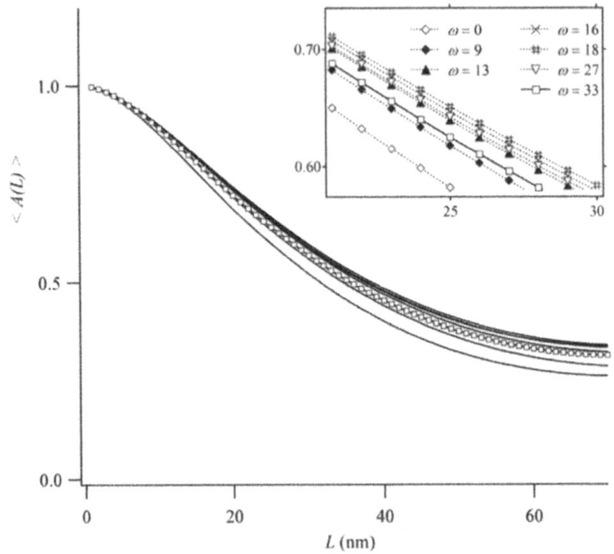

Fig. 12.9. Simulated Fourier coefficient curves of 311 diffraction line profiles for various ω tilts of the specimen, using contrast factors calculated for the elastically anisotropic case. The figure should be compared to Fig. 12.8

0.217551, 0.217551, 0.249228, 0.0950318, 0, 0.312617, 0.478336, 0.213461, 0.113117, 0.113117, 0.213461, 0.478336, 0.312617, 0.0681116, 0.0681116, 0.0927702, and 0.00289822 respectively. With these contrast factors, the curves have been recalculated, using only edge dislocations (again, similar results were obtained using other dislocation distributions), see Fig. 12.9. The figure should be compared to Fig. 12.8 in this chapter. Overall, the curves have shifted downwards. This is due to the somewhat higher average contrast factor for the anisotropic case (i.e. 0.2056) as compared to the isotropic case (0.1796). However, the curves follow the same trend as for the elastically isotropic case. This supports the conclusions drawn in this chapter.

Note added by J.-D. Kamminga and N. Armstrong.

References

1. P. Scardi, M. Leoni, Y.H. Dong: Eur. Phys. J. B **18**, 23 (2000)
2. G. Ribárik, T. Ungár, J. Gubicza: J. Appl. Cryst. **34**, 669 (2001)
3. A.C. Vermeulen, R. Delhez, T.H. de Keijser, E.J. Mittemeijer: J. Appl. Phys. **77**, 5026 (1995)
4. J.-D. Kamminga, R. Delhez: Mater. Sci. Eng. A **309–310**, 55 (2001)
5. R.W. Cheary, E. Dooryhee, P. Lynch, N. Armstrong, S. Dligatch: J. Appl. Cryst. **33**, 1271 (2000)

6. T. Ungár, M. Zehetbauer: Scripta Mat. **35**, 1467 (1996)
7. M. Wilkens: Phys. Status Solidi A **2**, 359 (1970)
8. I. Groma: Phys. Rev. B **57**, 7535 (1998)
9. B.E. Warren: *X-ray Diffraction* (Addison-Wesley, Reading MA 1969) pp. 251–275
10. J.-D. Kamminga, L.J. Seijbel: J. Res. Natl. Inst. Stand. Technol. (2003), in press
11. M. Wilkens: *Fundamental Aspects of Dislocation Theory*. Vol. II, ed. by J.A. Simmons, R. de Wit and R. Bullough (Nat. Bur. Stand. (US) Spec. Publ. No. 317 (1970)) pp. 1191–1193, 1195–1221
12. J.-D. Kamminga, R. Delhez: J. Appl. Cryst. **33**, 1122 (2000)
13. F.R.N. Nabarro: *Theory of Crystal Dislocations* (Clarendon, Oxford 1967) pp. 230–231
14. J.-D. Kamminga, T.H. De Keijser, E.J. Mittemeijer, R. Delhez: J. Appl. Cryst. **33**, 1059 (2000)
15. L.J. Seijbel, J.-D. Kamminga, R. Delhez: Mater. Sci. Forum 404–407, 715 (2002)

13 Line Profile Fitting: The Case of *fcc* Crystals Containing Stacking Faults

A.I. Ustinov, L.O. Olikhovska, N.M. Budarina, and F. Bernard

Monte Carlo simulation of the peak profiles in the powder XRD patterns of FCC structures containing a high density of stacking faults was performed using a general one-dimensional model. Features of the peak profiles have been found that allow the determination of the predominant type of stacking fault from an experimental XRD powder pattern with a high level of confidence. A non-monotonous change of the powder peak positions and their full widths at half maximum values depending on the density of the stacking faults was revealed. A procedure, based on the simulation of all powder peak profiles and comparison with experimental ones, is proposed to determine the structural characteristics of an *fcc* powder. The independent variables in the simulation procedure are the density of the stacking faults, the density of dislocations and the average crystallite size. As an example of application of the procedure, experimental XRD data are examined in the case of Cu powders produced by ball milling.

13.1 Introduction

The investigation of the microstructure of (in particular nanostructured) materials on the basis of X-ray diffraction (XRD) line broadening analysis, taking into account the finite size of coherently diffracting domains and microdistortions (often induced by the strain fields of dislocations) [1], is widely accepted. An application of the developed approaches in the case of a high-symmetry crystal could be complicated by the presence of stacking faults (SF) in the crystal. Indeed, in a single crystal, the XRD peaks of one family of *hkl* planes are differently affected by the presence of SF (i.e. some peaks are broadened and shifted whereas others not) as it was reported in many works [2–12]. For the purposes of this chapter, it is important to distinguish these two kinds of peaks; therefore, the peaks broadened and shifted by to the presence of stacking faults will be referred to as diffuse peaks in the following in order to distinguish them from the normal Bragg peaks.

Since the total intensity of each XRD peak is determined by the summation of the individual contributions of each family of *hkl* planes, the resultant XRD peak profile depends on the relationship between intensities and positions of the diffuse and Bragg components. As it will be demonstrated in this chapter in the case of a face centred cubic (*fcc*) structure, the main result of such dependence is that profiles and positions of peaks in the powder XRD patterns change non-monotonically versus the scattering angle and the SF concentration.

Earlier, many authors used several assumptions in order to find a correlation between the concentration of SF and the characteristics associated to the profile and the position of XRD peaks. Indeed, in the work reported by Warren [13], it was assumed that the concentration of SF was low and consequently the position of XRD peak could be defined as the arithmetic mean of the peak component positions.

However, in reality, the SF concentration is not always low [14–22]. In this case, from a numerical calculation for single crystals, the maximum intensity of the diffuse components of the XRD peaks decreases considerably when the SF concentration increases because of the intensity redistribution from the diffuse peak to the background [12]. Such a redistribution of the scattered intensity implies that the shapes and positions of peak profiles are mainly determined by their Bragg components.

It should be pointed out that the significant changes in the resultant XRD peak profiles caused by a difference between the intensities of the peak components appear also in the case when the polycrystalline specimen is textured or (and) the orientation of the SF differs from a random distribution [12].

From the preceding, it may be concluded that some corrections should be applied in the relationship between the SF concentration and the XRD peak parameters which have been formerly derived analytically [4, 13, 23].

In the present work, a new approach is proposed to analyze the XRD powder patterns. This latter is based on a fitting of the XRD peaks measured experimentally with those modelled taking into consideration the presence of the SF, the presence of dislocations, the finite size of the coherently diffracting domains and the instrumental broadening. This approach will be illustrated by an analysis of the XRD pattern of a set of Cu-nanostructured powders produced by ball milling.

13.2 Influence of Stacking Faults on XRD Powder Peak Profiles

13.2.1 Description of the Structures Containing Stacking Faults

The representation of a close-packed crystal using A, B, and C symbols is well known. Here, symbols A, B, and C correspond to three adjacent close-packed atomic layers differently packed relative to each other. Let us call this arrangement "*c*"-layer packing. The displacement of the atomic layers relative to each other resulting in a disruption of the regular ABC... sequence is considered as a SF [24]. The SF occurrence in a cubic structure results in the layer(s) being surrounded by two layers occupying identical positions

Table 13.1. Nomenclature of stacking faults in a *fcc* crystal

Description of the SF in "ABC" symbols	Description of the SF in "*hc*" symbols	Type of SF	
...ABCACABC...	...*ccc**hh**ccc*...	Intrinsic	Deformation
...ABCACBCABC...	...*ccc**hc'h**ccc*...	Extrinsic	Double deformation
...ABCACACB...	...*ccchc'c'c'*...	Twin	Growth

(...**AB**A..., for instance). Since such a stacking of layers is typical of a hexagonal close-packed structure, the layer packed in this manner could be referred to as "*h*"-layer. Note, that at each of the "*h*"-layers, the sequence changes its sign, that is the transition takes place from a "positive" arrangement of the layers, for example, ...ABCABC... (*cccc*...) to the "negative" arrangement ...ACBACB... (*c'c'c'c'*...), or vice versa.

In the literature, different terminologies concerning the types of SF are traditionally used. An *intrinsic* fault results from the displacement of one layer in a cubic structure, transforming the sequence ABCABCAB into, for example ABCA**C**ABC (*cchhcc*...); the bold letter indicates the layer where the displacement occurs. On the other hand, an *extrinsic* fault results when the displacement occurs in two successive layers of a cubic structure giving in such a way the sequence ABCA**CB**CABC (*ccchc'hccc*). Both *intrinsic* and *extrinsic* faults are often referred to as the *deformation* and *double deformation* faults, respectively, because these faults commonly occur during the deformation of a cubic crystal. In addition, another type of stacking faults exists; namely *twin* faults or *twins*. If we use A, B, C symbols, a *twin* fault can be described as, for example, ABCACACB (*ccchc'c'c'*). This type of stacking fault, which usually occurs during growth, is often referred to as a *growth* fault. The description and the nomenclature of these different types of SF are summarized in Table 13.1.

The alternation of the atomic layers in a crystal containing SF can be described by a probabilistic scheme in which all possible configurations of the atomic layers in the crystal are represented. The description of the "faulted" structure in terms of the "*hc*" symbols, considering the neighbourhood of each atomic layer, is really well suited to this work because the type of SF can be clearly determined without using a detailed analysis of the layer arrangements. One more advantage of such a description is a relative simplicity of the statistic schemes, as it will be shown later, especially when models of the crystals containing only one type of SF have to be investigated. Nevertheless, the sequence of "*h*" and "*c*" symbols can be easily arranged into the corresponding sequence of A, B, C symbols which allows a determination of the atomic layer coordinates and, in turn, the calculation of the scattered intensity.

The "A, B, C" symbols are convenient when a model of the faulted crystal is built using the probabilities of the *displacements* as it has been done by Paterson [3] and Warren [4]. On the contrary, if the faulted crystal structure has to be built with some type of *stacking faults*, the layer configurations corresponding to the stacking fault (h, or hh, or hch) seems to be preferable.

The difference between the faulted crystalline structures being formed by *intrinsic* faults and random displacements can be illustrated by an example. As discussed above, an *intrinsic* fault consists of adjacent layers having the hexagonal type of packing (hh) and it results from one displacement in a cubic crystal. For the occurrence of only hh-faults in a crystal, adjacent layers cannot be both displaced. However, if the case of random displacements is considered (model by Paterson and Warren (at $\beta = 0$), see schematic representation Scheme (13.1)), the next displacement would occur just in the layer adjacent to the layer where the previous displacement has occurred. This second displacement transforms the fault "hh" into the fault "h", a *twin* fault.

According to the model given in the schematic representation, the probabilities to find a SF hh having, for example, the ABAB configuration and a SF hc having, for example, the ABAC configuration are equal to $\alpha_{hh} = \alpha(1-\alpha)^2$ and $\alpha_h = \alpha^2(1-\alpha)$, respectively (see Scheme (13.1)). If the structure of such a crystal has to be characterized, other types of SF (*twins* in this case) are in the crystal, in addition to *intrinsic* faults, and must be considered in the model. Of course, the probabilities to find these types of SF may differ and, when the displacement concentration (α) is low, these other stacking faults can be neglected. The values of these two probabilities, α_{hh} and α_h, calculated for several values of α are given in Table 13.2.

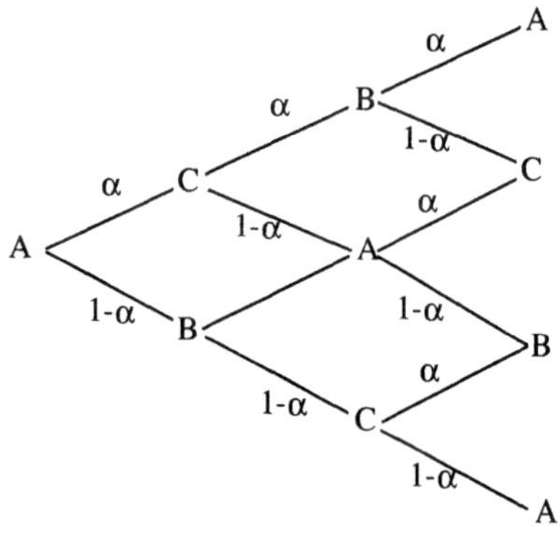

(13.1)

Another case is the model of a faulted crystal when only *intrinsic* stacking faults occur randomly in the cubic crystal. In such a crystal, the configuration hh occurs with a probability α at any possible position of the cubic crystal including arrangements where two and more hh configurations are placed alongside each other. The alternations of the layers in this case are represented by Schemes (13.2a) and (13.2b) using "ABC" and "hc" descrip-

Table 13.2. Probabilities of *intrinsic* fault and *twin* fault occurrences in the cubic crystal which is disordered by random displacements according to the Scheme (13.1)

Type of the SF or configuration of the layer sequences	Probabilities of the SF (configurations) (α is the probability of a random displacement) $\alpha = 0.01$	$\alpha = 0.05$	$\alpha = 0.1$	$\alpha = 0.2$
Intrinsic or *hh* (ABAB)	0.0098	0.0451	0.0810	0.128
Twin or *hc* (ABAC)	0.0001	0.0024	0.0090	0.032

tions, respectively. The comparison between Scheme (13.1) and Scheme (13.2) shows that the case of random *stacking faults* differs from the case of *random displacements* according to the following ways: (i) the "limiting" structure in the crystal with a random stacking faults are a regular cubic structure (at $\alpha = 0$) and a regular hexagonal structure (at $\alpha = 1$); (ii) the crystal does not contain another type of SF except for *intrinsic* faults.

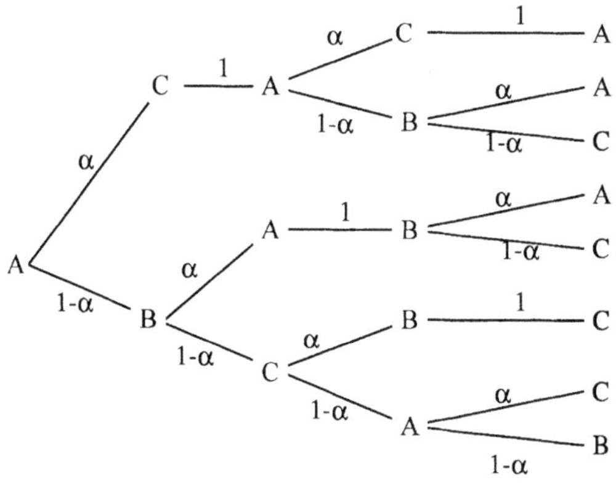

(13.2a)

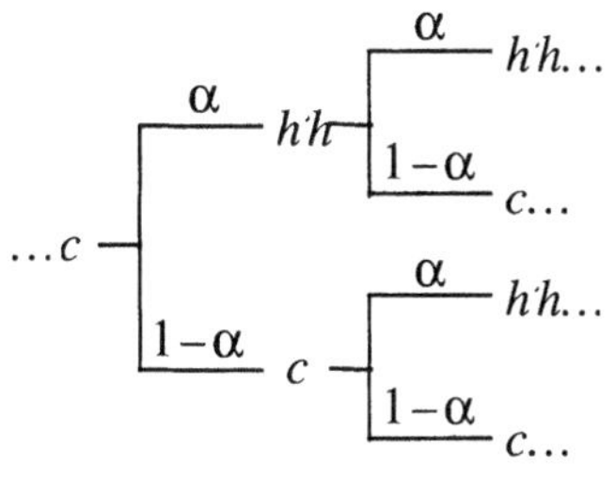

(13.2b)

One can see from Table 13.2 that an increase of the random displacement probability increases the probabilities to find other layer configurations besides the main one. Consequently, a difference would appear between the results obtained in the present chapter, where each type of SF are considered one by one, and the results derived from the model of a cubic crystal with random displacements. The higher the concentration of random dis-

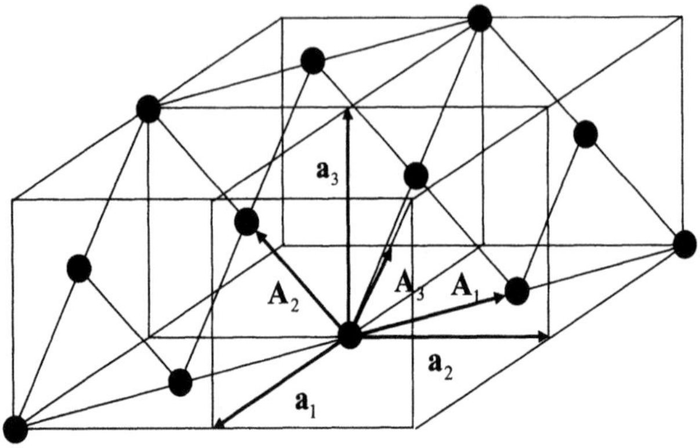

Fig. 13.1. Relationship between the cubic coordinate system ($\boldsymbol{a}_1$, $\boldsymbol{a}_2$, $\boldsymbol{a}_3$) and hexagonal coordinate system ($\boldsymbol{A}_1$, $\boldsymbol{A}_2$, $\boldsymbol{A}_3$)

placements (of stacking faults), the greater the difference between the XRD patterns calculated according to these two models.

The objective of this work is to determine, by means of the computer simulations, the basic features of the XRD peak profiles caused by each of the three types of stacking faults (i.e., *intrinsic* (hh), *twin* (h) and *extrinsic* ($hc'h$) SF) being randomly distributed in an FCC crystal having a *high concentration of SF*.

13.2.2 Calculation Procedure for the Intensity Scattered by a Single-Crystal

From the coordinate system described in Fig. 13.1, the structure of an *fcc* crystal can be represented as a packing of atomic layers along the axis $\boldsymbol{A}_3$ displaced by a vector $\boldsymbol{\delta} = (\boldsymbol{A}_1 - \boldsymbol{A}_2)/3$. This approach suggested by Wilson in the case of hexagonal close packed structure [25] was further used for other cases [26–30]. Then, the intensity of X-ray scattering by such a crystal, can be calculated using the probabilities $P_{\mathrm{AA}}(m)$, $P_{\mathrm{AB}}(m)$, $P_{\mathrm{AC}}(m)$ to find two atomic layers A–A, A–B, A–C, respectively, separated by m interplanar distances. The diffracted intensity by a crystal containing N atomic layers may be expressed as follows:

$$I_s(\mathrm{HK}\mathcal{L}) = C \sum_{m=-(N-1)}^{N-1} N(m) \exp\left(2\pi i \frac{\mathcal{L}m}{3}\right) \times \{P_{\mathrm{AA}}(m) + P_{\mathrm{AB}}(m)\exp(\psi) + [1 - P_{\mathrm{AA}}(m) - P_{\mathrm{AB}}(m)]\exp(-i\psi)\} \,, \tag{13.3}$$

where H, K are the indexes of the nodes of the reciprocal lattice along $\boldsymbol{A}_1^*$ and $\boldsymbol{A}_2^*$, respectively; $\mathcal{L}$ is the continuous coordinate along $\boldsymbol{A}_3^*$;

$$C = \frac{\sin^2 \pi N_1 \mathrm{H}}{\sin^2 \pi \mathrm{H}} \frac{\sin^2 \pi N_2 \mathrm{K}}{\sin^2 \pi \mathrm{K}}$$

is the two-dimensional Laue function; N_1 and N_2 are the numbers of atoms in the $\boldsymbol{A}_1$ and $\boldsymbol{A}_2$, directions, respectively; N is the number of atomic layers in the crystal in the $\boldsymbol{A}_3$ direction; $N(m)$ is the number of layer pairs spaced by m layers;

$$\psi = 2\pi \frac{\mathrm{H} - \mathrm{K}}{3} .$$

Considering that $N(m) = N - |m|$ and using an appropriate transformation, the expression (13.3) becomes

$$\begin{aligned} I_s(\mathrm{HK}\mathcal{L}) = CN + 2C \sum_{m=1}^{N} N - |m| \\ \times \{P_{\mathrm{AA}}(m) \cos m\phi + P_{\mathrm{AB}}(m) \cos(m\phi + \psi) \\ + [1 - P_{\mathrm{AA}}(m) - P_{\mathrm{AB}}(m)] \cos(m\phi + i\psi)\} , \end{aligned} \quad (13.4)$$

where $\phi = \pi \mathcal{L}$.

In order to calculate the pair correlation functions $P_{\mathrm{AA}}(m)$ and $P_{\mathrm{AB}}(m)$ for the specified SF concentration, a similar procedure which is described in detail by Berliner et al. [30] was used. For this purpose, a Monte Carlo method was applied to model the close packing of the atomic layers in *fcc* crystal containing different types of SF which are randomly arranged with a given concentration. The calculations were performed for such a model of crystals where "overlapping" of the SF is prevented. The known arrangement of the atomic layers allows the calculation of the values $P_{\mathrm{AA}}(m)$ and $P_{\mathrm{AB}}(m)$ to be easily made for any m. These correlation functions $P_{\mathrm{AA}}(m)$ and $P_{\mathrm{AB}}(m)$ have been calculated for $N = 500$ atomic layers; this number can be considered as the size of the coherently diffracting domains (i.e. crystallites) along the $\boldsymbol{A}_3$ direction. The calculated intensities have been averaged over an ensemble of 100 000 crystallites.

13.2.3 Calculation Procedure for the Intensity Scattered by a Powder

Following the above procedure, equation (13.4) can be used to calculate the intensity in each point of the reciprocal space along the rod with given H and K indices for an *fcc* crystal having a specified concentration of one SF type.

In order to calculate the XRD powder diagram, it is necessary to perform the summation of the intensities for all points of the reciprocal lattice which cross the Ewald sphere in point P defined by the angle θ, during the rotation of the reciprocal lattice around arbitrary axes passing through the zero node (point O, Fig. 13.2). The calculation can be simplified if the cross-sections of

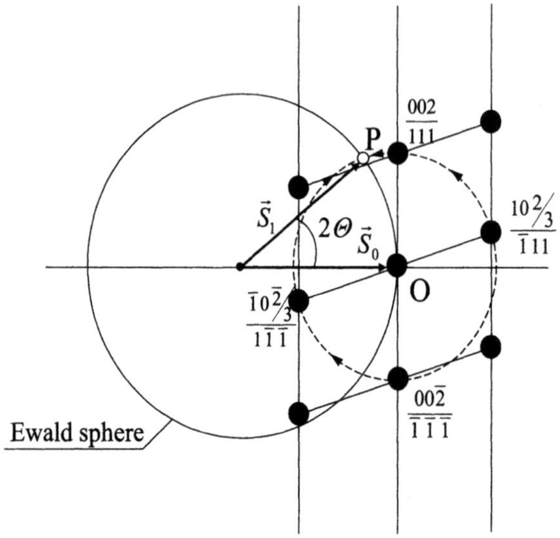

Fig. 13.2. Scheme of formation of the powder peak: $\boldsymbol{S}_0$ and $\boldsymbol{S}_1$ are vectors of the incident and scattered X-ray, respectively, $(|\boldsymbol{S}_0| = |\boldsymbol{S}_1| = 1/\lambda)$, λ is the X-ray wavelength; θ is Bragg angle; $\frac{\mathrm{HKL}}{\mathrm{hkl}}$ are indexes of the nodes of the reciprocal lattice in the $\frac{\text{hexagonal}}{\text{cubic}}$ coordinate systems

the reciprocal lattice rods with integer values of H and K, where the scattered intensity caused by the SF presence is localised, are negligibly small. This approximation is valid when the crystal dimensions in $\boldsymbol{A}_1$ and $\boldsymbol{A}_2$ directions are large enough. Then, only the reciprocal points located on the above rods will produce a significant effect on the XRD powder pattern. In such a way, the summation has to be performed only in the points of the reciprocal lattice located on the rods in which H and K indexes are integers. Then, in order to obtain the scattered intensity for one θ angle, coordinates $\mathcal{L}_i$ of the points in the $H_i^\theta K_i^\theta$ rods which cross the Ewald sphere, will be calculated from the expression

$$\frac{4\sin^2\theta}{\lambda^2} = (H_i^\theta \boldsymbol{A}_1^*)^2 + (K_i^\theta \boldsymbol{A}_2^*)^2 + (L_i^\theta \boldsymbol{A}_3)^2 \,, \tag{13.5}$$

where $\boldsymbol{A}_1^*$, $\boldsymbol{A}_2^*$ and $\boldsymbol{A}_3^*$ are the basis vectors of the reciprocal lattice.

For the given $H_i^\theta K_i^\theta \mathcal{L}_i^\theta$ coordinates, the scattered intensity at the angle θ can be determined as follows:

$$I_p(\theta) = \sum_i I_s(H_i^\theta K_i^\theta L_i^\theta) \,, \tag{13.6}$$

where $\sum_i$ is a summation over all points on the rods with indices $H_i^\theta K_i^\theta$ crossing the Ewald sphere at the scattering angle θ.

Realization of such a procedure using a defined step in θ results in the intensity distribution, which corresponds to the broadening of the XRD peaks due only to the presence of SF. So, the XRD peak profiles are calculated without taking into account the crystal distortions caused by the presence of other defects and the finite size of the coherently diffracting domains. In addition, these XRD peak profiles have to be further corrected by the instrumental function as well.

13.2.4 Intrinsic Stacking Faults (*hh*)

The probability to find an *intrinsic* SF in an *fcc* crystal at any distance from the previous one was expressed by α. Then, the possible sequences of the atomic layers in such a crystal can be represented in the terms of h- and c-layers using the Scheme (13.2b) which is given in Sect. 13.2.1.

Figure 13.3 shows some fragments of the XRD powder patterns calculated by the above procedure for an *fcc* crystal containing different concentrations

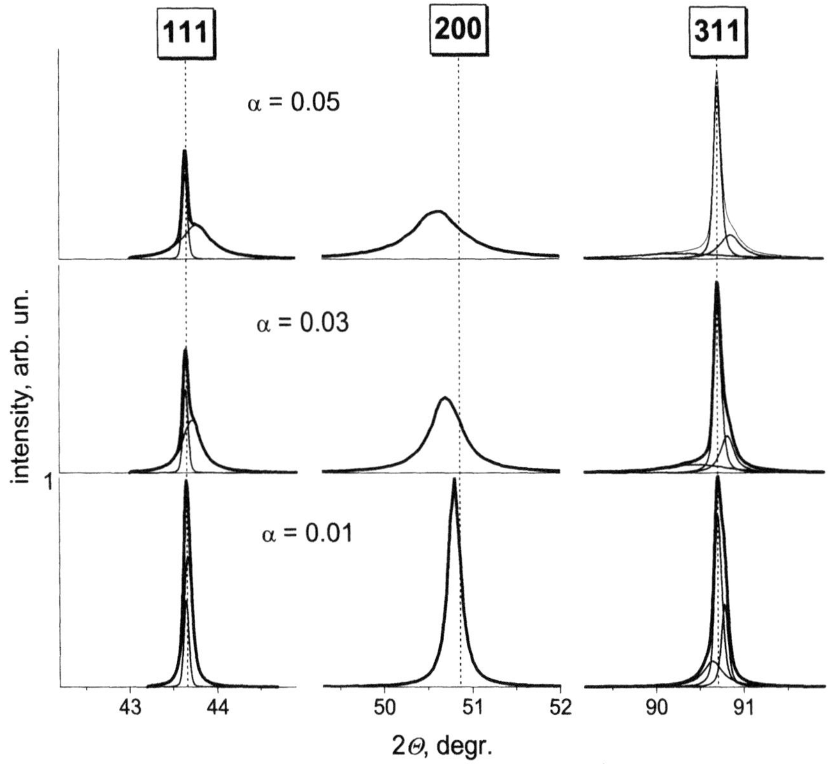

Fig. 13.3. Fragments of the calculated powder diagrams (peaks 111, 200, 311 and their components) of an *fcc* crystal with different concentrations of *intrinsic* SF: (**a**) $\alpha = 0.01$, (**b**) $\alpha = 0.03$, (**c**) $\alpha = 0.05$. The *dashe line* shows the positions of the 200 peak for $\alpha = 0$

of *intrinsic* faults. The introduction of a given concentration of the SF leads to a significant modification of the XRD peak profiles which depends on the *hkl* plane family. As an example, the resultant diffuse and Bragg components of the 111, 200 and 311 peaks are shown in Fig. 13.3. The 111 peaks is composed of eight components which can be divided into two groups: (i) two components are located on a rod where the condition H–K $= 3n$, $n = 1, 2, 3, \ldots$ holds i.e. where *intrinsic* faults do not affect their profiles; (ii) six components are located on the rods where the previous condition does not hold and, therefore, the components are broadened and also shifted towards larger reflection angles. Consequently, an asymmetry is observed on the right side of the resultant XRD peak maximum, but the peak remains nearly in Bragg position (The highest shift value of the peak maximum from its Bragg position is about 0.02 degrees when $\alpha = 0.01$).

On the contrary, the 200 peak contains six components, all located on the rods of the reciprocal lattice where H–K $\neq 3n$. Consequently, the broadening and shift of the XRD peak vary in a similar manner for all components. Hence, the peak maximum shifts towards smaller reflection angles remaining symmetrical whatever the concentration of SF.

Finally, the 311 peak is especially interesting because its diffuse components change in a different way when SF occur in a crystal. Indeed, among 12 diffuse components, 6 components are shifted towards smaller scattering angles and the 6 others are shifted towards larger scattering angles. Moreover, their broadening is also different. However, within each group, the diffuse components change their profiles and positions in the same manner. Therefore, considering also the existence of 6 Bragg components, the resultant 311 peak is composed of three different groups of components. The relationship between breadths, maximum intensities and positions of the peak maximums for these three groups is such that the resultant 311 peak does practically not change its position when the concentration of SF increases, whereas the profile becomes asymmetric on the right side.

The influence of the SF concentration on the full-width at half of various peaks is presented in Fig. 13.4. Fourier coefficients and distances between peaks demonstrate that this dependence is essentially non-monotonic whatever the *hkl*, except for the *h*00 peaks [31]. It should be noted that not only a shift of the diffuse components of the XRD peaks, but also a significant decrease of their maximum intensities accompanied by a considerable broadening, cause this non-monotonic change of the parameters characterizing the positions and the profiles of the peaks. Indeed, at high concentrations of SF, the FWHM of some peaks (cf. 311 peak in Fig. 13.3) decreases when the SF concentration increases. At the same time, background and asymmetry increase slightly.

Thus, a high concentration of just *intrinsic* faults in an *fcc* structure has the greatest impact on the peak shape (i.e. asymmetry and broadening), but the peak maximum positions practically do not change, although the gravity centres of these peaks are noticeably shifted when asymmetry of the

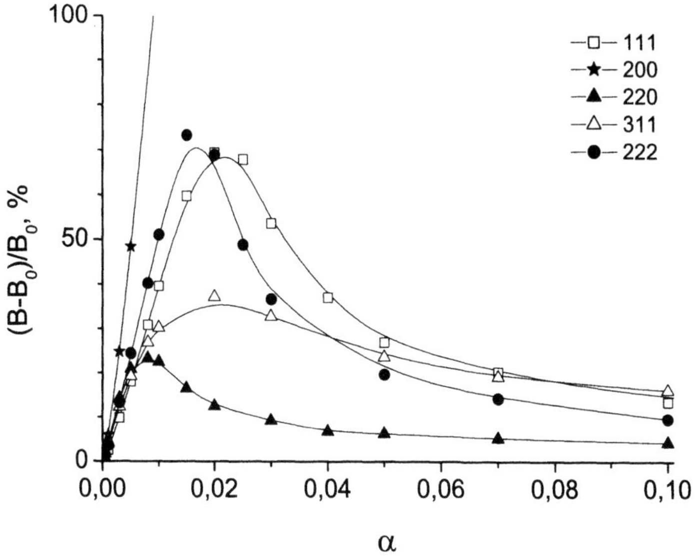

Fig. 13.4. Percentage changes of the FWHM (B) of different peaks calculated for various concentrations of the *intrinsic* faults (α); B_0 is the FWHM value of the corresponding peaks in the case $\alpha = 0$

peak profiles increases. However, this conclusion does not concern the profile and/or the position of the 200 peak. In the range of the studied concentrations, this peak was essentially broadened, but its profile remained practically symmetric. A gradual change of the 200 peak profile was accompanied by its monotonic shift from Bragg position.

13.2.5 Extrinsic Stacking Faults ($hc'h$)

The alternation of the atomic layers in a cubic structure where the *extrinsic* fault ($hc'h$) occurs with the probability β at any number of c-layers from the previous fault can be represented by the following probabilistic scheme (Scheme (13.7)).

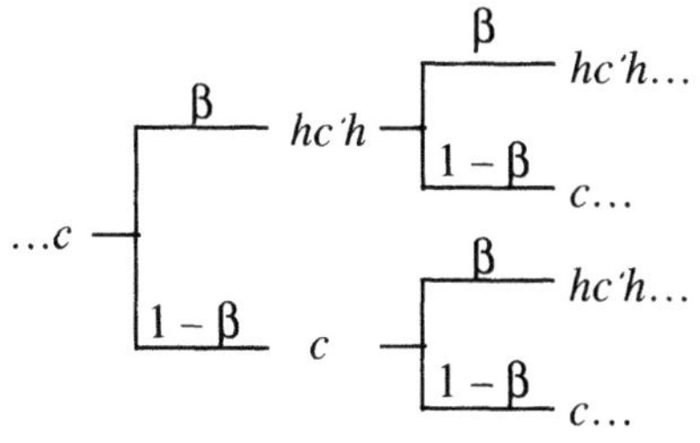

(13.7)

Two or more *extrinsic* faults can be arranged in such a way that a new structure (not cubic, not hexagonal) results. This so-called $9R_1$ polytype structure, is observed in some crystals.

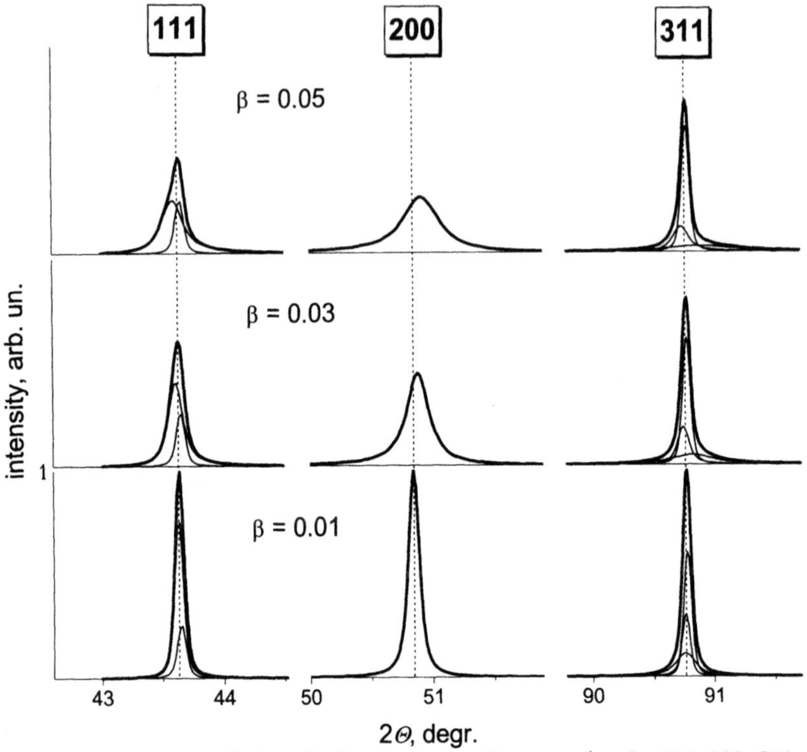

Fig. 13.5. Fragments of the calculated powder diagrams (peaks 111, 200, 311 and their components) of an *fcc* crystal with different concentrations of *extrinsic* SF: (**a**) $\beta = 0.01$, (**b**) $\beta = 0.03$, (**c**) $\beta = 0.05$. The *dashed line* shows the position of the 200 peak in the case $\beta = 0$

Let us consider peculiarities of the XRD peaks in the presence of the *extrinsic* faults (Fig. 13.5). As opposed to the previous case (i.e. *intrinsic* faults), 200 peak maximum is shifted towards larger scattering when *extrinsic* faults are present. In addition, the profile of the 111 peaks is asymmetric on the left side of the XRD peak maximum. Consequently, it is possible to discriminate between *intrinsic* and *extrinsic* faults by inspection of the 111 and 200 peak profiles. However, as the profile of the 311 peak is very close to that observed on Fig. 13.3, this peak can not be used to separate these two types of SF.

As in the case of the *intrinsic* faults, the broadening dependence of the peak profiles versus the concentration of *extrinsic* faults is non-monotonic (Fig. 13.6) except for the 200 peak.

13.2.6 Twin Stacking Faults (h)

This type of SF corresponds to an atomic layer packed as in the hexagonal close-packed structure. When a random distribution of *twin* faults only

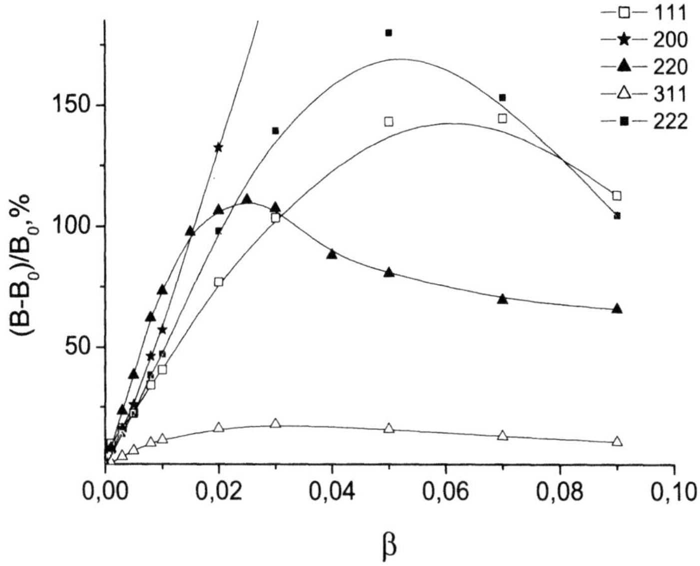

Fig. 13.6. Percentage changes of the FWHM (B) of different peaks calculated for various concentrations of *extrinsic* faults (β); B_0 is the FWHM value of the corresponding peaks for $\beta = 0$

is considered, the alternation of the atomic layers can be represented by Scheme (13.8) where γ is the probability to find a *twin* fault at a number of c-layers not less than two c-layers.

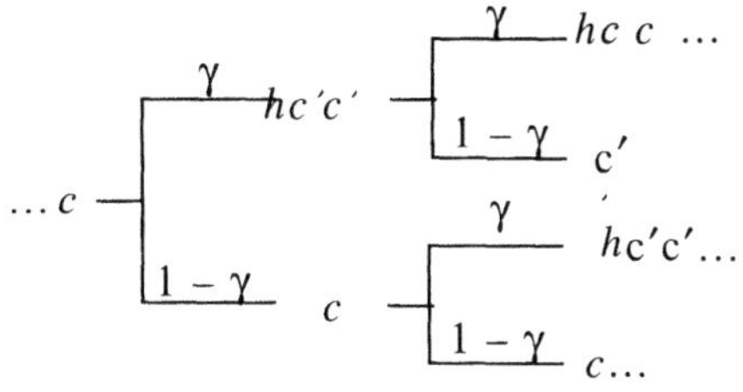

(13.8)

Figure 13.7 shows 111, 200 and 311 peaks for different concentration of *twin* faults. In analogy with the influence of *intrinsic* and *extrinsic* faults, the characteristics of the peaks change when the concentration of *twin* faults increases but, in this case, the shift from their Bragg positions is negligible, whatever their indexes. This is due to the fact that the XRD peak components affected by SF remain, in the case of *twin* faults, in the initial position whatever the *twin* concentration (see 111 and 311 peaks in Fig. 13.7).

When the concentration of *twins* increases, the change of the FWHM is non-monotonic for all peaks, except for $h00$ peaks (Fig. 13.8). At low *twin* concentration, peaks are noticeably wide, whereas, above a critical concentration, which depends on peak indices, the peaks become narrower with increasing *twin* fault concentration. Therefore, the determination of concen-

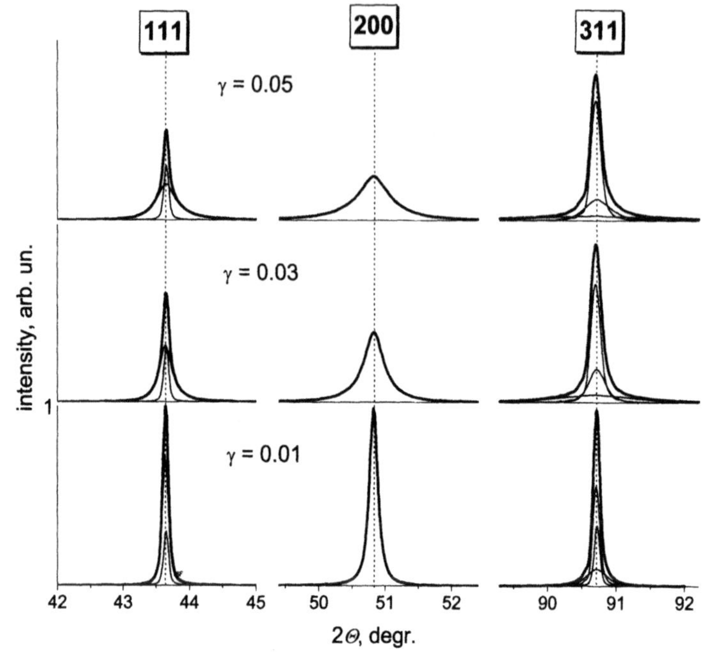

Fig. 13.7. Fragments of the calculated powder diagrams (peaks 111, 200, 311 and their components) of an *fcc* crystal with different concentrations of *twins*: (**a**) $\gamma = 0.01$, (**b**) $\gamma = 0.03$, (**c**) $\gamma = 0.05$. The *dashed line* shows the peak position of the 200 peak for $\gamma = 0$

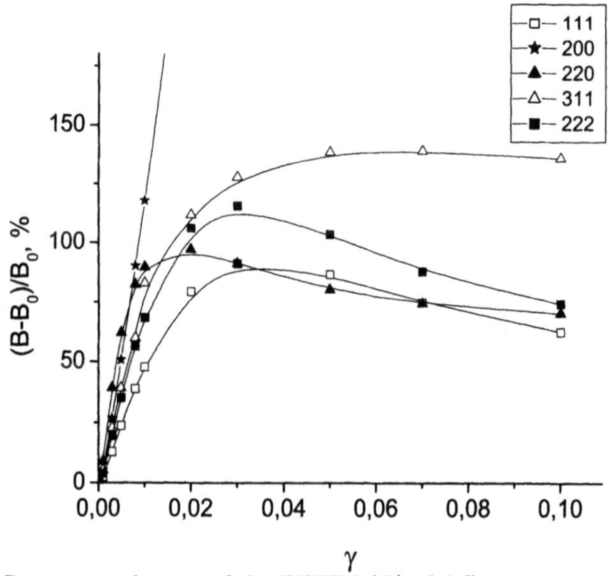

Fig. 13.8. Percentage changes of the FWHM (B) of different peaks calculated for various concentrations of *twin* faults (γ); B_0 is the FWHM value of the corresponding peaks for $\gamma = 0$

tration of *twin* faults based on the measurement of peak FWHMs with indices different from $h00$ seems to be problematic.

13.2.7 Determination of the Dominant Type of SF from the XRD Powder Diagram

The results presented above show that the qualitative characteristics of the XRD powder pattern (shift and profile asymmetry of the peaks) could be used to establish the presence of SF in an *fcc* crystal and to identify the SF type. Consequently, it is essential to compare the qualitative characteristics of the XRD peak profiles obtained experimentally with those simulated having different types of SF. The results of this comparative approach are given in Table 13.3.

In Table 13.3, symbols $\rightarrow$ and $\leftarrow$ indicate that the shift of XRD peaks occurs toward the smaller and larger scattering angles, respectively. Symbols ◭ and ◮ show the asymmetry direction of XRD peaks towards smaller and larger scattering angles, respectively. Symbol $\triangle$ corresponds to a symmetric profile.

So, the type of SF could be determined from the experimental XRD powder diagram by an analysis on one hand, of the XRD peak asymmetry for different *hkl* indices and, on other hand, of the asymmetry direction of the $h00$ XRD peaks. Unfortunately, the reliability of these results obtained from this qualitative analysis is not sufficient for a crystal containing several types of SF.

The XRD powder diagrams with a priori known concentrations of different SF were calculated in order to clarify the features of the XRD peak profiles caused by the simultaneous presence of different SF types. Figure 13.9 gives the evolution of the peak profiles calculated for an *fcc* crystal containing *twin* and *intrinsic* faults with different ratios of their concentration. The concentration of *intrinsic* faults was varied without changing the concentra-

Table 13.3. Characteristics of the peak profiles in the XRD powder patterns calculated for *fcc* crystal containing different types of SF

hkl	*Intrinsic* SF		*Extrinsic* SF		*Twin* SF	
	Shift direction	Peak profile asymmetry	Shift direction	Peak profile asymmetry	Shift direction	Profile asymmetry
111	no shift	◮	no shift	◭	no shift	△
200	→	△	←	△	no shift	△
220	no shift	△	no shift	△	no shift	△
311	no shift	◭	no shift	◮	no shift	△
222	no shift	◭	no shift	◮	no shift	△
400	←	△	→	△	no shift	△

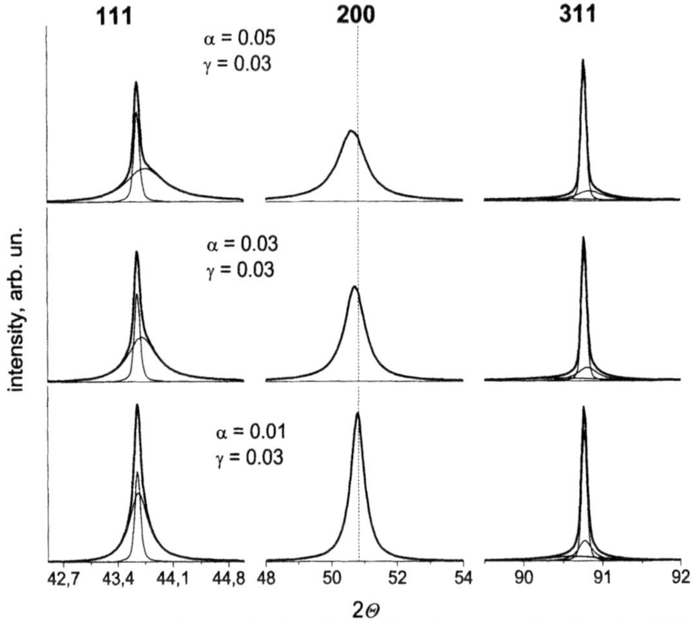

Fig. 13.9. Fragments of the calculated powder diagrams (peaks 111, 200, 311 and their components) of an *fcc* crystal containing a constant concentration of *twins* ($\gamma = 0.03$) and various concentrations of *intrinsic* SF: (**a**) $\alpha = 0.01$, (**b**) $\alpha = 0.03$, (**c**) $\alpha = 0.05$. The *dashed line* shows the position of the 200 peak for $\alpha = 0$, $\gamma = 0$

tion of *twin* faults ($\gamma = 0.03$). From this figure, the position of the 200 peak maximum is shifted towards lower scattering angles and the asymmetry of the 111 and 311 peak profiles similar to the case of *intrinsic* faults is observed (compare Figs. 13.9b,c and 13.3b,c, respectively).

The change of the FWHM value observed in Fig. 13.10 when simultaneously the *twin* faults (with an occurrence probability γ) and the *intrinsic* faults (with an occurrence probability α) are present, is interesting. Indeed, the FWHM of the 200 peak corresponds to a sum of the two contributions: (i) effect only due to the presence of *intrinsic* faults whose the occurrence probability is equal to α; (ii) effect only due to the presence of *twin* faults whose the occurrence probability is equal to γ. Thus, the features of XRD patterns resulting from the mixture of two different SF types correspond to an addition of these SF effects, at least, for the broadening of the 200 XRD peak.

In order to evaluate the predominant type of SF in the crystal, the ratio, ε, between the magnitudes of the shift and the FWHM of the 200 peak was calculated for different types of SF. Since both XRD peak parameters are directly proportional to the occurrence probability of a SF, ε is constant over the whole range of SF concentrations, but its value depends on the types of SF. When only *intrinsic* or *extrinsic* faults are present, ε is equal to 0.3 and 0.2 respectively. Consequently, if the value of ε obtained from the ex-

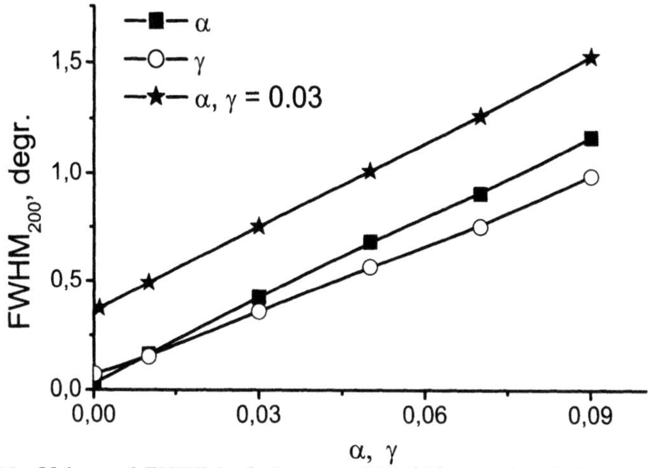

Fig. 13.10. Values of FWHM of the peak 200 (B) calculated for the *fcc* crystal containing various concentrations of the *twin* SF, γ, (curve 1), *intrinsic* SF, α, (curve 2), *intrinsic* SF at the constant concentration of *twin* SF ($\gamma = 0.03$) (curve 3)

perimental XRD patterns is equal to 0.3 (within the observation accuracy, $\varepsilon \cong 0.30 \pm 0.05$), only *intrinsic* faults are present: if $0.2 \leq \varepsilon \leq 0.3$, one can assume that *intrinsic* faults prevail over the rest of SF types. Finally, if ε is smaller than 0.2, it seems difficult to identify the predominant type of SF in this way. In this situation, a whole-powder-pattern fitting should be carried out. It has to be noted that knowledge of the real structure revealed by means of high-resolution electron microscopy would simplify the determination of the concentration of stacking faults and would improve the accuracy of results.

13.3 Simulation Procedure for "Real" XRD Powder Peak Profiles

The broadening of XRD peaks is known to be dependent not only on the presence of stacking faults, but also on the size of coherently diffracting domains and the mean value of microdistortions of the crystal lattice as well. Figure 13.11 describes schematically the change of the reciprocal node shape due to the presence of SF, microdistortions and finite size of crystallites in comparison with the reciprocal node shape in the perfect crystal. If crystallites are infinite and contain neither distortions nor SF, the reciprocal lattice nodes of the structure are point-like and all scattered intensity is localised in the node (i.e. the intensity distributions are like Dirac delta-functions (Fig. 13.11a). For small isotopic crystallites containing an isotropic arrangement of microdistortions, the point-like nodes become spherical (Fig. 13.11b). In contrast, the presence of SF produces an extension of the point-like nodes along the $\boldsymbol{A}_3^*$ direction as "needles" when the condition H–K $\neq 3n$ holds

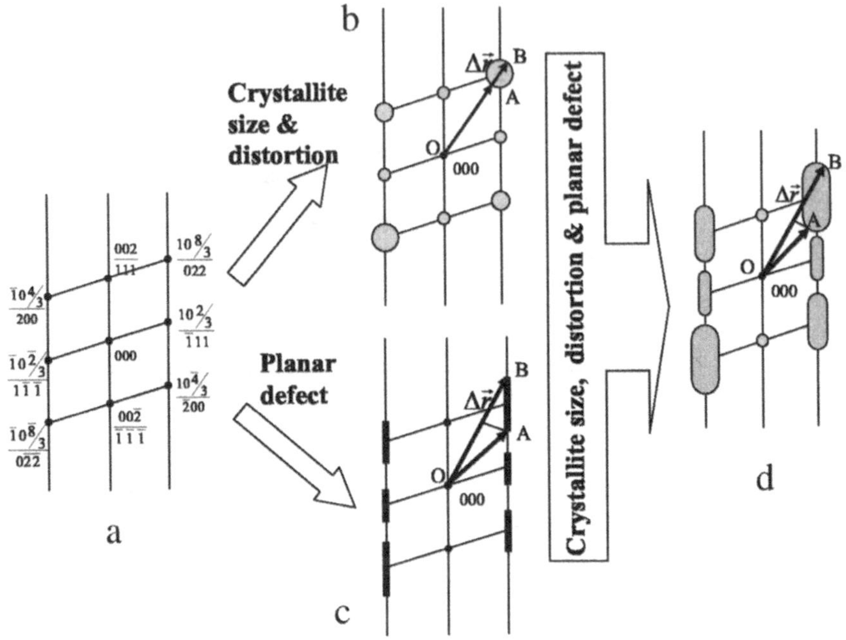

Fig. 13.11. Transformation of different nodes of the reciprocal lattice caused by the finite size of crystallites, microdistortions and stacking faults

(Fig. 13.11c). Consequently, when combined effects of SF, size and distortions are considered, the shape of the reciprocal lattice nodes located in the rod with indices H–K $\neq 3n$ looks like "sausages" (Fig. 13.11d).

The XRD peak profile is a convolution of the intensity distribution due to the presence of SF with the functions describing the influence of the size and distortion as well. In the general case, the function describing the peak profile due to the finite size of crystallites (coherently diffracting domains) can be complicated. Indeed, characteristics of crystallite shape and size distribution should be taken into account in this function which will be hereafter referred to as the size factor.

The function describing the change of the XRD peak profiles due to microstrains depends on many parameters such as the elasticity modulus, the dislocation structures (type of the dislocations, their density and spatial arrangement) etc. This function will be hereafter referred to as the microdistortion factor.

Consequently, each of the components of any XRD peak can be calculated as a convolution of the above-mentioned functions. It is clear the shapes of the functions depend on parameters of the crystal microstructure, namely on the type and concentration of SF, shape and size distribution of the coherently diffracting domains, types of dislocations and their distributions in the crystal, elastic properties of the crystal. Then, the XRD peak profiles calculated convoluting the functions have to be adjusted with those obtained experimentally in order to find the best agreement of both.

The goal of the present work is to demonstrate that such an approach can be used in practice. As a first step, the functions describing the size factor, microdistortion factor and instrumental broadening were considered in their simplest forms. Indeed, it was assumed that broadening due to finite size of the coherently diffracting domains is characterised by the integral breadth $B_s = \lambda/D\cos\theta$ (Scherrer formula), where λ is the wavelength of the X-ray radiation and D is the mean size of the coherently diffracting domains. Therefore, the size factor can be associated to a symmetric function $S(\theta)$ in which the integral breadth is equal to B_s.

In addition, distortions can affect the XRD peak profiles in a different way depending on the field of dislocations. Therefore, in many cases, it is assumed that both screw and edge dislocations are present and are randomly distributed in the crystal. In this case, the nodes of the reciprocal lattice will be broadened in an isotropic manner. From results reported by Ungár et al. [32], the integral breadth, B_d, of the XRD peak at the angle θ can be presented, in first approximation, as $B_d = 2Q\rho^{1/2}C_{hkl}tg\theta$, where $Q = (\pi A^2\boldsymbol{b}^2/2)^{1/2}$, A is a parameter which depends on the effective outer cut-off radius of dislocations, $\boldsymbol{b}$ is the Burgers vector and C_{hkl} is the mean contrast factor of dislocations for the hkl peak. (As it was shown by Wilkens [33], the value of A can be taken as $A = 10$ for a wide range of dislocation distribution.) Therefore, it is suggested in the calculations that the microdistortion factor is a symmetric function $D(\rho)$ in which the integral breadth is equal to B_d.

It should be remembered that the presence of SF always leads to an isotropic change of some nodes and non-isotropic for the others. These phenomena are opposed to those caused by the size and microdistortion factors, which produce, in principle, a symmetric change of all nodes (cf. Fig. 13.11). In addition, a change of the nodes located along rods satisfying to the condition H–K $\neq 3n$ is attributed not only to SF influence, but also to the crystallite size in the direction perpendicular to the SF plane (in the case of an *fcc* crystal this plane is the 111). Then, position of such nodes in the reciprocal space has to be considered when the projections of these nodes on Ewald's sphere are calculated. This is schematically given in Fig. 13.12.

To evaluate the value of the projection of the "sausage"-like shape of nodes on the Ewald's sphere, $\Delta\mathbf{r}$ may be represented (in a first approximation) by $\Delta\mathbf{r}' = \Delta\mathbf{r} + 2|\boldsymbol{r}_{s-s}|$ (Fig. 13.12), where $\Delta\mathbf{r}$ is the contribution due to SF and crystallite size in the direction perpendicular to the SF; the vector $\boldsymbol{r}_{s-s}$ is the contribution due to microdistorsions and the finite crystallite size along other directions. Considering that the microdistortion factor affects all nodes in the similar manner, its contribution to $\Delta\mathbf{r}_{s-s}$ is not dependent on the position of a reciprocal node relative to the origin. On the contrary, the broadening of the node due to the size factor increases as the angle between the diffraction vector and $\boldsymbol{A}_3^*$ axis decreases.

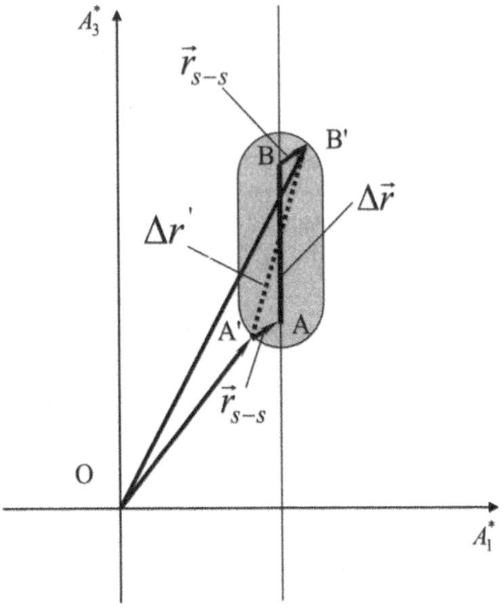

Fig. 13.12. To the calculation of the size of the reciprocal lattice node

Finally, considering all discussed above, the XRD intensity may be expressed as follows:

$$I(\theta) = \sum I_s(H_i^\theta K_i^\theta L_i^\theta) \otimes S_s(\theta) \otimes D(\theta) \otimes A(\theta)\,, \tag{13.9}$$

where $I_s(H_i^\theta K_i^\theta L_i^\theta)$ is the intensity distribution of the s-component due to the presence of SF in the crystallites; $\otimes$ means a convolution procedure; $S_s(\theta)$ and $D(\theta)$ are the functions describing the XRD peak broadening due to the finite size of the crystallites and microdistortions of the crystal lattice, respectively; $A(\theta)$ describes the peak broadening caused by the instrumental factors. To a first approximation, the peak profiles of a standard powder obtained under the same conditions as the XRD pattern of the specimen under investigation could be used as the instrumental function.

Based on the expression (13.9), the calculation procedure for microstructure characterisation from XRD powder patterns is the following: (1) Introduction of initial values characterising the SF, size and microdistortion factors; (2) calculation of all s components of m peaks in the scattering angle range where experimental XRD diagrams are recorded. This operation and all subsequent operations will perform for each m-peak; (3) correction of each s-component of each m-peak by the size factor taking into account the location of reciprocal nodes; (4) addition of all corrected for the size factor components (Point (3)) and then correction of the resulting XRD peak profile by the microdistortion factor; (5) correction of these latter peaks by the instrumental function; (6) subtraction of the background and, normalisation of peaks by the maximum peak intensity; (7) comparison m peak profiles with XRD peaks obtained experimentally after normalisation and background subtraction (8) modification of the characteristic parameters of the microstructure and come back to the point (2) until the best fit of the modelled XRD peaks to the experimentally determined peaks is achieved.

As an example for testing the proposed procedure, a set of Cu-nanostructured powders produced by ball milling were studied by XRD.

13.4 Defect and Microstructure Analysis of Ball-Milled Cu Powders by XRD

13.4.1 Preparation and Characterization of the Powders

Pure elemental copper powders were milled with a planetary mill referred to as G5 milling machine [34–36]. 125 ml of powders and 5 stainless steel balls (15 mm in diameter, 14 g in weight) were sealed under air, into the stainless steel vials. The ball to powder weight ratio is 7/1. The rotation speed of its vials (ω) which are fixed onto a rotating disk and the rotation speed of the disk (Ω) can be set independently. Each milling condition is characterised by 3 essential parameters ($\Omega(rpm)/\omega(rpm)/\tau(h)$) where τ is the duration of the milling process. The XRD patterns of three powder specimens produced under conditions 150/50/24, 250/50/24 and 350/50/24 that correspond to three different powers of shocks were selected to test the performance of the proposed XRD approach.

To identify the as-milled Cu powders and determine their microstructural parameters, XRD analyses were performed with a D5000 Siemens high resolution diffractometer using a monochromatic Cu–K_β beam ($\lambda = 0.1392$ nm) obtained by a curved graphite monochromator in the diffracted beam.

The morphology and size of the crystallites (shape and size of the coherently diffracting domains) and the structural imperfections (microstrains, stacking faults, etc.) in the as-milled powders were determined using the XRD line profile analysis described by Langford and Louër [37]. The procedure for performing this analysis, using the Pseudo-Voigt function, is as follows: (i) check that the Pseudo-Voigt function is appropriate to determine peak breadths for all profiles involved, then examine their shape factors to determine the Lorentzian–Gaussian ratio character, (ii) correct the data for instrumental effects, (iii) make a Williamson–Hall plot [38] ($\beta^* = \beta \cos\theta/\lambda$ is plotted as a function of $d^* = 2\sin\theta/\lambda$) to be certain of the nature of the sample broadening and to determine the precise $[hkl]$ dependence of peak breadths, and (iv) carry out a detailed analysis of peak breadth in accordance with the information from step (i).

The average particle diameter was estimated as being about 0.3 to 0.5 µm, from BET specific surface measurements and assuming a spherical shape. The large difference between particle size and crystallite size (20 ... 120 nm from

Williamson–Hall plot) is easily explained by the combined effect of fracture, agglomeration and consolidation processes that materials undergo during the severe mechanical treatment. Besides, aggregates formation induced by the ball milling process, was confirmed by scanning electron microscopy observations. Transmission electron microscopy was carried out using a TOPCON 002B microscope (working at 200 KV, with a point resolution of 0.18 nm) on the powders as-deposited on a carbon grid. The possible contamination by milling media and tools was evaluated by EDX analyses on pressed pellets. Analysis has shown slight pollution by chromium and iron, never exceeding 1 at%.

13.4.2 Full Profile Fitting

A preliminary analysis of the experimental XRD diagrams and subsequent fitting of the diagrams have been performed according to the procedure discussed above.

In the Williamson-Hall plot presented in Fig. 13.13, a non-monotonous behaviour is observed. This suggests that SF could play a role in the diffraction line broadening of the investigated specimens.

The XRD data indicate that the positions of XRD peaks correspond to their Bragg positions with a high degree of accuracy. In addition, experimental XRD peak profiles are symmetric (see Fig. 13.14). In concordance with the modelling results, these features evidence that the powders contain neither *intrinsic* nor *extrinsic* faults, and the *twin* SF are most likely present in the structure of the materials.

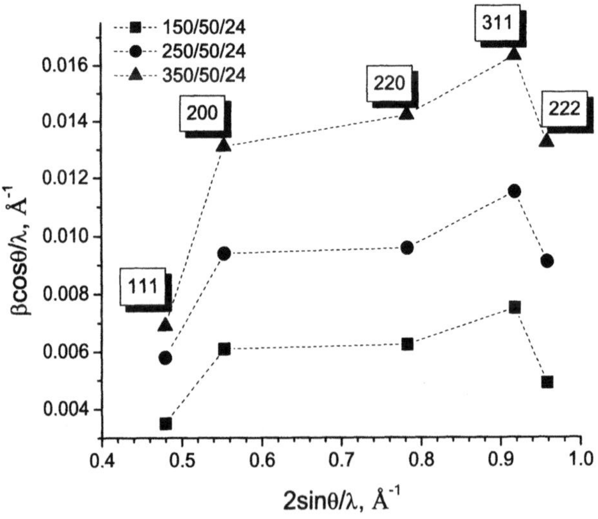

Fig. 13.13. Williamson-Hall plots (with indications of Miller indices) for Cu powders prepared employing three different milling conditions

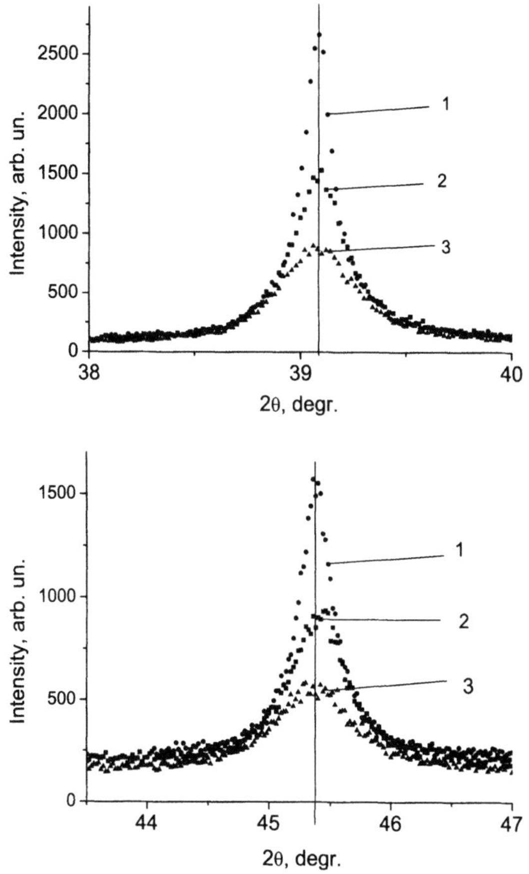

Fig. 13.14. The 111 and 200 peak profiles in the experimental XRD diagrams of Cu powders prepared under different conditions ($\Omega/\omega/\tau$) of ball milling: 1) – 150/50/24; 2) – 250/50/24; 3) – 350/50/24

In addition to XRD data, HRTEM observations were performed. For example, a high-resolution image of a Cu-nanograin (diameter 35 nm) is presented in Fig. 13.15. The centre of the particle is metallic copper oriented with a $\langle 110 \rangle$ zone axis parallel to the electron beam. The linear contrasts crossing the particle are associated to $\{111\}$ *twin* boundaries. These *twin* planes are also observed in grains of copper nano-powder synthesised by the cryomelting process [39]. Moreover, Huang et al. [40] reported that multiple *twins* and high-order *twins* are also frequently observed in grains of copper nano-powder milled during 20 h. From these different observations, only the presence of *twin* SF was revealed and therefore taken into account in the modelling. Based on the preliminary analysis of the experimental XRD peak profiles, TEM observations and the assumptions of the copper powder structures, three parameters (size, distortions and *twin* SF) have been investigated in this work.

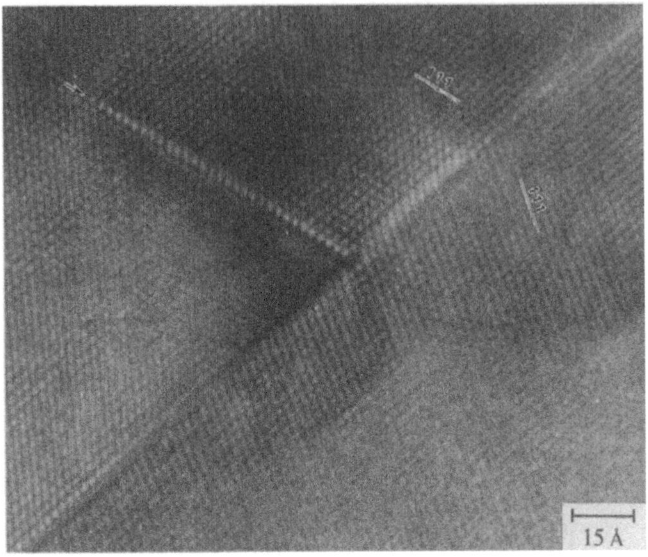

Fig. 13.15. HREM image of a particle of Cu-powder

To calculate the XRD peak profiles in accordance with (13.9), the parameters of the instrumental function $A(\theta)$ for each peak have been extracted from the resolution curve obtained from the broadening of BaF_2 XRD powder peaks (as standard) at different scattering angles. The XRD peak profiles of the standard have been also described by a symmetric pseudo-Voigt function in which parameters vary in a monotonous way versus the scattering angle. The magnitudes of these parameters were used to model the instrumental function at a given interval of the scattering angles.

The Gauss and Lorentz functions with integral breadths B_s and B_d, respectively, were used as the functions describing the size factor and distortion factor. The best agreement between the experimental and the calculated XRD peak profiles for different *hkl* indices was achieved by the variation of the parameters α, D and ρ at the condition that both the size factor and microdistortion factor are Lorentz functions.

As an illustration, experimentally determined peak profiles for a Cu powder (250/50/24) are compared in Fig. 13.16 with the corresponding peak profiles modelled by the proposed approach. A satisfactory agreement can be observed (cf. Fig. 13.16).

Results of the experimental data processing according to the procedure described above are shown in Table 13.4. Such changes of the Cu microstructure parameters are in agreement with those expected when the materials are subjected to the mechanical treatment.

The mean value of the crystallite size obtained by the XRD peak profiles fitting is nearly equal to the value determined from the Williamson–Hall method in the case when the SF concentration is low, as is shown in Ta-

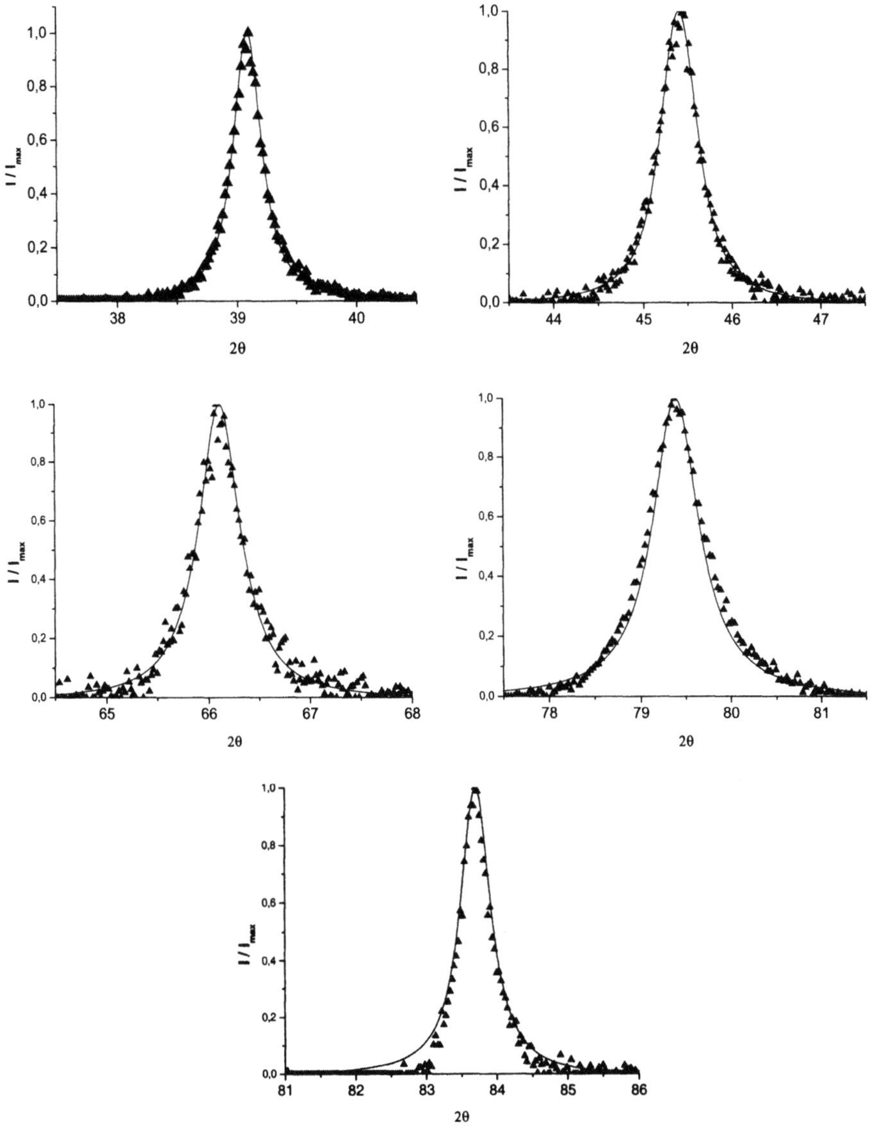

Fig. 13.16. Comparison between the peak profiles in the experimental XRD diagrams of the powders produced at the 250/50/24 regime of the ball milling and the corresponding peak profiles simulated according to the proposed approach

ble 13.4, in which the results from both procedures (our and Williamson–Hall) are presented. However, an increase in the *twin* SF concentration leads to an increase in the crystallite size difference determined by these two methods.

Table 13.4. Microstructural characteristics of the Cu powders produced by ball milling

		Results of the fitting			Results of Williamson–Hall method	
No.	$\Omega/\omega/\tau$	Concentration of the *twin* SF γ (%)	Mean size of crystallites D (nm)	Density of dislocations ρ (m^{-2}) 10^{14}	Mean size of crystallites D (nm)	Micro-distorsions %
1.	0/0/0	0.9	120	1.9	121.7	0.07
2.	150/50/24	2.7	87	7.7	41.3	0.16
3.	250/50/24	4.7	50	19.0	24.9	0.25
4.	350/50/24	5.5	32	46.1	20.7	0.41

13.5 Conclusion

Analysis of the XRD powder patterns calculated in the case of a *fcc* polycrystalline metal, containing a high concentration of SF, shows that shapes and positions of the XRD peaks are qualitatively dependent on the type of SF. The established features of transformation of the XRD peak profiles, as the concentration of the stacking faults increases, can be used to find the type of SF and, in such a way to reduce the number of parameters used in the fitting process.

On the basis of the developed approach, a successful fitting of the experimental data was performed by the simulation of powder diffraction patterns of *fcc* crystals taking into account the presence of SF, the finite size of crystallites and microdistortions caused by the presence of dislocations.

However, more information concerning the microstructure of these materials should be obtained, for example, by means of transmission and/or secondary electron microscopy in order to optimise the fitting procedure. This optimisation concerns, on one hand, the reduction of the varied number parameters by an identification of stacking fault types and, on the other hand, the improvement of the calculation accuracy taking into account, for example, the known shape of the crystallites or their size distributions.

The proposed XRD approach to study the *fcc* polycrystalline materials will be applied to materials with another high symmetry crystal lattice containing stacking faults.

Acknowledgements

The authors would like to thank Dr E. Gaffet (CNRS/Belfort) for performing the ball milling experiments and Dr Y. Champion (CNRS/Vitry) for TEM investigations.

References

1. R.L. Snyder, J. Fiala, H.J. Bunge (eds.): *Defect and Microstructure Analysis by Diffraction.* (University Press, Oxford 1999) pp. 785
2. L. Landau: Sow. Phys. **12**, 579 (1937)
3. M.S. Paterson: J. Appl. Phys. **23**, 805 (1952)
4. B.E. Warren: Prog. Met. Phys. **8**, 147 (1959)
5. J. Mering: Acta Cryst. **2**, 371 (1949)
6. G.W. Bridley, J. Mering: Acta Cryst. **4**, 441 (1951)
7. J. Kakinoki: Acta Cryst. **23**, 875 (1967)
8. W. Cochran, E.R. Howells: Acta Cryst. **7**, 412 (1954)
9. C.A. Jonson: Acta Cryst. **16**, 490 (1963)
10. D.A. Mirzaev, S.V. Rushchic: Fizika metallov i metallurgiya **37**, 912 (1974) (in Russian)
11. D.A. Mirzaev, S.V. Rushchic: Crystallografiya **21**, 670 (1976) (in Russian)
12. A.I. Ustinov, in *Defect and Microstructure Analysis by Diffraction.* (University Press, Oxford 1999) pp. 264–317
13. B.E. Warren, B.L. Averbach: J. App. Phys. **21**, 595 (1950)
14. G. Barret: Met. Trans. Soc. AIME **188**, 123 (1950)
15. Z. Nishiyama et al.: J. Phys. Soc. Jpn. **20**, 1192 (1965)
16. S. Kajiwara, H. Fujita: J. Phys. Soc. Jpn. **21**, 400 (1966)
17. S. Kajiwara: J. Phys. Soc. Jpn. **22**, 795 (1967)
18. L.I. Lysak, A.I. Ustinov: Phys. of Metals and Metallurgy **44**, 1050 (1977)
19. A.R. Tholen: Philos. Mag. A **53**, 259 (1986)
20. V.G. Pushin et al.: Phys. of Metals and Metallurgy **75**, 96 (1993)
21. F. Gauzzi, R. Montanari: Mater. Scien. and Eng. A **273–275**, 524 (1999)
22. N. Bergeon et al.: J. Phys. IV France **7**, C5 (1997)
23. L. Velterop, R. Delhez, T.H. de Keijser, E.J. Mittemeijer, D. Reefman: J. Appl. Cryst. **33**, 296 (2000)
24. J.P. Hirth, J. Lothe: *Theory of Dislocation* (Atomizdat, Moscow 1972) pp. 213–599
25. A.J.C. Wilson: Proc. Roy. Soc. A **180**, 277 (1942)
26. H. Jagodzinski: Acta Cryst. **2**, 201 (1949)
27. H. Jagodzinski: Acta Cryst. **2**, 208 (1949)
28. H. Jagodzinski: Acta Cryst. **7**, 17 (1954)
29. R. Gevers: Acta Cryst. **5**, 518 (1952)
30. R. Berliner, S.A. Werner: Phys. Rev. B **34**, 3586 (1986)
31. A.I. Ustinov, N.M. Budarina, L.O. Olikhovska: Metallofiz. Noveishie Tekhnol. **22**, 28 (2000) (in Russian)
32. T. Ungár, S. Ott, P.G. Sanders, A. Borbély, J.R. Weertman: Acta Mater. **46**, 3693 (1998)
33. M. Wilkens: Phys. Stat. Sol. (a) **2**, 359 (1970)
34. E. Gaffet, M. Abdellaoui, N. Malhouroux-Gaffet: Mater. Trans. JIM **36**, 198 (1995)
35. M. Abdellaoui, E. Gaffet: Acta Mater. **44**, 725 (1996)
36. E. Gaffet, L. Yousfi: Mater. Sci. Forum **88–90**, 51 (1992)
37. J.I. Langford, D. Louer: Rep. Prog. Phys. **59**, 131 (1996)
38. G.K. Williamson, W.H. Hall: Acta Metall. **1**, 22 (1953)
39. Y. Champion, J. Bigot: Scripta Mater. **35**, 517 (1996)
40. J.Y. Huang, Y.K. Wu, H.Q. Ye: Acta Mater. **44**, 1211 (1996)

Part V

Grain Interaction

14 Diffraction Elastic Constants and Stress Factors; Grain Interaction and Stress in Macroscopically Elastically Anisotropic Solids; The Case of Thin Films

U. Welzel, M. Leoni, and E.J. Mittemeijer

Thin films and surface-adjacent material of bulk polycrystals cannot generally be considered as macroscopically elastically isotropic; at most transverse isotropy can occur. However, in this case, the usually applied models for grain interaction (as those proposed by Voigt, Reuss, Neerfeld–Hill and Eshelby–Kröner) erroneously predict macroscopic elastic isotropy. The present chapter presents an overview of recent work on the diffraction analysis of elastic grain interaction in polycrystalline material. It has been demonstrated for the first time, by X-ray diffraction (XRD) analysis of stresses in thin films, that the elastic grain interaction can lead to macroscopically elastically anisotropic behaviour (as exhibited by non-linear so-called $\sin^2\psi$-plots), even in the absence of texture. A rigorous treatment has been given for the diffraction stress analysis of both macroscopically elastically isotropic specimens and macroscopically elastically anisotropic specimens. A new model has been proposed for the direction-dependent elastic grain interaction which predicts macroscopically anisotropic behaviour of thin films. The proposed model has been successfully applied in the X-ray diffraction study of the residual stresses in thin films. The notion surface anisotropy of bulk specimens has been revisited as a special case of direction-dependent grain interaction.

14.1 Introduction

Diffraction stress analysis is a well-established tool for the analysis of load and residual stress states in polycrystals (see, for example [1,2]) and is applied to a broad variety of specimens ranging from thin films to large mechanical components.

Traditionally, one *hkl* diffraction line is recorded at various specimen tilt (and possibly, dependent on the complexity of the stress state, rotation) angles. From the measured peak positions, diffraction (lattice) strains can be calculated. Seldom, different reflections are simultaneously considered (as has been done, for example, for the analysis of stress gradients [3,4]).

To obtain the components of the mechanical stress tensor from measured lattice strains, suitable elastic "parameters", which generally differ from the mechanical elastic constants, are required. For the case of macroscopically elastically isotropic specimens, these are the so-called (X-ray) diffraction elastic constants S_1^{hkl} and $\frac{1}{2}S_2^{hkl}$. The so-called (X-ray) diffraction stress factors $F_{ij}(\psi, \varphi, hkl)$, introduced by Dölle and Hauk [5, 6], are used for specimens exhibiting crystallographic texture. Only recently it has been demonstrated that the diffraction stress factors $F_{ij}(\psi, \varphi, hkl)$ are applicable to any macroscopically elastically anisotropic specimen [7].

For a given specimen these diffraction (X-ray) elastic constants (or respectively, stress factors) can be measured by (X-ray) diffraction by applying a known load stress under simultaneous diffraction strain measurement in a diffractometer. As the measurement is cumbersome, if possible at all, a theoretical route for obtaining the parameters can be very useful for practical purposes.

Only for macroscopically elastically *isotropic* specimens, measured or calculated diffraction (X-ray) elastic constants can be found for various materials in the literature (see, for example, [2]). For macroscopically elastically *anisotropic* specimens, a calculation algorithm for obtaining diffraction (X-ray) elastic constants and stress factors, from the single crystal elastic constants of the material composing the specimen, is generally required.

If the single crystallites composing the specimen are intrinsically elastically anisotropic, neither the mechanical elastic constants (relating mechanical (macroscopic) strains to mechanical (macroscopic) stresses) nor the diffraction elastic constants, relating (diffraction) lattice strains to mechanical stresses, can be derived straightforwardly from the single-crystal elastic constants of the crystallites composing the specimen. A so-called grain-interaction model is needed to describe the distribution of stresses and strains over the crystallographically differently oriented crystallites in the specimen subjected to a mechanical stress/strain state. This can be understood as follows: only for a single crystal, the elastic strain tensor can be calculated for any applied stress by employing Hooke's law. If elastically anisotropic crystallites constitute a polycrystal, the interaction of neighbouring grains imposes constraints on stresses and strains of an individual crystallite when the polycrystal is loaded. Such constraints can be expressed as grain-interaction conditions in grain-interaction models. By adopting a grain-interaction model, mechanical and diffraction (X-ray) elastic constants (and diffraction (X-ray) stress factors for macroscopically elastically anisotropic specimens) can be calculated from single-crystal elastic data, as will be shown in this chapter.

Common grain-interaction models are those due to Voigt [8], Reuss [9] and Eshelby–Kröner [10–13], which were devised for the calculation of *mechanical*, i.e. macroscopical, elastic constants of bulk materials from single-crystal elastic constants. Such models can also be used for the calculation of *diffraction* elastic constants from single-crystal elastic data.

The Voigt and Reuss models present extremes for grain interaction: either the strain tensor (Voigt model) or the stress tensor (Reuss model) for each crystallite is assumed to be equal to the macroscopic strain or stress tensor, respectively. These models are incompatible with physical reality because of the corresponding stress, respectively strain discontinuities at the grain boundaries that they imply. Hill showed theoretically that the Voigt and Reuss models set bounds for macroscopic elastic constants [14]. First Neerfeld [15] and later Hill [14] suggested on an empirical basis to use the arithmetic (or geometric,[1] [14]) averages of diffraction (X-ray) [15] and macroscopic [14, 15] elastic constants calculated according to the models of Voigt and Reuss. Values calculated on this basis for elastic constants come close to the experimentally determined. The mathematically more tedious Eshelby–Kröner model gives numerical values close to those obtained by the Neerfeld–Hill approach. The above models have been devised for bulk polycrystals, but have been applied to thin films as well.

Only recently it has been demonstrated experimentally by van Leeuwen et al. [16] for an untextured Ni film that diffraction effects can be observed (curved $\sin^2\psi$-plots; see further) that are incompatible with the traditional models for grain interaction. Van Leeuwen et al. [16] demonstrated that a grain-interaction model adapted to the microstructure of a columnar thin film, can explain these experimental findings. To this end, grain-interaction assumptions proposed by Vook and Witt [17, 18] were, for the first time rigorously, employed for the analysis of diffraction-stress measurements by assuming that all crystallites exhibit the same in-plane strain whereas they can deform freely in a direction perpendicular to the film surface (in correspondence with a columnar grain structure for the thin film).

Leoni et al. [19] extended this Vook–Witt based model to fibre-textured specimens and showed by simulations that errors as large as 50% for the stress can occur when traditional models are employed for the analysis of transversely isotropic specimens.

In the following, a grain-interaction model involving different grain-interaction assumptions along different directions in the specimen will be named *direction-dependent grain-interaction model.* Owing to this, the Vook and Witt model (and its subsequent modifications) can be considered a direction-dependent grain-interaction model. Direction-dependent grain-interaction assumptions give rise to elastic anisotropy on the macroscopic scale (see van Leeuwen et al. [16] (Vook–Witt model) and Welzel and Mittemeijer [7] (general case)).

Even though the Vook–Witt grain-interaction model can be considered as some "intermediate" model of grain interaction (with reference to the extreme Voigt and Reuss models), it still imposes extreme constraints and in particular identical in-plane strain and completely unconstrained deformation perpendicular to the surface for all grains. As previously said, ex-

[1] The geometrical average of n numbers is defined as the n-th root of the product of the numbers. For Young's modulus, as an example, the geometric average of the moduli according to the Reuss (E_R) and Voigt (E_V) models E_{GA} is defined as $\sqrt{E_R E_V}$.

treme grain-interaction assumptions are in general not compatible with the true behaviour of real polycrystals. Only recently, a more general approach to direction-dependent grain interaction, overcoming the unrealistic grain-interaction assumptions involved in the application of the Vook–Witt model, has been proposed by Welzel et al. [20].

The purpose of this chapter is to review the concept of direction dependent grain-interaction models and recent developments in this field.

14.2 Theoretical Background

14.2.1 Frames of Reference

Since stress and strain are tensorial quantities, it is convenient to consider three different Cartesian frames of reference, which are either fixed to an individual crystallite (frame of reference C), the specimen (frame of reference S) and the experimental apparatus (frame of reference L), respectively:

- The *crystal frame of reference* (C): in general, a convention for the definition of an orthonormal crystal system, such as the one given by Nye [21] (for a detailed treatment, see also [22]) has to be adopted. For cubic crystal symmetry, the axes are chosen to coincide with the a-, b- and c-axes of the crystal lattice.
- The *sample frame of reference* (S): the S_3-axis is oriented perpendicular to the specimen surface and the S_1- and S_2-axes lie in the surface plane.
- The *laboratory frame of reference* (L): this frame is chosen in such a way that the L_3-axis coincides with the diffraction vector of the (X-ray) diffraction experiment. For the relative orientation of laboratory and specimen frames of reference and the definition of angles φ and ψ, where ψ is the inclination angle of the sample surface normal with respect to the diffraction vector and φ denotes the rotation of the sample around the sample surface normal, consider also Fig. 14.1. For $\varphi = \psi = 0°$, the L and S frames coincide.

In the following a superscript (C, S or L) indicates the frame of reference adopted for the representation of tensors.

Transformations of tensors from one to another frame of reference can be accomplished by suitable rotation matrices (for a general introduction on the use of transformation matrices in the context of diffraction stress analysis, see for example, [2]; for the transformation matrices relevant in the context of this chapter, the reader is referred to van Leeuwen et al. [16] and Leoni et al. [19]).

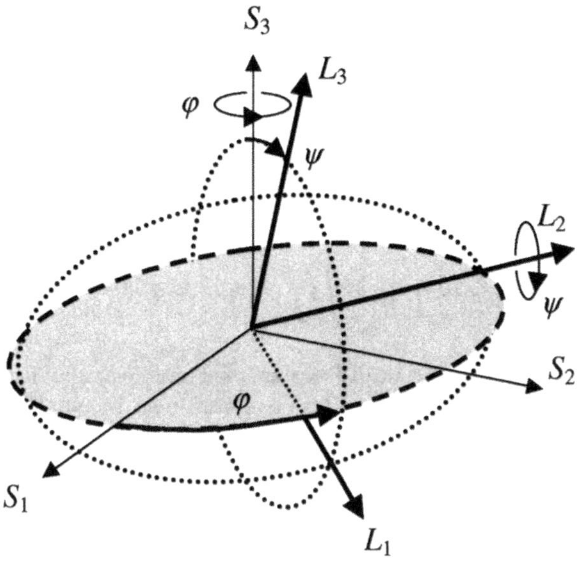

Fig. 14.1. Definition of the angles ψ and φ and relations between the sample (S) and laboratory (L) reference frames

14.2.2 Euler Angles and the Crystallographic Orientation Distribution Function (ODF)

The orientation of each crystallite in the specimen frame of reference S can be identified by three Euler angles α, β and γ (the convention of Roe and Krigbaum [23] is adopted).

A set of Euler angles is associated to a vector $\boldsymbol{g} = (\alpha, \beta, \gamma)$ in the three-dimensional orientation (Euler) space G (for a detailed introduction to the topic the reader is referred, for example, to Bunge [24]). Each point in the orientation space G represents a possible orientation of the frame of reference C with respect to S. In the absence of (crystallographic) texture, the volume fraction of crystallites having an orientation in the infinitesimal orientation range $d^3g = \sin(\beta)d\alpha d\beta d\gamma$ around $\boldsymbol{g}$ is independent of $\boldsymbol{g}$.

Crystallographic texture can be quantified by introducing the orientation distribution function (ODF) $f(\alpha, \beta, \gamma)$, specifying the volume fraction of crystallites having an orientation in the infinitesimal orientation range d^3g around $\boldsymbol{g}$:

$$\frac{dV(\boldsymbol{g})}{V} = \frac{f(\boldsymbol{g})}{8\pi^2} d^3g = \frac{f(\alpha, \beta, \gamma)}{8\pi^2} \sin(\beta) d\alpha d\beta d\gamma\,. \tag{14.1}$$

The ODF is normalised:

$$\iiint\limits_G \frac{f(\boldsymbol{g})}{8\pi^2} d^3g = 1\,. \tag{14.2}$$

14.2.3 Calculation of Averages of Tensors

In the context of diffraction stress analysis, mechanical and diffraction averages of stress, strain and elastic tensor averages are usually calculated in the Euler (orientation) space. The tensor $\Omega(\alpha,\beta,\gamma)$ (e.g. the strain tensor for the crystallites of orientation (α,β,γ)) has to be considered, in view of the diffraction experiments, as an average for all spatially distributed crystallites with this particular orientation:

$$\Omega(\alpha,\beta,\gamma) = \iiint_V \Omega(\boldsymbol{r},\alpha,\beta,\gamma)d^3r\,. \tag{14.3}$$

V is the volume sampled in a diffraction measurement and $\boldsymbol{r}$ is the position vector. Hence, the tensor $\Omega(\boldsymbol{r},\alpha,\beta,\gamma)$ can vary over the spatially distributed crystallites with a given orientation (α,β,γ).

In the following, angular brackets $\langle\,\rangle$ will be used to denote the volume-weighted average (i.e. mechanical average) of a tensor Ω (e.g. the strain tensor) for all crystallites in the aggregate considered:

$$\begin{aligned}\langle\Omega\rangle &= \frac{1}{8\pi^2}\iiint_G \Omega(\alpha,\beta,\gamma)\,f(\boldsymbol{g})\,d^3g \\ &= \frac{1}{8\pi^2}\int_{\gamma=0}^{2\pi}\int_{\beta=0}^{\pi}\int_{\alpha=0}^{2\pi} \Omega(\alpha,\beta,\gamma)f(\alpha,\beta,\gamma)\sin(\beta)\,d\alpha\,d\beta\,d\gamma\,. \end{aligned}\tag{14.4}$$

Similarly, braces $\{\,\}$ will indicate the average of a tensor Ω over the set of crystallites in diffraction condition only (i.e. the diffraction average). A diffraction line contains data of only a subset of the crystallites for which the diffracting planes are perpendicular to the chosen measurement direction. Because only the direction of the diffraction vector is defined, a degree of freedom occurs for the diffracting crystallites: the rotation around the diffraction vector (denoted by the angle λ in the following). In a diffraction measurement, the subset of diffracting crystallites is selected by hkl of the reflection considered and by the orientation of the diffraction vector with respect to the specimen frame of reference, i.e. by the two angles φ and ψ. These quantities, selecting the ensemble of crystallites for a diffraction average are attached to the corresponding average as sub- (φ,ψ) and superscripts (hkl):

$$\{\Omega\}_{\varphi,\psi}^{hkl} = \frac{\int_0^{2\pi}\Omega(hkl,\lambda,\varphi,\psi)f^*(hkl,\lambda,\varphi,\psi)d\lambda}{\int_0^{2\pi}f^*(hkl,\lambda,\varphi,\psi)d\lambda}\,. \tag{14.5}$$

$f^*(hkl,\lambda,\varphi,\psi)$ is the representation of the ODF in terms of the measurement parameters and the rotation angle λ. The ODF as defined in (14.1) cannot be directly used in (14.5) in analogy to (14.4) since the angles λ, φ, ψ are *not* Euler angles representing a rotation of the C system with respect to the S system (they provide the rotation of the system L with respect to the system S instead). However, the values of α, β, γ and thus $f(\alpha, \beta, \gamma)$ at every λ can be calculated from hkl, λ, φ and ψ, to be finally substituted for $f^*(hkl, \lambda, \varphi, \psi)$ in (14.5) (for a more detailed treatment of the necessary transformations, see also [19]). Thus, the diffraction strain $\varepsilon_{\varphi\psi}^{hkl}$ can be calculated as the average strain $\left\{\varepsilon_{33}^{L}\right\}_{\varphi\psi}^{hkl}$, i.e. the strain parallel to the L_3-axis:

$$\varepsilon_{\varphi\psi}^{hkl} = \left\{\varepsilon_{33}^{L}\right\}_{\varphi\psi}^{hkl} = \frac{\int\limits_{0}^{2\pi} \varepsilon_{33}^{L}(hkl, \lambda, \varphi, \psi) f^*(hkl, \lambda, \varphi, \psi) d\lambda}{\int\limits_{0}^{2\pi} f^*(hkl, \lambda, \varphi, \psi) d\lambda} . \tag{14.6}$$

14.2.4 Mechanical Elastic Constants

For simplicity, the analysis will be focused on a thin film subjected to a plane, rotationally symmetric state of stress. The film will be assumed being at least transversely isotropic i.e. either untextured or fibre-textured (fibre texture is frequently observed in thin films grown by physical or chemical vapour deposition). Thus, both the symmetry axis of the elastic properties and the symmetry axis of the stress field are parallel to the specimen normal.

The rotationally symmetric, plane state of stress is characterised by the following stress tensor:

$$\langle \sigma^S \rangle = \begin{pmatrix} \sigma_{\parallel}^{S} & 0 & 0 \\ 0 & \sigma_{\parallel}^{S} & 0 \\ 0 & 0 & 0 \end{pmatrix} . \tag{14.7}$$

Owing to the transverse isotropy, the stress field induces a rotationally symmetric strain field:

$$\langle \varepsilon^S \rangle = \begin{pmatrix} \varepsilon_{\parallel}^{S} & 0 & 0 \\ 0 & \varepsilon_{\parallel}^{S} & 0 \\ 0 & 0 & \varepsilon_{\perp}^{S} \end{pmatrix} . \tag{14.8}$$

Only two mechanical elastic constants are required to describe fully the relation between the stress and strain tensors for a transversely isotropic body subjected to a plane state of stress, as only two strain tensor components and one stress tensor component are independent and non-zero [see (14.7) and (14.8)]. Those mechanical elastic constants will be called A and B:

$$\varepsilon_{\parallel}^{S} = A\sigma_{\parallel}^{S} , \tag{14.9}$$

$$\varepsilon_{\perp}^{S} = B\sigma_{\parallel}^{S} . \tag{14.10}$$

14.2.5 Diffraction Elastic Constants and Stress Factors

For a macroscopically isotropic specimen the following equation relates the lattice strain measured by X-ray diffraction (diffraction strain) to the components of the mechanical stress tensor $\sigma_{\parallel}^{S}$ (see, for example, [2]):

$$\begin{aligned}\varepsilon_{\varphi,\psi}^{hkl} &= \frac{1}{2}S_2^{hkl}\sin^2\psi\left(\langle\sigma_{11}^S\rangle\cos^2\varphi+\langle\sigma_{12}^S\rangle\sin(2\varphi)+\langle\sigma_{22}^S\rangle\sin^2\varphi\right)\\ &+\frac{1}{2}S_2^{hkl}\left(\langle\sigma_{13}^S\rangle\cos\varphi\sin(2\psi)+\langle\sigma_{23}^S\rangle\sin\varphi\sin(2\psi)+\langle\sigma_{33}^S\rangle\cos^2\psi\right)\\ &+S_1^{hkl}\left(\langle\sigma_{11}^S\rangle+\langle\sigma_{22}^S\rangle+\langle\sigma_{33}^S\rangle\right). \qquad (14.11)\end{aligned}$$

For the case of a plane, rotationally symmetric state of mechanical stress ($(\langle\sigma_{12}^S\rangle=\langle\sigma_{13}^S\rangle=\langle\sigma_{23}^S\rangle=\langle\sigma_{33}^S\rangle=0)$, $\varepsilon_{\varphi,\psi}^{hkl}$ is independent of φ and (14.11) can be simplified leading to

$$\varepsilon_{\psi}^{hkl}=\left(2S_1^{hkl}+\frac{1}{2}S_2^{hkl}\sin^2\psi\right)\sigma_{\parallel}^{S}. \qquad (14.12)$$

In this case, the stress $\sigma_{\parallel}^{S}$ can be deduced from the slope of a plot of the lattice strain ε_{ψ}^{hkl} versus $\sin^2\psi$ ("$\sin^2\psi$-plot"). The constants S_1^{hkl} and $\frac{1}{2}S_2^{hkl}$, introduced on an empirical basis by Möller and Martin [25], are independent of ψ and φ and are called (X-ray) diffraction elastic constants. Decades after the introduction, Stickforth [26] proved on the basis of symmetry considerations, that the concept of (X-ray) diffraction elastic constants is justified for any macroscopically elastically isotropic aggregate, thereby confirming the validity of (14.12) for such a case.

Diffraction stress analysis of macroscopically elastically anisotropic specimens is also possible. Traditionally, only crystallographic texture, leading to macroscopically anisotropic elastic properties for a polycrystalline aggregate (if intrinsic (single crystal) elastic anisotropy occurs), was considered for diffraction stress analysis. Within this context the (X-ray) diffraction stress factors, F_{ij}, have been introduced [5,6], relating the diffraction strain to the mechanical stress tensor expressed in the specimen frame of reference S (the Einstein convention, i.e. summation over indices appearing twice in a formula, is adopted):

$$\varepsilon_{\varphi\psi}^{hkl}=F_{ij}(\psi,\varphi,hkl)\langle\sigma_{ij}^S\rangle. \qquad (14.13)$$

For the case of a plane, rotationally symmetric state of stress (14.13) can be simplified:

$$\varepsilon_{\varphi\psi}^{hkl}=\left[F_{11}(\psi,\varphi,hkl)+F_{22}(\psi,\varphi,hkl)\right]\sigma_{\parallel}^{S}. \qquad (14.14)$$

For the case of transverse elastic isotropy, the diffraction strain is independent of φ, thus φ can be arbitrarily set to zero and, as an index, suppressed:

$$\varepsilon_{\psi}^{hkl} = [F_{11}(\psi, 0, hkl) + F_{22}(\psi, 0, hkl)]\, \sigma_{\parallel}^{S} \,. \tag{14.15}$$

Note that the $F_{ij}(\psi, \varphi, hkl)$ due to their definition are not components of a tensor.

Only recently, it has been demonstrated by van Leeuwen et al. [16] that direction-dependent grain interaction leads to macroscopically elastically anisotropic behaviour, *even in the absence of texture.* In this case, an analysis on the basis of (14.12) is not possible, but (14.13) can be still employed, as has been shown by Welzel and Mittemeijer [7].

14.2.6 Principle of Diffraction Stress Analysis

Provided that the (X-ray) diffraction elastic constants or stress factors are known (measured or calculated from single crystal elastic data by adopting a grain-interaction model), the unknown stress tensor components can be obtained as fitting parameters in a (least-squares) minimisation of χ^2

$$\chi^2 = \sum_i \omega_i^2 \left[\varepsilon_i^{\text{calc}}(\langle\sigma^S\rangle, hkl, \varphi, \psi) - \varepsilon_i^{\text{meas}}(hkl, \varphi, \psi)\right]^2 , \tag{14.16}$$

where $\varepsilon_i^{\text{meas}}$ represents the measured lattice strains. The index i stands for all measured lattice strains (for all different ψ, φ and/or hkl employed). The ω_i are weighing factors which can be used to reflect the statistical relevance of the measured strains $\varepsilon_i^{\text{meas}}(hkl, \varphi, \psi)$ in the result of the minimisation and could correspond to the inverse of the standard deviations of the measured strains. The diffraction strains, $\varepsilon_i^{\text{calc}}$, have to be calculated following (14.13) (general case) or (14.15) (transversely elastically isotropic specimen subjected to a plane, rotationally symmetric state of stress). Of course, diffraction stress analysis on the basis of (14.16) can also be done for macroscopically elastically isotropic specimens. In this case, the diffraction strains, $\varepsilon_i^{\text{calc}}$, have to be calculated following (14.11) (general case) or (14.12) (specimen subjected to a plane, rotationally symmetric state of stress). Usually, analysis strategies based on fitting straight lines to measured lattice strain data (or sums and/or differences of lattice strains for positive and negative ψ, respectively) are employed for macroscopically elastically isotropic specimens (e.g. the so-called "Dölle–Hauk method", see, for example, [2]). However, employing the strategy outlined above for macroscopically elastically anisotropic specimens maybe useful also in this case, e.g., lattice strains measured using multiple reflections can be analysed simultaneously. Equations (14.11) and (14.12), employing φ- and ψ-independent (X-ray) diffraction elastic constants, have to be considered as simplifications of (14.13) and (14.15), respectively. For a review of different methods of diffraction stress analysis, see [37].

14.2.7 Effective Grain Interaction

The (X-ray) diffraction stress factors, the (X-ray) diffraction elastic constants and the mechanical constants of a polycrystal can be calculated from single crystal elastic data by adopting a grain-interaction model. The basics common to all grain-interaction models considered here are as follows:

For each crystallite, the stress and strain tensors in the specimen reference frame satisfy Hooke's law:

$$\varepsilon_{ij}^S = s_{ijkl}^S \sigma_{kl}^S \,. \tag{14.17}$$

The s_{ijkl}^S denote the single crystal elastic compliances transformed to the specimen frame of reference. Equation (14.17) represents a system of nine equations for eighteen unknowns, but as the strain and stress tensors are symmetric (i.e. $\varepsilon_{ij} = \varepsilon_{ji}$ and $\sigma_{ij} = \sigma_{ji}$), (14.17) thus is a short notation for six independent equations for twelve independent unknowns. If six components of the twelve unknowns are known, the other components for each crystallite can be calculated by solving the system of equations (14.17).

In the type of grain-interaction models considered here, a total of six stress and/or strain tensor components in the S system are taken equal to the mechanical averages for all crystallites and thus the other six (unknown) components can be calculated from the system of equations (14.17).

In the case of a plane, rotationally symmetric state of stress [cf. (14.7)], the elastic constants A and B for the case of transverse elastic isotropy can thus be calculated from [cf. (14.9) and (14.10)]

$$A = \frac{\varepsilon_{\parallel}^S}{\sigma_{\parallel}^S} = \frac{\langle \varepsilon_{11}^S \rangle}{\langle \sigma_{11}^S \rangle} = \frac{\iiint\limits_G \varepsilon_{11}^S f(\boldsymbol{g}) d^3 g}{\iiint\limits_G \sigma_{11}^S f(\boldsymbol{g}) d^3 g} \tag{14.18}$$

and

$$B = \frac{\varepsilon_{\perp}^S}{\sigma_{\parallel}^S} = \frac{\langle \varepsilon_{33}^S \rangle}{\langle \sigma_{11}^S \rangle} = \frac{\iiint\limits_G \varepsilon_{33}^S f(\boldsymbol{g}) d^3 g}{\iiint\limits_G \sigma_{11}^S f(\boldsymbol{g}) d^3 g} \,. \tag{14.19}$$

The diffraction strain can be calculated following (14.6) and the diffraction stress factors follow from (14.15):

$$[F_{11}(\psi, 0, hkl) + F_{22}(\psi, 0, hkl)] = \frac{\varepsilon_{\psi}}{\sigma_{\parallel}^S} \,. \tag{14.20}$$

Evidently, in this case, only the sum of the two (X-ray) diffraction stress factors indicated is needed. This sum will be abbreviated as

$$F(\psi, hkl) \stackrel{\text{def}}{=} F_{11}(\psi, 0, hkl) + F_{22}(\psi, 0, hkl) \,. \tag{14.21}$$

Frequently, the so-called Neerfeld–Hill model [14, 15] is employed for the calculation ((X-ray) diffraction and macroscopic) elastic constants. To this end, (X-ray) diffraction and macroscopic elastic constants are calculated as arithmetic averages of the constants calculated according to the Voigt and Reuss models. In a similar way, (X-ray) diffraction stress factors can be calculated as the arithmetic average of the stress factors calculated according to the Reuss (R) and Voigt (V) models (see, for example, [28,29]). Even though a generally accepted physical basis for mixing the models of Reuss and Voigt is lacking, a few authors have tried to give a physical basis of such a averaging procedure [29,30].

The background of any averaging of (extreme) grain-interaction models to represent physical reality could be described as follows. A real sample is conceived to be constituted from separate volume fractions of crystallites, each of which obeys a certain type of grain interaction. For the following discussion an average of (extreme) grain-interaction models will be called an effective grain-interaction model. It will be shown that, depending on the type of specimen considered, appropriate choices of the grain- interaction models "to be averaged" can be made.

Under the assumption that a volume fraction f_i of the polycrystalline aggregate obeys the (extreme) grain-interaction model i (in Sect. 14.3.2 four grain-interaction models will be considered), the average macroscopic strain of the aggregate parallel to the substrate $\langle \varepsilon_{\parallel}^{S} \rangle$ can be calculated as an arithmetic average:

$$\langle \varepsilon_{\parallel}^{S} \rangle = \left(\sum_{i} f_i A_i \right) \sigma_{\parallel}^{S} , \tag{14.22}$$

where the subscript i indicates that the averaging has to be performed with respect to the various volume fractions representing the various, extreme models. Thus, the averaged mechanical constant $\langle A \rangle_i$ reads in this case:

$$\langle A \rangle_i = \sum_{i} f_i A_i . \tag{14.23}$$

and, analogously, the average mechanical constant $\langle B \rangle_i$ is given by

$$\langle B \rangle_i = \sum_{i} f_i B_i . \tag{14.24}$$

14.3 Direction-Dependent Grain Interaction

In the inner part of a bulk specimen, each crystallite is surrounded by neighbouring crystallites in three dimensions. This is not the case for the surface-adjacent material of the specimen, and generally, this does not hold for thin films in particular. Thus, not all directions are equivalent for surface adjacent crystallites, as can be the case for crystallites located inside a bulk polycrystal. As a consequence, thin films or surface adjacent material cannot generally be considered as macroscopically elastically isotropic. Rather, at most transverse elastic isotropy, reflecting a rotational symmetry of the system with respect to the surface normal, can occur.

Direction-dependent grain-interaction assumptions adapted to the columnar microstructure frequently encountered in evaporated films have been formulated by Vook and Witt [17]. A rigorous adaptation to diffraction stress analysis based on $\sin^2\psi$-plots, taking the full elastic compliance tensors of individual crystallites into account was not done until recently by van Leeuwen et al. [16] for untextured specimens and Leoni et al. [19] for (fibre-) textured specimens. An approach similar to the Vook–Witt model was followed by Rafaja et al. [31], but in this case the full single-crystal elastic compliance tensor was not taken into account.

Stickforth [26] postulated independently of Vook and Witt [17] that the elastic behaviour of surface-adjacent material of a polycrystal may differ from the elastic behaviour of material remote from the surface and named the effect "surface anisotropy". In Sect. 14.5, the relation of surface anisotropy to direction-dependent grain interaction is discussed.

14.3.1 The Vook–Witt Model

The analysis is, for simplicity, focused on an at least transversely isotropic polycrystal subjected to a plane, rotationally symmetric state of stress.

The grain interaction assumptions adopted here are: (i) the strain is rotationally symmetric parallel to the surface and (ii) the in-plane strain components are equal for all crystallites and (iii) the stresses perpendicular to the surface are zero for all crystallites. These assumptions fix parts of the stress and strain tensors:

$$\varepsilon^S = \begin{pmatrix} \varepsilon_{\parallel}^S & 0 & \diamond \\ 0 & \varepsilon_{\parallel}^S & \diamond \\ \diamond & \diamond & \diamond \end{pmatrix}, \tag{14.25}$$

$$\sigma^S = \begin{pmatrix} \diamond & \diamond & 0 \\ \diamond & \diamond & 0 \\ 0 & 0 & 0 \end{pmatrix}. \tag{14.26}$$

The tensor components marked by $\diamond$ are not explicitly specified for every crystallite by the grain interaction assumptions. Following the procedure outlined in Sect. 14.2.7, these components can be calculated from Hooke's law (for every crystallite) according to (14.17). Thus, for a particular value of the strain parallel to the substrate $\varepsilon_{\parallel}^S$, all non-zero strain and stress tensor components can be calculated by solving the system of (14.17) (for explicit equations, see van Leeuwen et al. [16] (untextured case) and Leoni et al. [19] (textured case)). The elastic constants A and B can thus be calculated from (14.18) and (14.19). The diffraction strain can be calculated following (14.6) and the stress factors follow from (14.20) and (14.21).

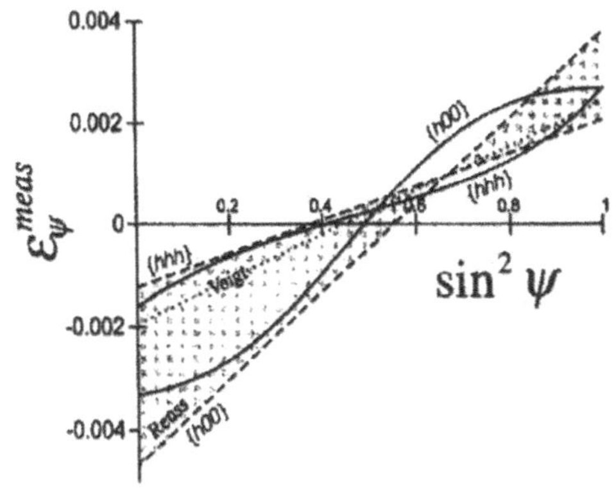

Fig. 14.2. $\sin^2\psi$-plots calculated accorded to the Voigt (dotted line, all reflections), Reuss (dashed lines, $00l$ and hhh reflections) and Vook–Witt (continuous lines, $00l$ and hhh reflections) grain interaction model for a nickel film imposing a plane, rotationally symmetric state of tensile stress of 800 MPa. All $\sin^2\psi$-plots for different reflections calculated according to the Reuss model fall within the lightly shaded area bounded by $00l$ and hhh plots. Taken from [16]

An example is presented in Fig. 14.2 for an untextured nickel layer. A striking result is the non-linearity of the $\varepsilon_\psi^{hkl} - \sin^2\psi$ diagrams according to the Vook–Witt based model, whereas the traditional models yield linear $\sin^2\psi$-plots. The non-linearities are due to the macroscopic elastic anisotropy induced by the direction-dependent grain interaction.

Experimentally, the model based on the grain-interaction assumptions proposed by Vook and Witt was applied to the stress analysis of an untextured nickel layer by van Leeuwen et al. [16]. Experimental findings not compatible with the traditional models, i.e. curved $\sin^2\psi$-plots in the absence of texture, became understandable as a consequence of the direction-dependent grain interaction; see Fig. 14.3.

Even though the Vook–Witt based grain interaction model is the first grain interaction model compatible with transverse elastic isotropy, the Vook–Witt assumptions represent extreme grain-interaction assumptions for the in-plane direction (crystallites are *tightly* connected together and exhibit *identical* strains) and the direction perpendicular to the surface of the film (crystallites can deform *freely*, independent from neighbouring crystallites and exhibit *identical* stresses). With reference to the traditional (bulk) models, a Voigt behaviour is attributed to the in-plane directions, whereas a Reuss behaviour is attributed to the direction perpendicular to the film surface. For this reason, the Vook–Witt model has to be considered as an extreme grain interaction model and will, in general, not be able to describe the true elastic behaviour of a real thin film.

An example is given in Fig. 14.4, showing $\sin^2\psi$-plots for the 200 reflection of a fibre-textured copper *thin film* with a thickness of 500 nm. An

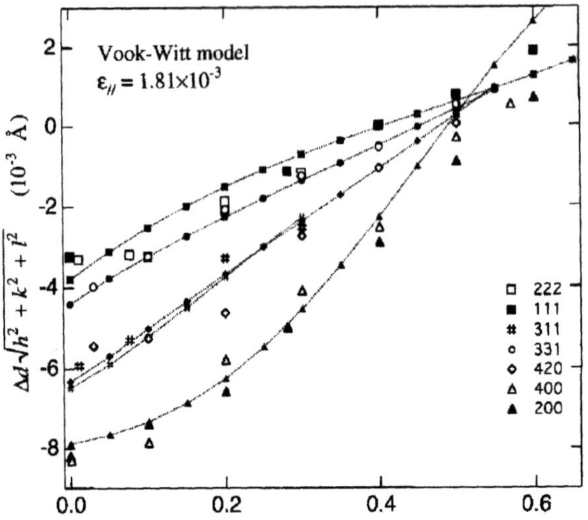

Fig. 14.3. Fit of the Vook–Witt model (continuos line) to the lattice strains measured from an evaporated nickel film (open and closed symbols) for different reflections. The open and closed symbols represent measurements on a ω- and a ψ-diffractometer, respectively. The symbols on the continuos lines correspond to the symbols representing measured strains. Figure taken from [16]

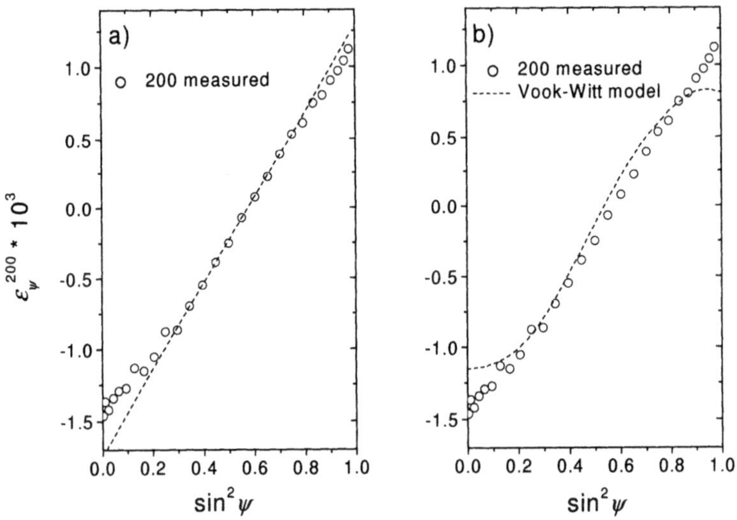

Fig. 14.4. $\sin^2\psi$-plots for the 200 reflection of a fibre textured copper film. (**a**) A straight line has been fitted to the linear part of the $\sin^2\psi$-plot. Traditional grain interaction models are incompatible with a non-linear $\sin^2\psi$-plot for this particular reflection. (**b**) The Vook–Witt models describes the observed deviations qualitatively, but overestimates the curvature

analysis on the basis of the traditional grain-interaction models is impossible, as the traditional models are incompatible with a non-linear $\sin^2\psi$-plot for the 200 reflection (see Fig. 14.4a). Even though the Vook–Witt based model qualitatively describes the experimentally observed deviation from the linear behaviour expected according to the traditional models (compare Figs. 14.4a and b), the observed deviations from a linear $\sin^2\psi$-dependence are overestimated.

Similar observations, i.e. an overestimation of the curvature in $\sin^2\psi$-plots, have been reported in the literature for *bulk* polycystals when using an extreme grain-interaction model (as the Reuss model, see, for example, [28]). A much more consistent analysis can be possible by using an average, effective grain interaction model. In particular, the Neerfeld–Hill model, a linear combination of the Reuss and the Voigt models, was frequently used (cf. also [29]).

Therefore, the development of a suitable "average grain-interaction model" combining "extreme" direction-dependent grain interaction models can be useful, as the experimental data presented below show.

14.3.2 Recent Developments

For bulk materials all directions for grain interaction can be equivalent. For columnar thin films, two principal directions can be indicated, that exhibit possibly different types of grain interactions; the in-plane direction (all in-plane directions are assumed to be equivalent) and the direction perpendicular to the surface of the film. Hence, the number of grain-interaction models based on extreme types of grain-interaction assumptions is two for bulk materials: Reuss and Voigt, provided that all directions within the bulk are equivalent for grain interaction. On the basis of similar reasoning, four types of extreme grain-interaction models can be formulated for columnar thin films, as the two principal directions each can be associated with two possible extreme grain interactions, namely Voigt and Reuss behaviours. Three of these models have already been mentioned in this chapter; the Voigt and Reuss models and the Vook–Witt model:

1. The Voigt Model

All crystallites are tightly connected together and exhibit identical strains in the specimen frame of reference.

2. The Reuss Model

All crystallites exhibit identical stresses in the specimen frame of reference but deform independently from neighbouring crystallites under these stresses.

3. The Vook–Witt Model

All crystallites exhibit identical in-plane strains (they are tightly connected together or to a common substrate with respect to the in-plane directions) in the specimen frame of reference, whereas they can deform freely in the perpendicular direction and thus exhibit identical stresses with respect to this direction in the specimen frame of reference.

The fourth extreme model has been presented recently for the first time [20]:

4. The Inverse Vook–Witt Model

All crystallites exhibit identical in-plane stresses (and can deform freely in in-plane directions) in the specimen frame of reference, whereas they are tightly connected together with respect to the perpendicular direction and thus exhibit identical strains in this direction in the specimen frame of reference.

The Inverse Vook–Witt Model

The inverse Vook–Witt model is a new grain-interaction model. It is required for the construction of an effective grain-interaction model (see below). The grain-interaction assumptions for a transversely isotropic film subjected to a plane, rotationally symmetric state of stress for this model are as follows: (i) the in-plane stress is rotational symmetric and (ii) equal for all crystallites and (iii) the strain perpendicular to the film surface is equal for all crystallites. Like in the (regular) Vook–Witt model, these assumptions fix certain strain and stress tensor components for all crystallites [cf. (14.25) and (14.26)]:

$$\varepsilon^S = \begin{pmatrix} \diamond & \diamond & 0 \\ \diamond & \diamond & 0 \\ 0 & 0 & \varepsilon_\perp^S \end{pmatrix}, \tag{14.27}$$

$$\varepsilon^S = \begin{pmatrix} \sigma_\parallel^S & 0 & \diamond \\ 0 & \sigma_\parallel^S & \diamond \\ \diamond & \diamond & \diamond \end{pmatrix}. \tag{14.28}$$

The missing strain and stress tensor components for each crystallite, marked by $\diamond$, can be calculated from the system of equations (14.17).

For the calculation of the mechanical elastic constants, A and B, it is necessary to assure that $\sigma_\perp^S = \langle\sigma_{33}^S\rangle$ is zero (with the Vook–Witt approach this is already assumed by the boundary conditions). To this end, a procedure as described in [20] can be followed.

When it has been realised that $\sigma_\perp^S$ is zero for a given combination of $\sigma_\parallel^S$ and $\varepsilon_\perp^S$, the elastic constants A and B can be calculated from (14.18) and (14.19). The diffraction strain can be calculated following (14.6) and the stress factors follow from (14.20) and (14.21) (cf. also Sect. 14.2.1).

Constructing an Effective Grain-Interaction Model

An effective grain interaction model for a thin film or, possibly (see Sect. 14.4), surface adjacent material can be constructed (see discussion in Sect. 14.2.7). The following example is relevant to the present work.

The Reuss approach and the Vook-Witt approach can be combined by assuming for the mechanical behaviour of the specimen that it is composed of an effective volume fraction f_R of the sample obeying the Reuss model and an effective volume fraction f_{VW} obeying the Vook–Witt model. Such a combined model leads, effectively, to partial in-plane Reuss-behaviour: whereas the crystallites are tightly connected together with respect to the in-plane direction in the Vook-Witt approach, the combination with the Reuss constraint gives the crystallites some degree of freedom for in-plane deformation independent of the neighbouring crystallites, by relieving the (unrealistic) constraint of identical planar strain for every crystallite.

A quantification of this effect of model mixing on the grain interaction can be achieved by defining interaction parameters for the two principal directions:

- in-plane interaction parameter $w_{\|}$: equals 1 for a Voigt-type interaction (crystallites are tightly connected together) and zero for a Reuss-type interaction (crystallites can deform independently of neighbouring crystallites but their stresses are identical);
- the out-of plane interaction parameter $w_{\perp}$: definition analogous to the in-plane parameter $w_{\|}$, i.e. $w_{\perp}$ equals 1 for Voigt-type interaction and zero for a Reuss-type interaction.

Combining the models for extreme types of grain interaction results in average values $\langle w_{\|} \rangle$ and $\langle w_{\perp} \rangle$ for the interaction parameters:

$$\langle w_{\|} \rangle = \sum_i f_i w_{\|,i} \,, \tag{14.29}$$

$$\langle w_{\perp} \rangle = \sum_i f_i w_{\perp,i} \,. \tag{14.30}$$

In (14.29) and (14.30), the f_i represent, like in (14.22)–(14.24), effective volume fractions of grains obeying the grain-interaction model i:

$$\sum_i f_i = 1 \,. \tag{14.31}$$

The four models for the extreme types of grain interaction are summarised, together with the in-plane and out-of-plane interaction parameters, $w_{\|}$ and $w_{\perp}$ respectively, in Table 14.1. For a given combination of grain-interaction models, the average interaction parameters can be calculated following (14.29) and (14.30).

Table 14.1. Grain interaction models (first column) with grain-interaction assumption for the in-plane direction (second column, V...Voigt type of grain interaction, i.e. crystallites are tightly connected, R...Reuss type of grain interaction, i.e. crystallites can deform freely) and for the direction perpendicular to the sample surface (third column) and the corresponding grain-interaction parameters $w_{\parallel}$ and $w_{\perp}$ (fourth and fifth columns)

	in-plane direction	perpendicularto sample surf.	$w_{\parallel}$	$w_{\perp}$
Voigt	V	V	1	1
Reuss	R	R	0	0
Vook–Witt	V	R	1	0
inverse Vook–Witt	R	V	0	1

Table 14.2. Single crystal elastic constants [32] of copper and gold used in the calculations. The compliance matrix components s_{ij} have been given together with the anisotropy factor A_i. The Voigt two-index notation (see, for example, [2]) is used; to change to the tensorial notation, the indices must be replaced following the scheme: $s_{ij} \to s_{mnop}$, $i \to mn$, $j \to op$ with $1 \to 11$, $2 \to 22$, $4 \to 12$ and recognising that $s_{ij} = 4s_{mnop}$ if $i, j > 3$

	s_{11} (TPa^{-1})	s_{12} (TPa^{-1})	s_{44} (TPa^{-1})
Cu (fcc)	14.98	−6.29	13.26
Au (fcc)	23.63	−10.814	24.096

Table 14.3. Macroscopic, mechanical elastic constants A and B according to the separate grain-interaction models and two effective models ($f_R = f_V = 0.5$ and $f_{VW} = f_{IVW} = 0.5$) for an untextured copper film. Single crystal elastic data taken from [32] (see Table 14.2)

	$f_V = 1$	$f_R = 1$	$f_{VW} = 1$	$f_{IVW} = 1$	$f_R = f_V = 0.5$ $w_{\parallel} = w_{\perp} = 0.5$	$f_{VW} = f_{IVW} = 0.5$ $w_{\parallel} = w_{\perp} = 0.5$
A (TPa^{-1})	4.65	5.77	5.24	4.92	5.21	5.08
B (TPa^{-1})	−4.50	−6.73	−5.68	−5.03	−5.62	−5.36

Of course, the same values of the interaction parameters can be realised by different combinations of the grain-interaction models. As an example, $w_{\parallel} = 0.5$ and $w_{\perp} = 0.5$ can be realised by the effective model with $f_R = 0.5$ and $f_V = 0.5$ and by the effective model with $f_{VW} = 0.5$ and $f_{IVW} = 0.5$ as well.

Numerical evaluations of the mechanical elastic constants A and B exhibit, that their values are practically independent of the choice of combination of models as long as the same values of the interaction parameters $\langle w_{\parallel} \rangle$ and $\langle w_{\perp} \rangle$ hold (see last two columns of Table 14.3 for an example). However,

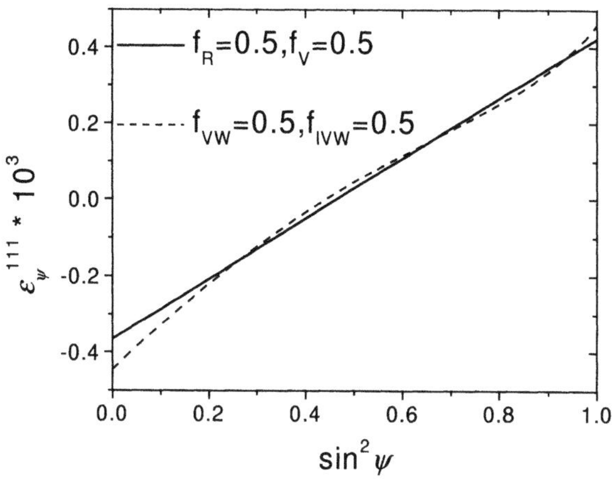

Fig. 14.5. Calculated diffraction strain for the 111 reflection of an untextured copper film subjected to a plane, rotationally symmetric, tensile stress of 100 MPa for two effective grain-interaction models: a combination of the Voigt and the Reuss model ($f_R = f_V = 0.5$, *continuous line*) and a combination of the Vook–Witt and the inverse Vook–Witt model ($f_{VW} = f_{IVW} = 0.5$, *dashed line*). Figure taken from [20]

this does not hold for the diffraction strain (or the stress factors), as can be seen in Fig. 14.5 (calculation using single crystal elastic constants for copper taken from Table 14.2): noticeable differences (large enough to be detected in a lattice strain measurement) in the diffraction strain occur as function of $\sin^2\psi$ for different combinations of grain-interaction models, although the same set of values holds for the parameters $\langle w_{\|}\rangle$ and $\langle w_{\perp}\rangle$. This observation is not in contradiction to the (approximate) independence of the elastic constants A and B for the type of grain interaction: the diffraction strain stems only from a fraction of all crystallites, whereas the mechanical elastic constants are calculated as averages for all crystallites. Therefore, diffraction strain analysis is a sensitive method to investigate direction dependent grain interaction.

Examples

Diffraction Strain Simulation on the Basis of the Various Grain-Interaction Models

Gold, a material exhibiting pronounced elastic anisotropy, was selected for the simulations. For gold, the $\langle 00l\rangle$ direction is much softer than the $\langle hhh\rangle$ direction. The single-crystal elastic compliances corresponding to the crystal frame of reference C have been summarised in Table 14.2. The (X-ray) diffraction strain was calculated and has been presented as a function of $\sin^2\psi$ (in $\sin^2\psi$-plots), for a number of reflections. Calculations were performed assuming an average stress $\sigma_{\|}^S = 100\,\text{MPa}$ for the film. In addition,

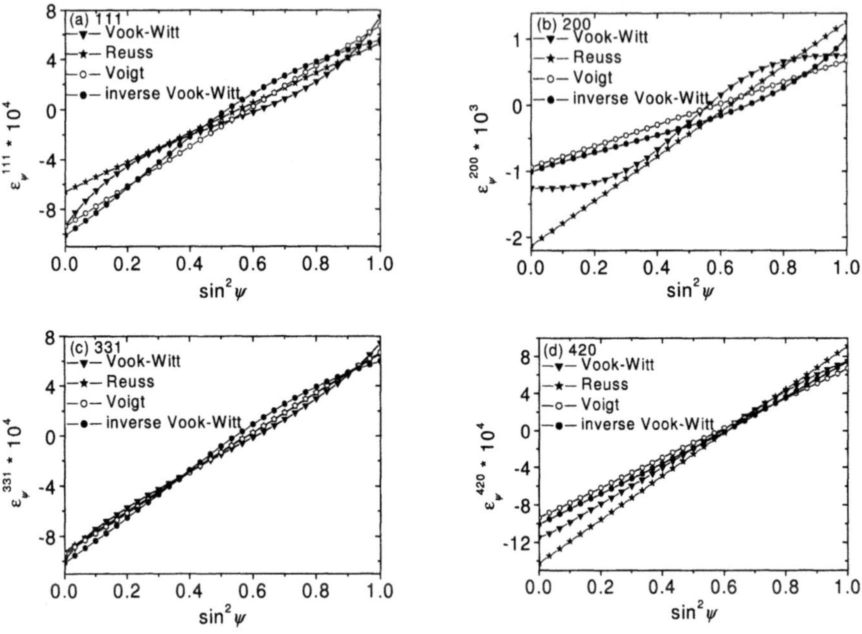

Fig. 14.6. Untextured gold film subjected to a plane, rotational symmetric state of stress (tensile stress, 100 MPa). Simulations of the diffraction strain ε_ψ^{hkl} versus $\sin^2\psi$ ("$\sin^2\psi$-plot") for (**a**) the 111, (**b**) the 200, (**c**) the 331 and (**d**) the 420 reflection. A legend indicating the symbols used for the different grain-interaction models has been given in the figure

the mechanical elastic constants A and B have been calculated using (14.18) and (14.19). The results for an untextured gold film are shown in Fig. 14.6.

As follows from Figs. 14.6a and b, in general the results for the various types of grain interaction do not coincide. In the absence of texture, linear $\sin^2\psi$-plots are obtained for all reflections from the traditional Reuss and Voigt models; this is not the case for the direction-dependent grain-interaction models (Vook–Witt, inverse Vook–Witt; see Figs. 14.6a–d; note that, although hardly visible, also in Fig. 14.6d a curved behaviour is present). This curvature is a direct consequence of the *macroscopic* elastic anisotropy of the body, caused by the direction-dependent grain-interaction model.

Four $\sin^2\psi$-plots for a textured gold specimen (Gaussian {111}-fibre-texture with a pole width of 5° half width at half maximum) are shown in Fig. 14.7. The volume fraction of diffracting crystallites, as represented by the integrated intensity of a diffraction line in a measurement, is indicated in the figures by a continuous line. As a consequence of the sharp texture, the intensity is zero in some ψ ranges (see Figs. 14.7a–d) and therefore the corresponding strain cannot be calculated neither measured. Major differences become evident upon comparing Figs. 14.6 and 14.7.

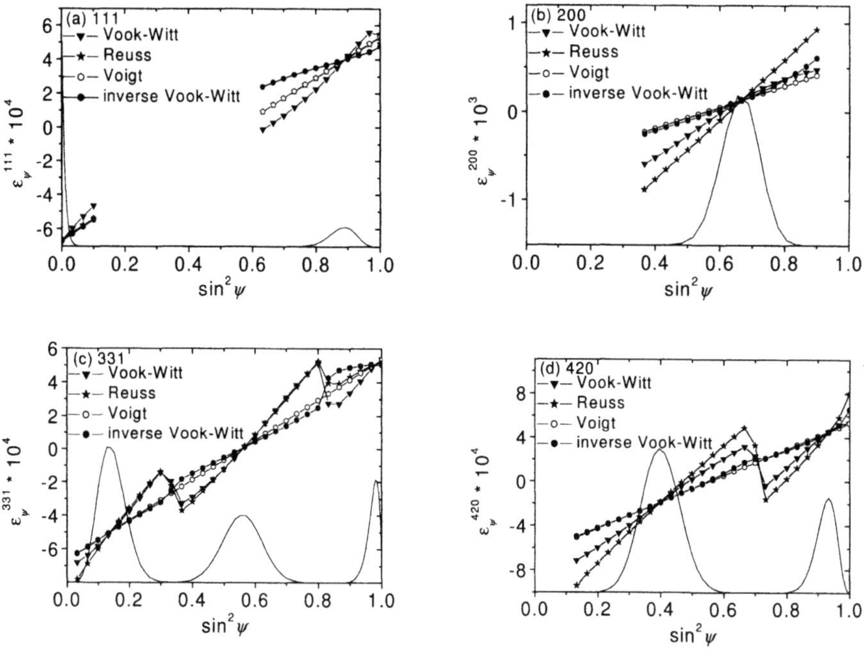

Fig. 14.7. Textured gold film (Gaussian 111-fibre texture, 5° pole half width at half maximum) subjected to a plane, rotational symmetric state of stress (tensile stress, 100 MPa). Simulations of the diffraction strain ε_{ψ}^{hkl} versus $\sin^2\psi$ ("$\sin^2\psi$-plot") for (**a**) the 111, (**b**) the 200, (**c**) the 331 and (**d**) the 420 reflection. A legend indicating the symbols used for the different grain-interaction models has been given in the figure

The Reuss model yields non-linear $\sin^2\psi$-plots for the 331 and the 420 reflections (Figs. 14.7c and d), whereas the 111 and the 200 reflections still exhibit linear behaviour (Figs. 14.7a and b). This non-linearity observed for the Reuss model for the 331 and 420 reflections is a consequence of the texture, leading to an anisotropic (i.e. transversely isotropic) mechanical behaviour of the film. For the Voigt model, linear $\sin^2\psi$-plots are obtained independently of the texture. Note that a textured polycrystal obeying the Voigt model is macroscopically elastically anisotropic, even though linear $\sin^2\psi$-plots are obtained.

A general observation is, that the $\sin^2\psi$-plots of the Vook–Witt model and those of the inverse Vook–Witt model exhibit curvatures with opposite signs for large portions of the $\sin^2\psi$-range.

The macroscopic, mechanical elastic constants A and B are summarised (for both cases: no texture and fibre texture) in Table 14.4 for Voigt, Reuss, Vook–Witt, and inverse Vook–Witt grain-interaction models. As expected [14], all constants lie between the extremes dictated by the Voigt and Reuss models. Due to the pronounced intrinsic elastic anisotropy of gold, the macro-

Table 14.4. Macroscopic, mechanical elastic constants A and B of a gold film (untextured and fibre textured; {111} fibre texture with a pole width of 5° half width at half maximum) according to the grain-interaction models Voigt, Reuss, Vook–Witt, inverse Vook–Witt and Neerfeld–Hill. Single crystal elastic data taken from [32] (see Table 14.2)

Au		R	V	VW	IVW	NH
no	A (TPa^{-1})	8.24	6.69	7.47	7.06	7.46
texture	B (TPa^{-1})	−12.48	−9.39	−10.94	−10.11	−10.93
fibre-	A (TPa^{-1})	5.42	5.34	5.38	5.35	5.38
texture	B (TPa^{-1})	−6.84	−6.68	−6.76	−6.71	−6.76

scopic elastic constants of the textured film differ significantly from those for an untextured film.

An Experimental Example

The above outlined theoretical considerations will be illustrated on the basis of an experimental example. The specimen investigated was a copper thin film of thickness 500 nm deposited by dc-magnetron sputtering (100 W power input) from a pure copper target (purity 99.95 at.%) onto a piece of an thermally oxidised ⟨510⟩-silicon wafer with a size of 14 mm by 14 mm. Details on the deposition can be found in [20].

The microstructure was investigated by using a Focused Ion Beam (FIB) workstation (FEI Fib 200) allowing cutting cross-sections and their imaging with the help of a Ga-ion beam. The specimen shows a columnar microstructure, and most grains are equiaxed and extend from the interface to the surface. The lateral grain size is of the order of the film thickness.

X-ray diffraction measurements were carried out using a Philips X'Pert MRD Pro diffractometer equipped with an Eulerian cradle (4-circle goniometer) and Cu K_α radiation (sealed X-ray tube operated at 45 kV and 40 mA) paralleled by an X-ray lens and monochromatised by a diffracted beam flat graphite monochromator.

Peak positions, heights, half widths at half maximum (HWHM), asymmetries and shape parameters were extracted from the measured data by means of a custom non-linear least squares fitting program based on Pearson VII functions [33].

No significant differences in peak position and integrated intensity were observed for symmetric and asymmetric peak fits. The $K_{\alpha 2}$ component was taken into account by simultaneous fitting of two Pearson VII functions, one for each radiation component; a fixed 2 : 1 ratio between the areas of the $K_{\alpha 1}$ and $K_{\alpha 2}$ components (as experimentally observed) was imposed.

The specimen exhibited a fibre texture. From the integrated areas of six diffraction lines (see Table 14.4) determined by peak fitting as a function of the specimen tilt angle ψ, pole figures for the 111 and the 200 reflections were generated. Following the correction scheme proposed by Welzel and Leoni [27], the pole figures were corrected for absorption, increase of illuminated sample area upon specimen tilting and instrumental intensity loss (occurring when a fraction of the incident beam misses the sample surface or a fraction of the diffracted beam misses the detector). The ODF was calculated from the corrected pole figures (111 and 200 reflections) using the X'Pert Texture (Philips Analytical, Almelo, The Netherlands) software package on a grid with 3° spacing for the three Euler angles and subsequently used in the calculation of elastic constants [see (14.18) and (14.19)], diffraction strain [see (14.6)] and X-ray stress factors [see (14.20)].

Six measured $\sin^2\psi$-plots (for the 111, 200, 220, 311, 331 and 420 reflections) were simulated simultaneously on the basis of the effective grain-interaction model (cf. Sect. 14.3.2). As the parameters f_i (i.e. the type of grain interaction cf. Sect. 14.3.2) are not known a priori, they were used as fitting parameters in addition to the unknown stress tensor components and to the strain-free lattice spacing d_0^{hkl} for each $\sin^2\psi$-plot. By using the strain-free lattice spacing d_0^{hkl} of each $\sin^2\psi$-plot as additional fit parameter, no accurate determination of the *absolute* peak position is required. Ten fit parameters were determined by fitting simultaneously six $\sin^2\psi$-plots, each corresponding to one hkl reflection: six strain free lattice spacings d_0^{hkl}, the stress parallel to the substrate $\sigma_{\parallel}^S$ and three (independent) volume fractions pertaining to the effective grain-interaction model involved (i.e. the Voigt, Reus, Vook–Witt and inverse Vook–Witt contributions). This is therefore a variant of the stress analysis method described in Sect. 14.2.5. The six $\sin^2\psi$-plots containing measured and simulated lattice strain data are shown in Fig. 14.8.

Note that the strain free lattice spacings have no direct physical meaning and cannot be used for the calculation of the lattice constant a, as they may be biased by instrumental errors. Another variant of the analysis, based on only five parameters can be realised and leads to identical results for the stress and the volume fractions pertaining to the different grain interaction models involved (by replacing the six "apparent" strain free lattice spacings by one lattice constant and a zero-offset of the 2θ-scale of the goniometer; for details, see [20]).

Evidently, the traditional models cannot describe the measured $\sin^2\psi$-plots of the copper film. In particular, it should be noted that only the direction-dependent grain-interaction models are compatible with the curved behaviour in the $\sin^2\psi$-plots observed for 111 and 200 (and in general hhh and $00l$) reflections; any combination of the Reuss and Voigt models (e.g. the traditional Neerfeld–Hill model) predicts linear $\sin^2\psi$-plots for the hhh and $00l$ reflections.

From the determined effective volume fractions pertaining to the different grain-interaction models, $f_{VW} = 0.6$, $f_R = 0.3$, $f_{IVW} = 0.1$, $f_V = 0.0$, the interaction parameters can be calculated. The in-plane interaction parameter,

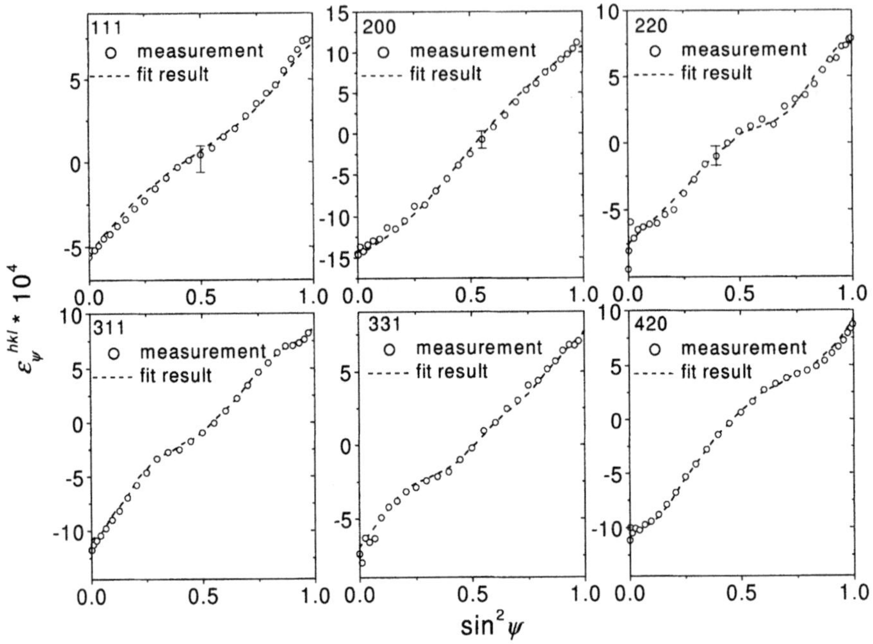

Fig. 14.8. Textured copper film. Measured (*open circles*) and simulated (*dashed line*) diffraction strain data for six different reflections as function of $\sin^2\psi$ ($\sin^2\psi$-plots). For details of the simulation, see text. Figure taken from [20]

$w_{\|}$ [cf. (14.29)], equals 0.6 whereas the out-of plane interaction parameter, $w_{\perp}$ [cf. (14.30)], equals 0.1. In view of the discussion given in Sect. 14.3.2, this indicates that the in-plane crystallite coupling is comparable to a bulk material whereas the out-of plane crystallite coupling allows relatively unconstrained crystallite deformation (note that $w_{\perp} = 0$ would correspond to completely unconstrained crystallite deformation). These findings are compatible with the microstructure of the layer, exhibiting a more or less columnar microstructure with grain boundaries extending from the layer-substrate interface to the surface.

14.4 Surface Anisotropy as a Special Case of Direction Dependent Grain Interaction

For crystallites adjacent to a surface of a polycrystal not all directions are equivalent and the (mechanical and diffraction elastic) properties of these crystallites can be at most transversely isotropic, reflecting the rotational symmetry of the aggregate with respect to the surface normal.

Stickforth [26] was the first to consider this “surface anisotropy” as a source of macroscopic (mechanical) elastic anisotropy. In this context, the consequences of macroscopic elastic anisotropy on diffraction stress analysis were analysed and, for a principal, plane state of stress, an expression relating the lattice strain to the components of the mechanical stress tensor was given for the first time:

$$\begin{aligned}\{\varepsilon_{33}^{L}\}_{\varphi\psi}^{hkl} &= t_1(hkl,\varphi,\psi)(\langle\sigma_{11}^{S}\rangle+\langle\sigma_{22}^{S}\rangle)\\ &+\frac{1}{2}t_2(hkl,\varphi,\psi)\sin^2\psi(\langle\sigma_{11}^{S}\rangle\cos^2\varphi+\langle\sigma_{22}^{S}\rangle\sin^2\varphi)\\ &+t_3(hkl,\varphi,\psi)\sin 2\varphi(\langle\sigma_{11}^{S}\rangle-\langle\sigma_{22}^{S}\rangle)\,. \end{aligned}\qquad(14.32)$$

The factors t_1, $\frac{1}{2}t_2$ and t_3 might be named "diffraction (X-ray) elastic constants" with reference to the traditional $\sin^2\psi$-law [cf. (14.11)]: t_1 takes the place of S_1^{hkl} and $\frac{1}{2}t_2$ takes the place of $\frac{1}{2}S_2^{hkl}$; for t_3 there is no equivalent in the traditional $\sin^2\psi$-law. For the case of macroscopic elastic isotropy t_3 equals zero and t_1 and $\frac{1}{2}t_2$ are then equal to the diffraction (X-ray) elastic constants S_1^{khl} and $\frac{1}{2}S_2^{hkl}$. The factors t_1, $\frac{1}{2}t_2$, and t_3 depend in addition to φ, ψ and hkl, on the single-crystal elastic constants and the grain interaction occurring in the specimen [7].

Whereas the X-ray elastic constants S_1^{hkl} and $\frac{1}{2}S_2^{hkl}$ are independent of φ and ψ, this is in general not the case for t_1, $\frac{1}{2}t_2$ and t_3. As a consequence, curvature occurs in the $\sin^2\psi$-plots. Note that (14.32) is valid also for textured specimens [7] and (for a plane, principal state of stress) is a fully equivalent, alternative formulation of (14.13).

Experimental investigations on the possible occurrence of surface anisotropy were performed by, for example, Nikolin [34] and Hauk et al. [35]. Hauk [2] summarised various publications and concluded (see also [34, 36]) that, if surface anisotropy is a genuine physical phenomenon, the surface region of a polycrystal affected by surface anisotropy must have a thickness of less than 400 nm.

The apparent lack of convincing proof of the relevance of surface anisotropy in the diffraction analysis of the surface region of bulk specimens may be understood as follows: straightforward experimental evidence of surface anisotropy is obtained if curvature occurs in $\sin^2\psi$-plots for untextured specimens in a plane, principal state of stress [cf. (14.11) and (14.32)]. This is the line of reasoning followed by van Leeuwen et al. [16] in their work on a stressed thin film, providing evidence of direction-dependent grain interaction in the diffraction stress analysis of an untextured, polycrystalline thin film in a plane state of stress. However, deviations from straight line behaviour in $\sin^2\psi$-plots can also be due to more complicated states of stress, texture and/or the occurrence of stress depth gradients. It is not unlikely that the unavailability of an appropriate direction dependent grain-interaction model (as, for example, the Vook–Witt based model proposed for diffraction stress analysis by van Leeuwen et al. [16]; see also [19]), and the tendency to either ignore non-linearities in measured $\sin^2\psi$-plots or to focus on interpretations based on shear components, texture effects and/or stress depth gradients, has obstructed the recognition of the significance of surface anisotropy for (also) bulk specimens.

14.5 Conclusion

- The traditional grain-interaction models due to Voigt, Reuss, Neerfeld–Hill and Eshelby–Kröner can be inappropriate for the stress analysis of thin films and, in general, surface adjacent material by X-ray diffraction. These models are based on isotropic grain interaction assumptions, which can be incompatible with the microstructure of the material, as the columnar microstructures frequently observed in thin films.
- The Vook–Witt approach to grain interaction, i.e. equal in-plane strain (parallel to the substrate) and zero stress perpendicular to the film surface for all crystallites, is the first direction-dependent grain interaction taking the microstructure of a (columnar) thin film into account. The model can explain the occurrence of curved $\sin^2\psi$-plots in the absence of texture and implies that a polycrystal is (at most) transversely elastically isotropic. In analogy with the pendants formulated by Voigt and Reuss for grain interaction, the inverse Vook–Witt grain-interaction model has been recently proposed by the authors, where all crystallites in a (thin layer) specimen can deform freely with respect to the in-plane direction under identical stresses, whereas they are tightly connected together with respect to the direction perpendicular to the surface.
- The four models, Voigt, Reuss, Vook–Witt and inverse Vook–Witt imply extreme grain-interaction assumptions, incompatible with physical reality. An "effective" grain-interaction model has been proposed that consists of a linear combination of basic, extreme models with effective volume fractions as weighting factors. This approach allows the description of "intermediate" grain interactions. Four basic grain-interaction models have to be taken into account for surface-adjacent polycrystals as a (columnar) thin film, recognising the occurrence of two principal directions (in-plane and perpendicular to the (film) surface) and two extremes of grain interaction (fully unconstrained (Reuss) and fully constrained (Voigt)).
- The direction-dependent grain-interaction model developed originally for diffraction stress analysis of thin films has relevance also for the phenomenon of surface anisotropy of bulk specimens. Observations of non-linearities in $\sin^2\psi$-plots may need a reinterpretation recognising the role of surface anisotropy.
- In case of macroscopic elastic *isotropy*, diffraction stress analysis can be performed employing the concept of diffraction (X-ray) elastic constants.
- In case of macroscopic elastic *anisotropy*, i.e. in the presence of preferred orientation and/or direction-dependent grain interaction, diffraction stress analysis can be performed employing the concept of diffraction (X-ray) stress factors.

References

1. I.C. Noyan and J.B. Cohen: *Residual stress. Measurement by diffraction and interpretation.* (Springer Berlin Heidelberg New York 1987)
2. V. Hauk (Ed.): *Structural and residual stress analysis by nondestructive methods.* (Elsevier, Amsterdam 1997)
3. M. Leoni, Y.H. Dong and P. Scardi: Mat. Sci. Forum **321–324**, 439 (2000)
4. P. Scardi and Y.H. Dong: J. Mater. Res. **16** [1], 233 (2000)
5. H. Dölle and V. Hauk: Z. Metallkde. **69**(6), 410 (1978)
6. H. Dölle and V. Hauk: Z. Metallkde. **70**(10), 682 (1979)
7. U. Welzel and E.J. Mittemeijer: J. Appl. Phys. **93** [11], 9001 (2003)
8. W. Voigt: *Lehrbuch der Kristallphysik.* (Teubner, Leipzig Berlin 1910)
9. A. Reuss: Z. Angewandte Mathematik und Mechanik **9**(1), 49 (1929)
10. J.D. Eshelby: Proc. Roy. Soc. A **241**, 376 (1957)
11. E. Kröner: Z. Physik **151**, 504 (1958)
12. G. Kneer: Phys. Stat. Sol. **9**, 825 (1965)
13. F. Bollenrath, V. Hauk and E.H. Müller, Z. Metallkde. **58**(1), 76 (1967)
14. R. Hill: Proc. Phys. Soc. London **65**, 349 (1952)
15. H. Neerfeld: Mitt. K.-Wilh.-Inst. Eisenforschg. **24**, 61 (1942)
16. M. van Leeuwen, J.-D. Kamminga and E.J. Mittemeijer: J. Appl. Phys. **86** [4], 1904 (1999)
17. R.W. Vook and F. Witt: J. Appl. Phys. **36** [7], 2169 (1965)
18. F. Witt and R.W. Vook: J. Appl. Phys. **39** [6], 2773 (1968)
19. M. Leoni, U. Welzel, P. Lamparter, E.J. Mittemeijer and J.-D. Kamminga: Phil. Mag. A **81**(3), 597 (2001)
20. U. Welzel, M. Leoni and E.J. Mittemeijer: Phil. Mag. **83**(5), 603 (2003)
21. J.F. Nye: *Physical properties of crystals.* (Oxford University Press, Oxford 1957)
22. C. Giacovazzo, H.L. Monaco, D. Viterbo, F. Scordari, G. Gilli, G. Zanotti and M. Catti: *Fundamentals of Crystallography.* (Oxford Science, Oxford 1998)
23. R.-J. Roe and W.R. Krigbaum: J. Chem. Phys. **40** [9], 2608 (1964)
24. H.-J. Bunge: *Texture analysis in materials science.* (Butterworths, London 1982)
25. H. Möller and G. Martin: Mitt. K.-Wilh.-Inst. Eisenforschg. **21**, 261 (1939)
26. J. Stickforth: Tech. Mitt. Krupp – Forsch.-Ber. **24** [3], 89 (1966)
27. U. Welzel and M. Leoni: J. Appl. Cryst. **35**, 196 (2002)
28. C.M. Brakman: Phil. Mag. A **58** [4], 635 (1988)
29. W. Serruys, P. van Houtte and E. Aernoudt, in *Residual Stresses in Science and Technology.* Ed by E. Macherauch and V. Hauk (Deutsche Gesellschaft für Metallkunde, Oberursel 1987) pp. 417–424
30. W. Serruys, F. Langouche, P. van Houtte, and E. Aernoudt, in *Proceedings of ICRS2.* Ed. by G. Beck, S. Denis and A. Simon (Elsevier Applied Science, London New York 1989) pp. 166–171
31. D. Rafaja, V. Valvoda, R. Kuzel, A.J. Perry and J.R. Treglio, Surf. Coat. Tech. **86–87**, 302 (1996)
32. M.A. Meyers and K.K. Chawla: *Mechanical Metallurgy, Principles and applications.* (Prentice-Hall, Englewood Cliffs, New Jersey 1984) pp. 57–58
33. E.J. Sonneveld, R. Delhez, T.H. de Keijser and E.J. Mittemeijer: Mat. Sci. Forum **79–82**, 85 (1991)

34. H.-J. Nikolin: Diploma thesis, Institut für Werkstoffkunde, RWTH Aachen (1983)
35. V. Hauk, H.-J. Nikolin, H.Z. Weisshaupt: Z. Metallkde. **76**(3), 226 (1985)
36. U.R. Hartmann: PhD thesis, Universität Karlsruhe (TH). 1973
37. U. Welzel, J. Ligot, P. Lamparter, A. Vermeulen and E.J. Mittemeijer: (2002) in preparation

15 Interaction between Phases in Co-deforming Two-Phase Materials: The Role of Dislocation Arrangements

R.E. Bolmaro, H.-G. Brokmeier, J.W. Signorelli, A. Fourty, and M.A. Bertinetti

This chapter discusses the problem of interaction among crystals in a polycrystalline material. The chosen strategy is to present experiments and simulations of textures in two-phase materials. By comparison of the fine texture structure with the simulated results some clues are obtained about the behavior of phase mixtures in relation with crystal shape, crystal symmetry, volume fraction, etc. A co-spin interaction scheme is presented and incorporated in self consistent models as a way to include dislocation arrays in texture simulations. The influence of dislocation structures over texture development and the needs for better models for X-ray diffraction pattern analysis are discussed.

15.1 Introduction

The problem of the so called "plastic grain interaction" in a polycrystalline material has been addressed in the past by several authors comprising a few view points. Usually the interaction among grains in a polycrystal is taken into account by providing a constitutive equation, either analytic or numeric, reflecting the main assumptions of the model. Interaction among crystals is more difficult to study than the interaction among the many molecules of a gas. Some advantages are available on top of the main disadvantage of having to consider always very strong interactions. The main advantage is that the influence of the crystals, over the behavior of some other particular one, is fading away with distance, following roughly an inverse proportionality law. However even this commonly accepted behavior has exceptions. When a hard crystal, as in a poly-phase material, or a "cluster" of them, as in a composite material, largely differs from the behavior of what is considered the matrix, the interaction effects can be large even at long distances (i.e. a few grains away). In that case the hard inclusions or second phases should actually be considered better as internal tools locally deforming the material and Finite Element Methods (FEM) seem to be mandatory [1].

In the meantime, and with the goal of considering just co-deforming materials, it is worthy to quote A. Cottrell:

"Such general understanding is a form of extreme simplification in which the fine detail is smothered in order to reveal a general dom-

> *inating feature. Simplification is an art rather like that of the cartoonist who captures the key features of a familiar face in a few deft strokes to make it instantly recognizable"* [2].

One of the most simple assumptions, the Taylor model, regards the crystals as equally deforming and the interaction behavior relies on that rule [3]. In the opposite extreme, the Sachs model, considers the grains deforming subject to a uniform stress field [4]. Those two cases are considered bounds or limits of the real plastic behavior of materials. In between those bounds, the most widely used model is the Self Consistent (SC) one, in its many variants [5]. The 1-site SC model assumes the crystals embedded in a matrix with elastic and plastic properties calculated as an average of the rest of the grains or Homogeneous Effective Medium (HEM) [6]. n-sites SC models incorporate an interaction scheme between each grain and a given number of preferential neighboring grains [7,8]. Grain interaction is a way of describing the strain and stress fields in terms of "what happens to a grain due to the properties of the grain itself" and "what happens to a grain because of the properties of its neighbors". Regarding the meaning of "grain interaction" special attention deserves the paper of U.F. Kocks [9] where general rules are established about the behavior of interacting grains. An original contribution of this chapter will be the analysis of grain interaction from the point of view of crystal spin, The analysis of experimental data can be valuable in comparison with simulation terminology, as it will be done in the second paragraph. The third paragraph will present experiments and simulation results obtained for Al-Cu and Cu-Fe powder composites deformed at Von Mises strains of 3.00. Following some previous work in the field [10, 11] a discussion of interaction strengths will be performed by direct comparison of experiments and simulations. The discussion, in the fourth paragraph, will address the problem of the influence of dislocation arrays on texture development. Some expected collaborations between the fields of texture and X-ray diffraction pattern analysis will be discussed.

15.2 Grain Interaction Assessment

15.2.1 Strain Interaction

U.F. Kocks presented a discussion of the "plastic interaction problem" that, more or less, we will follow in the next paragraph [9]. The discussion will be limited to the case of Visco Plastic (VP) behavior in order to deal with large deformations and avoiding the elastic regime.

Figure 15.1 shows a schematic elasto-plastic behavior of materials as usually plotted to evaluate strain rate sensitivity. The increment of strain rate results in an increment of stress, highly exaggerated in the figure. The analysis is usually performed through the equation

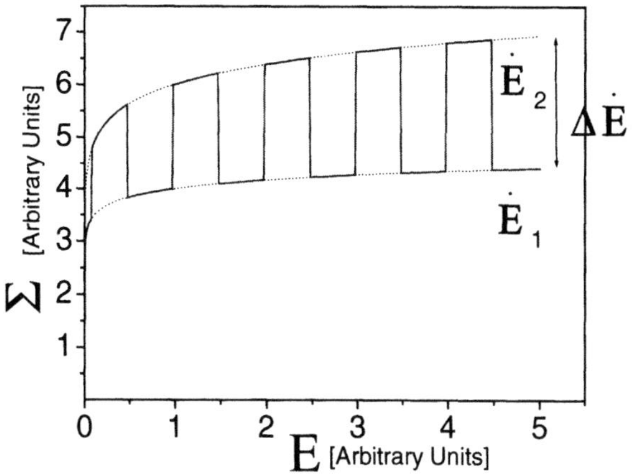

Fig. 15.1. Schematic behavior of a strain sensitive deforming material showing the usual technique to calculate strain rate sensitivity

$$m = \frac{\Delta \log \Sigma}{\Delta \log \dot{E}}, \tag{15.1}$$

where the finite increment of the logarithm of both variables, macroscopic stress Σ and strain rate $\dot{E}$, is computed for each step and m is the strain rate sensitivity. Metals are mostly strain rate insensitive and m values are usually in the order of 0.01 [12, 13].

Let us consider a viscoplastic constitutive equation

$$\dot{E} = \boldsymbol{k}\Sigma^n ; \quad \Sigma = \boldsymbol{c}\dot{E}^{1/n} = \boldsymbol{c}\dot{E}^m , \tag{15.2}$$

where n is the stress exponent or inverse of the strain rate sensitivity, $\boldsymbol{k}$ and $\boldsymbol{c}$ are the viscoplastic compliance and stiffness tensors. The tangent or instantaneous modulus can be defined as

$$\boldsymbol{M}^{(\mathrm{tg})} = \frac{\partial \dot{E}}{\partial \Sigma} = n\boldsymbol{k}\Sigma^{n-1} \tag{15.3}$$

and the secant modulus

$$\boldsymbol{M}^{\mathrm{s}} = \dot{E}\Sigma^{-1} . \tag{15.4}$$

The strain rate sensitivity can be calculated as

$$\log \Sigma = \log \boldsymbol{c} + m \log \dot{E} , \quad m = \frac{\partial \log \Sigma}{\partial \log \dot{E}} , \tag{15.5}$$

from which a simple connection can be obtained between the different moduli

$$\boldsymbol{M}^{(\mathrm{tg})} = n\boldsymbol{M}^{(\mathrm{s})} . \tag{15.6}$$

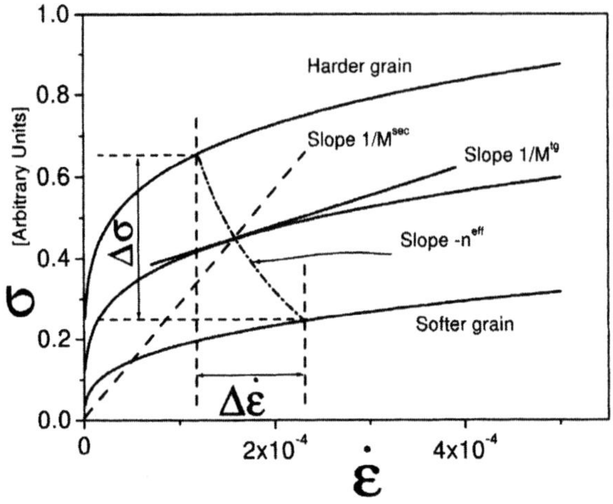

Fig. 15.2. Behavior of extreme cases of a collection of crystals in a polycrystal. The slope $1/M^{tg}$ measures the material stiffness. The slope $-n^{eff}$ measures the interaction between different crystals

This equation shows how to connect the tangent and secant moduli for a VP material. In a plot σ vs. $\dot{\varepsilon}$ the different variables and moduli can be defined. From here on we will use lower case symbols when we want to address the possibility of having microscopic (crystal) variables instead of macroscopic (average) ones. Figure 15.2 shows the secant and tangent moduli, keeping their meaning despite they had to be inverted and all variables are regarded as scalars. For easing the plotting, the curves represent strain rate sensitivities, m, between 1/3 (for the lowest one) and 1/5 (for the highest one) although real values are more between 1/30 and 1/300.

Both, upper and lower bounds, schematically represent the behavior of the "hardest" and "softest" grains of a two phase material or "stiffest" and "most compliant" orientations of a heterogeneous single phase polycrystal. In a sort of spatial derivative, by moving from the softest to the hardest phase, and regardless of their actual position, the interaction between both grains is grossly represented by the dot curve, $-n^{eff}$. The slope is expected to be negative because a harder grain will, both, deform at lower rate and larger stress, when compared with the average behavior, and vice versa. The softer (the harder) grain can represent the matrix behavior to which the harder (the softer) grain is being compared when a 1 site SC model is used. In such case the interaction coefficient is actually the interaction of each grain with the matrix. If they represent individual grains they provide the interaction coefficient between two neighbors (usually the closest ones) in a multi-site SC model. In any case the equation can be written:

$$\Delta\dot{\varepsilon} = -n^{eff}\Delta\sigma\,, \tag{15.7}$$

where, as it has been noticed by several authors, it becomes clear that n and n^{eff} are not necessarily the same. This equation just reflects the variation of strain rate when we move from one grain to other, with the consequent stress change.

Regarding SC calculations, the reader is assumed to be familiar with the VPSC model and just a brief summary is outlined with the double purpose of completeness and nomenclature uniformity.

In the following discussion the problem becomes more complex because we have to simultaneously deal with microscopic and macroscopic variables. The crystal behavior follows a potential law relating the applied shear stress τ^s in each plane and the Critical Resolved Shear Stress (CRSS) of that system τ_c^s. The total strain rate in each grain is obtained adding the shear strain rates of each active slip system and/or twining:

$$\dot{\varepsilon} = \left\{ \dot{\gamma}_0 \sum_s \frac{\boldsymbol{m}^s \otimes \boldsymbol{m}^s}{\tau_c^s} \left(\frac{\boldsymbol{m}^s : \sigma'}{\tau_c^s} \right)^{n-1} \right\} : \sigma' = \boldsymbol{M}^{c(\mathrm{sec})}[\sigma'] : \sigma' \,, \quad (15.8)$$

where $\dot{\varepsilon}$ is the crystalline microscopic strain rate, σ' is the microscopic deviatoric stress, $\boldsymbol{m}^s = 1/2(\boldsymbol{n}^s \otimes \boldsymbol{b}^s + \boldsymbol{b}^s \otimes \boldsymbol{n}^s)$ is the Schmid tensor describing the geometry of the slip system s in the single crystal, $\boldsymbol{n}^s$ is the slip plane normal, $\boldsymbol{b}^s$ is the slip direction, $\dot{\gamma}_0$ is a reference strain rate velocity, n is the inverse of the strain rate sensitivity, $\boldsymbol{M}^{c(\mathrm{sec})}$ is the secant viscoplastic compliance crystal modulus and $\otimes$ means the tensor product of involved quantities.

The material behavior under a viscoplastic regime is described through the following expression:

$$\Sigma' = \boldsymbol{M}^{(\mathrm{sec})^{-1}}[\dot{E}] : \dot{E} \,, \quad (15.9)$$

where Σ' is the macroscopic stress, $\dot{E}$ is the macroscopic deviatoric strain rate tensor and $\boldsymbol{M}^{\mathrm{sec}}$ is the secant material modulus. Currently $\boldsymbol{M}^{\mathrm{sec}}$ is calculated by means of a Self-Consistent scheme leading to the interaction equation that relates stress and strain rate deviations at crystal level [6–8].

$$\Delta\dot{\varepsilon} = \dot{\varepsilon} - \dot{E} = -\alpha\tilde{\boldsymbol{M}}^c : (\sigma' - \Sigma') = -\alpha\tilde{\boldsymbol{M}}^c : \Delta\sigma \quad (15.10a)$$

$$\tilde{\boldsymbol{M}}^c = n^{\mathrm{sec-tg}}(\boldsymbol{I} - \boldsymbol{S}^c)^{-1} : \boldsymbol{S}^c : \boldsymbol{M}^{(\mathrm{sec})} \,, \quad (15.10b)$$

where $\tilde{\boldsymbol{M}}^c$ is the accommodation tensor, α is a tuning factor and S is the Eshelby tensor. The $n^{\mathrm{sec-tg}}$ is a scalar parameter that fixes the interaction as follows:

$$\boldsymbol{M}^{(\mathrm{sec})} = \langle \boldsymbol{M}^{c(\mathrm{sec})} : \boldsymbol{B}^c \rangle : \langle \boldsymbol{B}^c \rangle^{-1} \quad (15.11a)$$

$$\boldsymbol{M}^{(\mathrm{tg})} = \langle \boldsymbol{M}^{c(\mathrm{tg})} : \boldsymbol{B}^c \rangle : \langle \boldsymbol{B}^c \rangle^{-1} \,, \quad (15.11b)$$

where

$$\boldsymbol{B}^c = (\boldsymbol{M}^{(\mathrm{sec})} + \alpha\tilde{\boldsymbol{M}}^c)^{-1}(\boldsymbol{M}^{c(\mathrm{sec})} + \alpha\tilde{\boldsymbol{M}}^c), \tag{15.12}$$

where $\boldsymbol{M}^{c(\mathrm{tg})}$ is the tangent viscoplastic crystal modulus. Simultaneously solving equations (15.11ba) and (15.11bb) leads to add a supplementary iteration level to the classic SC formulation. When different kinds of slip systems are active or different phases are interacting a general secant-tangent relation can be written as

$$\boldsymbol{M}^{(\mathrm{tg})} = f(\alpha)\boldsymbol{n}^{(\mathrm{sec-tg})}\boldsymbol{M}^{(\mathrm{sec})}, \tag{15.13}$$

where $\boldsymbol{n}^{\mathrm{sec-tg}}$ is now a 5×5 matrix and $f(\alpha)$ is a numerically calculated function of the tuning factor α. Particularly, when all matrix components n_{ij} are equal, $\boldsymbol{n}^{\mathrm{sec-tg}}$ reduces to a diagonal matrix with all components equal to the scalar parameter $n^{\mathrm{sec-tg}}$ as in (15.10b). Otherwise Hutchinson's relation between tangent and secant viscoplastic compliances does not longer hold [14]. Turner et al. [15] solved the mixed creep problem that we currently extend to a viscoplastic regime. The physical meaning and implications of this extension will be discussed later but notice that α has been added in (15.10ba) to allow different interactions among grains and between grains and matrix.

C.N. Tomé [11] proposed a model that allows the calculation of interaction coefficients by what he calls a Relative Directional Compliance Criterion (RDC). He is able to expand the options in order to include the possibility of having $1 < n^{\mathrm{eff}} < n$. By the different models we may have n^{eff} alternatively equal to

$$\boldsymbol{n}^{\mathrm{eff}} = \begin{cases} 0 & \text{Taylor} \\ 1 & \text{secant} \\ n & \text{tangent} \\ [1 - 0.4(1/n - 21)^2]^{-1} & \text{FE tuned} \\ 1 < n^{\mathrm{eff}} < n & \text{RDC}. \end{cases} \tag{15.14}$$

The one shown as "FE tuned" is the expression calculated by Molinari and Tóth [10]. Its limits are in between 1 and 10/6 for n in the interval $[1, \infty]$. For $0 < n < 1$ the expression gives large unreasonable variations and it is not expected to be applicable. Thus, the RDC model does not allow $n^{\mathrm{eff}} < 1$ and in any case extensions to poly-phase materials are yet to be developed.

15.2.2 Spin Interaction

One of the main problems of continuum mechanics has been the lack of a constitutive equation for the mechanical spin. Even in materials with no substructure a long discussion has been risen about the development of spin constitutive equations.

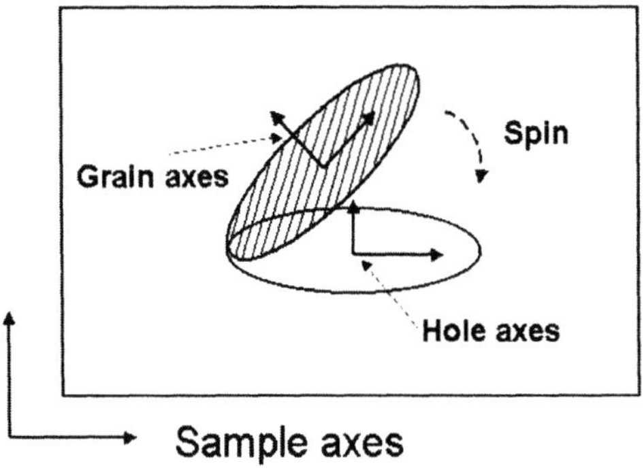

Fig. 15.3. Schematic representation of the reorientation mechanism in plastically deforming crystalline materials. The available slip systems are activated to fulfill average and local requirements. Compatibility is achieved by further reorientation of the crystallites

When a material substructure is present, Dafalias [16, 17], Mandel [18, 19] and Kratochvil [20, 21] have proposed the existence of preferential reference axes sets connected with the microstructure. Connected with that reference system the material will develop anisotropy, detected as new material properties.

Texture simulation research has followed an independently developed but similar approach. An isolated crystal does not develop any preferential orientation by plastic slip. However, subject to boundary conditions, either by being surrounded by other grains or solicited by external tools, it needs to comply with the continuity condition of not leaving any hole neither overlap with the vicinity. The hole in which a crystal is originally located must lodge the crystal even after plastic deformation. The spin produced by plastic slip is counteracted in such a way that texture arises because of that reorientation. Figure 15.3 shows schematically the reorientation produced by plastic slip in a single slip system.

The problem of interaction between crystals has been addressed, as we did in the former paragraph, just in terms of stress-strain interactions. Self Consistent models are a kind of mean field theories that regards a crystal as embedded in a mean stress-strain field. 1-site or general n-sites SC models take account of the stress-strain interaction between each crystal and the surroundings as a homogeneous medium or individually with each one of the n-sites.

Not surprisingly, and due perhaps to the lack of spin constitutive equations, not too much attention has been paid to interaction from the point of view of the expected influence of the spin of one crystal over the spin of the others. To introduce that quite different kind of interaction between crystals it is necessary to discuss further the problem of strain and spin heterogeneity.

Briefing the sources of heterogeneity, in the same way as the residual strains are usually classified, we can identify three main types: (a) First kind are those heterogeneities comprising many grains in a macroscopic arrangement reaching even the level of the sample. FEM are usually applied for dealing with this level of heterogeneity. Residual stress community knows that this kind of heterogeneity will make itself evident by the peak shift phenomena. (b) Second kind are those heterogeneous strains spanning over a few neighboring grains. They are usually approached by SC models and/or FEM working on the basis of one grain corresponding to one element of the simulation. Variations from grain-to-grain acquire special importance by the presence of two phases. Heterogeneities are larger and the models should not be stressed beyond their limits. From the view point of residual stresses development these kinds of heterogeneity are shown by peak broadening and, sometimes, by sample orientation dependent peak shift. (c) Third kind heterogeneities are developed internally in each grain by dislocation arrangements built up and consequent fragmentation. They are approached by the FEM scheme "one grain-many elements". Peak broadening, sometimes also dependent on sample orientation, is the phenomena observed in X-ray diffraction detection of this kind of heterogeneity. In the understanding of the current authors the fragmentation process, and its influence on texture development, is well simulated by using a spin sharing approach, technique that will be assessed in this chapter [22, 23].

Macroscopic heterogeneities (1st kind), those stemming from the test constrains, are among the mechanisms responsible for texture smoothing processes. Different regions of a sample are subject to different strain paths due to interaction with the machine tools and geometrical constraints imposed by the desired final shape. In rolled materials the sample surface undergoes a process of shearing and counter-shearing due to the roll action that is absent in the middle part of the sample. The superposition of texture patterns characteristic of each deformation path confers to the composed texture a lower severity or sharpness, comparing with homogeneously deforming materials. The main reasons for heterogeneous deformation are usually combined: geometrical as well as friction effects, which are not easily separable. In extruded or wire drawn materials the characteristic of the die (friction coefficient), geometry (die angle) and speed are combined to give lower sharpness textures. Die angle is the mayor modifier of the uni-axial velocity gradient. The most appropriate technique to evaluate strain heterogeneity at that level is FEM [24, 25].

Regarding the ability of SC models to simulate grain-to-grain heterogeneity it is worthy to emphasize the many neighbors surrounding a single crystal: 12 in an ideal compact regular arrangement and from 5 to 7 in a real polycrystal. Those surrounding grains should actually be representative of a local cluster reproducing in average the behavior of the matrix in which SC models assume the main grain is embedded. Second kind heterogeneity is quite likely well simulated by having each crystal embedded in a matrix with viscoplastic properties averaged over the rest of the crystals. The main variations from grain to grain are due mainly to the grain orientation and the statistical homogeneity of closest neighbor orientations represented locally by the SC scheme.

On the opposite end of the scale, the individual grains actually deform very heterogeneously, achieving strain compatibility and equilibrium by developing different internal dislocation structures (3rd kind). Those dislocation structures are grain orientation dependent and interrelated, at different deformations, in a not completely well known way. Notwithstanding they bear the common feature of being able to accumulate dislocations of the kind that are activated by the general Taylor rules. Minimum energy principles for the accumulated dislocations are the master rules to decide which dislocations are indeed active. The experiments conducted in this field are quite profuse. It is by now pretty clear that which were in principle assumed to be the main regions of a polycrystal acting as dislocation sink areas, the grain boundaries, are actually not. The dislocations are accumulated in regions internal to the grains and, there, they develop quite complex interactions and consequently carry the local deformations and crystal lattice spin (rotation rate). The experiments also show quite large misorientations among the many fragments developed in a grain. Some dislocation arrangements, known as Geometrically Necessary Dislocation Boundaries (GNDB), introduce larger misorientations than others, called Incidental Dislocation Boundaries (IDB) or statistically accumulated dislocations [26–28]. Those intra-grain misorientations, and consequent texture development, are simulated in this chapter by using a co-spin scheme between two close neighbor grains in a VPSC model. Details of the model have been published elsewhere [22] and will be briefed and discussed here. A detailed justification from the microscopic viewpoint and an application to a single phase material can be found in Bolmaro et al. [23].

The interaction effect between grains and fragments will be assessed by a combination of an n-sites Visco-plastic Self-Consistent (nS-VPSC used alternatively in its 1 and 2-sites versions) micromechanical model and Finite Element Methods (FEM). No details will be given about FEM implementations. The interested reader can consult references or Bolmaro et al. [29].

Let us go back to the spin interaction problem. For each grain k it holds, step by step, that:

$$\Omega_{ij} = \Omega_{ij}^{k} = \Omega_{ij}^{k^*} + \Omega_{ij}^{kP}, \tag{15.15}$$

where Ω_{ij}^{k} is the external macroscopic spin, assumed constant and equal to Ω_{ij} for every grain k, $\Omega_{ij}^{k^*}$ is the crystal system spin or spin of the substructure and Ω_{ij}^{kP} is the spin of the grain shape due to plastic crystal slip.

That assumption hides a more general assumption about the average spin of all grains in the sample. That is to say that

$$\Omega_{ij} = \langle \Omega_{ij}^{k} \rangle = \langle \Omega_{ij}^{k^*} + \Omega_{ij}^{kP} \rangle, \tag{15.16}$$

what means that, essentially, even self consistent models use a Taylor like approach to spin sharing among grains. Even modern versions of SC models,

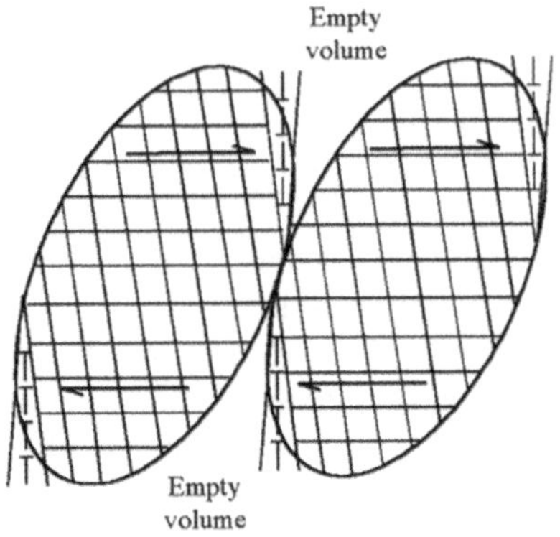

Fig. 15.4. Possible behavior of two similarly oriented neighbor grains. Dislocations accumulated in grain boundaries are responsible for grain reorientation. Except probably in the current case when the empty volumes left by the companion grain are readily occupied by dislocation arrays

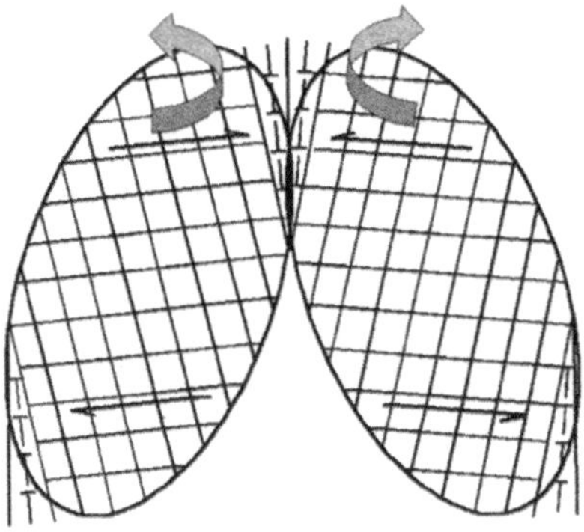

Fig. 15.5. Possible behavior of oppositely spinning grains. Overlap between dislocation arrays is forbidden and counter plastic spinning is enforced in the usual fashion

which consider a local spin of the hole different for each hole, do not consider, in the two-sites version, both coupled grains as a unique entity that has to fit that hole considering plastic spin too.

Figures 15.4 and 15.5 shows two limiting cases to get some insight on the spin behavior of one grain when the closest neighbor is, besides the influence of the matrix surrounding the grain, the largest influence in terms of plastic spin. Figure 15.4 shows the hypothetical case of having two neighboring grains oriented in the same direction. Because each grain is leaving behind, by plastic slip, empty volumes in the regions where the neighbor needs an empty region, driven by the sole interaction with the closest neighbor neither grain would counter spin to fit in the original hole. Figure 15.5 shows two crystals oriented in such way that the plastic spin is equal but of opposite signs. The counter plastic spin is enforced now because of the absence of overlapping between the dislocation arrays.

In some special situations we can assume that there is no need for single grain spin, Ω^{k^*}, if two grains can complementary fit the two-grains hole that has been apportioned to them (Fig. 15.4). In some other cases, the grains behave pretty much like in the usual model of each grain fitting its own hole (Fig. 15.5). Many different situations can be imagined but all of them require the perfect knowledge of geometrical and topological features. The relative strain rates and spins between both grains are straight consequence of them.

A less detailed vision, always keeping in mind the conditions assumed in the beginning of the paragraph about strain rates and spins, will be worked out in the following paragraphs. However, despite the simplicity, the model will become operative in that way.

Let us write the equation (15.16) for just two grains of equal volume

$$\Omega_{ij} = \langle \Omega_{ij} \rangle = \frac{(\Omega_{ij}^{1^*} + \Omega_{ij}^{2^*} + \Omega_{ij}^{1P} + \Omega_{ij}^{2P})}{2}, \tag{15.17}$$

where $\langle\,\rangle$ means the average magnitude. If the average spin Ω_{ij}^{*} is associated with the spin of the "average hole" apportioned to each pair of grains, then

$$\langle \Omega_{ij}^{*} \rangle = \frac{(\Omega_{ij}^{1^*} + \Omega_{ij}^{2^*})}{2} = \Omega_{ij} - \frac{(\Omega_{ij}^{1P} + \Omega_{ij}^{2P})}{2} \tag{15.18}$$

and we should use an average grain shape spin (produced by the plastic deformation), for both grains, subtracted from the macroscopic spin. The direct consequence is that both grains in a 2-sites SC approach will keep constant relative orientation. That counter-intuitive result is rather puzzling and can be understood schematically as follows.

Whenever an extra crystal plane is created, in the form of an edge dislocation, a reorientation of crystal lattices at both sides is produced. That reorientation is approximately equal in magnitude and opposite in sign in both sides of the just created array. And that regardless of being those extra planes located either on original grain boundaries or newly created GNDBs or IDBs. Those dislocation arrangements must cope with a mass imbalance in such way that a reorientation of a grain or a fragment will induce similar reorientation not in its closest neighbor (which is actually counter rotated) but in its second closest vicinity. If we understand the process of crystal reorientation as stemming from mass accumulation creating more grain boundary surfaces, the lattice spin canceling out at medium ranges is a natural consequence, probably valid regardless of crystal symmetry or slip system nature.

Figure 15.6 shows a schematic representation, simplified by using only edge dislocations, of the short distance anti-correlation plus long range correlation of crystal lattice spins. As well as strains, spins have to be compensated in a short range. Canceling out spins grain by grain has been used by many models but has not been taken further until now.

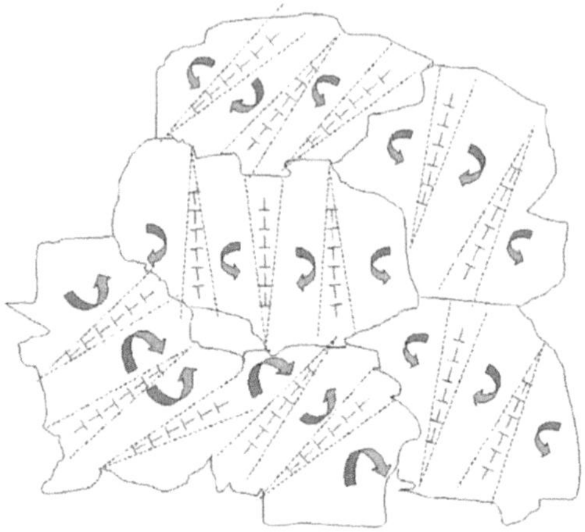

Fig. 15.6. Schematic representation of grain fragmentation. The newly created GNDB are arranged alternatively compensating the spin of the second neighbor. The first neighbor is spinning driven by its own slip systems and the required spin compensation is provided by the creation of misorientations

It is clear from experiments that different parts of a grain spin to different orientations. That would suggest that we should move to lower simulation scales. Instead, the opposite can also be true if we learn how to deal with statistic correlation between differently spinning volumes. This last approach is worked out in the current model.

Assuming that the current statistical assumption holds for calculated misorientations between two parts of a crystal we simulated a channel die simile rolling test for FCC materials. Calculated misorientations showed a distribution function in quite good agreement with EBSD (Electron Back Scattering Diffraction) experiments [26–28]. The results were published elsewhere [23]. Figure 15.7 shows the distribution of misorientation angles for different deformation stages. The curve follows a log-normal distribution that resembles quite well experimental values. Figure 15.8 shows experimental misorientations at GNDBs obtained by Hughes and Hansen [28] by Transmission Electron Microscopy for Al., Ni and Ta.

Confirmation of agreement at a lower scale (grain to grain and not just statistical) is in the process of being simulated.

15.3 Results

The purpose of this chapter is to apply the model to the behavior of very demanding co-deforming materials. A collection of different volume fractions of Al-Cu and Cu-Fe powder composites are available and have been previously published. The model should take account of the different behaviors of

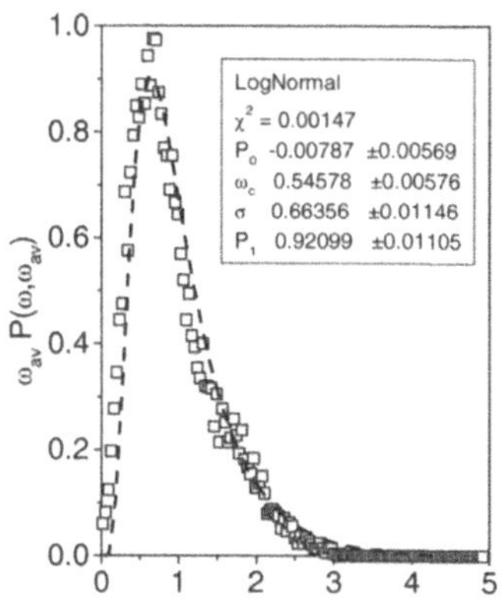

Fig. 15.7. Simulated probability density functions of the intra-grain misorientation angles normalized by the average misorientation angle, $\varepsilon_{vM} = 0.25$–2.0. All misorientation angles for every simulated deformation fitted by a log-normal distribution

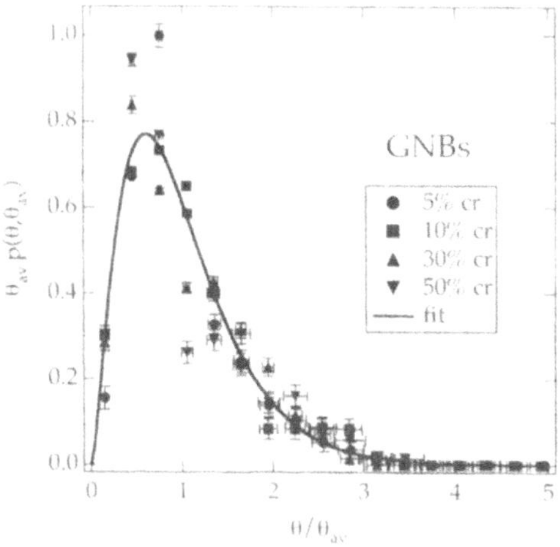

Fig. 15.8. Probability density function of the experimental GNDBs misorientation angles normalized by the average misorientation angle. Reprinted from [26] with permission from Elsevier

each mixture and crystal symmetry. The experimental data consists of textures measured at Tex-2 neutron diffractometer at GKSS Forschungszentrum Geesthacht, Germany. Highly precise measurements are allowed by neutron diffraction for low volume fractions.

Despite the model was inspired by co-deforming two-phase materials, most arguments are also valid for single phase materials. A torsion test simulation for FCC materials will be also shown, together with X-ray diffraction experimental textures measured for a Cu torsion test.

15.3.1 Extruded Al–Cu

The volume fraction of both phases in Al-Cu samples span from 0% through 100% in 10% steps. The samples were deformed to Von Mises equivalent strains of 3.0 starting from pre-forms of slightly compacted mixed powders.

Slip systems for Al and Cu were the usual $\{111\}\langle 110\rangle$ and the stress exponents were 100 for Cu and 50 for Al. The hardening law was taken linear with slopes of 1.8 for Cu and 1 for Al in arbitrary units. The Critical Resolved Shear Stress was considered 2 for the Cu with respect to the Al that was taken equal to 1 in the same arbitrary units.

Morphology and phase distribution observations show that the interaction between grains should be dependent on the volume fraction of each phase. Highly dependent in the case of strain interaction and only slightly dependent in the case of spin interaction [30]. We considered the presence of 6 phases alternatively behaving as co-spinning, strongly interacting Al-Al, Al-Cu and Cu-Cu pares. Al usually flows around Cu for being the softer phase. For high Al volume fraction the morphology shows Cu grains not having Cu-Cu interfaces. The simulation is performed in that case considering only 4 phases and interactions Al-Al and Al-Cu. Figure 15.9 shows the sharpness of texture components for Al-Cu powder composites vs. Al volume fraction together with the simulated values. Complete experimental details have been given elsewhere [31]. A sample of results for some volume fractions is shown in Fig. 15.10 as inverse pole figures. A quite close agreement of experiments and simulations is evident and only some differences appear for the Al $\langle 100\rangle$ component due probably to dynamic recrystallization that takes that component to the range $\langle 103\rangle$–$\langle 113\rangle$. The agreement has been obtained by using the

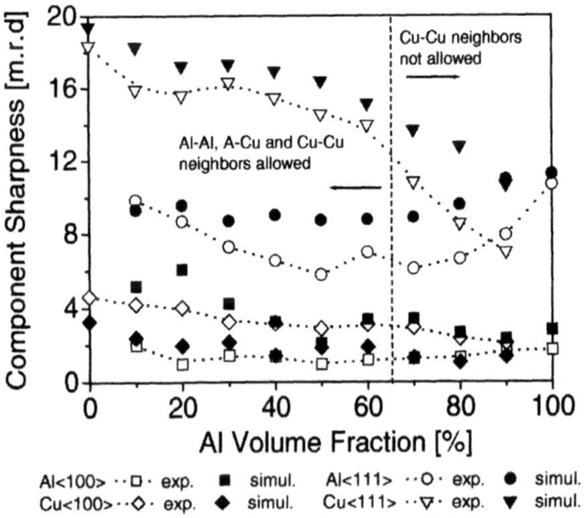

Fig. 15.9. Experimental and simulated component sharpness for Al and Cu vs. Al volume fraction

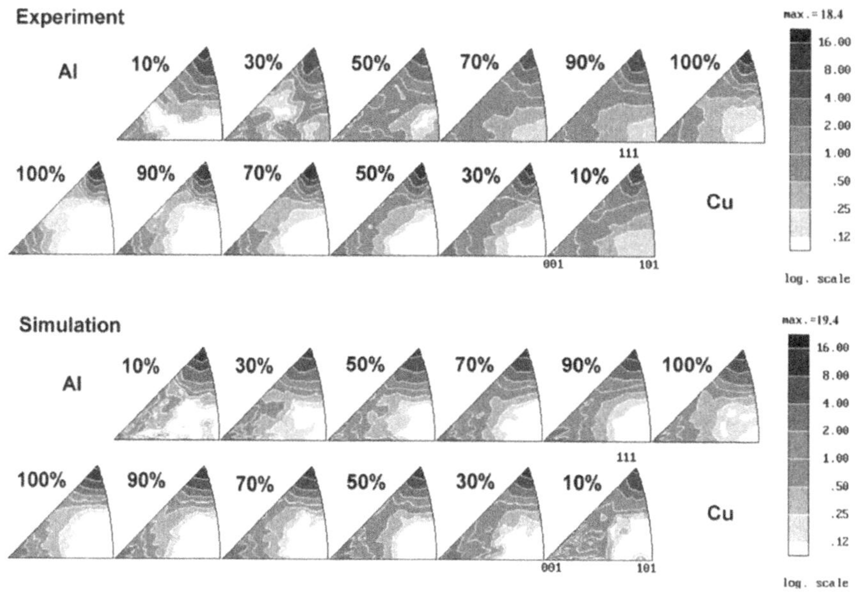

Fig. 15.10. Inverse pole figures for some of the volume fractions of extruded Al-Cu deformed samples. (**a**) Experiments. (**b**) Simulations

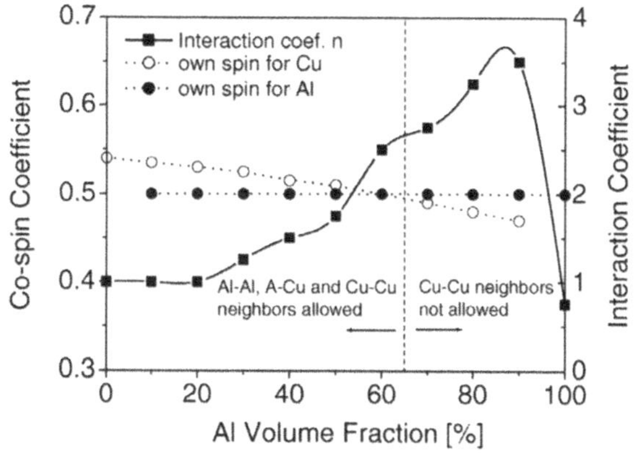

Fig. 15.11. Interaction coefficients and spin sharing for Al and Cu vs. Al volume fraction. The interaction coefficients are very close to the FEM tuned values of 1.6. Spin sharing slightly varies with volume fraction

interaction coefficients shown in Fig. 15.11. Small variations between 1 and 3.5 confirm previous tuning performed by Molinari and Tóth ($\approx n^{\text{eff}} = 1.6$) combining SC and FEM models [10].

15.3.2 Extruded Cu–Fe

Cu-Fe experimental textures were published by Brokmeier et al. [32]. The current simulations are applied to a set 75%Cu25%Fe extruded until Von Mises equivalent strains of 3.0. Intermediate strains (every 0.25 Von Mises strains) were also produced and measured but not analyzed in this chapter. Previous simulations performed in Cu-Fe extruded composites have shown that the "curling effect", usually exhibited by extruded Fe, enforces a similar behavior in the Cu phase [29]. That interaction effect means a much larger interaction coefficient than for FCC-FCC mixtures. It reached a value of 67 for 75%Cu25%Fe. The main effect is the enforcing of plane strain over Cu, which is characteristic of Fe deformation. To achieve compatibility "curling" is a second order effect. Spin sharing varies slightly at around 50% for each phase almost independently of the volume fraction. Figure 15.12 shows experimental and simulated textures as inverse pole figures. They show a quantitative agreement never reached before by any simulation. In this case we also included FEM calculations of the different velocity gradients imposed at varying distances from the sample center.

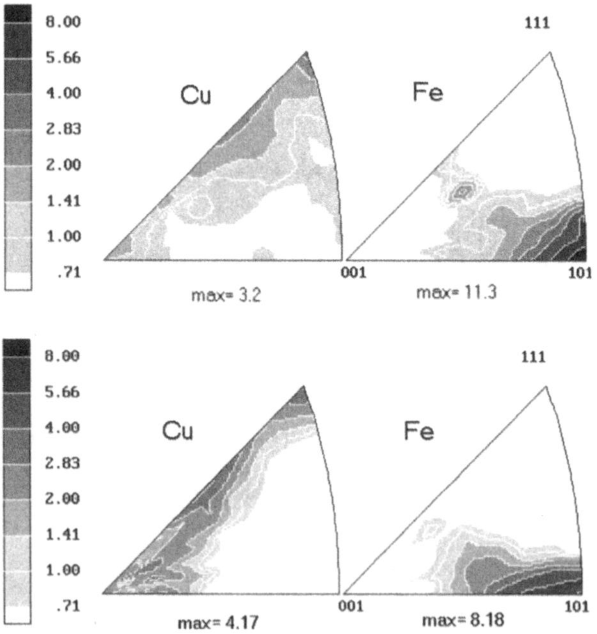

Fig. 15.12. Inverse pole figure for Cu and Fe. (**a**) Experiments. (**b**) Simulations

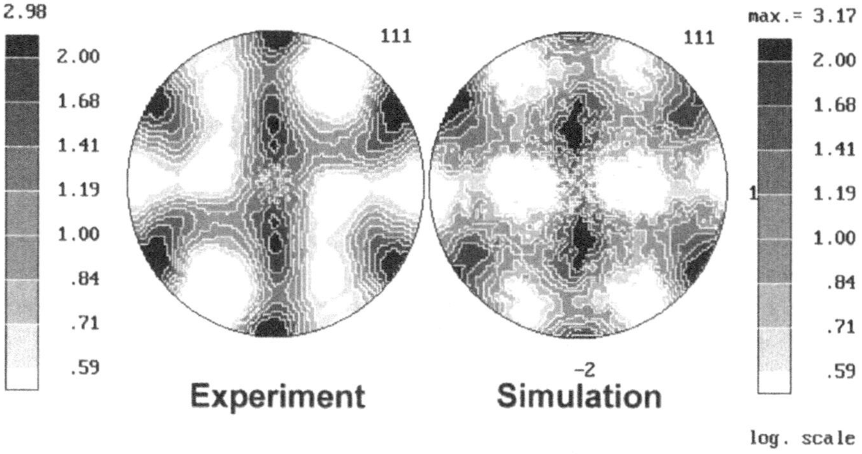

Fig. 15.13. (111) pole figures for simile Cu deformed under torsion to a Von Mises equivalent strain of 2.0. (**a**) Experiment . (**b**) Simulation

15.3.3 Cu Torsion

Simulation of FCC torsion (pure shear) deformation is very difficult. We proceeded to simulate large deformation torsion tests by using the same material parameters used for Cu in Al-Cu and Cu-Fe to equivalent Von Mises strains of 2.0. We used a co-spin scheme and same hardening parameters. Experimental textures were taken by X-ray diffraction techniques. The experimental and simulated results are shown in Fig. 15.13 as (111) pole figures.

15.4 Discussion and Conclusions

15.4.1 What does Our Model Provide?

The main conclusion is that it is possible to interpret the texture development of co-deforming mixtures by a combined action of individual phase behavior and a separated interaction effect represented by interaction matrices. Those interaction effects are currently well understood in terms of physical variables as volume fraction and phase distribution and morphology. Of paramount importance is the assumption of local compensation of crystal spins. It allows, by a simple "ad hoc" rule, to reach texture sharpness close to experimental ones and even get some insight on microscopic behavior. An interpretation of the current results would be that long range orientational order (texture) can be achieved only by producing local orientational disorder through the creation of misorientation regions guaranteeing spin compensation at short distances. The first neighbor is forced to deform driven by its own orientation and available slip systems. That deformation builds up stresses, because of strains and spins, that can be relieved only by fragmenting and counter-spinning of the newly created fragment.

Worthy to be emphasized is the ability of the current approach of simulating textures with high accuracy, not only regarding components but also with respect to the relative strength of simulated pole figures. No smoothing is necessary as it is usually performed for texture simulations, which come out much stronger than experiments. Also some small components are better simulated by incorporating the co-spin scheme. The simultaneous inclusion of different stress exponents and interaction coefficients allows another parameter to be fit but the simulations were performed with the same parameters regardless of material composition or mechanical test.

15.4.2 What Do We Need?

What do we need from X-ray diffraction analysis? This chapter shows a collection of simulation-experiments texture studies that proves the necessity of having a better understanding of grain sizes, dislocation distributions and arrangements. The model deals with the problem of dislocation accumulation in an indirect way. Just by formulating a general rule about local spin canceling out or compensation, the micromechanical models (regardless of being Taylor, Sachs or Self-Consistent) provides a surprisingly good agreement with experimental textures. Moreover, some more detailed studies about misorientation distributions among different texture components show results corresponding well with EBSD and Transmission Electron Microscopy (TEM) experiments.

It is clear however that a better understanding of dislocation arrangements is necessary to improve the behavior of models and derived information. Just to consider the results presented in the current conference, it is worthy to mention the following works as contributions to the field:

1. Y. Perlovich, M. Isaenkova and co-workers presented a few works [33–36] showing the large dependence of peak broadening on crystal orientations. Generalized Pole Figures measure different mechanical and physical properties in function of the orientation of crystals in a textured polycrystal. Peak broadening and peak shift are particularly instructive in showing the great challenge presented to physically based peak profile analysis. Large efforts should be devoted to the understanding of crystal orientation dependence for grain size, dislocation arrays and residual strains. By no considering the texture, measured peak broadening and peak shift are integrating magnitudes summing up over grains embedded in different environments and having different dislocation arrays. That may preclude proper modeling and understanding.
2. S. Lomine, P. Klimanek and M. Masimov [37] showed rather conflictive comparisons between EBSD and X-rays misorientation measurements. EBSD misorientations come almost an order of magnitude larger than X-ray determinations. Besides the possible higher low limit presented by

EBSD technique determinations, also the models relating peak broadening with misorientation should be revised. In any case, whichever the model, a more detailed information will be necessary and peak broadening is manifestly orientation dependent.

3. I. Groma [38,39] presented a model for the evaluation of X-ray line broadening due to heterogeneous dislocation distribution. Correlation between characteristics of dislocation arrays and grain orientations can help to fill the gap between dislocation modeling of internal stresses and experimental peak shift and peak broadening.

Taking the risk of being redundant it is worthy to emphasize the many dislocation structures stemming from plastic deformation and the close interrelation between them and grain size. Dislocation arrangements are always described as frozen structures that are most probably continuous variations going from uniformly distributed dislocations (Taylor structures) through cell boundaries (or Incidental Dislocations Boundaries) until they reach the complex GNDB (Geometrically Necessary Dislocation Boundaries) structure. This last ones are likely to be structurally equivalent to regular grain boundaries and more probably detected as grain size peak broadening. However, models are usually particular cases with no continuous connections between the different dislocation array varieties. The task of developing wide span models, comprehensive of all possible strain fields, seems to be rather difficult. Extra information can come from texture – peak shift – peak broadening experiments.

Being texture development by plastic strain so dependent on dislocation structures, it should reveal itself both as a powerful tool for model checking and as a demanding field for more careful X-ray line profile analysis.

Acknowledgements

This project was partially funded by the German-Argentine Intergovernmental Agreement No. AL/A99-EXII/18 and CONICET-Argentina. One of the authors (R.E. Bolmaro) is specially grateful for the partial support received from the Organizing Committee of Size-Strain III to attend the Conference.

References

1. R.E. Bolmaro, R.V. Browning, F.M. Guerra, U.F. Kocks, P.R. Dawson, J.D. Embury and W.J. Poole: Acta metall. mater. **41**, 1893 (1993)
2. A. Cottrell: *Von Hippel Award address.* MRS Fall Meeting, 1996, or MRS Bulletin **22**, 15 (1997)
3. G.I. Taylor: J. Inst. Metals **62**, 307 (1938)
4. G. Sachs: Zeitschrift Verein Deutscher Ingenieure **72**, 734 (1928)
5. R.A. Lebensohn and C.N. Tomé. Acta metal. Mater. **41**, 2611 (1993)
6. A. Molinari, G.R. Canova and S. Ahzi: Acta metall. **35**, 2983 (1987)

7. G.R. Canova, H.-R. Wenk and A. Molinari: Acta metall. mater. **40**, 1519 (1992)
8. R.A. Lebensohn and G.R. Canova: Acta mater. **44**, 3687 (1997)
9. U.F. Kocks: Mat. Sci. Eng. A **175**, 49 (1994)
10. A. Molinari and L.S. Tóth: Part I. Acta metall. mater. **42**, 2453 (1994); Part II. Acta metall. mater. **42**, 2459 (1994)
11. C.N. Tomé: Model. Simul. Mater. Sci. Eng. **7**, 723 (1999)
12. Z.S. Basinski: Phil. Mag. **4**, 393 (1959)
13. H. Mecking and U.F. Kocks: Acta Metall. **29**, 1865 (1981)
14. J.W. Hutchinson: Proc. Roy. Soc. London A **348**, 101 (1976)
15. P.A. Turner, C.N. Tomé, N. Christodoulou and C.H. Woo: Phil. Mag. A **79**, 2505 (1999)
16. Y.F. Dafalias: *Colloq. Int. du CNRS 351*. Ed. J.P. Boehler (Paris 1983)
17. Y.F. Dafalias: Mech. Mater. **3**, 223 (1984)
18. J. Mandel: *Courses and Lectures*. Int. Center for Mechanical Sciences, Udine. Vol. 97 (Springer, Berlin Heidelberg New York 1971)
19. J. Mandel: Int. J. Solids Struct. **9**, 725 (1973)
20. J. Kratochvil: J. Appl. Phys. **42**, 1104 (1971)
21. J. Kratochvil: Acta Mech. **16**, 127 (1973)
22. R.E. Bolmaro, R.A. Lebensohn and H.-G. Brokmeier: Comp. Mat. Sci. **9**, 237 (1997)
23. R.E. Bolmaro, A. Fourty, A. Roatta, M.A. Bertinetti and J. Signorelli: Scripta mater. **43**, 553 (2000)
24. P.R. Dawson and A.J. Beaudoin: J. of Metals **49**, 34 (1997)
25. T. Aukrust, S. Tjøtta, H.E. Vatne and P. Van Houtte: Int. J. of Plast. **13**, 111 (1997)
26. D.A. Hughes and N. Hansen: Acta mater **45**, 3871 (1997)
27. Q. Liu, D. Juul Jensen, and N. Hansen: Acta mater **46**, 5819 (1998)
28. N. Hansen and D.J. Jensen: Phil. Trans. R. Soc. Lond. A **357**, 1447 (1999)
29. R.E. Bolmaro, A. Fourty, J.W. Signorelli and H.-G. Brokmeier: *Development of Wire Drawing Textures in Cu-Fe. The Influence of Macroscopic and Microscopic Heterogeneities*. To be published.
30. H. Gertel: Technischen Universitat Clausthal als Dissertation angenommene Arbeit, GKSS, 1992
31. H. Gertel-Kloos, H.-G. Brokmeier and H.J. Bunge: Mat. Sci. Forum **157–162**, 685 (1994)
32. H.-G. Brokmeier, U. Zink, R. Schnieber, B. Witassek: Mat. Sci. Forum **273–275**, 277 (1998)
33. Y. Perlovich and M. Isaenkova: *Proceedings of Size – Strain III., December 2–5.* (Trento 2001) p. 49
34. M. Isaenkova and Y. Perlovich: *Proceedings of Size – Strain III. December 2–5.* (Trento 2001) p. 92
35. M. Isaenkova, Y. Perlovich, H.J. Bunge and V. Fesenko: *Proceedings of Size – Strain III. December 2–5.* (Trento, 2001) p. 94
36. M. Isaenkova, Y. Perlovich and V. Fesenko: *Proceedings of Size – Strain III. December 2–5.* (Trento, 2001) p. 119
37. S. Lomine, P. Klimanek and M. Masimov: *Proceedings of Size – Strain III. December 2–5.* (Trento 2001) p. 106
38. I. Groma: *Proceedings of Size – Strain III. December 2–5.* (Trento 2001) p. 22
39. I. Groma and A. Borbély: this volume, Chap. 11

Part VI

Surface and Interface Effects

16 Grain Surface Relaxation Effects in Powder Diffraction

M. Leoni and P. Scardi

Most nanocrystalline materials, in powder or bulk form, show unit cell parameters sensibly different from the coarse-grain or single-crystal counterpart. Even if the phenomenon is well known, a sound explanation is not always available: stoichiometric effects (e.g. related to vacancies or oxygen in excess) are frequently invoked to interpret the experimental evidence. Grain-surface relaxation is a possible alternative for data interpretation: due to the presence of the surface itself, atoms next to a surface are permanently displaced from their equilibrium lattice positions. The consequences of this, unveiled when the number of atoms in the surface layers starts to supersede the number of atoms in the rest of the grain (as in the case of nanocrystalline materials), are usually not considered in powder diffraction and are a main issue in this chapter.

A general Fourier-based method is therefore proposed to account for grain-surface relaxation effects on powder diffraction profiles (line shift and broadening). The model can be easily incorporated into the Whole Powder Pattern Fitting (WPPF) and Whole Powder Pattern Modelling (WPPM) algorithms, recently introduced for a non-destructive microstructural analysis of materials based on X-ray diffraction.

The specific case of fcc materials is considered in detail. The proposed method accurately describes the diffraction patterns of nanocrystalline ceria produced by sol-gel and calcinated at various temperature, and provides microstructural parameters (crystallite size and distribution, dislocation content, etc.) consistent with those refined by WPPF and WPPM. The average unit cell parameter after calcination at four different temperatures is correctly modelled by using only two grain surface relaxation parameters, kept constant for the studied sample series. The results of the non-destructive X-ray diffraction measurement are in good agreement with the direct Transmission Electron Microscopy (TEM) observations, giving a further validation of the proposed model.

16.1 Introduction

In recent years, nanostructured and nanocrystalline materials have earned a growing interest both from the scientific community and from the industry. This is mainly due to their unique physical and mechanical properties not easily found in their coarse-grain counterparts [1–4]. As an example, the phenomenon of superplasticity, i.e. the possibility of extremely high deformation before fracture, is one of the new properties often associated to the nanostructured materials world.

This differences between the nanocrystalline and the coarse-grained world are due to the hyperbolic trend of the surface-to-volume ratio with respect to the grain diameter, diverging for very small grain sizes (like in finely dispersed powders, supported catalysts, heavily deformed and nanocrystalline materials, etc.). It will be shown how surface effects can be observed with traditional powder diffraction techniques. After a brief survey of the state-of-art, a novel method of analysis, based on a Fourier formalism, will be presented. Fundamental feature of the proposed approach is the possibility of implementation within the recently developed Whole Powder Pattern Fitting (WPPF) [5, 6] and Whole Powder Pattern Modelling (WPPM) [6, 7] frames, opening the doors to the possibility of a complete microstructural refinement from powder diffraction data.

As an example of application of the method to fcc materials, the case of nanocrystalline cerium oxide will be presented in detail.

16.2 Surface Relaxation

The problem of surface relaxation and surface reconstruction is not a recent one. The existence of near-surface relaxation was in fact theorised in the late seventies and clearly observed in the last years (vide infra).

The simplest model for a perfect infinite crystal accounting for its atomic structure, assumes a perfect lattice and harmonic forces between neighbouring atoms. This implicitly states that (possible) permanent displacements of the atoms from their equilibrium positions are very small [8].

The picture changes when defects are considered; we should not limit our definition of "defects" to the traditional point and line ones, but also include more complex defects like surfaces, interfaces and grain boundaries. Atoms nearby a defect are more free to move or to shift from their lattice position, because of a reduced number of nearest neighbours; as a consequence, the amplitudes of atomic vibrations (phonons) can be quite different from those of the infinite crystal (even one order of magnitude higher [8–11]).

A surface is thus a source of anharmonicity in lattice vibrations (effect known as phonon confinement) and can lead to a permanent displacement of some (namely the near-surface) atoms from their "regular" *unrelaxed* lattice positions.[1] The displacement is permanent but not random: Allan and

[1] In this chapter the term unrelaxed will be used for a structure in its ideal state where all atoms reside on lattice positions (i.e. the structure is free from any relaxation phenomena).

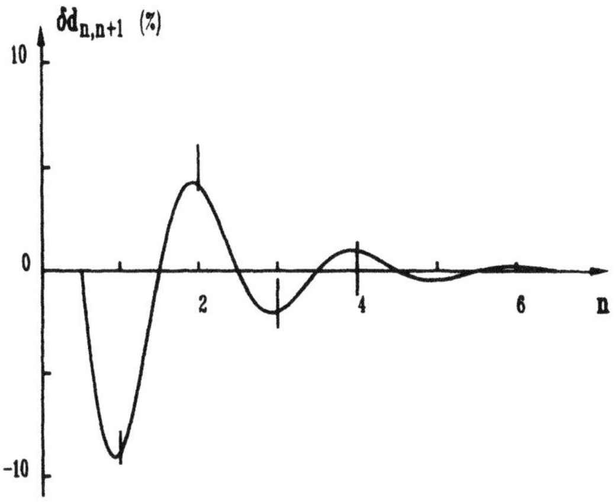

Fig. 16.1. Difference $\delta d_{n,n+1}$ between the lattice spacing of a couple of layers n and the subsequent one $(n + 1)$ from the surface towards the bulk of aluminium (from [12]. Copyright (1988) by the American Physical Society. Reprinted with permission)

Lannoo [12], for instance, observed an oscillatory damped relaxation (see Fig. 16.1) as asymptotic trend for a semi-infinite crystal. This oscillatory trend is easily obtained by evaluating the stable configurations for the elastic bulk phonons normal to the considered planes (e.g. those parallel to the surface), altered because of the energy variation of the outer atomic layers. The oscillatory damped behaviour is experimentally confirmed by a discrete number of semi-infinite crystals, including Ni(110), Al(110) and Fe(111) [12].

A simple generalisation of the anharmonicity problem to crystals showing a higher degree of phonon confinement (e.g. nano-sized grains) is not possible. More complex cases have been studied and analytical formulations of the problem, proposed. The case of a finite-size crystal has been examined by Żabiński and Zieliński who provided analytical expressions for the static response functions [13]. Besides these equations, the authors present the interesting case of a crystal composed of alternating A and B planes (with two different atomic species), met in a variety of simple compounds. In all cases, a permanent displacement of the surface atoms from their unrelaxed positions is observed and the structure appears superficially relaxed.

The extent and magnitude of surface relaxation is not a constant among the materials: incredibly high values are shown by the $(10\bar{1}0)$ surface of Beryllium (*hcp*) where the top layer moves inward of 25% [14]. This is not the rule, but relaxation as high as few hundredths of nanometre seems quite common in metal oxides (see e.g. [15] on TiO_2 or [16] on SnO_2).

There is no rule of thumb for estimating a priori the extent of surface relaxation; ab initio Molecular Dynamics (MD) calculations, however, can quite accurately describe and quantify the phenomenon. Following the progress in computers and calculation schemes, MD permits nowadays the evaluation of bulk and surface properties of small atomic clusters and polycrystalline ag-

Fig. 16.2. Displacement field calculated by Molecular Statics methods and necessary for the relaxation of the structure in an aggregate of iron grains (from [22]. Copyright (1995) by Elsevier. Reprinted with permission)

gregates (see e.g. [15,17–22]). It is interesting to observe that the calculation of a stable configuration of all atoms results in unbalanced forces at the surface of a cluster; a Molecular Statics (MS) method is then often applied in order to adjust the atomic positions as to minimise the energy of the whole system [21]. In this step, surface relaxation is therefore modelled.

In [21,22] Chen presents the example of a bulk aggregate of iron nanocrystals (maximum diameter of about 2 nm); Fig. 16.2 shows the displacement field, calculated by MS, necessary for relaxing the structure obtained from a MD algorithm. It is apparent that atoms close to the core of the grains are less displaced than those on the boundaries and that surface effects can be appreciated even on the nanometre scale. This is of key importance since X-ray diffraction (XRD) techniques are particularly suited for the study of microstructural features up to a few hundreds nanometres.

Another key observation stemming from the results of Chen [21] is that, although the crystalline lattice is slightly distorted in the various crystallites, the average unit cell parameter for the overall structure seems to be expanded with respect to the equilibrium value. Even if this is a by-product of Chen's calculations, this result has a number of consequences in materials science, especially concerning nanometre-sized materials. In fact, this effect is more a rule than an exception in nanocrystalline powders; a large number of publications related to severe plastic deformation show an evolution (increasing or decreasing trend) of the unit cell parameter with the grain size [1–3, 23–26]. Analogous effects are observed in finely dispersed catalysts and crystalline powders produced from amorphous precursors [27,28].

In many cases this variation is not given an explicit interpretation, especially when XRD techniques are used for the determination of unit cell parameters and microstructural features of the material (such as defect content and crystallite size). However, few exceptions exists [29–32]: Nunes and

Yang [29] present the case of a ferrite nanocrystal simulated ab initio and show that the resulting diffraction pattern (taken as the squared, spherically averaged Fourier transform of the intensity scattered by a crystal) possesses some of the features observed experimentally (such as peak position shift). This work confirms that relaxation phenomena, generally measured on extended flat surfaces, can be appreciated also in nanometre particles.

Transferring the results from the surface science to the XRD field is *not* straightforward and presents serious difficulties because of the different environment in which surface atoms are posed. Nunes and Yang [29] discuss on the fact that their results are in reasonable but not in perfect agreement with data relative to (111)-oriented films of the same material and argue that the discrepancy can be ascribed to the different neighbours of the atoms lying on a semi-infinite surface with respect to the surface of a small cluster.

The disagreement is not a general rule. Muliu and Pakkanen [33] found for instance that the values calculated for very small crystallites of zinc sulphide (ZnS) can be extrapolated to larger values (i.e. larger surface) and that the displacement of the surface atoms depends linearly on the crystallites' edge-to-surface ratio. Ab initio Hartree–Fock Molecular Orbital model calculation shows that the picture is more complex than a simple collective inward or outward movement of the surface or sub-surface layers. In small ZnS crystallites, in fact, the outer layer tends to move inward whereas the sub-surface layer shows an opposite trend [33].

Sometimes, all these indirect observations of the surface relaxation phenomenon and of its (rather complex) nature can be checked by High Resolution Transmission Electron Microscopy (HRTEM). A good example of this is provided by [34] and concerns the observation of gold nanorods obtained by bioreduction of $KAuCl_4$. The data presented in [34] clearly show grain surface relaxation associated with extended faceting (Au (110) surface).

This further adds complexity to the problem because surface relaxation is accompanied by (and intimately related to) surface reconstruction. However, only the former aspect will be dealt with in the present work.

16.3 Surface Relaxation Models in Powder Diffraction

16.3.1 State of the Art

Even if the literature on surface relaxation (and reconstruction) phenomena is quite rich, scarce data exist on their possible effects on traditional diffraction measurements. In the following we describe two recent models accounting for surface relaxation phenomena in powder diffraction. Both of them deal only with size and relaxation problems, therefore excluding a priori other possible sources of line profile broadening such as line or plane defects, quite common in nano-sized particles.

Model of Nunes and Lin (1995)

Nunes and Lin [30] observed asymmetric tails in the Bragg profiles of ball milled (nanometre-sized) ferrite powders. These powders show a reduction in magnetisation with the grain size; the phenomenon is often associated to a core-shell structure: each grain is composed of a ferritic core bounded by a magnetically anomalous surface whose thickness is of the order of a nanometre. This dependence of properties on grain size has been observed in ferroelectrics by many authors [31,35–37].

In their model for ferrite grains, Nunes and Lin consider each crystallite as spherical and made of $N+1$ lattice planes (Fig. 16.3). Surface relaxation is introduced as a graded variation in the spacing of the outermost layers (the case of expansion is explicitly considered, and the occurring curvature of the lattice planes is neglected). According to the kinematical theory, the scattering power for a system of N (unrelaxed) planes arranged symmetrically as in Fig. 16.3, is

$$A(R,\delta) = a\left[R^2 + 2\sum_{n=1}^{N/2} G_n \cos(H_n\delta)\right] , \qquad (16.1)$$

where

- $\delta = (4\pi \sin\theta)/\lambda$ is the transferred momentum (2θ is the diffraction angle and λ the wavelength);
- R is the sphere radius;
- H_n is the distance between the sphere centre and the n-th plane;
- G_n is a geometrical factor proportional to the area of the n-th diffracting plane $G_n = R^2 - H_n^2$, accounting for the spherical crystallite shape;

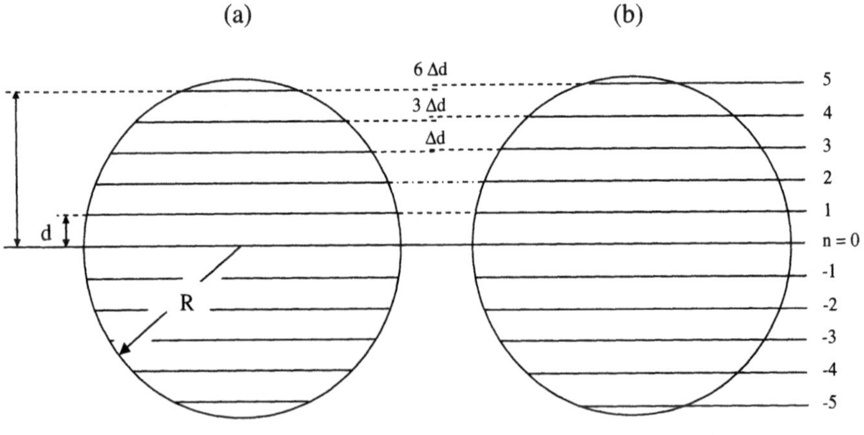

Fig. 16.3. Spherical crystallite unaffected (**a**) and affected (**b**) by surface relaxation. The variation in the spacing of the three outermost layers, following the treatment of Nunes and Lin, is shown (adapted from [30])

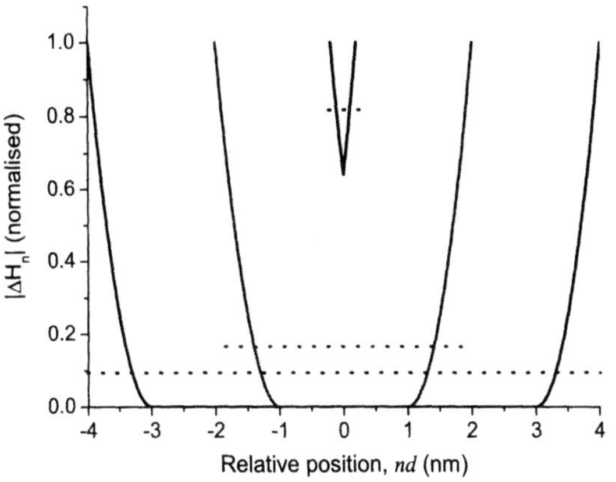

Fig. 16.4. Normalised relaxation function ($|\Delta H_n| = (H_n - nd)/(M(M+1)\Delta d/2)$, cf. (16.2)) employed by Nunes and Lin for different crystallite diameters ($D = 1$, 4, 8 nm, respectively). The number of relaxed planes is $M = 10$, whereas $\Delta d = 0.01$ nm, $d = 0.1$ nm (N changes correspondingly). The average shift in the lattice spacing is indicated, for each case, by horizontal dotted lines

- a is proportional to the scattering amplitude per unit area of each scattering plane.

Equation (16.1) modifies by inclusion of the relaxation effect since the function H_n changes from linear to piecewise linear (M is the number of relaxed planes):[2]

$$H_n = \begin{cases} nd & 1 \leq n < \dfrac{N}{2} - M \\ nd + \Delta d \sum\limits_{j=1}^{n-N/2-M} j & \dfrac{N}{2} - M \leq n \leq \dfrac{N}{2} \,. \end{cases} \tag{16.2}$$

This function is antisymmetrical, i.e. $H_{-n} = -H_n$, therefore only the $1 \leq n \leq N$ range is defined. The part of this function describing the relaxation is shown in Fig. 16.4 for spheres of 1, 4 and 8 nm in diameter, respectively. The graph is normalised to the maximum displacement allowed $M(M+1)\Delta d/2$. Since M is constant for all spheres considered, a notably large increase in the average displacement is observed for the smaller spheres (cf. horizontal dotted lines in Fig. 16.4).

If $p(R)dR$ is the normalised crystallite size distribution, the final expression for the diffracted intensity becomes:

[2] The proposed equation, taken from [30], can be simplified noting that

$$\sum_{j=1}^{n-N/2-M} j = \frac{1}{2}\left(n - \frac{N}{2} - M\right)\left(n - \frac{N}{2} - M + 1\right)$$

$$I(\theta) = K \int_0^\infty \frac{A^2(R,\theta)}{R^2} p(R) dR \,, \tag{16.3}$$

where K contains all instrumental, geometric and polarization factors.

As this approach does not use a rigorous Fourier formalism (as in the traditional methods of LPA, cf. e.g. [38]) for the description of the diffraction profiles, different broadening sources cannot be easily considered within the same model (in this sense, the possible generalisation of the problem was not considered by Nunes and Lin).

Model of Uemori et al. (1998)

In order to explain the observed trend of the average unit cell parameter and line profile broadening with respect to the average "grain diameter" for a ferroelectric material, Uemori et al. [31] proposed a relaxation model based on an exponential change of the interatomic spacing with the distance from the surface:

$$\Delta a_n = \Delta a_{\max} \exp\left[-\frac{(m-n)a_\infty}{\kappa}\right] , \quad 0 < n \le m \,, \tag{16.4}$$

where, in accordance with Fig. 16.5, a_∞ is the unrelaxed interatomic spacing, $\Delta a_{\max}$ the maximum interatomic spacing shift and κ is a relaxation constant describing the rate (in terms of depth below the grain surface) at which the relaxation effect vanishes. The relaxation factors $\Delta a_{\max}$ and κ are supposed

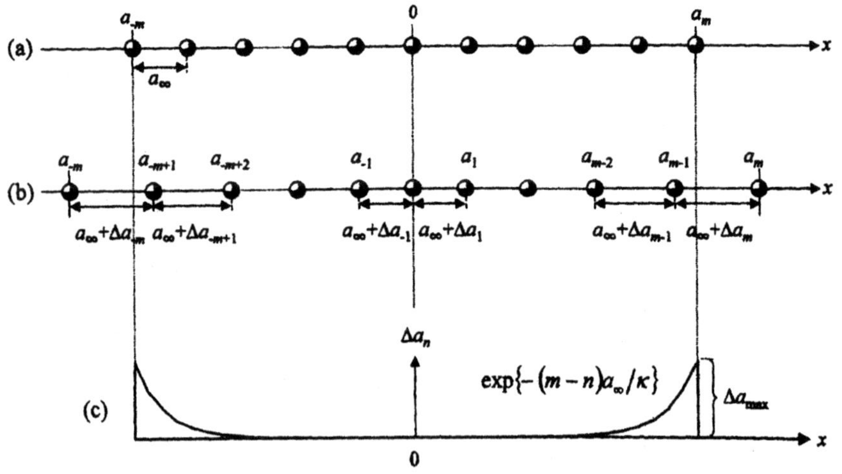

Fig. 16.5. Surface relaxation: the unrelaxed atomic positions (**a**) are changed if surface relaxation occurs as in (**b**). The displacement of each atom from its unrelaxed position follows an exponential law, shown in (**c**) (from [31]. Copyright (1998) by the University of Shizuoka. Reprinted with permission)

independent of the grain size, i.e. related only to the material. Even if defined for a single atomic row, (16.4) can be extended to lattice planes considering each atom as representative of the atomic plane it belongs to. Due to the symmetry of the problem with respect to the atom lying in the origin, an equation analogous to (16.4) can be written for the lattice planes lying in the range $-m \leq n < 0$; the magnitude of the relaxation is the same for a_k and a_{-k} atoms, but the sign is different, otherwise the net relaxation would be zero.

The choice of an exponential relaxation function [cf. (16.4)] is certainly arbitrary, but physically plausible [12] and gives a smooth trend with only two refinable parameters. However, there is no a priori criterion to prefer (16.4) over the parabolic model proposed by Nunes and Lin [30] [cf. (16.2) above].

The position of each atom within a given crystallite is obtained from the shift of the n-th atomic layer with respect to its unrelaxed state (16.4) as

$$x_n = na_\infty + \Delta a_{\max} \exp\left[-\frac{(m-1)a_\infty}{\kappa}\right] \frac{1-\exp(na_\infty/\kappa)}{1-\exp(a_\infty/\kappa)} . \qquad (16.5)$$

The diffracted intensity from a cube-shaped crystallite is then calculated by summing the phase differences of each couple of scatterers (for more details, see [31]).

Because of this calculation scheme, spherical crystallites are more difficult to model: a suitable clipping algorithm is necessary to properly select the atom lying inside the grain (clipping would introduce a certain degree of terracing of grain surface). This may explain the choice of the cubic grain shape, implicitly operated by these authors [31].

The effective crystallite size of the powder under investigation, used for comparing model and experiment, is evaluated from the measured diffraction pattern using the Scherrer formula

$$D_{\text{Scher}} = \frac{K\lambda}{\beta \cos\theta} , \qquad (16.6)$$

relating the (volume weighted apparent) crystallite size D_{Scher} to the integral breadth of the profile β and accounting for the cubic crystallite shape using a factor $K = 0.9$ (for the justifications of this choice see e.g. [39–41]).

Calculations are rather fast in the case proposed in [31] because a unique grain size is used. However, in the general case a distribution of grain diameters has to be considered and calculation speed can be a serious limit of this approach (the number of calculations scales with the cube of the grain size).

Besides neglecting possible effects of a grain size distribution, as already pointed out commenting on Nunes and Lin's model, Uemori et al. did not consider the presence of other sources of line profile broadening, nor any instrumental effect. However, as shown in Fig. 16.6, despite its simple nature the model describes with good accuracy the observed trend of the unit cell parameter versus the average Scherrer crystallite size.

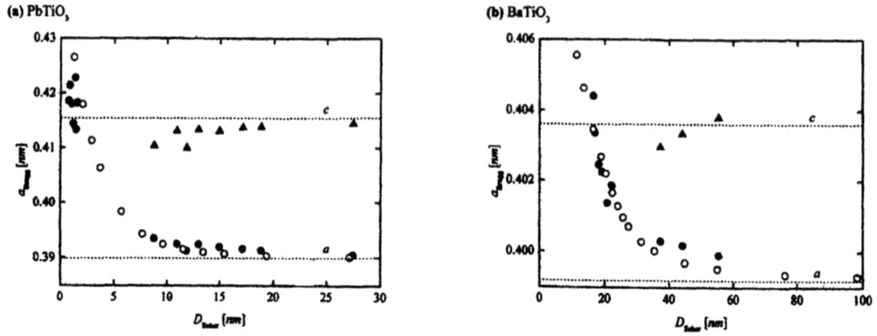

Fig. 16.6. Effective lattice parameter a_{Bragg} of (**a**) $PbTiO_3$ and (**b**) $BaTiO_3$ nanoparticles as a function of the effective grain diameter D_{Scher} along the a-axis (full dots) and c-axis (triangles). Open circles are obtained by the model proposed by Uemori et al. [31] whereas the dotted lines denote the bulk lattice constant (figure from [31]. Copyright (1998) by the University of Shizuoka. Reprinted with permission))

More interestingly, the authors show a clear correlation between the extent of the relaxed region (κ in the proposed model) and the extrapolation length, a measure of the local change of the electric polarisation at the surface of a ferro-electric grain. As already pointed out previously, this is not an isolated result and seems a common trend among ferroelectrics. It is also an indication that the proposed relaxation function is a good candidate for the description of an *effective* relaxation on the surface of (at least ferro-electric) nanometre particles.

16.3.2 Proposed Model of Effective Grain Surface Relaxation

A Fourier treatment of the surface relaxation problem is proposed here to allow for a direct integration with existing models of line broadening, from the oldest and most common Warren–Averbach [42,43] or Williamson–Hall [44] analyses to the more recent WPPF [5,6], WPPM [6,7] and even Rietveld [41–46] methods. The Fourier series coefficients are calculated for a single coherently-diffracting spherical crystallite subject to the static deformation field induced by the grain surface relaxation. The treatment follows the formalism of Warren [38], suitably extended to account for shape and size distribution of the diffracting crystallites (see for instance Scardi and Leoni [41]). The attention will be focused on cubic materials with spherical crystallites whose diameter is distributed according to a lognormal function, a case often met when dealing with nanocrystalline powders (see [7] and references therein). The extension to lower lattice symmetries and different crystallite shapes and/or crystallite size distributions is straightforward.

The traditional approach to the calculation of the Fourier coefficients for a generic crystallite along the scattering direction $\boldsymbol{g} = [hkl]$ follows the idea

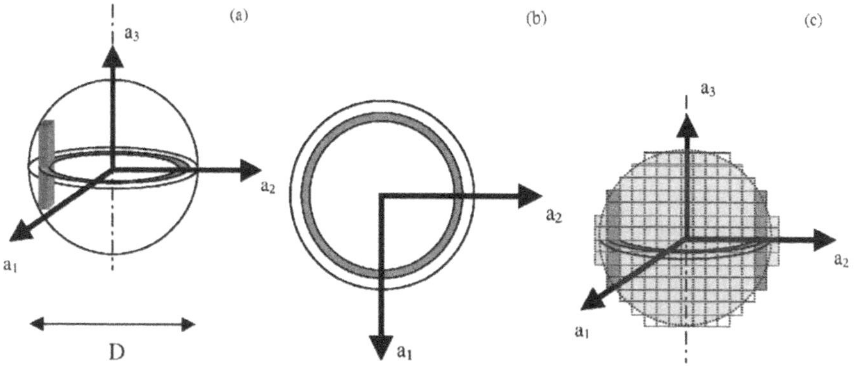

Fig. 16.7. Subdivision, along the scattering direction $[00L]$ in the orthorhombic reference frame $\{\boldsymbol{a}_1, \boldsymbol{a}_2, \boldsymbol{a}_3\}$, of a spherical crystallite of diameter D into columns of cubic blocks (**a**). Single column drawn on one of the annuli in which the cross section of the sphere is subdivided. The annulus is visible in the projection along $\boldsymbol{a}_3$ (i.e. on the plane $a_3 = 0$) (**b**). Due to the finite size of the blocks, the effective shape of the crystallite is not exactly spherical and some terracing on the surface occurs (**c**)

of Bertaut [49, 50], where the crystallite is subdivided into columns of cells[3] along $\boldsymbol{g}$. The expressions leading to the Fourier coefficients can be simplified by introducing an orthorhombic co-ordinate system, such that the columns are oriented along the $[00L]$ direction (the treatment loses no generality if this representation is used [38, 47, 48]). From this definition it follows that $[00L]$ is equivalent to $[hkl]$ in the corresponding original cubic notation.

In our particular case, the symmetry of the crystallite further simplifies the problem. As a spherical crystallite certainly possesses cylindrical symmetry, it can be subdivided into concentric cylinders about the orthorhombic $[00L]$ scattering direction (annular sections, see Fig. 16.7a and b), in such a way that all columns belonging to one annulus (ring) have the same height. Annuli of infinitesimal width can be chosen such that the effect of terracing (cf. Fig. 16.7c) is minimised. Mathematically, this allows a simplification in the calculations and permits an averaging over different terracing configurations, but it could be seen as an oversimplification of the problem since the phenomenon of terracing is known and recurrent in the field of surface science.

As columns of a given annular section have equal length, actual calculations can be done along a radius of the sphere and extended by symmetry to the whole crystallite. For a given radius r, the number of columns $n_{\mathrm{col}}(r)$ of a given length $2z_{\max}(r) = \sqrt{D^2 - 4r^2}$ can be calculated from the (known) diameter of the crystallite D as $n_{\mathrm{col}}(r) = 8z_{\max}(r)/D^2$.

[3] In the following, these cells will be called blocks to distinguish them from lattice unit cells.

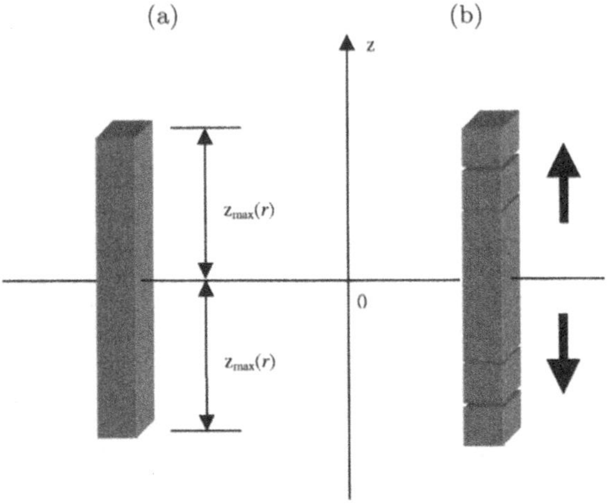

Fig. 16.8. A column of blocks in (**a**) unrelaxed (traditional) and (**b**) relaxed case. If atomic planes passing through the centre of the blocks are considered, then we observe a net increment of the interplanar spacing when going from the centre of the crystallite to its surface. The horizontal line indicates the position of the maximum diameter of the sphere

The problem of calculating the intensity diffracted from a relaxed crystallite is thus reduced to the calculation of the diffraction behaviour of the columns lying on its radius. Actually, only the centre of mass of each of the blocks composing the columns, i.e. the orthorhombic $(00L)$ diffraction plane passing through the centre of each block, is considered for calculation purposes.

For each column (see Fig. 16.8a) we introduce a relaxation function analogous to that proposed by Uemori et al. [cf. (16.4) above]. The choice is certainly arbitrary, but can be regarded as a first order approximation of the problem (for instance, oscillatory effects (cf. Sect. 16.2) are neglected). However, the treatment that follows is independent of the functional form of the relaxation function.

It is assumed that the orthorhombic $(00L)$ planes are shifted by a quantity Δz from their relaxed position according to

$$\Delta z(r,z) = \xi \frac{z}{|z|} e^{-\frac{z_{\max}(r)-|z|}{\kappa}} . \quad (16.7)$$

The variable z is the distance from the middle of the column (the origin of the z axis is on the $a_3 = 0$ plane) along $[00L]$. As in [31], the two refinable parameters ξ and κ represent, respectively, the maximum shift of the scattering planes from the unrelaxed position and the decay rate of the relaxation effect below the surface. Moreover, it is implicitly assumed that the grain surface relaxation effect is the same for all scattering directions.

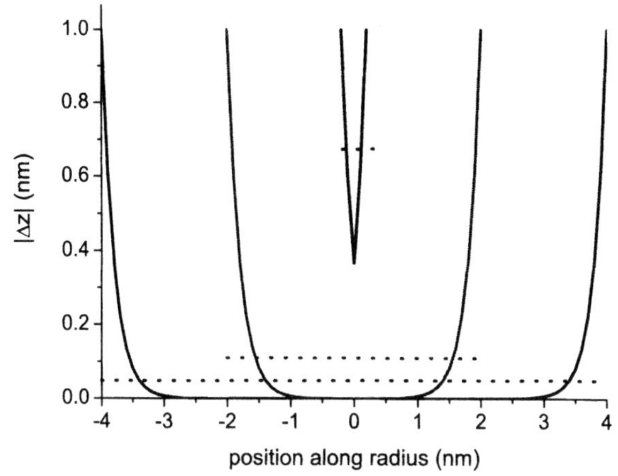

Fig. 16.9. Effect of surface relaxation (normalised absolute variation of the local unit cell parameter) and average increase of the unit cell parameter for different grain diameters (1 nn, 4 nm and 8 nm, respectively). The smaller the diameter, the higher the average unit cell parameter variation from the relaxed value (horizontal dotted lines are the average values of the respective curves)

To satisfy these conditions, the relaxation is scaled to be compatible with the chosen lattice. In our case, for each (hkl) plane, cubic lattice restrictions lead to

$$\Delta z(r, z(h,k,l)) = \frac{\xi}{\sqrt{h^2+k^2+l^2}} \frac{z}{|z|} e^{-\frac{z_{\max}(r)-|z|}{\kappa}} . \tag{16.8}$$

Accordingly, ξ is the maximum variation in the unit cell parameter (at the surface) allowed for any crystallite. A plot of (16.8) for different crystallite sizes (1, 4 and 8 nm, respectively) is shown in Fig. 16.9. Since the parameter ξ is supposed independent of crystallite diameter, smaller crystallites will exhibit a larger average unit cell parameter (the average shift of the unit cell parameter from its unrelaxed value increases with decreasing diameter as evident from the horizontal lines in Fig. 16.9).

Grain surface relaxation gives a monotonous change of interplanar distances (expansion or contraction), and as a consequence a peak shift is produced. To account for this effect, we consider two possible strategies, compatible with the Fourier approach:

(a) Integration of (16.8) over a single crystallite and then over the crystallite size distribution. Only one average peak profile is obtained for each spherical crystallite, shifted from the unrelaxed peak position (that can be calculated using Bragg's law from a_0, the unrelaxed unit cell parameter, measurable for instance on a polycrystalline specimen with sufficiently large crystallites). In this simple case, the average unit cell parameter $\bar{a}$ for a crystallite of diameter D can be obtained as

$$\bar{a} = a_0 + \Delta a = a_0 \int_0^{D/2} 2 \int_0^{z_{\max}(r)} \xi \frac{z}{|z|} e^{-\frac{z_{\max}(r)-|z|}{\kappa}} dz \frac{8 z_{\max}(r)}{D^2} dr \,, \tag{16.9}$$

and, analogously, the reciprocal relaxed lattice spacing is

$$d^*_{\{hkl\}} = d^*_{\{hkl\},\text{unrelaxed}} + \Delta d^*_{\{hkl\}} = \frac{\sqrt{h^2+k^2+l^2}}{\bar{a}}$$
$$\Rightarrow \Delta d^*_{\{hkl\}} = \left(\frac{a_0}{\bar{a}} - 1\right) d^*_{\{hkl\},\text{unrelaxed}} \,. \tag{16.10}$$

(b) Integration of (16.8) over columns of blocks (only the inner integral in (16.9); single columns are implicitly assumed to scatter independently, as in Bertaut's formulation [49, 50]). In this case we consider the interplanar distance shift and the corresponding line profile for each column (actually, each ring of equal length columns (Fig. 16.7)) in the crystallite. For each sphere we thus obtain a set of peak profiles, one for each ring (i.e. diameter) considered. The calculation should be carried out for each crystallite and for all crystallite diameters covered by the crystallite size distribution.

Equation (16.9) is common to the two approaches; what changes is the way the integration is carried out (analytically or numerically) and the level at which the peak position is assigned (crystallite or column, in (a) and (b), respectively).

In the following we will use approach (b): in this way diffraction peaks will show both the shift produced by the relaxation phenomenon and the asymmetry related to the uneven (and skewed) distribution of lattice distances along the scattering direction.

As shown in more detail in [51], the diffraction profile $I_{\{hkl\}}$ for the family of reflection $\{hkl\}$ in a cubic material can be calculated, in the reciprocal space, as weighted sum (w_{hkl} is the weight) of the various hkl family components $I_{hkl}(s_{hkl})$ (in the general cubic case, up to 12 independent ones) as

$$I_{\{hkl\}}\left(d^*, d^*_{\{hkl\}}\right) = k(d^*) \cdot \sum_{hkl} w_{hkl} I_{hkl}\left(s_{hkl}\right) \,, \tag{16.11}$$

where $s_{hkl} = d^* - (d^*_{\{hkl\}} + \delta_{hkl})$ is a reciprocal space coordinate ($s_{hkl} = 0$ on the peak centroid), δ_{hkl} is the displacement of the hkl component from the unrelaxed defect-free Bragg position $d^*_{\{hkl\}} = 2\sin\theta_{\{hkl\}}/\lambda$ and $k(d^*)$ accounts for all effects related to instrument, Lorentz, polarization and structure factors.

The profile of each component (possibly different, depending on the particular source of broadening involved) is in turn given by

$$I_{hkl}(s_{hkl}) = \int_{-\infty}^{\infty} \mathbb{C}_{hkl}(L) e^{2\pi i L \cdot s_{hkl}} dL$$

$$= 2 \int_{0}^{\infty} (A_{hkl}(L) \cos 2\pi L s_{hkl} - B_{hkl}(L) \sin 2\pi L s_{hkl}) dL \quad (16.12)$$

where the function $\mathbb{C}_{hkl}(L) = A_{hkl}(L) + iB_{hkl}(L)$ is the complex Fourier transform of the profile. Equations (16.11) and (16.12) are the basis of the WPPF and WPPM algorithms and, in general, of the Fourier methods of Line Profile Analysis (LPA).[4] Suitable expressions for $A_{hkl}(L)$ and $B_{hkl}(L)$ can be found in the literature [5–7, 38, 51] for the most common sources of broadening (mainly size effects, line and plane defects).

If all broadening sources are considered as independent, the transform $\mathbb{C}_{hkl}(L)$ can be written as product of the transforms associated to each broadening source [38, 51].

Following the treatment proposed by Warren, it is possible to calculate the Fourier transform component $\mathbb{C}_{hkl}^{SR}(L)$ introduced by the presence of surface relaxation. To this end, however, we must step back and start from the Fourier series treatment [leading also to (16.12)] in order to find a convenient method for the calculation of the transform.

If we suppose only size and surface relaxation effects to be present (superscript SR), the diffracted intensity for each profile component can be calculated as

$$I_{hkl}^{SR}(d^*) \equiv I_{00L}^{SR}(d^*) \propto \sum_{n=-\infty}^{\infty} N_n(\text{sample}) \langle e^{2\pi i L Z_n} \rangle e^{2\pi i n h_3}, \quad (16.13)$$

where [38]:

1. N_n(sample) is the number of blocks in the sample having an n-th neighbour in the same column;
2. Z_n is the phase difference between two of these blocks;
3. $\langle e^{2\pi i L Z_n} \rangle$ an average over such pairs of n-th neighbours in the same column for the whole sample.

The variable h_3 is the coordinate referring to the peak centroid: a set of equivalent (but useful) definitions is proposed:

$$h_3 = \frac{2\sin\theta}{\lambda |\boldsymbol{b}_3|} = \frac{2\sin\theta}{\lambda} \frac{L\lambda}{2\sin\theta_{\{hkl\}}} = \frac{L\sin\theta}{\sin\theta_{\{hkl\}}} = \frac{Ld^*}{d^*_{\{hkl\}}}, \quad (16.14)$$

with the usual meaning for d^*, θ, $d^*_{\{hkl\}}$ and $\theta_{\{hkl\}}$ (the latter equal to d^*_{00L} and θ_{00L}, respectively, in orthorhombic notation). The vector $\boldsymbol{b}_3$ defines the

[4] All modern techniques of line profile analysis try to derive analytical expressions for $A_{hkl}(L)$ and $B_{hkl}(L)$, based on the nature of the broadening source being considered.

position in the reciprocal space of the $00L$ point. With the choices operated in (16.14), the peak will be centred at $h_3 = L$, where L is the diffraction order in the orthorhombic notation.

The summation in (16.13) should be limited to the planes composing the sample, but it has been extended to infinity, N_n being conventionally nil outside the grain boundaries [38].

For symmetry reasons (a purely complex function is odd), if we expand the exponential factors in (16.13) into their real and imaginary parts,

$$I_{hkl}^{SR}(d^*) \propto \sum_{n=-\infty}^{\infty} \frac{N_n}{N_3} \times \left\{ \begin{array}{c} \langle \cos 2\pi L Z_n \rangle \cos 2\pi n h_3 - \langle \sin 2\pi L Z_n \rangle \sin 2\pi n h_3 \\ +i\left(\langle \cos 2\pi L Z_n \rangle \sin 2\pi n h_3 + \langle \sin 2\pi L Z_n \rangle \cos 2\pi n h_3\right) \end{array} \right\} \tag{16.15}$$

and we define

$$A_n^{SR} = \frac{N_n}{N_3} \langle \cos 2\pi L Z_n \rangle \,, \quad B_n^{SR} = \frac{N_n}{N_3} \langle \sin 2\pi L Z_n \rangle \tag{16.16}$$

where N_3 is the average number of cells per column, we can reduce (16.13) to

$$I_{hkl}^{SR}(d^*) \propto \sum_{n=-\infty}^{\infty} \left\{ A_n^{SR} \cos 2\pi n h_3 + B_n^{SR} \sin 2\pi n h_3 \right\} \,. \tag{16.17}$$

The term N_n/N_3 in (16.16) depends only on size parameters and it is usually defined as (real) size Fourier term A_n^s; being n an integer, all Fourier coefficients are defined for integer values of n.

Equation (16.17) represents a periodic series of peak profiles; to be of practical use (i.e. to provide a single profile) it must be rewritten in terms of Fourier transform:

$$I_{hkl}^{SR}(d^*) \propto \int_0^{\infty} \left[A_{hkl}^{SR}(L) \cos 2\pi L s_{hkl} + B_{hkl}^{SR}(L) \sin 2\pi L s_{hkl} \right] dL \,. \tag{16.18}$$

For mathematical convenience [and uniformity with (16.12)], the peak defined by (16.18) is centred in $s_{hkl} = 0$; the diffraction order is therefore not indicated. This needs to be considered when comparing (16.17) with (16.18). Moreover, being the relaxation function (hence the entire surface relaxation treatment for cubic materials) considered independent of the family component, $\delta_{hkl} = 0$ and thus $s_{hkl} = d^* - d^*_{\{hkl\}}$ (therefore, $X_{hkl}^{SR} = X_{\{hkl\}}^{SR}$, where X stands for I, A, B).

The real and imaginary parts of the Fourier transform in (16.18) ($A_{\{hkl\}}^{SR}$ and $B_{\{hkl\}}^{SR}$) are equivalent to the coefficients introduced by (16.16) when n is considered as a continuous variable (and $L = n/d^*_{\{hkl\}} = nd_{\{hkl\}}$). This, in turn, means that from the Fourier transform (calculated from the continuous displacement fields relative to the studied case), the Fourier coefficients can be easily obtained.

We now recall that the Fourier transform $\mathbb{C}(L)$ is the product of the transforms associated to the various defects (consider for instance (16.16) presenting the product of the size and distortion coefficients); expressions are given in [51] for most broadening sources, and in the following for combined size and relaxation effects only.

Omitting for clarity the explicit dependence on L of each coefficient, we can write

$$\begin{aligned} A_{hkl}(L) &= T^{IP} A^F_{hkl} \\ &\quad \times \left[\left(A^D_{\{hkl\}} - B^D_{\{hkl\}} B^F_{hkl} \right) A^{SR}_{\{hkl\}} - \left(B^D_{\{hkl\}} + A^D_{\{hkl\}} B^F_{hkl} \right) B^{SR}_{\{hkl\}} \right] \\ B_{hkl}(L) &= T^{IP} A^F_{hkl} \\ &\quad \times \left[\left(B^D_{\{hkl\}} + A^D_{\{hkl\}} B^F_{hkl} \right) A^{SR}_{\{hkl\}} + \left(A^D_{\{hkl\}} - B^D_{\{hkl\}} B^F_{hkl} \right) B^{SR}_{\{hkl\}} \right] \end{aligned} \tag{16.19}$$

where T^{IP} is the (real part of the) Fourier transform of the instrumental profile, obtained by suitable modelling of the peaks of a standard powder (like the SRM 660a [52]). For a well aligned high-resolution diffractometer the instrumental contribution is symmetrical across a large fraction of the accessible 2θ range, thus the imaginary part of T^{IP} is zero or negligible.

The coefficients with superscript D in (16.20) refer to dislocations-induced broadening and those having a superscript F cope with mistakes in the stacking sequence of lattice planes (stacking and twin faults). If the sine distortion term $B^D_{\{hkl\}}$ is neglected (as is often the case [51]), the expressions are further simplified:

$$\begin{aligned} A_{hkl}(L) &= T^{IP} A^F_{hkl} A^D_{\{hkl\}} \left[A^{SR}_{\{hkl\}} - B^F_{hkl} B^{SR}_{\{hkl\}} \right] \\ B_{hkl}(L) &= T^{IP} A^F_{hkl} A^D_{\{hkl\}} \left[B^F_{hkl} A^{SR}_{\{hkl\}} + B^{SR}_{\{hkl\}} \right] . \end{aligned} \tag{16.20}$$

These expressions can be employed in place of the original ones both in the WPPF/WPPM algorithms [7,51] and in the traditional (simplified) methods of line profile analysis.

The main task of the surface relaxation algorithm is now the numerical calculation of the Fourier transform (i.e. $A^{SR}_{\{hkl\}}(L)$ and $B^{SR}_{\{hkl\}}(L)$) for the distributed crystallites, considering that the position of the various cells/blocks in each crystallite is not simply a multiple of the (unrelaxed) lattice spacing, but has to be corrected according to the chosen relaxation model.

We can easily calculate the Fourier coefficient [according to (16.16)] from the phase difference between couples of blocks along the columns composing each crystallite. For a crystallite of a given diameter D (such that the maximum dimension for a column lying on the circle of radius r is $z_{\max}(r)$) we

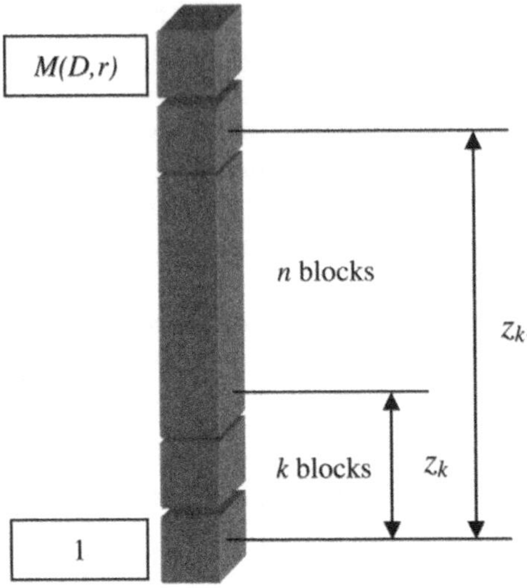

Fig. 16.10. A column of blocks used in evaluating the Fourier coefficients. The total number of blocks in the column ($M(D,r)$) and the coordinates of two blocks (z_{k+n}, z_k) are indicated

can introduce the partial coefficients:

$$\tilde{A}_n^{SR}(D,r) = \sum_{k=-z_{\max}/d_{\{hkl\}}}^{z_{\max}/d_{\{hkl\}}} \cos\left(\frac{2\pi(z_{k+n}-z_k)}{d_{\{hkl\}}}\right),$$

$$\tilde{B}_n^{SR}(D,r) = \sum_{k=-z_{\max}/d_{\{hkl\}}}^{z_{\max}/d_{\{hkl\}}} \sin\left(\frac{2\pi(z_{k+n}-z_k)}{d_{\{hkl\}}}\right), \qquad (16.21)$$

where the number of blocks in the column considered is $M(D,r) = 2z_{\max}(r)/d_{\{hkl\}}$ (for a sphere where the centre coincides with the corners of blocks; $M(D,r) = 2z_{\max}(r)/d_{\{hkl\}} + 1$ if the centre of a block coincides with the centre of the sphere). The variables z_{k+n} and z_k are the position of two blocks separated by n blocks along the same column (see Fig. 16.10) and can be calculated as $z_x = xd_{\{hkl\}} + \Delta z(r, z(h,k,l))$. It is worth noting that for pure size broadening the distance between blocks in the same column is an integer multiple of the lattice spacing, i.e. $z_{k+n} - z_k = nd_{\{hkl\}} = L$ [38]. Again, we should stress the fact that the result of (16.21) is a set of cosine and sine Fourier coefficients for a single column inside each single crystallite of the distribution.

If integration is substituted for the sum in (16.21), the transforms for a single column $\tilde{A}_{\{hkl\}}^{SR}(L,D,r)$ and $\tilde{B}_{\{hkl\}}^{SR}(L,D,r)$ are obtained; inserted into (16.20), they lead to $A_{\{hkl\}}(L,D,r)$ and $B_{\{hkl\}}(L,D,r)$, accounting for all considered sources of broadening.

The calculation for the whole system must be carried out by summing the scattered intensity for all columns in the crystallite and then for all crystallites in the distribution $g_i(D)$. As said above, since the Fourier coefficients describe only the profile shape (i.e. how the profile decays from its maximum), the peak position of each reflection is an input parameter for the WPPM algorithm, thus the evaluation of the integral over the crystallite size distribution can become quite difficult. The formulation will thus change according to the route followed in determining the peak position.

If the averaging of the relaxation function over the column distribution of a grain is executed, then a unique peak position determined by $\bar{a}$ [cf. (16.10)] will be available for each crystallite. In this case, the profile function becomes

$$I(d^*) \propto \int_0^\infty \left\{ \int_0^{D/2} \int_0^\infty \left[\begin{array}{l} A_{\{hkl\}}(L,D,r) \cos 2\pi L \left(s_{hkl} + \Delta d^*_{\{hkl\}} \right) \\ + B_{\{hkl\}}(L,D,r) \sin 2\pi L \left(s_{hkl} + \Delta d^*_{\{hkl\}} \right) \end{array} \right] \times \frac{8 z_{\max}(r)}{D^2} dL dr \right\} g_i(D) D^3 dD \,. \tag{16.22}$$

If the peak position associated to a single column is considered, then the final profile will be the sum of the profiles associated to all columns of all crystallites in the size distribution, each profile having a (possibly) different position (the difference will be negligible or small depending on the value of the relaxation parameters). The scattered power associated to each column will be

$$I(d^*) \propto \int_0^\infty \left\{ \int_0^{D/2} \int_0^\infty \left[\begin{array}{l} A_{\{hkl\}}(L,D,r) \cos 2\pi L \left(s_{hkl} + \Delta d^*_{\{hkl\}}(D,r) \right) \\ + B_{\{hkl\}}(L,D,r) \sin 2\pi L \left(s_{hkl} + \Delta d^*_{\{hkl\}}(D,r) \right) \end{array} \right] \times \frac{8 z_{\max}(r)}{D^2} dL dr \right\} g_i(D) D^3 dD \,. \tag{16.23}$$

where the now explicit dependence of the Fourier coefficients on D and r is indicated.

By using the Fourier transforms proposed here, the WPPM algorithm becomes of mixed nature: some terms are calculated very fast (all those contained in the original WPPM formulation [7,51]), whereas the effect of surface relaxation, given the highly non-linear nature of (16.21), is quite slow because of the triple summation involved. The data throughput is ca. 1 profile calculation/h on a 1.5 GHz Pentium-IV machine; highly optimised numerical integration schemes (e.g. those based on multiple range quadrature-interpolation) must be used to bring the calculation speed up to about 30 profiles/min.[5]

16.4 Experimental Validation: Nanocrystalline Cerium Oxide Powders

To illustrate the application of the surface relaxation algorithm we have chosen cerium oxide (ceria). Due to its high ionic conductivity, ceria is one of the candidates as oxygen ion sources in modern Solid Electrolyte Fuel Cells (SOFC) [53, 54]. It is also widely used in catalytic converters for automotive applications and as refractory material or as stabiliser for zirconia ceramics.

Sol-gel is a common route for the production of high-purity ceramics starting from metallorganic precursors. The process passes through an amorphous stage and the crystalline product is obtained after a suitable thermal treatment (calcination) aimed also at the combustion of the organic constituent. An evolution in the crystallite size can be clearly observed during the calcination process and sharpening diffraction peaks emerge from the starting amorphous pattern. Powders produced by the sol-gel route are often nanocrystalline and the size of the grains is frequently distributed according to a lognormal curve. Recent publications [5–7, 55, 56] have shown new routes for the analysis of the grain size distribution in nanocrystalline materials. Moreover, [55] and [56] confirmed (even if indirectly) the lognormal size distribution to be appropriate for the description these systems.

The number of publications on this subject is increasing. Among them, the work of Audebrand et al. [57] presents some features useful for the ongoing discussion. For a (hydrated) cerium oxide powder obtained from precipitation of cerium sulphate tetrahydrate ($Ce(SO_4)_2 \cdot 4H_2O$) in a solution of ammonia, the authors observed a unit cell parameter sensibly different from that considered as "reference" [58], $a = 0.541134\,\mathrm{nm}$. In particular, for crystallites of average size 1.9 nm, Audebrand et al. [57] found a unit cell parameter $a = 0.5428(7)\,\mathrm{nm}$ and compared it with the value of a CeO_2 standard (NIST, [59]) observing that the two agree within 3 esd's.

As a first observation, the authors deal with hydrated ceria so some discrepancy with respect to pure cerium oxide should be expected. However, water is not expected to be a substitutional or interstitial species for cerium oxide.

Oxygen defects (anionic vacancies) have been observed as the consequence of the possible tri- and tetra- valence of cerium: in an oxygen-defective material, the anions can form colour centres leading to a blackening of the material [58, 60]. An exhaustive study on the nonstoichiometry effects in cerium

[5] Note added in proofs: with the recent introduction (December 2002) of a Fast Fourier Transform (FFT) algorithm, calculation time is now comparable to WPPF procedures.

oxide has been proposed by Kim [61] and later completed by Mogensen et al. [58] and possible variations of the unit cell parameter discussed. The same authors claim that, because of the high mobility of Ce^{3+}, cerium oxide is often a solid solution of Ce_2O_3 in CeO_2. For small concentrations, Vegard's rule for mixtures can be assumed, thus the unit cell parameter of CeO_2 should be linearly dependent on the quantity of dopant (i.e. Ce_2O_3 present): the slope of this line is often called Vegard's slope. This is in fact valid and Kim found an empirical relationship between the unit cell parameter of ceria and the quantities of possible solutes [61]:[6]

$$a = 0.5413 + \sum_k (0.0220 \Delta r_K + 0.00015 \Delta z_k) m_k \tag{16.24}$$

where a (in nm) is the unit cell parameter of the ceria solid solution at room temperature, Δr_k (in nm) is the difference in ionic radius ($r_k - r_{\mathrm{Ce}}$) of the k-th dopant and the Ce^{4+}, Δz_k is the difference in the valences ($z_k - 4$), and m_k is the mole percent of the k-th dopant in the form of MO_x. This formula can possibly give misleading results since the unit cell parameter for pure ceria at room temperature $a = 0.5413$ nm is slightly larger than the widely accepted 0.541134 nm (see above).

From (16.24), considering $CeO_{1.5}$ as dopant for CeO_2, we can obtain the following relationship for the unit cell parameter:

$$a = 0.5413\,\mathrm{nm} + m \times 2.306 \times 10^{-4}\,\mathrm{nm\,mol\%^{-1}\,CeO_{1.5}}\,. \tag{16.25}$$

As m (in %) $= 200x$ (x in CeO_{2-x}), we get for homogeneous samples at room temperature:

$$a = 0.5413\,\mathrm{nm} + x \times 4.612 \times 10^{-2}\,\mathrm{nm}\,. \tag{16.26}$$

An analogous approach has been followed by Hong and Virkar [62] for fluorite-type lanthanide oxides, leading to a formula analogous to (16.24), where m is the mole content of the dopant:

$$a = 0.54112\,\mathrm{nm} + 0.023027(\mathrm{mol\%\,LnO_{1.5}})^{-1}(r_k - 0.1024\,\mathrm{nm})m\,. \tag{16.27}$$

Inserting the Ce^{3+} radius in the latter, we get

$$a = 0.54112\,\mathrm{nm} + m \times 2.740 \times 10^{-4}\,\mathrm{nm\,mol\%^{-1}\,CeO_{1.5}}\,. \tag{16.28}$$

Transferred to [57], Kim's relation (16.26) would provide at room temperature a fraction of vacancies $x = 0.034$, i.e. the composition of the powder should be $CeO_{1.966}$. Conversely, if the Hong and Virkar's [60] formula is used [cf. (16.28)], where the undoped unit cell parameter is closer to the widely accepted one, at room temperature, we obtain:

[6] This is the original formula of Kim [61]. The same formula, as reported in [58], is wrong.

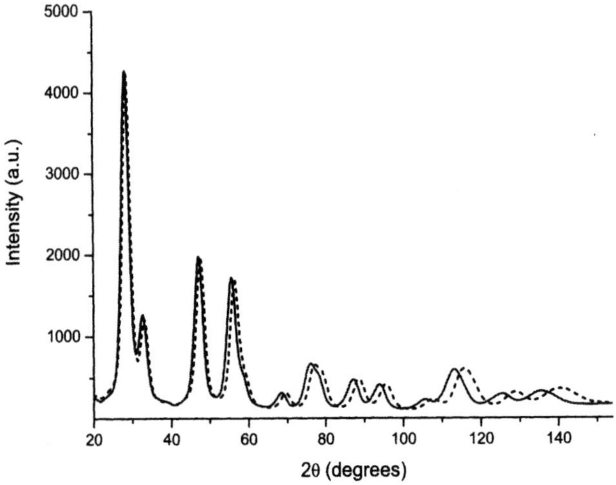

Fig. 16.11. Calculated patterns for ceria powder according to the traditional WPPM algorithm (i.e., no grain surface relaxation) (dashed curve) and according to the GSR-WPPM algorithm (i.e., with grain-surface relaxation) (full curve)

$$a = 0.54112\,\mathrm{nm} + x \times 5.580 \times 10^{-2}\,\mathrm{nm}\,. \tag{16.29}$$

According to this formula, the fraction of vacancies in the samples of Audebrand et al. would be slightly lower, $x = 0.030$ (and the composition would become $CeO_{1.970}$).

The presence of vacancies should not be regarded as the only possible explanation for the phenomenon: water adsorbed on the surface of the powder particles, hypothesised in [57], can locally change the atomic ordering and hence the interplanar spacing leading to a surface relaxation. This relaxation introduces into the system a fictitious strain, often termed *stress-free strain* since deformation is not accompanied by any residual stress.

In the following, calculated patterns and experimental data will be presented on sol-gel ceria powders. It will be shown how surface relaxation could possibly explain the phenomenon observed by Audebrand et al. accounting, within a self-consistent model, for the variation in the average unit cell parameter in powders calcinated at increasing temperature.

Calculation of XRD Powder Patterns by the WPPM Algorithm

Patterns calculated according to the traditional WPPM algorithm and the WPPM algorithm with grain-surface relaxation (GSR-WPPM) (introducing relaxation parameters in the latter, the other parameters being the same in both cases) are shown in Fig. 16.11. The integrated intensity for each reflection and the average "unrelaxed" unit cell parameter for cubic ceria ($a = 0.541134(12)$ nm) have been taken from the ICDD PC-PDF card No. 34-0394.

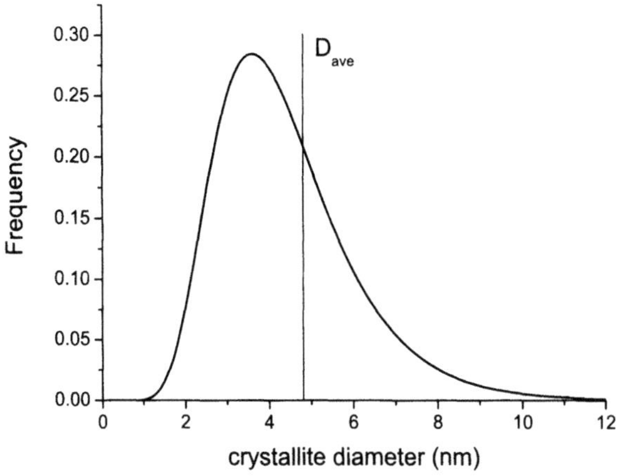

Fig. 16.12. Crystallite size distribution used in the calculated ceria powder diffraction pattern shown in Fig. 16.11

Calculations have been carried out for a lognormal distribution of spherical crystallites (3 nm average diameter, 0.3 lognormal variance, see Fig. 16.12) adding $\rho = 10^{16}\,\mathrm{m}^{-2}$ of dislocations (outer cut-off radius $R_e = 3\,\mathrm{nm}$), 1% of stacking faults and 2% of twin faults, and considering surface relaxation with parameters $\xi = 0.05\,\mathrm{nm}$ and $\kappa = 0.3\,\mathrm{nm}$.

The corresponding trend for the unit cell parameter is similar to that shown in Fig. 16.9. In this case, the surface zone influenced by lattice relaxation is roughly one third of the average crystallite diameter and the maximum variation in the distance of the influenced planes is quite limited (maximum stress-free strain of 0.055): a large effect is observable on the diffraction pattern (Fig. 16.11). This is particularly evident at high angle where small differences in the unit cell parameter are certainly amplified in reverting from reciprocal to angular space. Numerical values for the two parameters ξ and κ are well above the possible figures that can be found in practice: they have been selected in order to enhance the relaxation effect and make it predominant over the other sources of peak broadening.

Surface relaxation can thus mimic the effect of a variation in the unit cell parameter, like that observed in the cited paper by Audebrand et al. Additionally, peak shape changes, but since profiles tend to be quite broad, small variations in the peak shape cannot be easily appreciated.

The broadening introduced by surface relaxation is limited and mainly related to the shape of the crystallite diameter distribution function. The variation in the diffraction pattern induced by a variation in the mean crystallite diameter is shown in Fig. 16.13a.

The simulation is done for a dislocation density $\rho = 1.2 \times 10^{16}\,\mathrm{m}^{-2}$, in absence of planar faulting and considering a surface relaxation with $\xi = 0.04\,\mathrm{nm}$ and $\kappa = 0.4\,\mathrm{nm}$. The variance of the crystallite size distribution is kept constant, whereas the lognormal mean is changed; data are presented

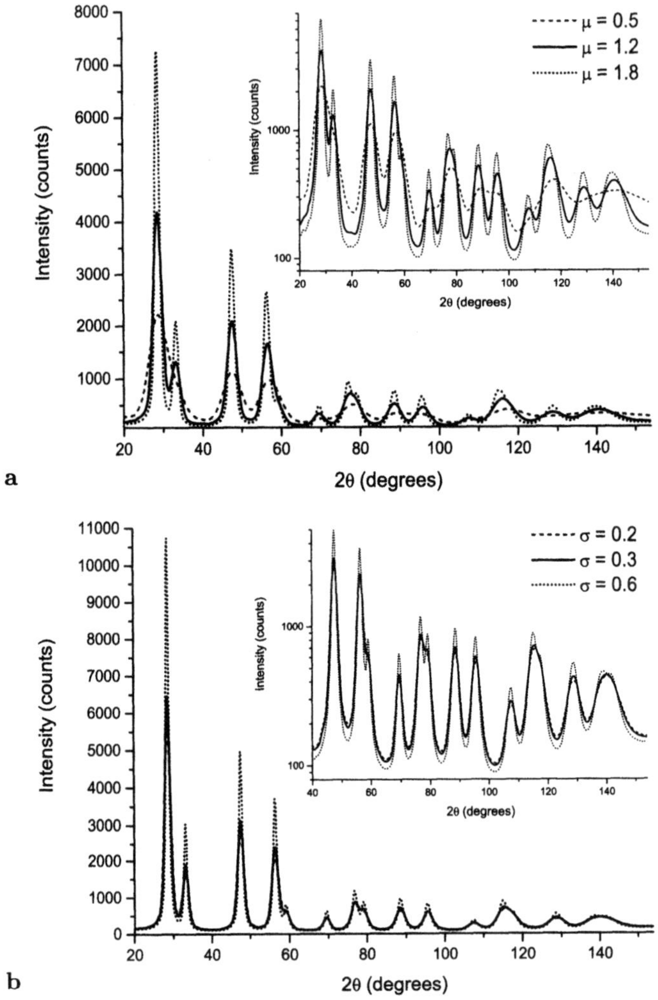

Fig. 16.13. (**a**) Powder patterns calculated with the GSR-WPPM algorithm for ceria powders with a lognormal crystallite size distribution at increasing lognormal mean (μ) and fixed variance ($\sigma = 0.2$). Clearly the relaxation effect is less evident in a coarse-grained powder sample (higher μ). Details can be appreciated in the log-scale inset. (**b**) As in (**a**) but at fixed lognormal mean ($\mu = 1.5$) and increasing variance (σ). The relaxation effect decreases by increasing the variance. Details can be appreciated in the log-scale inset (portion of the pattern)

also in log scale in order to make visible the small differences in the tails. Reducing the average crystallite size results in quite broad profiles and in an increase of the relaxation effects observable as shift of the high angle reflections (quite evident above ca. 100°). We should remember that in all patterns the same relaxation parameters are considered, i.e. the relaxation

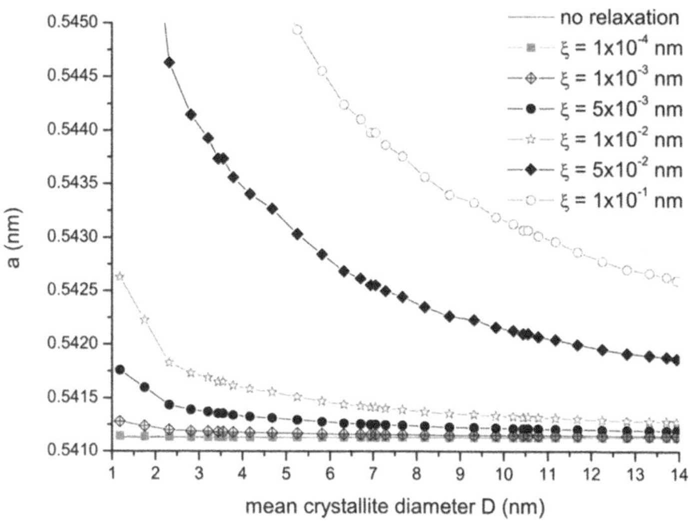

Fig. 16.14. Average unit cell parameter as a function of the mean crystallite diameter within a lognormal distribution of grains (mean diameter 4.4 nm, corresponding to $\mu = 1.5$, and $\sigma = 0.2$). The various plots correspond to an increasing surface relaxation

effect faints in bigger crystallites (because of the lower surface/volume ratio) as expected.

Less marked is the effect of the lognormal variance on a relaxed powder pattern simulation. Figure 16.13b shows the patterns for a powder having a mean crystallite size of 4.4 nm (lognormal mean $\mu = 1.5$) and all other parameters set as for the previous case. As observed for the average crystallite size, increasing the breadth of the distribution hides the effects of a surface relaxation because a larger fraction of "less relaxed" (large) crystallites is taken into consideration.

These hiding effects can be often met in common practice and they render the evaluation of the unrelaxed unit cell parameter quite difficult. This can also be a serious limitation in the production of a powder reference or in unit cell parameters determination from powder data. Particular care should be taken in the choice of a suitable grain size to avoid or, whenever this is not possible, to limit surface relaxation effects; in practice this would mean dealing with coarse grained powders. On the other hand, the presence of very big grains can be a serious limitation. When only a small number of grains is irradiated, orientation and grain size distribution statistics can be quite poor, and the diffraction patterns may even depend on the way the specimen is mounted.

By simulation we can determine the average unit cell parameter for a chosen crystallite diameter within a given size distribution. An example is presented in Fig. 16.14 for a lognormal crystallite size distribution (mean crystallite diameter 4.4 nm, lognormal variance $\sigma = 0.2$) and increasing relaxation (keeping the width of the influenced region, constant to $\kappa = 0.1$ nm).

As expected, the higher the relaxation, the larger the average unit cell parameter, in particular for the smaller crystallites, where the surface/volume ratio enhances the surface relaxation effect.

Results from Traditional WPPF and WPPM Analyses

Various samples of cerium oxide have been synthesised through a sol-gel route and calcinated for 1 h at increasing temperatures [63].

To clarify the significance of incorporation of grain-surface relaxation into the WPPM algorithm (GSR-WPPM), all specimens have been preliminarily characterised by WPPF and WPPM in their traditional form [7, 51], i.e., without incorporation of surface relaxation effects. Care has been taken in the choice of the starting parameters to limit the possibility for the nonlinear least squares Levenberg–Marquardt algorithm, on which WPPF and WPPM rely, to fall in local minima or to diverge. The refined parameters obtained after starting from different points (i.e. from a different set of trial parameters) have been checked for consistency and have been found to agree within the estimated standard deviations (esd's).

The data presented below have been chosen as the most representative to show the crystallisation process. In particular, samples calcinated at 300°C, 400°C, 500°C and 600°C [63] are compared in order to show a possible trend in the refined parameters and to explain the microstructural evolution of the cerium oxide. The results obtained for the sample calcinated at 400°C will be analysed in detail and the meaning of the various parameters assessed by comparison with data obtained by TEM.

Traditional techniques for the analysis of nanocrystalline powders, such as TEM, could give a good insight in the microstructure but, being quite localised (few micrometers) are not among the best candidates for a quantitative analysis of the average behaviour of the material. In this case diffraction could be complementary, but the observations must match on both scales.

CeO_2 Calcinated 1 Hour at 400°C

The result of the WPPF and WPPM analysis for the sample calcinated at 400°C is presented in Table 16.1 and the refined pattern in Fig. 16.15. Only some key parameters are listed in the table; for brevity and clearness, background, instrumental and emission profile data have been omitted.

The instrumental function has been measured using the NIST SRM 660a standard (lanthanum hexaboride) and parameterised according to the procedure proposed in [5–7] and illustrated in detail in [51].

The intensity of each reflection is also not shown since it is irrelevant for the goal of our analysis. The total number of microstructural parameters necessary for the description of the profile shape for the whole ceria pattern

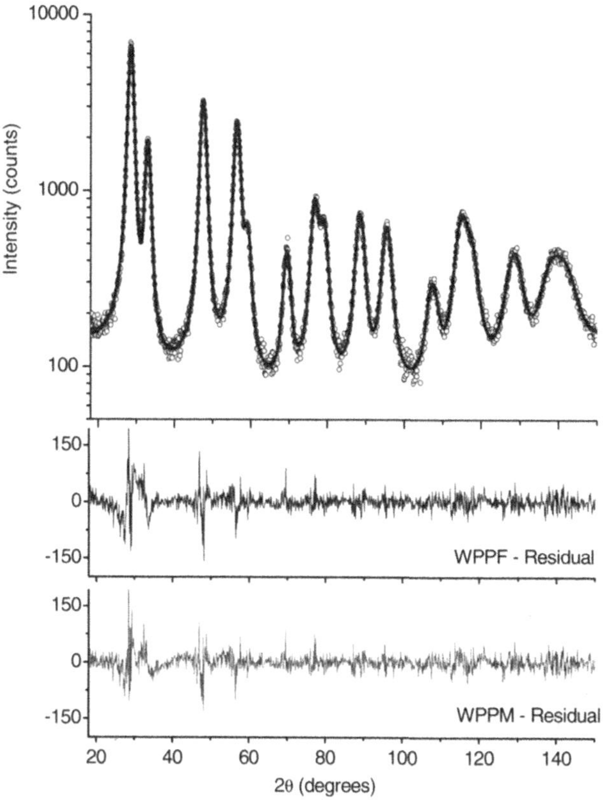

Fig. 16.15. Measured (*dots*) and modelled pattern according to the WPPF (continuous line) and WPPM (*dotted line*) algorithms. The residual (difference between measured and modelled data) is shown below, in linear scale

(16 reflections accessible with copper radiation) is 6. The other parameters shown in Table 16.1 are kept fixed or calculated from those shown in bold. A grand total of 27 parameters (5 for the background, 6 for the microstructure and 16 peak intensities) has been simultaneously refined, sufficient for the complete description of each of the analysed cerium oxide patterns.

The results of WPPF and WPPM are in good quantitative agreement as indicated by very close values of the R_{wp} and R_{exp} statistic indices [5–7, 46]. The Goodness of Fit (GoF) parameter obtained from WPPF is slightly better than that observed for WPPM. This should not be surprising since WPPM puts more restraints to the shape of the profiles.

The size distributions obtained from WPPF and WPPM, as visible in Fig. 16.16, are very close each other: the curves are slightly asymmetric and the mean crystallite size is roughly 4.5 nm.

To validate the results of WPPF and WPPM, a crystallite size distribution has been built from a set of TEM micrographs comprising a large number

Table 16.1. Results of the WPPF, WPPM and GSR-WPPM analyses on the ceria powder calcinated in air at 400°C for 1 h. Parameters in bold are those actually refined. The number in parentheses after each value is the estimated standard deviation referred to the least significant figure. In the GSR-WPPM algorithm, the cell parameter (indicated by a *) was *not* refined

		WPPF	WPPM	GSR-WPPM
Instrument				
sample displacement	(µm)	9.2(6)	9.1(6)	8.2(7)
Cell parameter				
cell parameter	(nm)	0.54150(3)	0.54153(3)	0.541134*
Size distribution (spherical grains)				
lognormal μ		1.49(2)	1.41(2)	1.42(1)
lognormal σ		0.322(7)	0.355(6)	0.364(6)
average diameter	(nm)	4.68(1)	4.37(6)	4.40(6)
Dislocations				
dislocation density	(m^{-2})	$2.0(10) \cdot 10^{16}$	$1.4(10) \cdot 10^{16}$	$1.08(4) \cdot 10^{16}$
edge/screw dislocation ratio		1 : 1	1 : 1	1 : 1
cutoff radius R_e	(nm)	2(1)	2(1)	3(1)
A (from el. constants)		0.1187	0.1187	0.1187
B (from el. constants)		0.1618	0.1618	0.1618
Wilkens parameter M		0.29(1)	0.25(8)	0.31(3)
Grain surface relaxation				
relaxation factor ξ	(nm)	–	–	0.008(3)
decay constant κ	(nm)	–	–	0.16(4)
Statistical estimators				
R_{wp}		5.29	5.51	5.58
R_{exp}		4.67	4.67	4.67
GoF		1.13	1.18	1.20

of clearly-distinct grains (ca. 800). One of these micrographs is presented in Fig. 16.17 and shows a cluster of cerium oxide grains. A 300 keV Jeol 3010 microscope, equipped by a slow-scan Gatan CCD camera (mod. 794), was used for the analysis.

Remarkably good TEM specimens have been easily prepared by dispersion of some powder on a graphite frame. Grains are spherical and well separated, limiting the possible doubts on the reliability of the distribution obtained by the TEM analysis and validating the chosen model based on the assumption of spherical grains.

The TEM distribution is presented in Fig 16.16 superimposed to the results of the WPPF and WPPM analyses. The agreement is remarkably good, especially considering the difference between the techniques employed. A

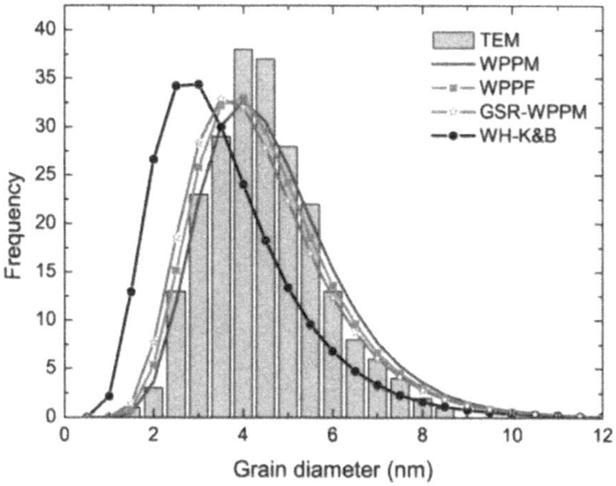

Fig. 16.16. Grain size distribution for the cerium oxide powder calcinated for 1 h in air at 400°C

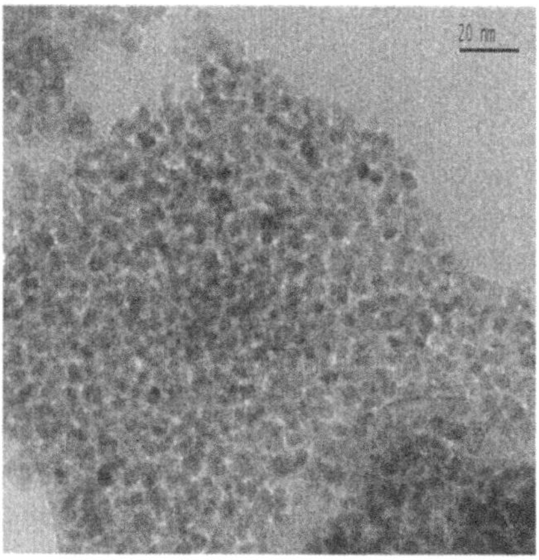

Fig. 16.17. Clustering of grains in cerium oxide powder calcinated for 1 h in air at 400°C (bright field (111) TEM micrograph)

small difference can be appreciated in the high-diameters tail of the curve; a higher number of "big" grains is expected with respect to what is actually observed.

This could be a problem of TEM statistics ("only" 800 grains have been considered, at least eight orders of magnitude less than those counted by XRD), or some limitation in the model employed. For instance, as shown in Fig. 16.18, the distribution obtained from the TEM analysis is only approximately lognormal (the relationship between the logarithm of the grain size and the cumulative probability should be linear for a lognormal distribution) as assumed for the WPPF and WPPM analyses.

An interesting comparison is that made between the results of the whole pattern methods, those of transmission electron microscopy and the results of

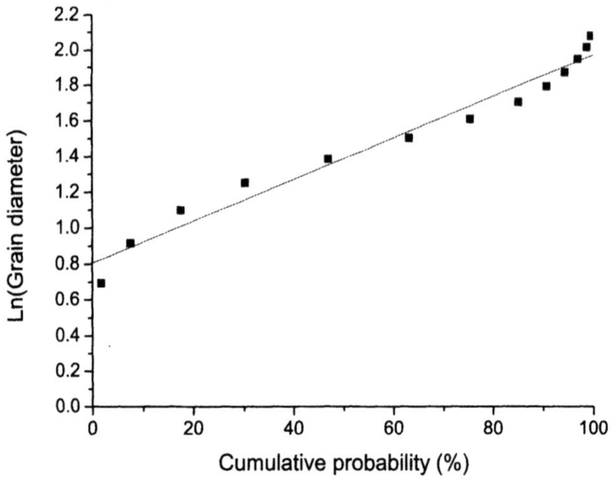

Fig. 16.18. Trend of the logarithm of the grain diameter versus the cumulative probability for the TEM grain-size distribution shown in Fig. 16.16 (*points*) and possible linear fit (*line*)

the traditional methods of line profile analysis. The latter gives somewhat deviating results: if we consider the crystallite size distribution obtained from the use of the analytical formulae proposed by Krill and Birringer [55] allowing a lognormal size distribution of spherical crystallite diameters to be obtained from area and volume-weighted average crystallite sizes, we obtain the curve presented in Fig. 16.16 as WH-K&B. The two mean crystallite sizes needed to build that curve, have been calculated using the traditional Warren-Averbach and Williamson–Hall methods. The disagreement can be explained considering that traditional methods do not directly deal with the measured data points, but rely on pre-processed data (the full width of the diffraction profiles), often affected by peak overlap and/or profile truncation errors, whereas WPPF and WPPM uses information contained in the whole pattern, including the statistical error associated to each of the experimental points [64].

Besides allowing the assessment of grain shape and dimensions, transmission electron microscopy permits dislocations to be directly observed: qualitative results can be obtained for dislocation densities up to 10^{14}–$10^{15}\,\mathrm{m}^{-2}$, about two orders of magnitude lower than the maximum detectable by XRD. A couple of examples, obtained on the specimen treated at 500°C, is shown in Fig. 16.19 where single dislocations are clearly visible. These are presumably edge dislocations spanning almost the entire width of the grain. The dislocation density estimated using the WPPF method is higher than that obtained using the WPPM approach even if both algorithms are based on Wilkens theory of the dislocations-induced line profile broadening [65, 66]. The parameters needed for the calculation of the average contrast factor for

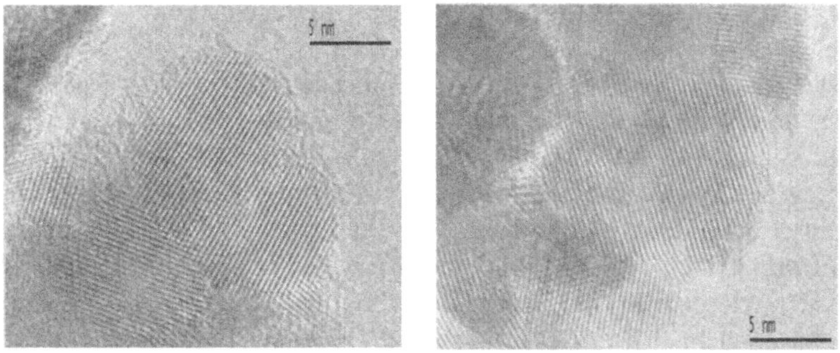

Fig. 16.19. Edge dislocations in two grains of cerium oxide powder calcinated for 1 h in air at 500°C (bright field (111) TEM micrograph)

all reflections observed in the pattern have been obtained from the literature elastic constants of ceria [67] and reported in Table 16.1 as A and B. A refinement of the values would have been possible, but would have introduced instability in the algorithm. As a test, the parameter B has been freed in the calculation of some patterns and it has been observed to vary slightly with respect to the imposed value. For the same reason, the ratio between screw and edge dislocations has been set to 1:1. The line broadening due to dislocations in ceria is certainly a minor contribution with respect to that of crystallite size even if, on an absolute scale, the dislocation density ρ has a relatively high numerical value.

An average number of dislocations per grain can be estimated from the dislocation density and the grain size. Assuming an average grain diameter $D_{\mathrm{ave}} = 4.5\,\mathrm{nm}$, i.e. a grain section of $\pi D_{\mathrm{ave}}^2/4$, and assuming a packing factor of 0.5 (the powder is not highly compacted), we reach a density of grains of about $3 \times 10^{16}\,\mathrm{m}^{-2}$. This would mean, roughly, an average of one dislocation in every 2 grains.

The density measurable from a direct TEM observation is certainly lower, about 50% of these figures; it is worth noting, however, that only part of the dislocations are actually visible in TEM pictures (i.e., properly oriented).

The Wilkens parameter M (see for instance [65]), defined as $M = R_e\sqrt{\rho}$ can also help in attaching some meaning to the dislocation density, giving some indications on their arrangement. Strong interaction (for instance the presence of dislocation clusters or walls) is expected when the parameter is close to unity. This is typical for ball milled samples where dislocations are preferentially locked on grain boundaries and the number of randomly distributed ones is limited.

In our case a value of $M \approx 0.25$ would indicate a strong interaction between dislocations. This seems not justified at least for the observed grains. However, extended defects like small angle grain boundaries, can be responsible for the low value of M [68].

CeO_2 Calcinated 1 Hour at 300, 500 and 600°C

The sample microstructure seems to be well described by the results of the WPPF and WPPM analyses and agrees reasonably with the TEM observation. Considering thus the WPPM data as representative for the microstructure of a given sample, we can compare the results from various samples calcinated at different temperatures. By increasing the calcination temperature from 300°C to 600°C, both mean crystallite size and breadth of the crystallite size distribution, increase (cf. Fig. 16.20). These features are further discussed in [63], on the basis of a thorough TEM investigation and thermal analysis study.

Figure 16.21 shows the trend of the mean crystallites diameter and log-normal variance for the four samples considered, data extracted from the plots of Fig. 16.20.

It is clear how a direct comparison of the various samples is not possible, especially between the one calcinated at 300°C and that at 600°C, since the variances of these distributions are very different. This is particularly valid for those features which are strongly dependent on the dimension of the grains like those related to the defects or those possibly dependent on surface or volume of the grains. For instance, Fig. 16.22 shows the trend of the dislocation density ρ as a function of the calcination temperature. Because of the different features in the crystallite size distribution, a direct comparison between the results of the samples calcinated at the lowest and highest temperature is not easy and only qualitative conclusions can be drawn.

By increasing the crystallites size, the unit cell parameter decreases towards the "standard" one (Fig. 16.23a). As evident in Fig. 16.23b, this decrease is in good agreement with the hyperbolic law of decrease of the area/volume ratio for a sphere, thus confirming that the effect is strongly

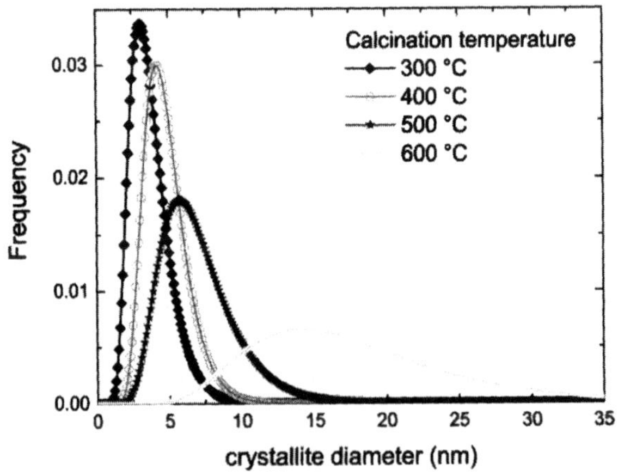

Fig. 16.20. Crystallite size distribution for cerium oxide calcinated at various temperatures

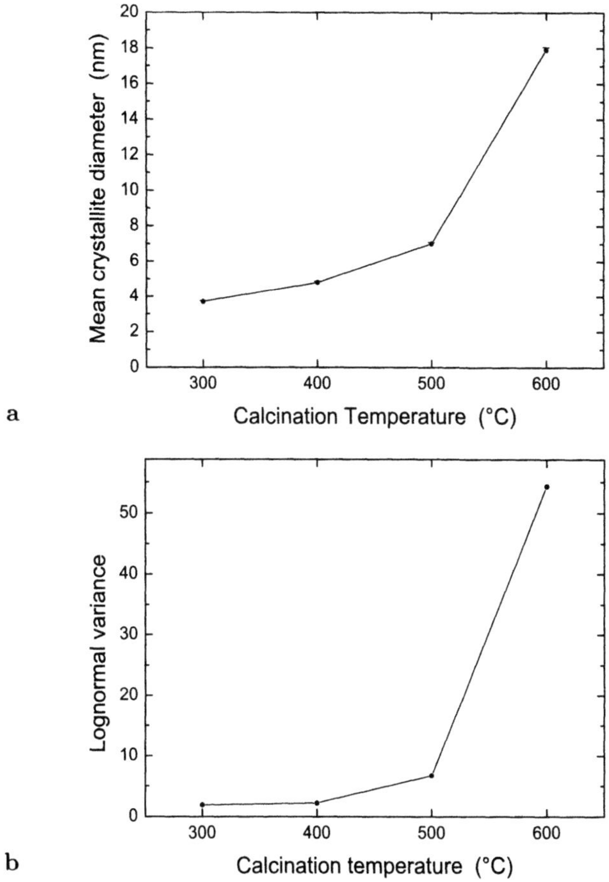

Fig. 16.21. Variation of mean crystallite diameter (**a**) and lognormal variance (**b**) for the cerium oxide samples as a function of the calcination temperature

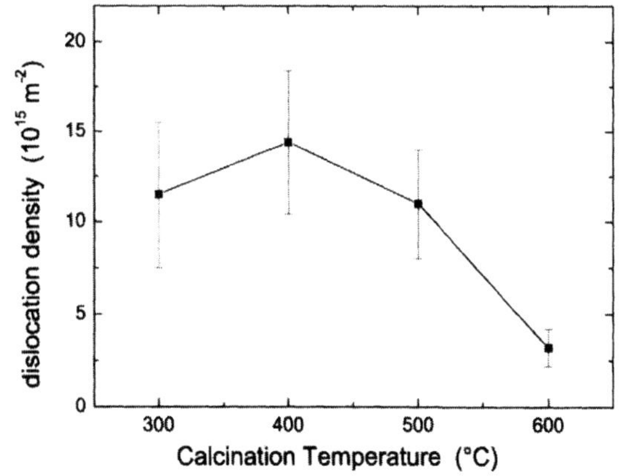

Fig. 16.22. Average dislocation density for the four cerium oxide samples as a function of the calcination temperature

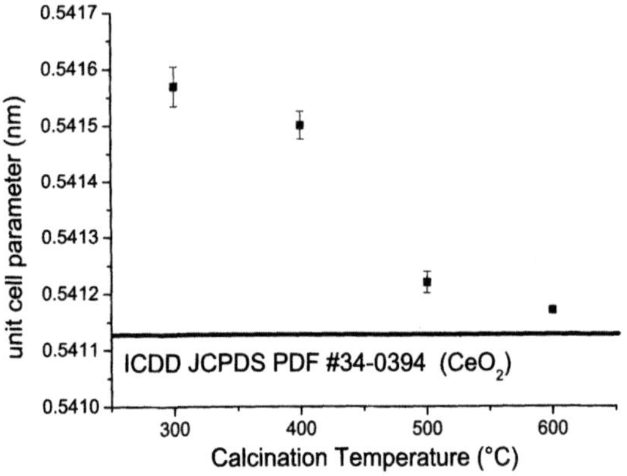

a

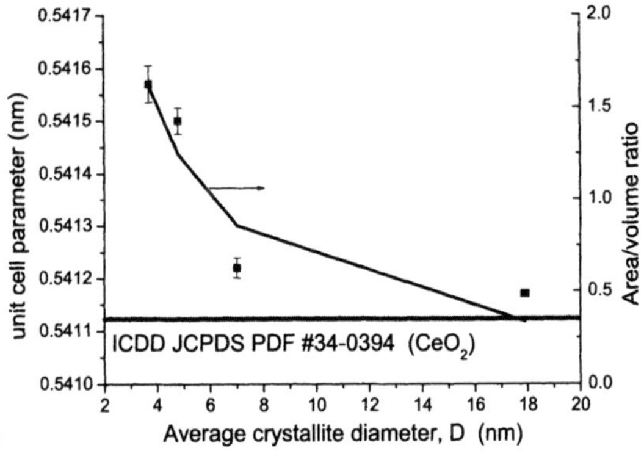

b

Fig. 16.23. Dependence of the average unit cell parameter (from a traditional refinement) on the calcination temperature (**a**) and on the corresponding mean crystallite diameter (**b**). In (**b**), the area/volume ratio corresponding to the 4 mean crystallite sizes is also shown (*line*)

correlated with the increase of bulk effects over surface ones. This is a strong indication that the observed trend of the unit cell parameter with the calcination temperature is mainly due to surface relaxation.

Vacancies alone, in fact, cannot account for the observed trends. Moreover, the equilibrium concentration of vacancies stable at room temperature should decrease by increasing the calcination temperature following a trend analogous to that of Fig. 16.23a. If the relationship of Kim [61] is supposed valid, the result is that shown in Fig. 16.24 (full dots), i.e. the powder calcinated at the highest temperatures should be over-stoichiometric. Even if (16.29) above is used (Fig. 16.24, open dots), then the powder calcinated at 300°C should be highly understoichiometric at room temperature. In principle this would be possible, but from the data of Fig. 2 in the cited paper of Mogensen et al. [58], the temperature needed to obtain the predicted vacancies concentration in ceria for the partial pressure of oxygen in air, is higher than 1500°C, a temperature never reached by any of the samples.

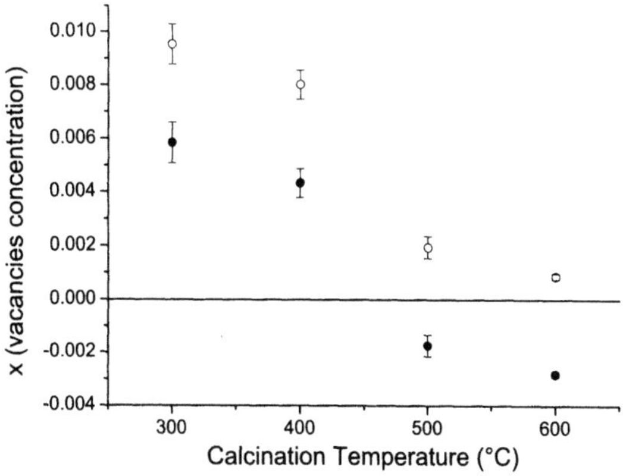

Fig. 16.24. Concentration of vacancies as calculated from the measured unit cell parameter according to Kim [61] (*full dots*) and using a modified formula based on the accepted unit cell parameter for pure ceria (*open dots*)

Comparison of WPPF and WPPM Results with Those of GSR-WPPM

The change of the unit cell parameter with the calcination temperature cannot be interpreted by the traditional models. To try to get more sound results, all powders have also been analysed using the grain surface relaxation algorithm implemented into the WPPM calculation frame (GSR-WPPM). The refined pattern and relative parameters for the powder calcinated at 400°C are shown in Table 16.1 and Fig. 16.25, respectively.

A first check of the results shows a quite good uniformity of values among the various models, indication that at least the three models are consistent. However some subtle but substantial differences can be spotted out. The GSR-WPPM algorithm keeps the unit cell parameter fixed to the reference value $a = 0.541134\,\mathrm{nm}$ [58] thus all variations in the position of the peaks are accommodated by surface relaxation and by the specimen displacement error.

Apparently, the introduction of surface relaxation brings an additional degree of freedom into the system (one more refinable parameter) with respect to the traditional WPPM. This is true if we analyse just this dataset alone. All four samples presented previously (calcinated at 300, 400, 500 and 600°C, respectively) can be modelled by the GSR-WPPM algorithm without any change in the ξ and κ relaxation factors with respect to the values shown in Table 16.1. Moreover, if the relaxation parameters are set free for all four samples, they tend to refine to the same value.

An example of the relaxation function refined for the sample calcinated in air at 400°C for 1 h is shown in Fig. 16.26. Figure 16.27 shows the refined crystallite size distribution function together with the variation of the unit cell parameter for some crystallite sizes randomly chosen from the distribu-

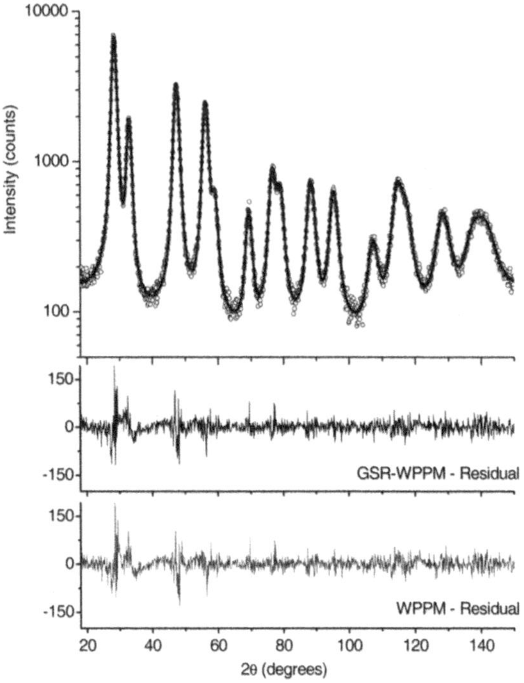

Fig. 16.25. Measured (*dots*) and modelled pattern according to the GSR-WPPM (*continuous line*) and WPPM (*dashed line*) algorithms. The plot is shown in linear and log scale to allow small details to be clearly seen

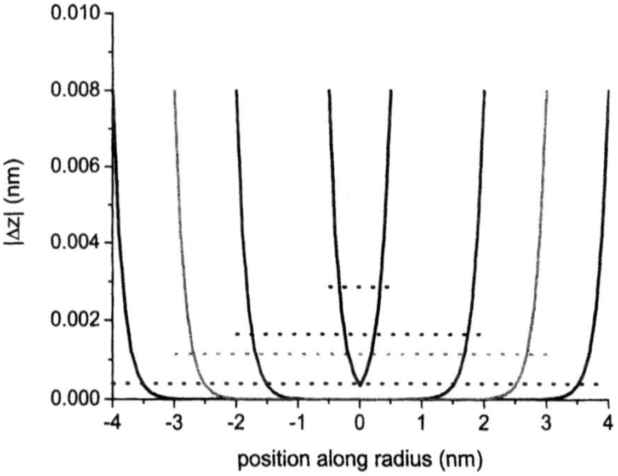

Fig. 16.26. Relaxation curves for the cerium oxide sample calcinated at 400°C for various crystallite sizes. *Horizontal lines* indicates the average unit cell parameter shift for the corresponding crystallite

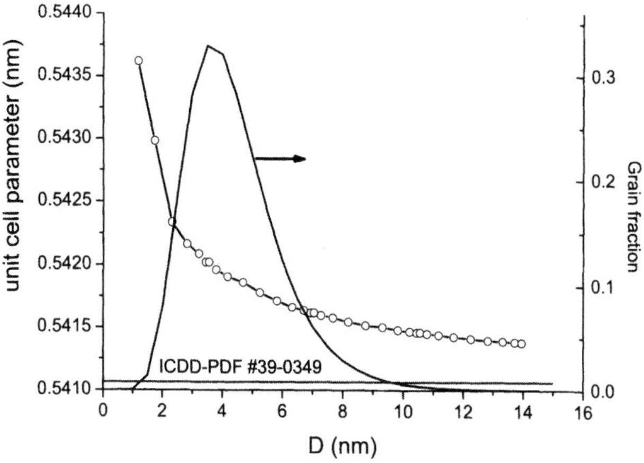

Fig. 16.27. Dependence of the unit cell parameter on the crystallite diameter for the ceria sample calcinated at 400°C for 1 h

tion. The largest variation in the unit cell parameter is associated to a very small fraction of grains; because of the small crystallite size, none of the grains having a statistical presence in the distribution possesses the unit cell parameter of bulk ceria. The value is reached asymptotically for crystallites of infinite diameter.

A comparison between the crystallite size distribution obtained above using the traditional WPPF/WPPM methods and that calculated by GSR-WPPM is shown in Fig. 16.16. A slight variation can be observed, well within the esd's of the parameters. The agreement with TEM data is still quite good.

The specimen displacement has changed, indicating that something in the shape of the profiles and in their position has been modified (especially at high angle). It cannot be established a priori if this effect is real: either a part of the specimen surface displacement in the traditional algorithms was due to some lacks in the modelling, or a local minimum has been found for the GSR-WPPM algorithm. To reduce the possibility of finding local minima, refinement was repeated more than once starting from a different point in the parameters space. The various program runs led to close values, within the esd's. In any case a displacement of few micrometers is certainly tolerable (we must remember that the powder specimens are almost self-supporting in the aluminium frame, thus some displacement, especially in the middle of the specimen, can be present).

Although only a slight change in the parameters related to size broadening is found, there is a large change in those dealing with line defects. The dislocation density halves with respect to the value refined using the WPPF algorithm, indication that the broadening due to surface relaxation can mimic the effect of a higher dislocation content. Owing to this reduction in the dislocation density, we expect a slight increase in the effective outer cutoff radius

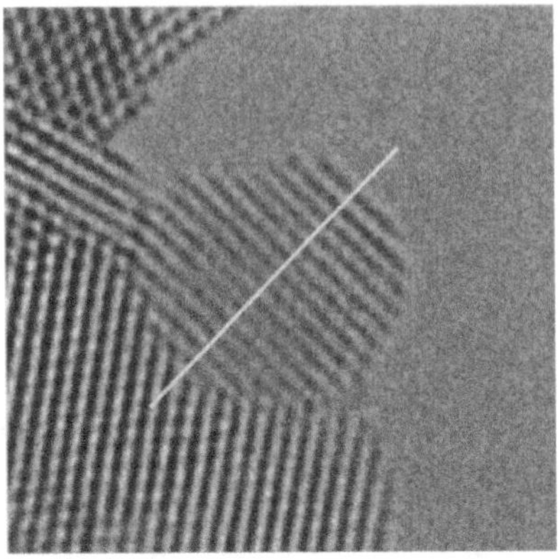

Fig. 16.28. Bright field (111) TEM micrograph of a ceria grain (400°C) clearly showing the surface relaxation effect both as variation in the lattice spacing at the surface and buckling of some lattice planes in the nearby of the surface. Quantitative measurement of surface relaxation have been conducted along the *white line* indicated on the micrograph

R_e demonstrating some correlation between the refinement of the unit cell parameter and the dislocation model (a relationship does exist: the modulus of the Burgers vector is in fact proportional to the unit cell parameter). A lower dislocation density is also more compatible with the TEM observations.

Trying to do on cerium oxide the kind of accurate lattice spacing measurement done by Yacamán et al. [34] on gold nanorods, measurements have been conducted on calibrated TEM pictures of ceria grains in order to search for some traces of surface relaxation. Good candidates were selected among the subset of small and isolated grains (where the effect should be larger). Figure 16.28 shows an example of those particles.

By means of a commercial software for TEM micrographs analysis [69], all lattice spacings have been measured along the line drawn in Fig. 16.28. Figure 16.29 shows the result; the peaks and valleys correspond to the dark/white lines of Fig. 16.28, i.e. to the (111) atomic planes of ceria. By taking a fixed-width marker (multiple of the "unrelaxed" grain spacing) and moving along the line drawn in Fig. 16.28 (i.e. from side to side in Fig. 16.29), a variation in the lattice spacing can be appreciated. In particular, while the marker spans exactly 9 lattice planes in the middle of the grain (Fig. 16.29b), planes seems more distant at the edges of the grain (Fig. 16.29a and c). The choice of a multiple of the interplanar distance has been chosen in order to minimise the evaluation errors and to maximise the effect (the precision and accuracy of the result certainly scales with the length to be measured).

The absolute maximum variation in the distance is of the order of magnitude of the relaxation parameter ξ. It is difficult to give a direct measurement of the agreement of the parameter κ, since it refers to a continuous function.

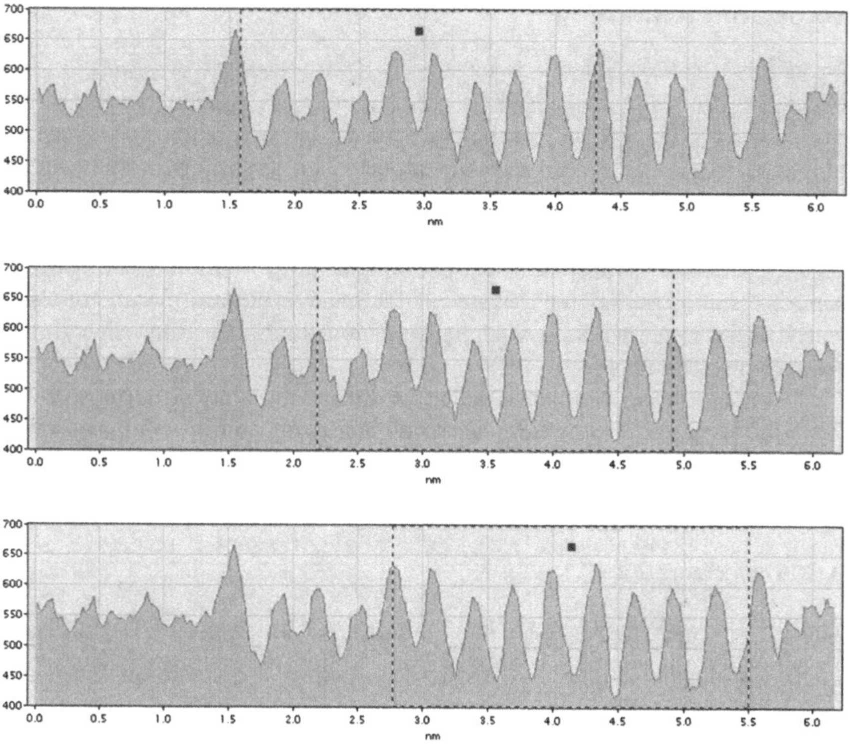

Fig. 16.29. Scan along the line drawn in Fig. 16.28. The dashed marker rectangle (fixed span width) is moved along the line in Fig. 16.28; due to surface relaxation, the distance of the outermost planes is slightly increased and the marker does not intercept them

It should in a certain way, be related to the extension of the relaxed zone, thus limited to one or two lattice planes.

A more accurate quantitative evaluation of TEM data would be possible on a better isolated grain and highly calibrated images; even if the result of Fig. 16.29 is considered as merely qualitative, it certainly confirms GSR-WPPM results.

Both GSR-WPPM and TEM results find a further confirmation in a recent paper by Wu et al. [70] where relaxation effects are observed in nanometric cerium oxide particles produced by a microemulsion method. EXAFS measurements conducted on those ceria particles can be explained only by recurring to a surface-core model, where the surface region of the particles shows longer bonding distances (i.e. increased unit cell parameter). Owing to the presence of a surfactant in the ceria powder of [70], a quantitative comparison between GSR-WPPM and EXAFS results cannot be made. However, the order of magnitude of the GSR effect in cerium oxide is confirmed by the EXAFS study.

16.5 Conclusions

Grain surface relaxation effects have been incorporated into a general frame of Whole Powder Pattern Modelling. Simulations and modelling of the diffraction pattern of an *fcc* nanocrystalline powder (cerium oxide) have been employed to show the features of the model. The variations of the unit cell parameter of ceria with respect to the bulk value, have been interpreted in terms of surface relaxation. This is of great importance since four datasets, showing different average unit cell parameter, have been modelled using the same relaxation parameters. More reliable microstructural parameters seem to result from the whole powder pattern modelling algorithm when surface relaxation is introduced.

The results from the non-destructive diffraction analysis have been validated by means of transmission electron microscopy and by comparison with the available literature. The relaxation parameters are shown to be qualitatively and quantitatively sound.

Acknowledgements

The authors wish to thank Dr Rosa Di Maggio (University of Trento, Italy) for providing the samples and Dr Stefano Polizzi (University of Venezia Cà Foscari, Italy) for TEM imaging.

References

1. A.S. Edelstein, J.S. Murday, B.B. Rath: Progr. Mater. Sci. **42**, 5 (1997)
2. R.Z. Valiev, R.K. Islamgaliev, I.V. Alexandrov: Progr. Mater. Sci. **45**, 103 (2000)
3. H. Gleiter: Acta Mater. **48**, 1 (2000)
4. C. Delerue, G. Allan, M. Lannoo: J. Luminescence **80**, 65 (1999)
5. P. Scardi, M. Leoni, Y.H. Dong: Eur. Phys. J. B **18**, 23 (2000)
6. P. Scardi, M. Leoni, Y.H. Dong: Mat. Sci. Forum. **378–381**, 132 (2001)
7. P. Scardi, M. Leoni: Acta Cryst. A **58**, 190 (2002)
8. P. Zieliński, Z. Łodziana, T. Srokowski: Prog. Surf. Sci. **59**, 265 (1998)
9. P. Zieliński: Phys. Rev. B **38**, 12338 (1988)
10. S.Y. Tong, G. Xu, W.N. Mei: Phys. Rev. Lett. **52**, 1693 (1985)
11. J. Bohr, R. Feidenhans'l, M. Nielsen, M. Toney, R.L. Johnson, I.K. Robinson: Phys. Rev. Lett. **54**, 1275 (1985)
12. G. Allan, M. Lannoo: Phys. Rev. B **37**, 2678 (1988)
13. K. Żabińska, P. Zieliński: Phys. Rev. B **40**, 12071 (1989)
14. O. Hjortstam, J. Trygg, J.M. Wills, B. Johansson, O. Eriksson: Surf. Sci. **355**, 214 (1996)
15. X. Yin, R. Miura, A. Endou, I. Gunji, R. Yamauchi, M. Kubo, A. Stirling, A. Fahmi, A. Miyamoto: Appl. Surf. Sci. **119**, 199 (1997)
16. T.T. Rantala, T.S. Rantala, V. Lannto: Surf. Sci. **420**, 103 (1999)

17. H.-G. Fritsche: Zeitsch. Phys. Chem. **192**, 21 (1995)
18. J. Huang, L.S. Bartell: J. Mol. Struct. **567–568**, 145 (2000)
19. H. Van Swygenhoven, D. Farkas, A. Caro: Phys. Rev. B **62**, 831 (2000)
20. H. Van Swygenhoven, A. Caro, D. Farkas: Mat. Sci. Eng. A **309–310**, 440 (2001)
21. D. Chen: Comp. Mat. Sci. **3**, 327 (1995)
22. D. Chen: Mat. Sci. Eng. A **190**, 193 (1995)
23. K. Lu, Y.H. Zhao: Nanostr. Mater. **12**, 559 (1999)
24. X.D. Liu, H.Y. Zhang, K. Lu, Z.Q. Hu: J. Phys. Condens. Matter **6**, L497 (1994)
25. J.A. Eastman, M.R. Fritzsimmons, L.J. Thompson: Phil. Mag. B **66**, 667 (1992)
26. H. Bakker, G.F. Zhou, H. Yang: Progr. Mat. Sci. **39**, 159 (1995)
27. P. Scardi, P.L. Antonucci: J. Mater. Res. **8**, 1829 (1993)
28. K. Kinoshita, in *Modern aspects of electrochemistry*. Ed. by J.O'M. Bockris, B.E. Conway, R.E. White (Plenum, New York 1982) pp. 557–637
29. A.C. Nunes, L. Yang: Surf. Sci. **399**, 225 (1998)
30. A.C. Nunes, D. Lin: J. Appl. Cryst. **28**, 274 (1995)
31. T. Uemori, K. Nagareda, K. Ishikawa: Bull. Res. Inst. Electr. Shizuoka University **33**, 91 (1998). In Japanese. For an English version, refer to K. Ishikawa, T. Uemori: Phys. Rev. B **60**, 11841 (1999)
32. Z. Kaszkur: J. Appl. Cryst. **33**, 1262 (2000)
33. J. Muliu, T.A. Pakkanen: Surf. Sci. **364**, 439 (1996)
34. M.J. Yacamán, J.A. Ascencio, G. Canizal: Surf. Sci. **486**, L449 (2001)
35. K. Ishikawa, K. Yoshikawa, N. Okada: Phys. Rev. B **37**, 3852 (1988)
36. A. Hartridge, A.K. Bhattacharya, M. Sengupta, C.K. Majumdat, D. Das, S.N. Chintalupudi: J. Mag. Mag. Mat. **176**, L89 (1997)
37. W.L. Zhong, Y.G. Wang, P.L. Zhang, B.D. Qu: Phys. Rev. B **50**, 698 (1994)
38. B.E. Warren: *X-ray diffraction.* 2nd edn. (Dover, New York 1990) [1st edition, Addison Wesley, Reading 1969]
39. P. Scherrer: Nachr. Ges. Wiss. Göttingen, 26. September (1918) p. 98
40. J.I. Langford, A.J.C. Wilson: J. Appl. Cryst. **11**, 102 (1978)
41. P. Scardi, M. Leoni: Acta Cryst. A **57**, 604 (2001)
42. B.E. Warren, B.L. Averbach: J. Appl. Phys. **21**, 595 (1950)
43. B.E. Warren, B.L. Averbach: J. Appl. Phys. **23**, 497 (1952)
44. G.K. Williamson, W.H. Hall: Acta Metall. **1**, 22 (1953)
45. H.M. Rietveld: J. Appl. Cryst. **2**, 65 (1969)
46. R.A. Young (Ed.): *The Rietveld Method.* (Oxford University Press, Oxford 1993)
47. B.E. Warren: Acta Cryst. **8**, 483 (1955)
48. A.R. Stokes, A.J.C. Wilson: Proc. Camb. Phyl. Soc. **40**, 197 (1944)
49. E.F. Bertaut: C. R. Acad. Sci. Paris **228**, 492 (1949)
50. E.F. Bertaut: Acta Cryst. **3**, 14 (1950)
51. P. Scardi, M. Leoni: this volume, Chap. 3
52. J.P. Cline, R.D. Deslattes, J.-L. Staudenmann, E.G. Kessler, L.T. Hudson, A. Henins and R.W. Cheary: *Certificate SRM 660a.* (NIST Gaithersburg 2000)
53. M.S. Dresselhaus, I.L. Thomas: Nature **414**, 332 (2001)
54. V.V. Kharton, F.M. Figueiredo, L. Navarro, E.N. Naumovich, A.V. Kovalevsky, A.A. Yaremchenko, A.P. Viskup, A. Carneiro, F.M.B. Marques, J.R. Frade: J. Mater. Sci. **35**, 1105 (2001)

55. C.E. Krill, R. Birringer: Phil. Mag. A **77**, 621 (1998)
56. I.J. Langford, D. Louër, P. Scardi: J. Appl. Cryst. **33**, 964 (2000)
57. N. Audebrand, J.P. Auffrédic, D. Louër: Chem. Mater. **12**, 1791 (2000)
58. M. Mogensen, N.M. Sammens, G.A. Tompsett: Solid State Ionics **129**, 63 (2000)
59. Standard Reference Material 674, National Institute of Standards and Technology (formerly Nat. Bur. of Stand.) (Gaithersburg 1983)
60. R. Dal Maschio, P. Scardi, L. Lutterotti, G.M. Ingo: J. Mat. Sci. **27**, 5591 (1992)
61. D.-J. Kim: J. Am. Ceram. Soc. **79**, 1415 (1989)
62. S.J. Hong, A.V. Virkar: J. Am. Ceram. Soc. **78**, 433 (1995)
63. M. Leoni, R. Di Maggio, S. Polizzi, P. Scardi: J. Amer. Ceram. Soc. (2003) Submitted
64. P. Scardi, M. Leoni, R. Delhez: J. Appl. Cryst. (2003) Submitted
65. M. Wilkens: Phys. Stat. Sol. (a) **2**, 359 (1970)
66. M. Wilkens: *Fundamental Aspects of Dislocation Theory*. Ed. by J.A. Simmons, R. de Wit, R. Bullough. Vol. II. Nat. Bur. Stand., (US) Spec. Publ. No. 317 (Washington 1970) pp. 1195–1221
67. A. Nakajima, A. Yoshihara, M. Ishigame: Phys. Rev. B **50**, 13297 (1994)
68. J.G.M. van Berkum, PhD Thesis. Delft University of Technology (Delft 1994)
69. Gatan Digital Micrograph software version 3.4.1
70. Z. Wu, L. Guo, H. Li, Q. Yang, Q. Li, H. Zhu: Mat. Sci. Eng. A **286**, 179 (2000)

17 Interface Stress in Polycrystalline Materials

R. Birringer, M. Hoffmann, and P. Zimmer

Based on a generalization of a capillary equation for solids, we develop a method for measuring the absolute value of grain-boundary stress in polycrystalline samples having a large interface-to-volume ratio. The grain-boundary stress in nanocrystalline Pd is calculated from X-ray diffraction measurements of the average grain size and the residual-strain-free lattice spacings, yielding a value of $1.2 \pm 0.1\,\mathrm{N/m}$. The random distribution of crystallite orientations in the sample suggests that this value is characteristic of high-angle grain boundaries in Pd.

17.1 Introduction and Basic Ideas

The modelling of stress generation in metallic systems and semiconductor heterostructures, the effect of stresses on microstructural evolution and the influence of internal stresses on physical properties have developed into a major research topic in the field of materials science [1]. Our interest in this area concerns the internal stresses that are generated by the presence of internal interfaces in solids. When a solid with a fine-scale microstructure is elastically deformed, work is performed against volume (bulk) forces and interface forces. Even if there is no external load applied, mechanical equilibrium prescribes that stresses in the bulk are required to compensate interface stresses. In a conventional polycrystalline material, the bulk stresses arising from interface stresses are negligible; however, for a solid with a nanometer-scale microstructure, interface stresses can no longer be neglected, because they induce bulk stresses on the order of f/L, where f is the scalar interface stress and L denotes a characteristic length scale of the microstructure (such as the average crystallite size). Following [2–5], Cahn [6, 7] has pointed out that the interface stress should be defined by the reversible work needed to elastically deform a preexisiting interface of area A. If interface area A is measured in coordinates of the undeformed solid, then, the interface stress $\boldsymbol{f}_{ij}$, a 2×2 tensor with entries that may have either sign, is defined as

$$\boldsymbol{f}_{ij} = \frac{\partial \gamma_0}{\partial \boldsymbol{e}_{ij}}, \tag{17.1}$$

where γ_0 is the always positively valued interfacial free energy per unit reference area A_0 of the undeformed interface, and $\boldsymbol{e}_{ij}$ represents the mean tangential strain along the interface. Using the material points in an undeformed configuration as reference state – Lagrangian coordinates –, it is implied that the area of interface A_0 is taken constant, independent of the presence of finite values of $\boldsymbol{e}_{ij}$. As a consequence, the true area A_{true} of the deformed interface and the Lagrangian measure of interface area A_0 differ by

$$A_{\text{true}} = A_0(1 + \boldsymbol{e}_{ii}) , \tag{17.2}$$

where $\boldsymbol{e}_{ii}$ is the trace of the interface strain (summation convention used). Interface stress can also be expressed in terms of laboratory coordinates [6] to obtain

$$\boldsymbol{f}_{ij} = \delta_{ij}\gamma + \frac{\partial \gamma}{\partial \boldsymbol{e}_{ij}} , \tag{17.3}$$

where γ is the interfacial free energy per unit area of the deformed interface A_{true}. Since the free energy of a given piece of interface should be the same in both descriptions, the identity $\gamma_0 A_0 = \gamma A_{\text{true}}$ still holds. Therefore, both definitions of $\boldsymbol{f}_{ij}$ may be used to account for energy changes associated with elastically deforming an interface. Since they describe the same phenomenon, both are equally correct, and they lead to the identical numerical value for $\boldsymbol{f}_{ij}$. Besides its formal simplicity, (17.1) also unambiguously illustrates that the entries of $\boldsymbol{f}_{ij}$ are independent thermodynamic quantities which are, in contrast to what is suggested by (17.3), unrelated to the value of γ by magnitude as well as sign, and in addition (17.1) reflects the formal analogy between the definitions of stress in bulk elasticity $\boldsymbol{\sigma}_{ij} = V_0^{-1}(\partial U/\partial \boldsymbol{\varepsilon}_{ij})$ and the interface stress $\boldsymbol{f}_{ij} = A_0^{-1}[\partial(\gamma_0 A_0)/\partial \boldsymbol{e}_{ij}]$; the change in strain energy U for an arbitrary uniform small strain (Lagrangian strain) is measured with reference to the undeformed ($\boldsymbol{\varepsilon}_{ij} = 0$) body of volume V_0. A more general notion of interface stress permitting a discontinuity in the displacement across an internal interface has been discussed by Gurtin et al. [8], and suggests stresses which are conjugated to the normal and tangential components of the jump in displacement and jump in the displacement gradient at the interface.

The rapid development in our theoretical and experimental understanding of stresses at solid surfaces and the role they play in phenomena like surface reconstruction, epitaxial growth, adsorbate-induced changes of surface stress or restructuring of surfaces, as well as the influence of inhomogeneous surface stress on the equilibrium structure of surfaces or on the stabilization of mesoscopic structures has been reviewed by Ibach [9]. Much less complete, however, is our understanding of the influence of internal interfaces on intrinsic stresses and their consequent impact on the properties of materials with a large interface-to-volume ratio, such as multilayered thin films or nanostructured materials [10]. A review of the current understanding of the effect of interfaces on the intrinsic stresses in polycrystalline thin films has been given recently by Spaepen [11]. The thermodynamic aspects of surface and interface stress are discussed in a review article by Cammarata et al. [12]. Equations for calculating the volume stresses induced by interface stresses have been derived for special sample geometries [13,14] and for general microstructures subjected to a hydrostatic load [15]. Several recent experiments have taken advantage of such relations to obtain experimental values for f in multilayered thin films [16–19]. Surprisingly, the research conducted on thin films provided rather different values for the magnitude of f, although, experiments have been carried out on comparable multilayered thin film systems, and it is still unclear why the experimentally determined interface stresses are compressive (negative) while atomistic calculations using the embedded atom method [20] predict that positive (tensile) interface stresses should be observed in experiment. Regarding polycrystalline materials, the relative change of grain-boundary stress upon hydrogen loading of nanocrystalline Pd has been investigated [21]. Nevertheless, many aspects of the interface stress remain largely unexplored in experiment, such as the hitherto unknown absolute value of interface stress related to grain boundaries, manifesting itself by its direct mechanical action on the crystallites adjoining a grain boundary. Therefore, in this study we are aiming to demonstrate that the absolute value of grain-boundary stress in polycrystalline materials can be experimentally determined.

17.2 Theory and Methodology

In what follows, we describe a method for determining the absolute value of the scalar grain-boundary stress in a polycrystalline sample made up of nanometer-sized grains (crystallites), i.e. a so-called nanocrystalline (nc) material [22]. Our approach is based on a generalized capillary equation derived by Weissmüller and Cahn [15] using the principal of virtual work [23]. Their central result was obtained by relating the hydrostatic components of the volumetric stress tensor $\boldsymbol{\sigma}_{ij}$ averaged over all of the grains in a polycrystalline solid to the integrated (average) stresses of the interfaces contained within the sample

$$\langle \Delta P \rangle_V = \frac{2}{3} \left(\frac{A}{V} \right) \langle f \rangle_A \, . \tag{17.4}$$

In (17.4), $\langle \Delta P \rangle_V$ denotes the volume-averaged difference between the local pressure P^{int} in the sample and an externally applied hydrostatic pressure P^{ext}, i.e. $\Delta P = P^{\text{int}} - P^{\text{ext}}$. The quantity $\langle f \rangle_A$ is the area-averaged interface stress; in analogy to the scalar pressure P in the theory of bulk elasticity, in which $P = -\text{Tr}\,(\boldsymbol{\sigma}_{ij}/3)$, the scalar interface stress (interface pressure) f is defined as $f = \text{Tr}\left(\boldsymbol{f}_{ij}/2\right)$. Finally, A/V represents the total interface area divided by the total grain (bulk) volume. In (17.4), contributions to the volume stress arising from interface junctions, which are one-dimensional elements of microstructure and second order in A/V, are neglected. We see from this equation that the mean pressure is independent of the spatial arrangement, orientation and curvature of the interfaces.

It seems worthwhile to point out that the interface-induced stress (pressure) in solids as described by (17.4) has an only formal analogy in the capillary pressure in liquids which obeys the Young–Laplace equation [24,25]

$$\Delta P = \gamma \left(\frac{1}{R_1} + \frac{1}{R_2} \right) , \tag{17.5}$$

where the pressure jump ΔP at each point on a fluid interface – e.g. the jump in hydrostatic pressure on the interface of a liquid droplet in a microemulsion – is related to the mean curvature, given by the principal radii of curvature R_1 and R_2, and to the surface or interface free energy γ (per unit area in some reference state). Note that the quantity γ is often also referred to as the surface or interfacial *tension*. This nomenclature may have caused confusion with the concept of surface or interface stress as defined in the introductory paragraph or by (17.4); as a consequence, (17.5) has been frequently misused to explain size-dependent changes of lattice constants in *solid* nanometer-sized particles or similar-sized grains embedded in solid matrices. There are several reasons why the Young–Laplace equation is not applicable to analyze such phenomena, but the capillary equation for solids (17.4) has to be preferred. The curvature of solid surfaces or interfaces can vary in an arbitrary manner depending on the position along the surface, whereas in fluids, the mean curvature is a constant along each surface. Therefore, (17.5) fails when applied to solids, since surfaces and interfaces in solids do not constitute manifolds of constant mean curvature; interface curvature in solids behaves rather like a locally varying scalar field quantity. Furthermore, measurable quantities like weighted averages of particle or grain size are non-local stereological quantities, which are – if at all – non-trivially related to the concept of curvature, again implying that the use of (17.4) is appropriate for solids rather than the misuse of (17.5). Finally, the application of (17.5) for analyzing stress effects in solids is even locally incorrect because the relevant capillary parameter is the tensor $\boldsymbol{f}_{ij}$ with entries that may have either sign and not the positive-valued scalar interface free energy γ.

In order to devise a method for measuring the interface stress in a nanocrystalline sample, we express (17.4) in terms of experimentally accessible quantities. The stereological identity $A/V = 2/\langle L \rangle$ relates A/V to the average grain intercept length $\langle L \rangle$, which represents an area-weighted average grain size and can be obtained directly from a Fourier analysis of Bragg-peak profiles, for more details see [26]. Concerning $\langle \Delta P \rangle_V$, we can derive a relation between the mean pressure in the bulk and the sample lattice parameter, which can be measured with high accuracy in a diffraction experiment. For the sake of argument, let us consider two nc-solids that differ by their total interface area A^β and A^α, respectively, but are otherwise identical. Inserting the definition of the isothermal bulk modulus $1/K = -V^{-1}(\partial V/\partial P) \mid_T$ in (17.4), and making use of the relation $\langle \Delta P \rangle_V = \langle P^{\mathrm{int}} \rangle_V - P^{\mathrm{ext}}$ (see Eq. (8) in [15]), we obtain

$$-K\left(\frac{V^\beta - V^\alpha}{V^\alpha}\right) = \frac{2}{3}\langle f\rangle_A\left(\frac{2}{\langle L\rangle^\beta} - \frac{2}{\langle L\rangle^\alpha}\right), \tag{17.6}$$

where V^β and V^α measure the total grain volume of the nc-solids having a grain boundary area A^β and A^α, respectively.

We remind the reader that the relative change in volume given in (17.6) is different from the macroscopic change in sample volume as shown in references [21, 22]. We also note that for polycrystalline microstructures the effective bulk modulus $K^\star$ should appear in (17.6) instead of K. However, for solids with cubic symmetry, K is isotropic and the identity $K^- = K^\star = K^+ = K = 1/3(C_{11} + 2C_{12})$ holds [27], where K^- and K^+ are the lower and upper bounds, respectively, of $K^\star$, and the C_{ij} are elastic constants in Voigt notation. For non-cubic systems $K^- < K^\star < K^+$, implying that $\langle f\rangle_A$ can only be approximated by its lower and/or upper bounds. Statistical continuum mechanics provides concepts for computing upper and lower bounds of moduli of heterogenous systems [27–29].

The isotropy of K and consideration of only small affine deformations allow to express $\Delta V/V$ (lhs of (17.6)) by the volume-averaged change of the unit cell volume, $\langle\frac{\Delta v}{v}\rangle_V$, where the identity $v = a^3$ holds for the lattice parameter a of cubic systems. As a result, we obtain $\Delta V/V = 3\langle\frac{\Delta a}{a}\rangle_V$, with $\langle a\rangle_V$ being a non-local, volume-averaged lattice parameter. If we further consider an interface-free solid (i.e. $\langle L\rangle \to \infty$) as our reference state, (17.6) simplifies to

$$\frac{\langle a_{\mathrm{nc}}\rangle_V - a_0}{a_0} = -\frac{4}{9}\frac{1}{K}\langle f\rangle_A\frac{1}{\langle L\rangle}, \tag{17.7}$$

where $\langle a_{\mathrm{nc}}\rangle_V$ denotes the lattice parameter of the nc-solid with average intercept length $\langle L\rangle$, and $\langle a_0\rangle_V = a_0$ is the (ground state) lattice parameter of the (coarse-grained) reference state. In Fig. 17.1, the magnitude of the relative change in lattice parameter versus $\langle L\rangle$ is depicted for a set of typical parameter values $\langle f\rangle_A$ and K.

Obviously, the relative change in lattice parameter is rather small for a wide variety of possible parameter combinations. However, the pressure $\langle P^{\mathrm{int}}\rangle_V$ (see (17.4)) associated with such small changes in $\Delta a/a_0$ can easily amount to large values of one or even several GPa as shown in Fig. 17.2.

Finally, we conclude this paragraph by emphasizing that the interface stress $\langle f\rangle_A$ can be obtained from the slope of a plot of $\langle a_{\mathrm{nc}}\rangle_V$ versus $\langle L\rangle$ as indicated by (17.7).

Before attempting to carry out such measurements, it is important to realize that (17.4) and (17.7) are valid only when the volume average extends over the entire specimen rather than merely a portion of it. Equation (17.4) was derived using the principle of virtual work by summing up the individual amounts of work performed by external, internal and interface stresses during a homogeneous deformation of a body of a single component nanocrystalline material. This principle requires that for sufficiently small deformations, the total work must vanish in mechanical equilibrium. In general, the internal

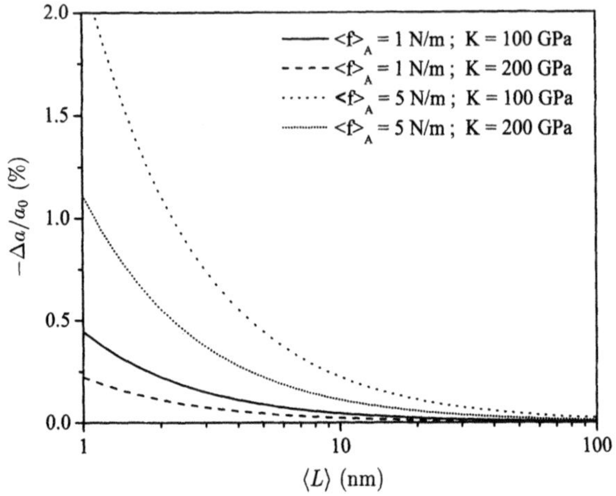

Fig. 17.1. Lattice parameter change $\Delta a/a_0$ plotted as a function of the area-weighted average grain size $\langle L \rangle$ for reasonable parameter values of $\langle f \rangle_A$ and K (see (17.7))

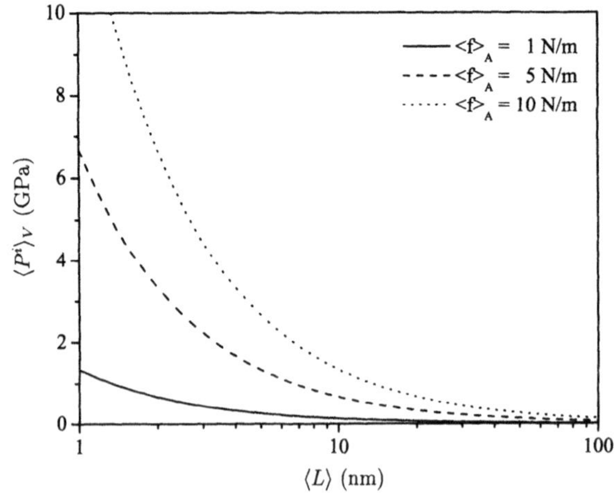

Fig. 17.2. Interface stress-induced pressure $\langle P^{\mathrm{int}} \rangle_V$ experienced by crystallites adjoining an internal interface, plotted as a function of the area-weighted average grain size $\langle L \rangle$; the parameter $\langle f \rangle$ is supposed to take on positive values of plausible magnitude. As indicated by (17.4), contributions to $\langle P^{\mathrm{int}} \rangle_V$ from P^{ext} are not considered

stress field $\boldsymbol{\sigma}_{ij}(\boldsymbol{r})$ of a heterogeneous solid is a linear superposition of two stress fields with different physical origins: (i) the stress conjugated to the strain $\boldsymbol{\varepsilon}_{ij}(\boldsymbol{r})$ resulting from an externally applied load, and (ii) the stress conjugated to the residual strain $\boldsymbol{\varepsilon}_{ij}^{\mathrm{res}}(\boldsymbol{r})$ arising from the elastic response of a material to an inhomogeneous distribution of non-elastic strains, such as plastic strains (e.g. due to processing) or those induced by thermal expansion, phase transformations, etc. [30]. As a consequence, the work W_V performed by the total internal stress can formally be written as

$$W_V = \int_V \sigma_{ij}(\boldsymbol{r})\varepsilon_{ij}(\boldsymbol{r})dV + \int_V \sigma_{ij}^{\mathrm{res}}(\boldsymbol{r})\varepsilon_{ij}(\boldsymbol{r})dV \tag{17.8}$$

with summation over repeated indices. Owing to the self-equilibrating nature of the residual stresses, the second integral in (17.8) vanishes as long as the range of integration extends over the entire sample volume [27], thus explaining why residual stresses do not appear in (17.4).

If X-ray diffraction is used to determine the lattice parameters appearing in (17.7), it may be impossible, depending on the specifics of specimen geometry, beam cross-section and available radiation energy, to carry out a measurement that averages over the entire sample volume. In such a case, the second integral in (17.8) no longer vanishes, and a derivation analogous to that of (17.7) would contain work integrals extending over the irradiation volume V_{irr}. As a result, we would obtain an expression for $\langle f \rangle_{A_{\mathrm{irr}}}$ that depends on the typically unknown value of the residual strain, rendering a determination of the interface stress from lattice-parameter and grain-size data impossible.

Although determination of the residual-strain tensor of a given sample is a challenging task, standard methods for residual-stress analysis [31] allow one to identify scattering geometries in which residual-strain-free lattice spacings can be measured by diffraction, despite probing a confined scattering volume V_{irr}. In these geometries, the residual-stress contribution to W_V vanishes even when $V_{\mathrm{irr}} < V$, because the integrand of the second integral in (17.8) is identically zero. Consequently, we can still use (17.7) to compute the interface stress, provided the quantity $\langle a_{\mathrm{nc}} \rangle_V$ is replaced by the residual-strain-free lattice parameter $[\langle a_{\mathrm{nc}} \rangle_{V_{\mathrm{irr}}}]_0$.

The theory of stress analysis by X-ray diffraction provides an explicit protocol for extracting $[\langle a_{\mathrm{nc}} \rangle_{V_{\mathrm{irr}}}]_0$ values from measurements of hkl-dependent lattice spacings d^{hkl}. For a polycrystalline specimen, average values of the latter can be related to the average strain according to

$$[\langle \varepsilon'_{33} \rangle_{V_{\mathrm{irr}}}]_{\phi,\psi}^{hkl} = \frac{[\langle d \rangle_{V_{\mathrm{irr}}}]_{\phi,\psi}^{hkl} - d_{\mathrm{ref}}^{hkl}}{d_{\mathrm{ref}}^{hkl}} , \tag{17.9}$$

where the prime signifies a laboratory coordinate system oriented with z-axis parallel to the scattering vector and azimuthal (ϕ) and polar (ψ) angles defined relative to the plane of the sample surface. The term d_{ref}^{hkl} denotes the lattice spacings of a reference state, which may be chosen without loss of generality to correspond to the residual-strain-free state – i.e. strain contributions arise only from the homogeneous deformation of all grains by interface (grain boundary)-stress-induced forces. The applicability of such a reference state to a given experiment must be justified a posteriori. The scattering geometry (ϕ, ψ) permitting measurement of residual-strain-free (but interface-stress-modified) *hkl* lattice spacings is specified by the condition

$$[\langle \varepsilon'_{33} \rangle_{V_{\rm irr}}]^{hkl}_{\phi,\psi} \equiv 0\,. \qquad (17.10)$$

As demonstrated in [32], (17.10) may be solved by considering the ensemble-averaged (i.e. ϕ-averaged) strain

$$\left[\overline{\langle \varepsilon'_{33} \rangle_{V_{\rm irr}}}\right]^{hkl}_{\psi} = (1/4) \sum_{n=0}^{3} [\langle \varepsilon'_{33} \rangle_{V_{\rm irr}}]^{hkl}_{n\pi/2,\psi}\,,$$

which can be shown to vanish for ψ values equal to

$$(\psi_0)^{hkl} = \arcsin\left[2\nu^{hkl}/(1+\nu^{hkl})\right]^{1/2}\,.$$

The ν^{hkl} parameters are *hkl*-dependent Poisson numbers, which can be computed for a given material according to the theory of X-ray elastic constants [33]. This solution to (17.10) relies upon the applicability of Hooke's law and equations of mechanical equilibrium in conjunction with the supposition that the nanocrystalline solid is statistically homogeneous and isotropic within $V_{\rm irr}$. A sample characterization must be carried out to determine whether or not these requirements are met. We conclude that the lattice parameters appearing in (17.7) can be replaced by their irradiated-volume-averaged equivalents, provided residual-strain-free values $[\langle a_{\rm nc} \rangle_{V_{\rm irr}}]_0$ are determined from lattice spacings $\left[\overline{\langle d_{\rm nc} \rangle_{V_{\rm irr}}}\right]^{hkl}_{(\psi_0)^{hkl}}$, where an average is performed at each *hkl* over the scattering geometries $(n\pi/2, (\psi_0)^{hkl})$ for $n = 0, 1, 2, 3$.

17.3 Experiments and Discussion

We now demonstrate the applicability of this approach to the experimental determination of the interface stress $\langle f \rangle_A$ in a sample of nanocrystalline Pd. A disc-shaped specimen 8 mm in diameter and about 0.1 mm thick was prepared by the inert-gas condensation technique and subsequently preannealed at 60°C for 24 h in order to equilibrate the grain boundaries [34]; the specimen state set after such a relaxation anneal is denoted by the as-prepared state. Lattice constants were extracted from diffraction measurements performed on a Huber powder X-ray diffractometer equipped with an Eulerian cradle and stationary pinhole collimators. The Cu-$K\alpha_{1,2}$ ($\lambda = 1.54056$ Å) radiation scattered by a sample held at 20°C with a stability of ±0.01°C was detected by an energy-dispersive solid-state detector. In order to determine the peak maximum position of intensity peaks, we fitted single peak profiles. A pseudo-Voigt function which is a linear combination of Lorentzian and Gaussian functions was found to give a good description of the intensity peaks. Statistical isotropy (random texture) of the specimen was verified by the radial symmetry of the iso-intensity contours of (111) and (200) pole figures.

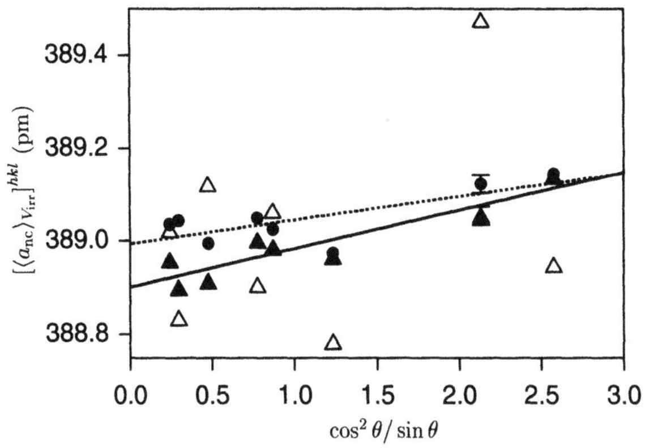

Fig. 17.3. Lattice parameter $[\langle a_{\mathrm{nc}}\rangle_{V_{\mathrm{irr}}}]^{hkl}$ of as-prepared nanocrystalline Pd plotted relative to the extrapolation function $\cos^2\theta/\sin\theta$, where θ denotes the Bragg angle. All $[\langle a_{\mathrm{nc}}\rangle_{V_{\mathrm{irr}}}]^{hkl}$ values were determined at the corresponding angles $(\psi_0)^{hkl}$ listed in Table 17.1 (△: raw data; ▲: data corrected for stacking faults). The y-intercept of the least-squares fit line (solid line) yields the lattice parameter $\langle a_{\mathrm{nc}}\rangle_V$ of the nanocrystalline solid. For comparison, we include the lattice parameters of a sample annealed at 600°C for use as a reference material (•, dashed line)

The average grain intercept length $\langle L\rangle$ of the as-prepared specimen was found to be ~ 10 nm by means of the standard Warren–Averbach analysis [35, 36]. The value for $\langle L\rangle$ was obtained from a plot of normalized cosine Fourier size coefficients against column length calculated from the (111) and (222) diffraction peaks. The rms inhomogeneous strain (microstrain), $\langle\varepsilon^2(L = 10\,\mathrm{nm})\rangle^{1/2}$, amounted to 0.4%. For a detailed discussion of the Fourier analysis of X-ray diffraction peak profiles (as implemented by Warren and Averbach), we refer the reader to our careful study [26] carried out on the same material. Annealing the sample in a sealed quartz tube filled with He (99.999%) induced grain growth [37, 38], corresponding to marked increases in $\langle L\rangle$ and a concomitant reduction of microstrain. Specimens with $\langle L\rangle$ values of 30 nm and 80 nm were prepared by annealing the sample at 120°C and 180°C, respectively, for 1 h. A coarse grained (micrometer-sized) reference sample was generated by annealing the sample at 600°C for 24 h.

Although the true lattice parameter of a crystalline material is independent of the Miller indices hkl, the same does not in general apply to the lattice constants a^{hkl} computed from individual lattice spacings d^{hkl}. A reliable estimate for the hkl-independent lattice parameter is obtained by plotting the a^{hkl} values against an appropriate extrapolation function [39]. Such a procedure is illustrated in Fig. 17.3 for nanocrystalline Pd in the as-prepared state. Each data point represents the average over four diffraction measurements carried out at the angles $\phi = n\pi/2$ $(n = 0, 1, 2, 3)$ and $(\psi_0)^{hkl}$, the values of which are given in Table 17.1.

Table 17.1. X-ray elastic constants ν^{hkl} of Pd, computed for the various hkl lattice planes according to the method of [33]. The angles $(\psi_0)^{hkl}$ were calculated from the ν^{hkl} constants using the formula given in the text

hkl	ν^{hkl}	$(\psi_0)^{hkl}$ (deg)
111	0.36	46.7
200	0.41	49.5
220	0.38	47.6
311	0.39	48.5
222	0.36	46.7
400	0.41	49.5
331	0.37	47.4
420	0.39	48.4

The strong scatter in the raw data points of Fig. 17.3 was found to arise from the presence of stacking faults in the specimen. After correcting the lattice constants for stacking faults as described in [35,40–44], we obtain a set of data points lying nearly along a straight line. The y-intercept of a linear fit to the latter gives the residual-strain-free lattice parameter $[\langle a_{\mathrm{nc}}\rangle_{V_{\mathrm{irr}}}]_0 = \langle a_{\mathrm{nc}}\rangle_V$ of the nanocrystalline solid. A revision of Warren's description of the intensity distribution of powder pattern reflections containing two-dimensional faults [35] has been recently given by Velterop et al. [45]. These authors removed some of the unrealistic assumptions contained in Warren's theory, especially, by taking into account higher order stacking- and twin-fault probabilities. In our specimen, we found rather small stacking fault probabilities on the order of 0.01 suggesting that Warren's approach should be applicable. Furthermore, the presence of stacking faults and/or twin boundaries was only observed in the as-prepared specimen. The heat treatment at elevated temperatures not only induced grain growth but also reduced the density of two dimensional faults (recovery) below the resolution limit of our equipment. For a sample containing two-dimensional faults, the Warren–Averbach analysis yields the average intercept length of the coherent scattering domains, the so-called effective intercept length $\langle L\rangle^{\mathrm{eff}}$, which is smaller than the true intercept length $\langle L\rangle^{\mathrm{true}}$, owing to the peak broadening induced by stacking faults incorporated in the individual nanocrystallites. The quantity $\langle L\rangle^{\mathrm{true}}$ was determined from $\langle L\rangle^{\mathrm{eff}}$ using the procedure of [35], thus permitting $\langle a_{\mathrm{nc}}\rangle_V$ to be plotted as a function of $1/\langle L\rangle^{\mathrm{true}}$ (Fig. 17.4).

Apparently, the data points in Fig. 17.4 can be fitted by a straight line. The slope of a fit of (17.7) to these data points gives $\langle f\rangle_A = 1.2\pm0.1$ N/m for $K = 192\pm24$ GPa [46]; possible small contributions of two-dimensional faults to $\langle a_{\mathrm{nc}}\rangle$ have been neglected. From (17.4) we compute a compressive bulk pressure of 0.17 ± 0.03 GPa in the as-prepared state. In the limit $\langle L\rangle^{\mathrm{true}} \to \infty$, $\langle a_{\mathrm{nc}}\rangle_V$ should tend toward the lattice parameter of a Pd single crystal,

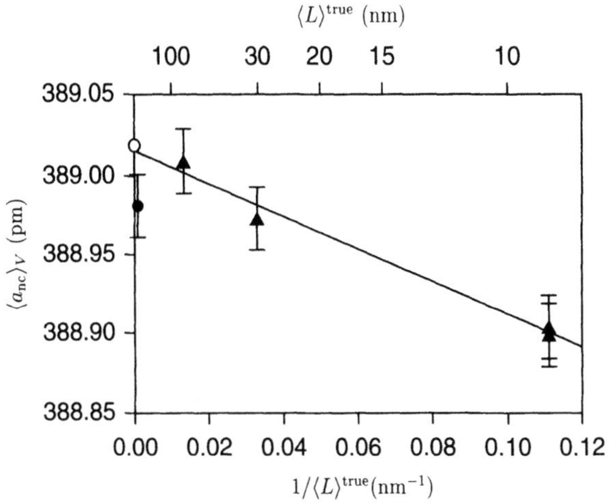

Fig. 17.4. Lattice parameter $\langle a_{\rm nc}\rangle_V$ of nanocrystalline Pd (▲) plotted as a function of the reciprocal average grain intercept length. The solid line is a least-squares fit to the data points. Also shown are the literature value for Pd (○) and the lattice parameter measured for the coarse-grained reference sample of Fig. 17.3 (•)

$a_0 = 389.019 \pm 0.002\,$pm [47]; extrapolation of the fit line in Fig. 17.4 yields a value of 389.02 pm for a_0, which is in remarkably good agreement with the literature value. Moreover, independent measurement of a disk-shaped coarse-grained reference sample of Pd found a lattice parameter of 388.99 pm. This consistency justifies both the assumption of statistical homogeneity as well as the supposition that interface stress generates a homogeneous deformation of the nanocrystallites upon which it acts.

A quantitative X-ray study [48] of the thermal properties of nanocrystalline Pd (prepared by the inert-gas condensation technique, $\langle L\rangle = 13\,$nm) from 16 K to 300 K revealed a lattice parameter $a_{\rm nc} = 389.00\,$pm. Since these measurements were not corrected for the presence of residual strain ($\psi = 0°$), a direct comparison with our result is not amenable.

In a preceding study to this work, we investigated the change of lattice parameter $a_{\rm nc}(\psi = 0°)$ in nanocrystalline Pt as a function of increasing grain size in Bragg-Brentano untilted geometry ($\psi = 0°$, no residual strain correction) in more detail. The result of this work is shown in Fig. 17.5.

Obviously, the change in lattice parameter $a_{\rm nc}(\psi = 0°)$ as a function of $1/\langle L\rangle$ cannot be fitted by a straight line. Depending on the selected $\langle L\rangle$-range, linearly fitted data points assume positive or negative slope even indicating a change of sign of $\langle f\rangle_A$. This finding is unphysical since the value of $\langle f\rangle_A$ as defined in the introduction must be independent of the size of interface area $A \propto 1/\langle L\rangle$. As a consequence, the residual strain correction as suggested in this study seems mandatory when determining absolute values of interface stress from changes of lattice parameter unless neutron or synchrotron radiation is used that allows to probe the total bulk volume of a sample.

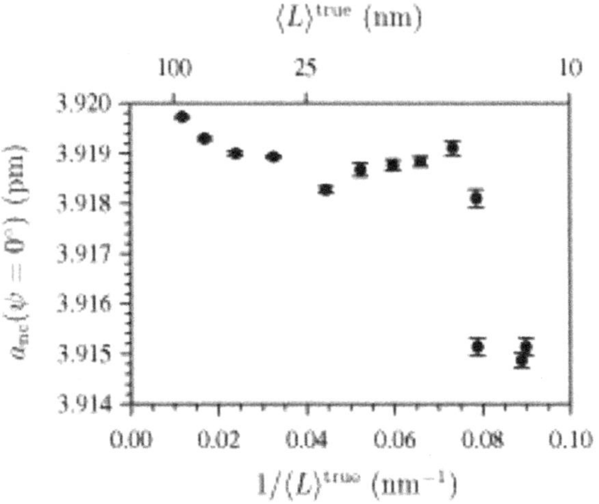

Fig. 17.5. Lattice parameter $a_{\rm nc}(\psi = 0°)$ of nanocrystalline Pt plotted as a function of the reciprocal average grain intercept length

Since $\langle f \rangle_A$ represents an orientational average over many different interfaces, we may assume that its value is characteristic of the predominant type of grain boundary – namely, high-angle [49, 50]) – in the specimen. This conclusion is supported by the fact that calorimetric measurements, performed on an as-prepared nanocrystalline Pd samples, yielded a value of $0.8 \pm 0.1\,\mathrm{J/m^2}$ for the enthalpic part of the area-averaged grain-boundary energy $\langle \gamma \rangle_A$ (compare details in [51]), which agrees well with typical values for random high-angle grain boundaries in Pd [52, 53].

Finally, we note that the value obtained in this study for the interface stress in nanocrystalline Pd differs significantly from $\langle f \rangle_A = 0.1 \pm 0.1\,\mathrm{N/m}$, as measured by Weissmüller and Lemier [21]. This discrepancy is most likely caused by the presence of irreversibly dissolved hydrogen in their specimens [54], as it was found that the sign and magnitude of the interface stress is significantly affected by the segregation of hydrogen to the core region of grain boundaries.

17.4 Outlook

The upsurge in research on nanostructured materials within the last decade has led to a renaissance in experimental and theoretical studies on macroscopic aspects of interfaces. With the ever-growing abilities to engineer surfaces and interfaces and using their impact to tailor the properties of materials with a high interface to volume ratio, the thermodynamic description of surfaces and interfaces has become appreciated again. The particular value of such a description using averaged state variables, lies in the fact that interface phenomena can be qualitatively studied – in many instances even on a quantitative basis – without requiring explicit knowledge about the complexity of structure, dynamics and processes involved on the atomic scale at or near interfaces. Although, it has been recognized for a long time that interface stress f, similarly than interface energy γ, is an important (independent of γ) thermodynamic quantity, it took, however, until only recently that interface stress has become amenable to experimental studies. Expectedly, the study of surface and interface stress has raised a number of not yet answered questions that are addressed in the review papers by Ibach [9] and Spaepen [11]. With regard to nanocrystalline materials, we may focus on a few pertinent issues that are of common interest: Firstly, it would be very useful to learn more about the impact of grain boundaries on thermodynamic equilibrium in nanocrystalline multicomponent materials and the resulting modifications of our e.g. binary phase diagrams; in fact, it would be desirable to make our phase diagrams also applicable to nanoscaled materials by completing them in specifying the state variable A/V (interface area per unit volume of crystal phase) and its effect on the phase boundary lines. Secondly, with the reduction in size of individual structures, e.g. crystalline entities in devices, we observe not only a concomitant decrease of their structural correlation length, but also a shrinkage in the spatial extension of interfaces. There is only little known about the structure and properties of such interfaces of finite extent and any crossover-behavior with reference to their counterparts of quasi-infinite extent. Thirdly, in nanocrystalline materials, the volume fraction of triple junctions – one-dimensional defects that are located where three or more interfaces meet – scales with $(\delta/D)^2$, where δ denotes the structural width of an interface and D represents the average intercept length of the adjoining crystalline phases, thus, indicating that the nature and behavior of materials characterized by D values on the order of a few nanometers will be inevitably influenced by triple junctions (of finite extent), that are as yet scarcely explored.

Acknowledgment

The authors are obliged to C.E. Krill, A. Tschöpe and J. Weissmüller for critical readings of the manuscript and helpful suggestions. We are indebted to Prof. H. Ruppersberg for his guidance in setting up the stress measurements and in interpreting the residual-strain data. This research was supported by the Deutsche Forschungsgemeinschaft (Grant Bi 385/13, Sonderforschungsbereich 277).

References

1. W. Pompe and M. Bobeth: Curr. Opin. Solid St. M. **3**, 269 (1998)
2. J.W. Gibbs: Trans. Conn. Acad. III, (1878)
3. J.W. Gibbs: *Scientific Papers of Gibbs, J. Williard.* Vol. 1 (Dover Publications, New York 1961)
4. C. Herring: *Structure and Properties of Solid Surfaces.* (University of Chicago, Chicago 1953)
5. R. Shuttleworth: Proc. Phys. Soc. A **63**, 444 (1950)
6. J.W. Cahn. In *Interfacial Segregation.* Ed. by W.C. Johnson and J.M. Blakely (ASM, Metals Park, OH 1987) pp. 3–23
7. J.W. Cahn: Acta metall. **28**, 1333 (1980)
8. M.E. Gurtin, J. Weissmüller and F. Larché: Phil. Mag. A **78**, 1093 (1998)
9. H. Ibach: Surf. Sci. Reports **29**, 193 (1997)
10. H.S. Nalwa: *Handbook of Nanostructured Materials and Nanotechnology.* (Academic Press, London 2000)
11. F. Spaepen: Acta Mater. **48**, 1899 (2000)
12. R.C. Cammarata and K. Sieradzki: Annu. Rev. Mater. Sci. **24**, 215 (1994)
13. D. Josell: Acta metall. mater. **42**, 1031 (1994)
14. J.A. Ruud, A. Witvrouw and F. Spaepen: J. Appl. Phys. **74**, 2517 (1993)
15. J. Weissmüller and J.W. Cahn: Acta mater. **45**, 1899 (1997)
16. K.O. Schweitz, H. Geisler, J. Chevallier, J. Bøttiger and R. Feidenhans'l: Mater. Res. Soc. Symp. Proc. **505**, 559 (1998)
17. A.L. Shull and F. Spaepen: J. Appl. Phys. **80**, 6243 (1996)
18. B.M. Clemens, W.D. Nix and V. Ramaswamy: J. Appl. Phys. **87**, 2816 (2000)
19. D. Josell, J.E. Bonevich, I. Shao and R.C. Cammarata: J. Mater. Res. **14**, 4358 (1999)
20. P. Gumbsch and M.S. Daw: Phys. Rev. B **44**, 3934 (1991)
21. J. Weissmüller and C. Lemier: Phys. Rev. Lett. **82**, 213 (1999)
22. R. Birringer, C.E. Krill and M. Klingel: Phil Mag. Lett. **72**, 71 (1995)
23. S. Timoshenko and J.N. Goodier: *Theory of Elasticity.* (McGraw Hill, New York 1951)
24. I. Young: Phil. Trans. Royal Soc. (London) **95**, 65 (1805)
25. P.S. Laplace: Méchanique Céleste **10**, Supp. 1, (1806)
26. C.E. Krill and R. Birringer: Phil. Mag. A **77**, 621 (1998)
27. W. Kreher and W. Pompe: *Internal Stresses in Heterogeneous Solids.* (Akademie, Berlin 1989)
28. S. Nemat–Nasser and M. Hori: *Micromechanics: Overall Properties of Heterogeneous Materials.* (Elsevier, Amsterdam 1993)
29. I. Alber, J.L. Bassan, M. Kantha, V. Vitek and G.J. Wang: Phil. Trans R. Soc. A **339**, 555 (1992)
30. T. Mura: *Micromechanics of Defects in Solids.* (Nijhoff, The Hague 1982)
31. I.C. Noyan and J.B. Cohen: *Residual Stress.* (Springer, Berlin Heidelberg New York 1987), in Japanese. For an English version, see: K. Ishikawa, T. Uemori: Phys. Rev. B **60**, 11841 (1999)
32. H. Ruppersberg, I. Detemple and J. Krier: Phys. Stat. Sol. (a) **116**, 681 (1989)
33. F. Bollenrath, V. Hauk and E.H. Müller: Z. Metallkde. **58**, 76 (1967)
34. A. Tschöpe, R. Birringer and H. Gleiter: J. Appl. Phys. **71**, 5391 (1992)
35. B.E. Warren: *X-Ray Diffraction.* (Dover Publications, New York 1990)

36. R. Delhez, Th.H. de Keijser, and E.J. Mittemeijer: *Accuracy in Powder Diffraction*, pp. 213–253. National Bureau of Standards Special Publication No. 567, edited by S. Block and C.R. Hubbard (Washington, DC: National Bureau of Standards 1980) p. 213
37. C.E. Krill III, L. Helfen, D. Michels, H. Natter, A. Fitch, O. Masson and R. Birringer: Phys. Rev. Lett. **86**, 842 (2001)
38. A. Michels, C.E. Krill, H. Ehrhardt, R. Birringer, and D.T. Wu. Acta Mater. **47**, 2143 (1999)
39. J.B. Nelson and D.P. Riley: Proc. Phys. Soc. **57**, 160 (1945)
40. C.N.J. Wagner. In *Local atomic Arrangements studied by X-ray Diffraction.* (Gordon & Breach, New York 1996) p. 219
41. M.S. Paterson: J. Appl. Phys. **23**, 805 (1952)
42. C.A. Johnson: Acta Cryst. **16**, 490 (1963)
43. C.N.J. Wagner, A.S. Tetelman, and H.M. Otte: J. Appl. Phys. **33**, 3080 (1962)
44. B.E. Warren: J. Appl. Phys. **34**, 1973 (1963)
45. L. Velterop, R. Delhez, Th.H. de Keijser, E.J. Mittemeijer, and D. Reefman. *J. Appl. Cryst.*, 33:296–306, 2000.
46. E. Kröner: *Statistical Continuum Mechanics.* (Springer, Wien New York Udine 1972)
47. JCPS file no. 46-1043. (International Centre for Diffration Data, Newtown Square, PA)
48. J.A. Eastman, M.R. Fitzsimmons and L.J. Thompson: Phil. Mag. B **66**, 667 (1992)
49. H.E. Schaefer, K. Reimann, W. Straub, F. Phillip, H. Tanimoto, K. Brossmann and R. Würschum: Mater. Sci. Eng. A **286**, 24 (2000)
50. J.K. Mackenzie: Acta metall. **12**, 223 (1964)
51. A. Tschöpe and R. Birringer: Acta Metall. Mater. **41**, 2791 (1993)
52. A.R. Miedema: Z. Metallkde. **69**, 287 (1978)
53. L.E. Murr: *Interfacial phenomena in metals and alloys.* (Addison-Wesley, Reading, Ma 1975)
54. J. Weissmüller, private communication (2001)

Part VII

Microstructural Gradients; Thin Films

18 Problems Related to X-Ray Stress Analysis in Thin Films in the Presence of Gradients and Texture

C. Genzel

18.1 Introduction

Surface processing such as grinding, shot-peening, case hardening or coating deposition, is well known today to give rise to complex residual stress fields in the near-surface region of polycrystalline materials, which may vary strongly with depth [1–3]. Because these stresses influence the properties of technical components, such as resistance to abrasive wear or corrosion, to a large extent, detailed knowledge of both their amount and distribution is of increasing interest. Therefore, strong efforts are made in order to develop and to improve suitable methods for their detection. In this connection, non-destructive diffraction methods play a role, because they yield the phase-specific lattice spacings and, thus, the lattice strain in a straightforward and unique way [4].

However, compared with most of the "non-diffraction methods" for residual stress analysis, such as optical, ultrasonic or magnetic methods, X-ray stress analysis (XSA) of polycrystalline materials is a quite long and demanding procedure with respect to both measurement and evaluation. This especially applies to thin films with a strong texture and/or steep residual stress gradients, the investigation of which has become an important topic in modern materials science research, because they influence the mechanical as well as the electronic properties of the film, either in a positive or in a negative way. So the high compressive stresses of TiN- or TiC-coated technical components, for example, contribute to the high hardness of the films, whereas tensile film stresses may lead to failure by cracking and fracture [5–7]. In microelectronics, the residual strain in superlattice structures is used, for example, in order to adjust the electronic band gap and, therefore, the properties of semi-conducting devices such as laser diodes [8]. Misfit dislocations, on the other hand, which are introduced in single-crystal epitaxy between the layer and the substrate in order to reduce the lattice mismatch, may lead to electronic damage, if they grow into the film [9] and penetrate the p–n junction.

The aim of the present chapter is to give a compact survey of the "state-of-the-art" concerning the analysis of residual stresses in thin films by means of X-ray diffraction methods. In consideration of the limited length of the chapter, the discussion will be restricted to polycrystalline films, the XSA of which can be treated within the framework of the kinematical theory of diffraction. But even here the number of experimental techniques developed so far is too large for us to give a detailed report on all of them. For a more comprehensive survey the reader is referred, for example, to [3] and [10].

This chapter deals with the special problem of the depth-resolved analysis of the residual stress/strain distribution with respect to the film thickness. After a short discussion of the possible origins of residual stress in thin films, it will be shown in Sect. 18.2 that steep residual (macro-) stress gradients $\sigma(z)$ may even occur in very thin layers. Section 18.3 addresses the basic problems in the depth-resolved thin-film XSA. Here, the appropriate variation of the X-ray penetration depth τ within the material plays a key role. Consequently, the concepts developed so far in order to analyse residual stress gradients in thin films, summarised in this section, are divided according to the method that is used for the variation of τ. In Sect. 18.4 some of the most interesting experimental methods are reported, which deal with very sophisticated (but unfortunately frequently occurring) problems of depth-resolved XSA in very thin and/or textured films.

18.2 Residual Stress in Thin Films: Its Origin and Possible Stress States

18.2.1 Extrinsic and Intrinsic Film Stresses

Long-range macro- or homogeneous phase-specific residual stress fields in polycrystalline materials usually arise from an inhomogeneous plastic deformation with respect to any sample-cross section, or at least to any macroscopic sample area [1,3,11]. In many cases, only the near-surface region after mechanical processing is of particular interest [2]. Macro-residual stresses may also arise from the elastic strain between individual parts of technical compounds. This especially applies to thin films deposited on any kind of substrate (amorphous, and single- or polycrystalline). According to their origin in the film evaporation process, it is convenient to distinguish between extrinsic and intrinsic residual stresses (Fig. 18.1).

Extrinsic residual stresses have been generally defined as the thermal stresses that are due to the difference in the coefficients of thermal expansion between the film and the substrate, $\Delta\alpha$ (Fig. 18.1a). If T_{dep} is the deposition temperature, the thermal stress σ_T at any temperature T is as follows [10]:

$$\sigma_T \propto (T_{\mathrm{dep}} - T)\Delta\alpha - \varepsilon^{pl} . \qquad (18.1)$$

Here, ε^{pl} is the plastic strain that arises, for large $\Delta\alpha$ or ΔT, if the film stress exceeds the yield point at any temperature. Therefore, the total stress that remains within the film at room temperature is reduced by an amount proportional to ε^{pl}.

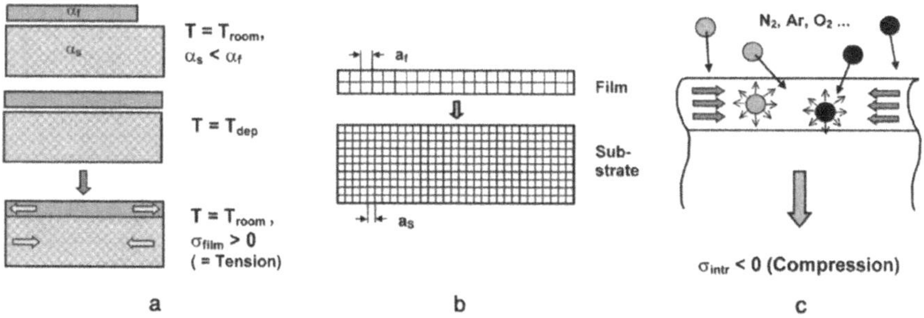

Fig. 18.1. Possible origins for the formation of extrinsic (**a**) and intrinsic (**b,c**) residual stresses in thin films. (**a**) The thermal stress due to the difference between the coefficients of thermal expansion of the film and the substrate, α_f and α_s, respectively. (**b**) The misfit-induced stress due to the mismatch between the lattice parameters of the film and the substrate, a_f and a_s, respectively. (**c**) Stresses due to defect incorporation during evaporation

Intrinsic residual stresses are defined as those stresses having lattice-inherent origins. In single-crystal hetero-epitaxy, for example, where the lattice parameter of the growing film a_f often differs only by a small amount from that of the substrate, a_s, the residual stress σ_M induced by this lattice misfit is proportional to the interfacial strain $\varepsilon_\parallel$ [12] (Fig. 18.1b):

$$\sigma_M \propto \frac{a_f - a_s}{1/2(a_f + a_s)} = \varepsilon_\parallel \,. \tag{18.2}$$

Equation (18.2) holds for pseudomorph film growth without taking into account bending effects of the substrate as well as misfit dislocations, which would reduce the interfacial strain by a certain amount. Further origins for intrinsic film stresses, which are of particular interest in the physical (PVD) or chemical vapour deposition (CVD) of polycrystalline films, are defect incorporation into the film during evaporation (Fig. 18.1c), the formation of non-equilibrium structures, re-crystallisation etc. (see, for example, [13, 14]). Compared with the thermal or misfit-induced stresses, which can be calculated (or at least estimated) by means of (18.1) and (18.2), respectively, a theoretical evaluation of the residual stresses of the last group is difficult, because their amount, as well as their distribution within the growing film, depend on the deposition kinetics in a very complex way.

18.2.2 Possible Macrostress States in Thin Films

In the following, the substrate is assumed to be much thicker than the film, so that bending effects of the complete system due to internal moments [15] can be neglected. Furthermore, let us assume, for the sake of simplicity, that the film should consist of a single-phase material. In this case, the residual

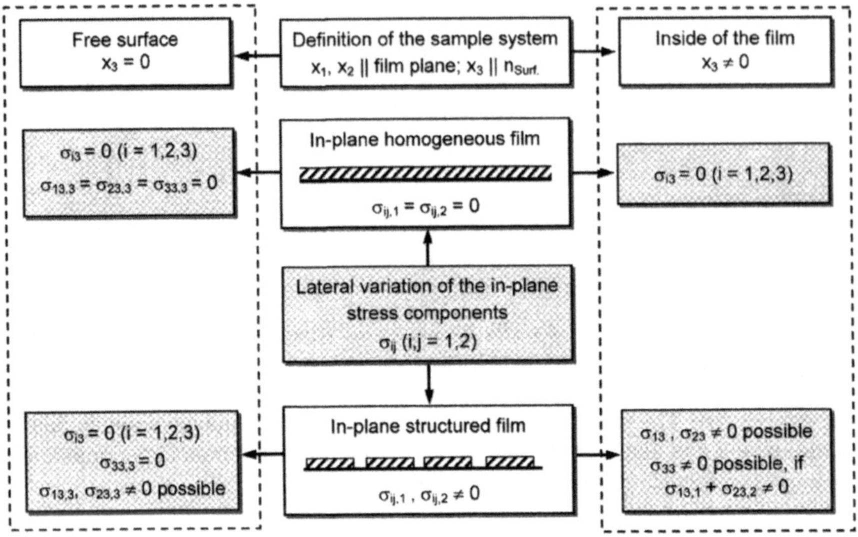

Fig. 18.2. Possible macrostress states in single-phase thin films. According to the Einstein suffix notation, a derivate of the term with respect to a coordinate is indicated after the comma and summation over repeated indices is used. Note that the existence of non-zero derivates of the σ_{i3}-components with respect to x_3 is a direct consequence of the differential conditions (18.3) and (18.4) and depends on the existence of lateral gradients of the in-plane stress components σ_{ij} $(i, j = 1, 2)$. See the text for further details

stresses analysed by means of X-ray diffraction, in which we are interested here, are the actual first kind or macrostresses, which are the mean value over a sufficiently large number of grains irradiated by the X-ray beam [16].

In order to get an idea of the possible macrostress states within thin films (Fig. 18.2), we have to consider the fundamental equations of elasticity [17–20]. They are given by the differential equilibrium conditions which, using the Einstein suffix notation, are

$$\sigma_{ij,j} = 0 \quad (i, j = 1, 2, 3) \tag{18.3}$$

and the boundary conditions for any external body surface that is free of external forces:

$$\sigma_{ij} n_j = 0 \, , \tag{18.4}$$

where n_j is the exterior unit normal vector on the free surface of the body. It should be noted that the conditions (18.3) must be fulfilled at any point, i.e. at the surface and in the interior of the body [18].

For example, consider the case of a thin homogeneous film on a substrate. Defining the outer surface normal as the x_3 direction, with x_1 and x_2 in the plane of the film, the boundary conditions (18.4) take the form:

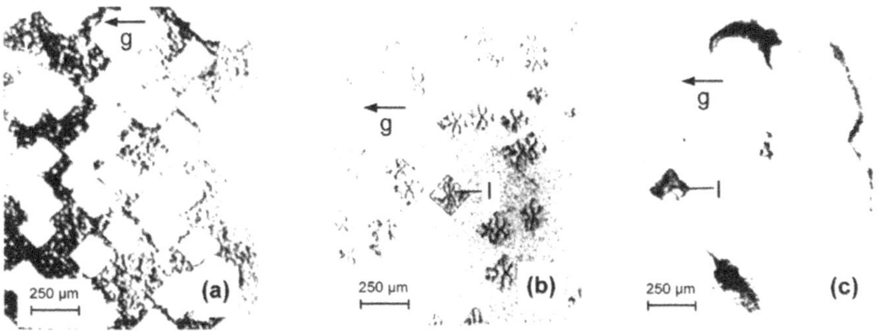

Fig. 18.3. X-ray topographs of small (100) single-crystalline $Pb_{0.8}Sn_{0.2}Te$ islands (edge length and height approximately 180 µm and 15 µm, respectively) deposited by liquid-phase epitaxy on a (100) PbTe substrate. Due to the lattice misfit of $m = -4.2 \times 10^{-3}$ the film islands and the substrate diffract at different Bragg angles. (**a**) A double crystal topograph of the substrate, 620-reflexion. (**b**) A double crystal topograph of the film islands (I), 620-reflection, superposition of four sample positions separated by $\Delta\theta = 0.03°$ on one film. (**c**) A transmission Lang topograph, 020-reflection of a film island after substrate removal (after [23])

$$\sigma_{13} = \sigma_{23} = \sigma_{33} = 0 \quad \text{at} \quad x_3 = 0\,. \tag{18.5}$$

Far from the edges of the film, (18.3) and (18.4) do not contain any restrictions for the in-plane stress components σ_{11}, σ_{22} and σ_{12}. However, the existence of macrostress fields of the out-of-plane stress components σ_{13}, σ_{23} and σ_{33}, respectively, within the film is attached to the variation of the in-plane stresses with respect to x_1 and x_2 (cf. Fig. 18.2). If the in-plane residual stresses σ_{ij} $(i, j = 1, 2)$ do not vary at least within the irradiated part of the film, the derivatives of the σ_{i3}-components with respect to x_3 are zero and therefore, the out-of-plane stresses have to vanish with respect to the total film thickness, because they cannot change their values at the free surface given by the boundary condition (18.5). Hence, the film stress state remains biaxial. In the case of an in-plane structured film, on the other hand, the relaxation of the in-plane stresses near the free edges will lead to complex triaxial stress states at least near the edges.

For example, let us consider a hetero-epitaxial system consisting of small $Pb_{0.8}Sn_{0.2}Te$ film islands deposited by liquid-phase epitaxy (LPE) on a (100) PbTe substrate (Fig. 18.3). For the investigation of the residual stress/strain state of such single-crystalline film structures, it is favourable to use X-ray topography [21, 22]. So the striking vertical stripe-like contrast of the film islands obtained in the superposition topograph (Fig. 18.3b) is due to a strong curvature of the crystal lattice. After removal of the substrate, the film islands show a nearly uniform contrast (Fig. 18.3c), which indicates the relaxation of the film stress. Assuming that the lattice misfit $m = -4.2 \times 10^{-3}$ between the substrate and the film islands is completely accommodated by pure elastic

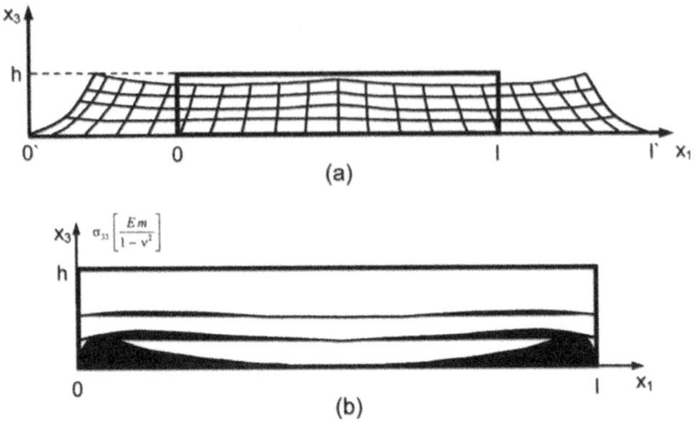

Fig. 18.4. The strain (**a**) and out-of-plane normal stress σ_{33} (**b**) within a heteroepitaxial film island on a single-crystalline substrate. The lattice misfit m (taken from the example in Fig. 18.3) between the film and the substrate is assumed to be completely accommodated by pure elastic deformation of the film. The displacements in (**a**) are enlarged by a factor $1/|m| \approx 240$. The stresses normal to the film plane shown in (**b**) refer to $x_3 = 0$ (interface), $h/4$ and $h/2$, respectively. They are given in arbitrary units $[Em/(1-\nu^2)]$ (E – Young's modulus, ν – Poisson ratio) (after [23])

deformation of the film lattice, the residual stress/strain state of the islands can be calculated by simple analytical models [23]. Figure 18.4 shows that the deformation of the film lattice as well as the stress component σ_{33} normal to the film plane, which is necessary to keep the interface flat, are especially large near the edges of the islands. In the literature, some hints are reported that such complex triaxial residual stress states can even occur in closed polycrystalline films. In [24], for example, the columnar grain structure in TiN coatings is conceived as an assembly of freestanding columns (Fig. 18.5). The authors assume a strong interaction, with full elastic accommodation of the thermal misfit at the column/substrate interface on the one hand, but a rather weak interaction between the individual columns. From this state of strain, they conclude relatively high shear stresses σ_{12} and normal stresses σ_{33} perpendicular to the interface, which can stimulate debonding of the coating. According to the authors, these stresses should exist at least near the edges of the columns, which would be similar to the case shown in Fig. 18.4.

18.3 Basic Principles of X-Ray Stress Gradient Analysis

18.3.1 Problems Related to Thin Films

The considerations and examples in Sect. 18.2 served the purpose of explaining that the macrostress state in thin films can take on very complex forms, the investigation of which by means of X-ray diffraction requires the application of rather sophisticated methods. The diffraction analysis of non-uniform

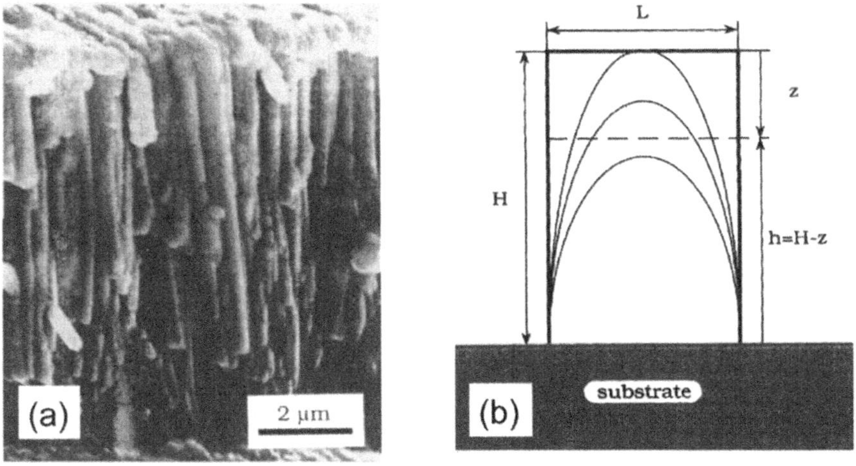

Fig. 18.5. The microstructure of a CVD TiN coating on a Mo substrate (**a**) and a schematic presentation of a freestanding column with iso-strain contours on a substrate (**b**) (after [24])

near-surface stress fields can be subdivided into individual steps, each of which involves a number of practical problems that depend on the material to be investigated itself, as well as on the applied experimental technique (Table 18.1) [25].

The first step is the determination of the interplanar lattice spacings $d_{\varphi\psi}(hkl)$ for various positions φ and ψ of the diffraction vector $\boldsymbol{g}$ with respect to the sample reference system $\{\boldsymbol{P}\}$ (Fig. 18.6). In the case of textured thin films, restrictions exist with regard to the accessible measurement range (φ, ψ), because the diffracted intensity is rather weak outside the texture poles. Another problem may arise due to very steep stress/strain gradients within the penetration depth of the X-rays [27]. They can be shown to cause an asymmetry in the diffraction line profiles, the disregarding of which could lead to considerable errors in the evaluation of the stress gradient [28].

Because of the finite penetration of X-rays into the material, an appropriate measure has to be defined for the depth ξ below the surface to which the lattice spacings can be assigned. Here, some problems may arise due to the arbitrary character of the definitions; for a detailed discussion the reader is referred to [29]. A suitable measure is given by the so-called penetration depth τ, which is based on the idea that the contribution to the recorded signal of each sublayer of thickness dz at a depth z below the surface must be weighted with the exponential attenuation factor according to Beer's law:

$$\langle z \rangle = \frac{\int_0^D z \exp(-\mu k z) dz}{\int_0^D \exp(-\mu k z) dz} . \tag{18.6}$$

Table 18.1. Problems involved in the evaluation of near-surface residual stress gradients by diffraction methods (after [25])

Step in the procedure of X-ray stress gradient analysis	Possible problems
Determination of the lattice spacings	Asymmetric line profiles due to steep strain gradients
by Bragg's equation:	Limited measurement range (φ, ψ) due to texture
$d_{\varphi\psi}(hkl) = \dfrac{\lambda}{2\sin\theta(hkl)}$	small irradiated volume in the case of very thin films
Assignment of $d_{\varphi\psi}(hkl)$ to a depth ξ below the surface: $d_{\varphi\psi}(hkl) \rightarrow d_{\varphi\psi}(hkl, \xi)$	Definition of the depth ξ
Calculation of the strain depth profiles:	Availability of the accurate strain-free lattice parameter $d_0(hkl)$
$\varepsilon_{\varphi\psi}(hkl, \xi) = \dfrac{d_{\varphi\psi}(hkl, \xi) - d_0(hkl)}{d_0(hkl)}$	$d_0(hkl, z)$ due to composition gradients
Evaluation of the residual stress gradient:	Determination of the stress factors $F_{ij}(hkl, \varphi, \psi)$
$\varepsilon_{\varphi\psi}(hkl, \xi) = F_{ij}(hkl, \varphi, \psi)\sigma_{ij}(\xi)$	Influence of the elastic anisotropy in the presence of texture
	Max. 6 independent stress components σ_{ij}
	Transformation $\sigma_{ij}(\xi) \rightarrow \sigma_{ij}(z)$

In (18.6), μ is the linear absorption coefficient of the material, D denotes the sample thickness and k is the geometry factor, which depends on the diffraction geometry (cf. Fig. 18.7):

$$k = \frac{\sin\alpha + \sin\beta}{\sin\alpha \sin\beta} . \tag{18.7}$$

For bulk materials, the upper integration limit D in (18.6) can be replaced by infinity, which yields the "conventional" penetration depth:

$$\tau = \langle z \rangle_{D\rightarrow\infty} = \frac{1}{\mu k} . \tag{18.8}$$

τ defines the thickness of a surface layer the diffracted integrated power P_τ of which is $(1 - 1/e)$ or 63% of that of an infinitely thick sample, given by P_∞. In the case of thin films with a thickness D smaller than the penetration

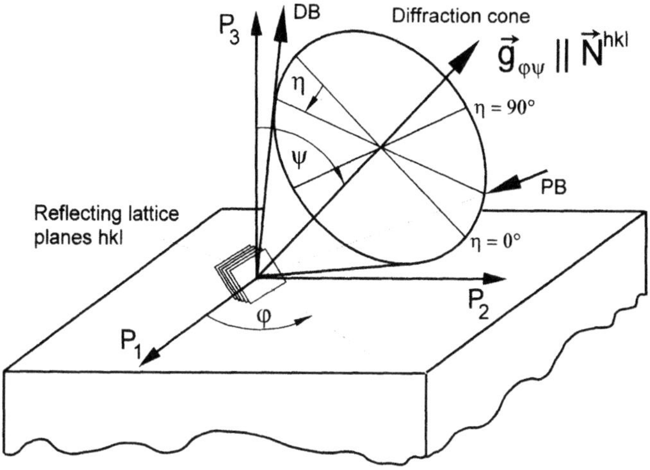

Fig. 18.6. Diffraction geometries in X-ray stress analysis. η denotes the rotation of the sample around the diffraction vector $\boldsymbol{g}_{\varphi\psi}$ for a fixed measurement direction (φ, ψ). The positions $\eta = 0°$ and $\eta = 90°$ correspond to the limiting cases of the Ω-mode (for $\psi < 0$) and the Ψ-mode, respectively. PB, DB – primary and diffracted beam, respectively

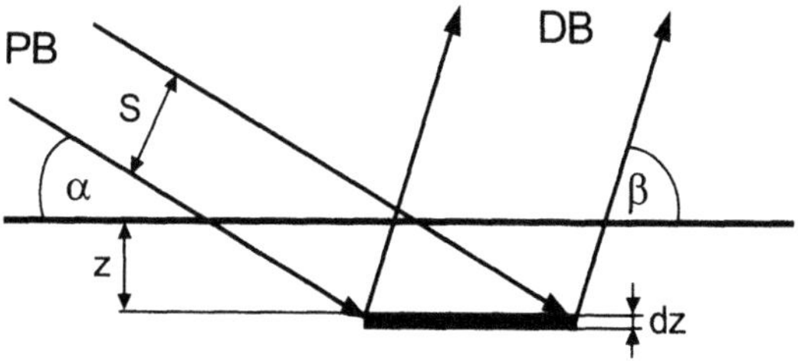

Fig. 18.7. A schematic view of the geometrical path of the X-rays within the material. α and β denote the angles of the primary beam (PB) and the diffracted beam (DB), respectively, to the sample surface

of the X-rays, the evaluation of (18.6) yields the so-called information [29] or effective penetration depth [28]:

$$\tau_{\text{eff}} = \frac{1}{\mu k} - \frac{D \exp(-\mu k D)}{1 - \exp(-\mu k D)} \,. \tag{18.9}$$

The second step in residual stress gradient analysis by diffraction methods concerns the calculation of the lattice strains $\varepsilon_{\varphi\psi}$. Especially in the case of triaxial stress states, even small uncertainties in the stress-free lattice parameter, d_0, are known to lead to large errors in stress evaluation [30]. Furthermore, gradients $d_0(z)$ due to variations in the chemical composition, for example, may produce "ghost stresses" which would also falsify the depth profiles $\sigma_{ij}(z)$ [29].

The final and probably most difficult step is the calculation of the individual stress profiles $\sigma_{ij}(z)$ from the strain distributions $\varepsilon_{\varphi\psi}(\xi)$. Assuming a linear correlation according to Hooke's law between the lattice strain $\varepsilon_{\varphi\psi}(hkl)$ of the diffracting crystallites and the mechanical stress σ_{ij} defined in the sample system $\boldsymbol{P}$, the corresponding formulation for quasi-isotropic bulk materials is given by the fundamental equation of XSA:

$$\begin{aligned}\varepsilon_{\varphi\psi}(hkl,\tau) = {} & \frac{1}{2}s_2(hkl) \\ & \times \left\{ \begin{array}{l} \left[\sigma_{11}(\tau)\cos^2\varphi + \sigma_{22}(\tau)\sin^2\varphi + \sigma_{12}(\tau)\sin 2\varphi\right]\sin^2\psi \\ + \left[\sigma_{13}(\tau)\cos\varphi + \sigma_{23}(\tau)\sin\varphi\right]\sin 2\psi + \sigma_{33}(\tau)\cos^2\psi \end{array} \right\} \\ & + s_1(hkl)\left[\sigma_{11}(\tau) + \sigma_{22}(\tau) + \sigma_{33}(\tau)\right] , \end{aligned} \quad (18.10)$$

where $s_1(hkl)$ and $1/2s_2(hkl)$ are the diffraction elastic constants (DEC). $\varepsilon_{\varphi\psi}(hkl,\tau)$ and $\sigma_{ij}(\tau)$ in (18.10) are the exponentially averaged "Laplace" profiles of the actual strain and stress depth profiles, $\varepsilon_{\varphi\psi}(hkl,z)$ and $\sigma_{ij}(z)$, respectively [31]:

$$\varepsilon_{\varphi\psi}(hkl,\tau) = \frac{\int \varepsilon_{\varphi\psi}(hkl,z)e^{-z/\tau}dz}{\int e^{-z/\tau}dz} ,$$

$$\sigma_{ij}(\tau) = \frac{\int \sigma_{ij}(z)e^{-z/\tau}dz}{\int e^{-z/\tau}dz} . \quad (18.11)$$

In the case of textured materials or thin films, it is convenient to rewrite (18.10) in the form

$$\varepsilon_{\varphi\psi}(hkl,\tau) = F_{ij}(hkl,\varphi,\psi)\sigma_{ij}(\tau) , \quad (18.12)$$

where the F_{ij} are the so-called stress factors [32]. They can be either measured experimentally, which is usually quite a long and demanding procedure, or calculated using different models for the grain interaction.

Often, however, it is difficult to decide which model fits best to a particular problem and, therefore, much research work has been done in this field in recent times. For a detailed discussion of these problems the reader is referred, for example, to [33] and to the comprehensive survey given elsewhere in this book [34].

The symmetry of (18.10) with respect to ψ allows only the separated evaluation of the shear components $\sigma_{13}(\tau)$ and $\sigma_{23}(\tau)$ [35]. The normal components $\sigma_{ii}(\tau)$, however, remain coupled, which may result in serious problems in stress gradient evaluation. So the number of unknown parameters needed for describing the individual depth profiles σ_{ii} in the z-space becomes large and, consequently, the systems of equations to be solved in order to determine them by a least-squares fit from (18.10) or (18.12) may become ill-conditioned.

18.3.2 Concepts in Depth-Resolved Thin-Film X-Ray Residual Stress Analysis

From the considerations of the last section it should have become clear that the appropriate variation of the X-ray penetration depth plays a key role in depth-resolved residual stress analysis. For this reason, many efforts have been made in order to develop suitable methods to record lattice spacing depth profiles, which take into account the problems listed above. These methods are based on the fundamental equations (18.6)–(18.12), but differ mainly in the method of defining the geometry factor k. Table 18.2 gives a survey of some of the most interesting methods for the variation of the X-ray penetration depth. It should be emphasised that the references in the last column refer to their application in thin-film stress gradient analysis; for a comprehensive survey of the use of the methods in general cases of depth-resolved XSA, see, for example, [3]. While the first two approaches are directly related to the $\sin^2\psi$ method [36] performed in the Ψ- and the Ω-geometry, respectively, the remaining measurement techniques were especially developed to meet the special demands of thin-film stress analysis. A more detailed discussion will be given in Sect. 18.4.

There is no method that solves all of the problems listed in Table 18.1 at once. So the choice of the most appropriate measurement technique always depends on the special question under consideration. In the case of untextured, or at least weakly textured, materials, for example, use can be made of those methods that are based on strain depth profiling with variable orientation (φ, ψ) of the diffraction vector with respect to the sample system. For strongly textured samples, on the other hand, strain depth profiling at a fixed orientation should be advantageous. In other cases, if one is interested in the residual stress state at a certain depth below the surface, the application of those methods that are based on strain analysis at a predefined depth is recommended.

It should be emphasised that some of the XSA methods given in Table 18.2 can be used in several of the modes listed below the table. Some of these possibilities will be discussed in the next section, in order to give the interested reader ideas for planning his or her own experiments in thin-film stress analysis.

Table 18.2. Methods applied in the depth-resolved XSA of thin films. The footnotes denote the mostly used/possible modes, respectively. See the text for details

Method	Penetration depth	Remarks	Ref.
$\sin^2\psi$-method / Ψ-mode	$\frac{1}{2\mu}\sin\theta\cos\psi$	Sample tilt around axis parallel to the diffraction plane[a,b]	[37, 38]
$\sin^2\psi$ method/ Ω-mode	$\frac{1}{2\mu}\frac{\sin^2\theta-\sin^2\psi}{\sin\theta\cos\psi}$	Sample tilt around axis perpendicular to the diffraction plane[a,b]	[39]
$\sin^2\psi$ method/ mixed Ω–Ψ mode	$\frac{1}{2\mu}\frac{\sin\alpha\sin(2\theta-\alpha)}{\sin\alpha+\sin(2\theta-\alpha)}\cos\psi_d$	Sample tilt in the Ω mode by $\lvert\alpha-\theta\rvert$ towards small incidence, followed by successive tilts in the Ψ mode by ψ_d with subsequent azimuth correction[a,b/d]	[40, 41]
Low incidence beam angle diffraction (LIBAD)	$\frac{1}{2\mu}\frac{\sin\alpha\sin(2\theta-\alpha)}{\sin\alpha+\sin(2\theta-\alpha)}$	2θ scan for constant (small) incidence angles α between the primary beam and the surface[b,d]	[42, 43]
Scattering vector method	$\frac{\sin^2\theta-\sin^2\psi+\cos^2\theta\sin^2\psi\sin^2\eta}{2\mu\sin\theta\cos\psi}$	Lattice spacing depth profiling by sample rotation η around the diffraction vector for fixed orientation (φ,ψ)[a,c/d]	[27, 44]
Grazing incidence diffraction (GID)	$\frac{\lambda}{4\pi}\left\{\frac{1}{2}\left[\sqrt{(\alpha^2-\alpha_c^2)^2+4\beta^2}+\alpha^2-\alpha_c^2\right]\right\}^{-\frac{1}{2}}$	Lattice spacing depth profiling at small incidence angles α near the critical angle of total reflection, α_c, either in symmetric ($\psi\approx 90^\circ$) or asymmetric ($\psi\approx\theta$) Bragg diffraction geometry[a,c]	[45, 46]

[a] Stress gradient analysis with direct conversion $\varepsilon(\tau)\rightarrow\sigma(\tau)$
[b] Strain depth profiling with variable orientation (φ,ψ) of the diffraction vector
[c] Strain depth profiling at fixed orientation (φ,ψ) of the diffraction vector
[d] A $\sin^2\psi$-based analysis at predefined penetration depths

18.4 Problem-Oriented Solutions for Special Cases in Thin-Film XSA

18.4.1 Measurement Techniques Using Grazing and Low Incident Angles

The Grazing Incidence Diffraction Method (GID)

From X-ray diffraction theory (see, for example, [47,48]), it is well known that the refractive index $n = 1 - \delta - i\beta$ (δ and β are the refractive index decrement and the absorption index, respectively) for most materials is slightly smaller than one. Therefore, the X-ray beam penetrates only the top few nanometres or less of the surface, if the angle between the incident beam and the surface, α, becomes smaller than the so-called critical angle of total external reflection, α_c, given by $\alpha_c = \sqrt{2\delta}$ [49]. Typical values for the critical angle are in the 0.2° to 0.6° range for Cu$K\alpha$ radiation with $\lambda = 0.154$ nm. Figure 18.8 shows that the $1/e$ penetration depth τ according to Beer's law [cf. (18.8)] is reduced drastically, by about two orders of magnitude, as the incidence angle α approaches α_c. The inclination ψ of the diffraction vector with respect to the sample system, however, remains nearly unchanged during strain depth profiling, because the slope of the $\tau(\alpha)$-curve is very steep for $\alpha \approx \alpha_c$.

Two basic geometries are possible for observing Bragg diffraction with the glancing-angle technique (Fig. 18.9). For asymmetric Bragg scattering geometry (Fig. 18.9a), a refraction correction of the observed Bragg angle is

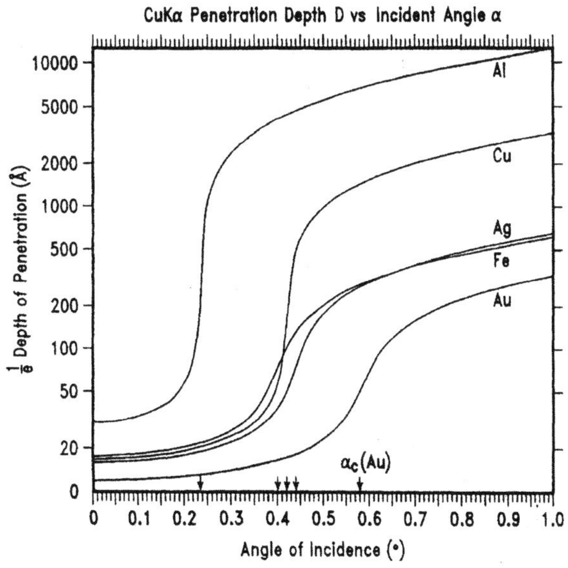

Fig. 18.8. Calculated Cu$K\alpha$ X-ray penetration depths for various metals (after [10])

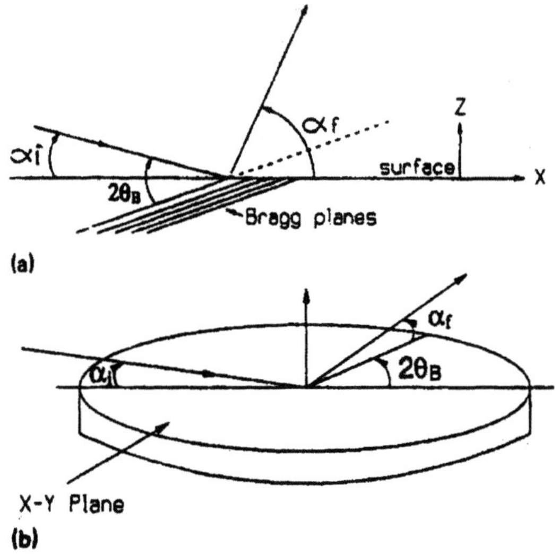

Fig. 18.9. Glancing-angle geometries: (**a**) asymmetric Bragg diffraction geometry and (**b**) symmetric Bragg diffraction geometry. α_i and α_f are the angles that the incident beam and the diffracted beam, respectively, make with the surface (after [46])

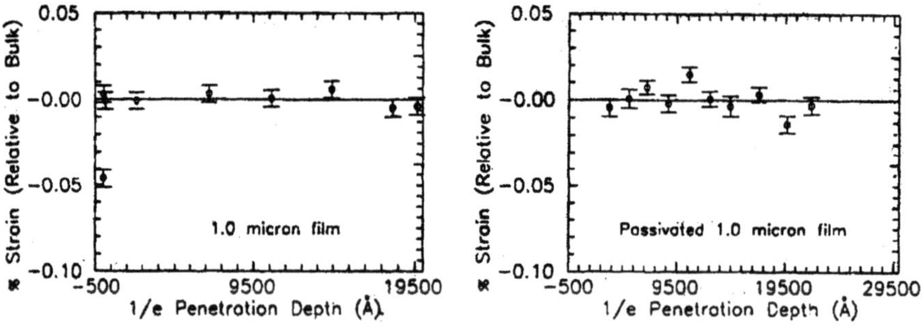

Fig. 18.10. Relative percentage strain (deviation from "bulk") versus penetration depth for Al – 2% Cu thin films on Si (001) wafers. The strain analysis was performed by means of the symmetric GID method at the {422} lattice planes, which are perpendicular to the ⟨111⟩ textured films (after [46])

needed [50]. In the second technique, which results in a symmetric Bragg geometry (Fig. 18.9b), this correction is not necessary.

For the analysis of the in-plane residual stress/strain state in thin films, in particular the symmetric mode, is of interest, because the inclination angle ψ between the surface normal and the reflecting lattice planes is close to 90° and, therefore, lattice spacing depth profiling yields approximately $d_{\psi=90^\circ}(hkl,\tau) = d_\perp(hkl,\tau)$. Figure 18.10 gives an example of a strain gradient analysis by means of the symmetric GID method [46]. By choosing the {422} lattice planes with their normals perpendicular to the ⟨111⟩ fibre axis of the films, the authors found that the residual stress relaxation in strongly ⟨111⟩ textured Al − 2% Cu films on (001) silicon wafers is restricted to a very small surface layer of about 35 Å.

The example demonstrates that the GID method is well-suited for detecting steep residual strain and stress gradients, even in very thin surface layers with a strong texture. However, it should be noted that the evaluation of the in-plane residual stress depth profiles, $\sigma_{\parallel}(\tau)$, from the measured lattice spacing profiles, $d_{\perp}(hkl, \tau)$, according to (18.10) or (18.12), respectively [51], requires knowledge of the exact value of the strain-free lattice parameter, $d_0(hkl)$, because even a small uncertainty would lead to a considerable shift of the $\sigma_{\parallel}(\tau)$-curves on the absolute scale, in the same sense as it is known from triaxial bulk stress analysis by means of neutron diffraction methods [52].

The Low-Incident-Beam-Angle-Diffraction Method (LIBAD)

Working in the asymmetric Ω-mode of diffraction with incident angles α larger than the critical angle α_c leads to penetration depths expected from normal absorption that start to appear on the right-hand side of Fig. 18.8. The so-called low-incident-beam-angle-diffraction method (LIBAD) [42, 43] is based on the idea that the depth of X-ray penetration τ can be controlled using a proper incident angle α that is kept fixed during the recording of the diffraction patterns (Fig. 18.11). Thus, the inclination angle ψ between the surface normal and the diffraction vector becomes a discrete function of α and the respective reflection hkl, given by $\psi(hkl) = \alpha - \theta(hkl)$. Figure 18.12 shows that the penetration depth τ in this case is dominated by α (cf. Table 18.2), but it is almost independent of the Bragg angle over a wide range of 2θ.

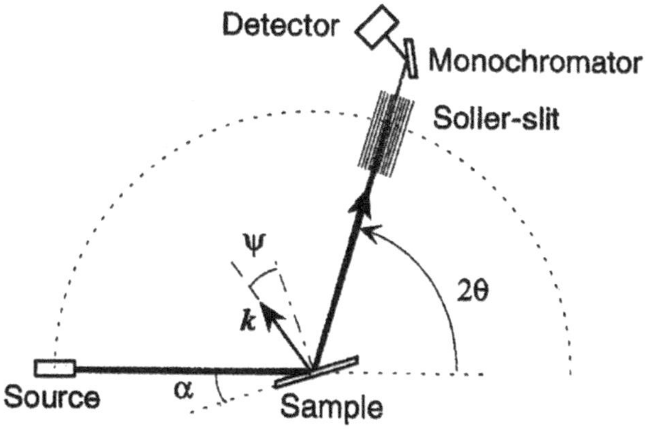

Fig. 18.11. The experimental set-up for the LIBAD method. The angle α between the incident beam and the surface remains fixed during the recording of the diffraction patterns by stepwise variation of 2θ (after [43])

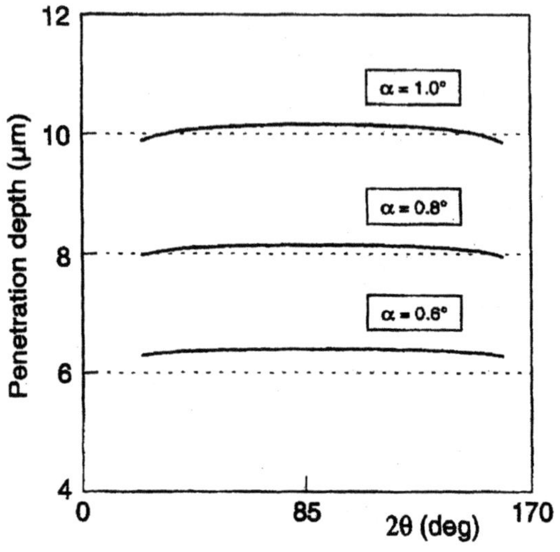

Fig. 18.12. The 2θ dependence of the X-ray penetration depth of Cu$K\alpha$ radiation into diamond in an Ω-diffractometer at various incident-beam angles (after [43])

For this reason, the method allows the analysis of the in-plane residual stress state for predefined values of τ adjusted by the incident angle α, which are much shallower than the penetration depths encountered in the conventional $\sin^2\psi$ technique. So the example in Fig. 18.13 demonstrates that the intensity of substrate peaks in the diffraction patterns can be effectively suppressed if the incident-beam angle α is reduced to such small values, at which the penetration depth approaches the film thickness [43]. Plotting the lattice strain data $\varepsilon(hkl)$ obtained for the individual reflections hkl versus $\cos^2\psi$, the in-plane residual stress $\sigma_{\|}$ at the penetration depth adjusted by α can be calculated by a least-squares fit from [53] (Fig. 18.14):

$$\varepsilon_{\psi}(hkl) = \left[2s_1(hkl) + \frac{1}{2}s_2(hkl)\right]\left[1 - \frac{1/2s_2(hkl)\cos^2\psi}{2s_1(hkl) + 1/2s_2(hkl)}\right]\sigma_{\|} \,. \quad (18.13)$$

Using an appropriate thin-film attachment consisting of a parallel beam soller slit (cf. Fig. 18.11) and a crystal monochromator in the diffracted beam in order to suppress problems due to defocusing, the LIBAD experiments can be performed even on a two-circle Ω-diffractometer. However, attention should be paid to some drawbacks of the method. The rather low accuracy in the peak determination at small 2θ angles may lead to considerable errors in the strain/stress evaluation. Furthermore, the DEC must be known for all measured reflections. An added complication is the distribution of micro- or pseudo-macrostresses and their dependency on the different reflections used. And, finally, it should be noted that the method is only applicable to untextured, or at least only weakly textured, films.

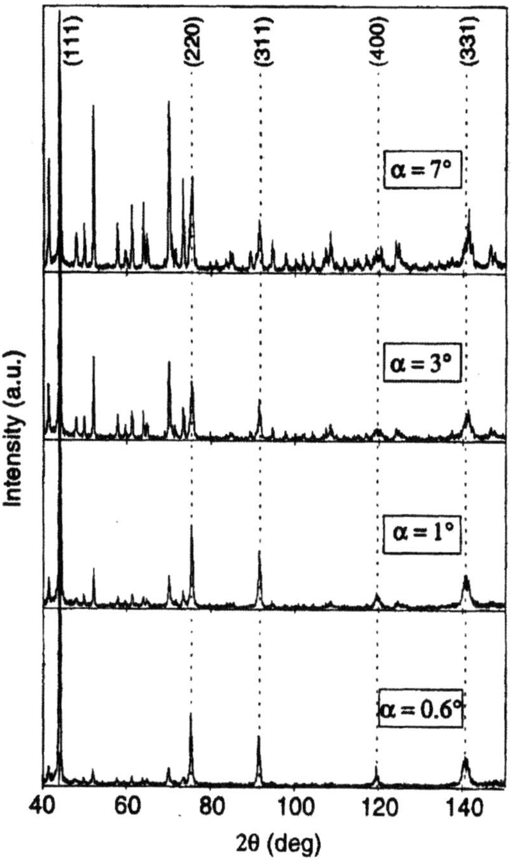

Fig. 18.13. The diffraction patterns of a 6 µm thick CVD diamond film on a SiAlON substrate obtained at various incident-beam angles α with Ku$K\alpha$ radiation (after [43])

18.4.2 Experimental Techniques Based on Sample Tilt

The "Mixed Ω–Ψ-Mode Technique"

In many cases, the average penetration depth achieved by the $\sin^2\psi$ technique is much larger than the thickness of the investigated film. This may result in a wrong interpretation of the $d_\psi - \sin^2\psi$ data, because the plots often remain linear up to large inclination angles y, even in the presence of steep in-plane residual stress gradients $\sigma_{\|}(z)$ in the film [28]. For this reason, efforts have been made in order to reduce the $\sin^2\psi$ penetration depth by an additional inclination of the sample system relative to the laboratory system [54].

A method developed especially for thin-film stress gradient analysis was suggested by [40, 41]. This "mixed Ω–Ψ-mode technique" is based on the adjustment of the penetration depth in two stages. In the first step, the sample

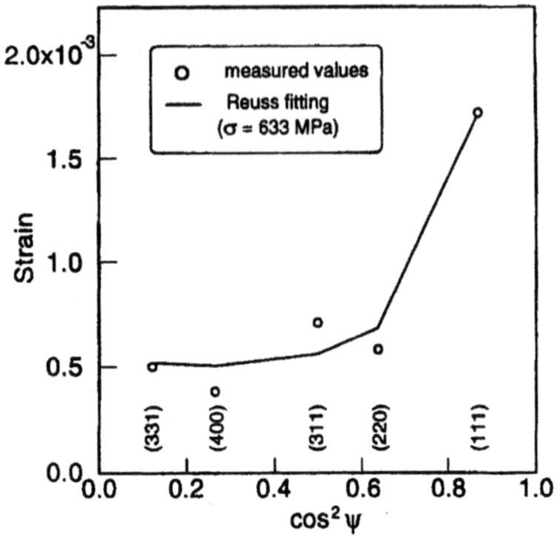

Fig. 18.14. The lattice strain versus $\cos^2\psi$ measured with Ku$K\alpha$ radiation on a CVD diamond film (thickness 6 µm) on a SiAlON substrate for an incident-beam angle of 0.6°. The least-squares fit based on the Reuss model is shown by a solid line (after [43])

is tilted in the Ω-mode around the Ω-axis perpendicular to the diffraction plane by an amount $|\alpha - \theta|$ in order to get a small angle α between the incident beam and the surface. X-ray stress analysis is then performed in the Ψ-mode, i.e. by recording the diffraction lines after stepwise sample tilt around an axis defined by the intersection of the sample surface and the diffraction plane. Denoting the tilt angle in the Ψ-mode by ψ_d, the actual measurement direction (φ, ψ) in the asymmetric mode is given by

$$\tan\varphi = \frac{\tan(\theta - \alpha)}{\sin\psi_d} \quad \text{and} \quad \cos\psi = \cos(\theta - \alpha)\cos\psi_d\,, \qquad (18.14)$$

respectively (Fig. 18.15) [40]. In this way, it becomes possible to perform $\sin^2\psi$-measurements on different τ-scales below the surface, which can be adapted to the film thickness. The example in Fig. 18.16 refers to the residual stress analysis of PVD-Ti(C,N) films [41]. One clearly realises that the slopes of the $2\theta_{\varphi,\psi}$ versus $\sin^2\psi$ plots are different for the different τ-ranges that are covered in the symmetric and the asymmetric mode, respectively (Fig. 18.16a). This indicates steep stress gradients within the films (Fig. 18.16b), which were quantitatively evaluated by the Piecewise–Polynomial Method [55]. In this method, the depth distribution of the in-plane stress, $\sigma_{\|}(z)$, is described by polynomial spline functions of low degree. Their Laplace transforms, given by (18.11), are then determined by means of a least-squares fit from the $2\theta_{\varphi,\psi}$ versus $\sin^2\psi$ data. In principle, it is even possible to evaluate the gradients of the strain-free lattice parameter a_0 using this procedure (Fig. 18.16c). However, it should be noted that a large number of fit parameters often leads to numerical instabilities of the solutions.

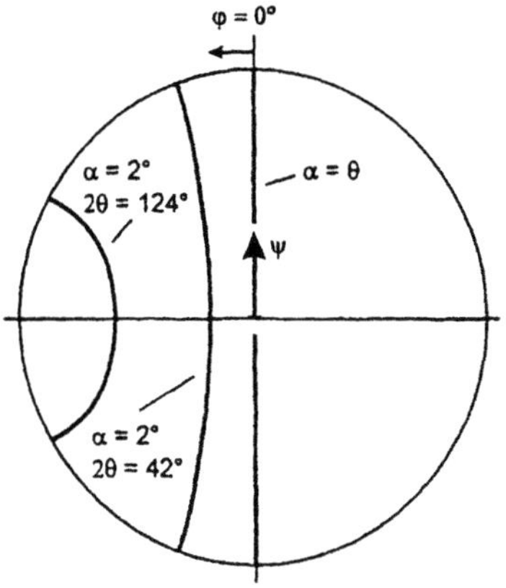

Fig. 18.15. The stereographic projection of the measurement direction (φ, ψ) for the symmetric configuration with $\alpha = \theta$ and for the asymmetric configuration with $\alpha = 2°$ and peak positions $2\theta = 42°$ and $124°$, respectively (after [40])

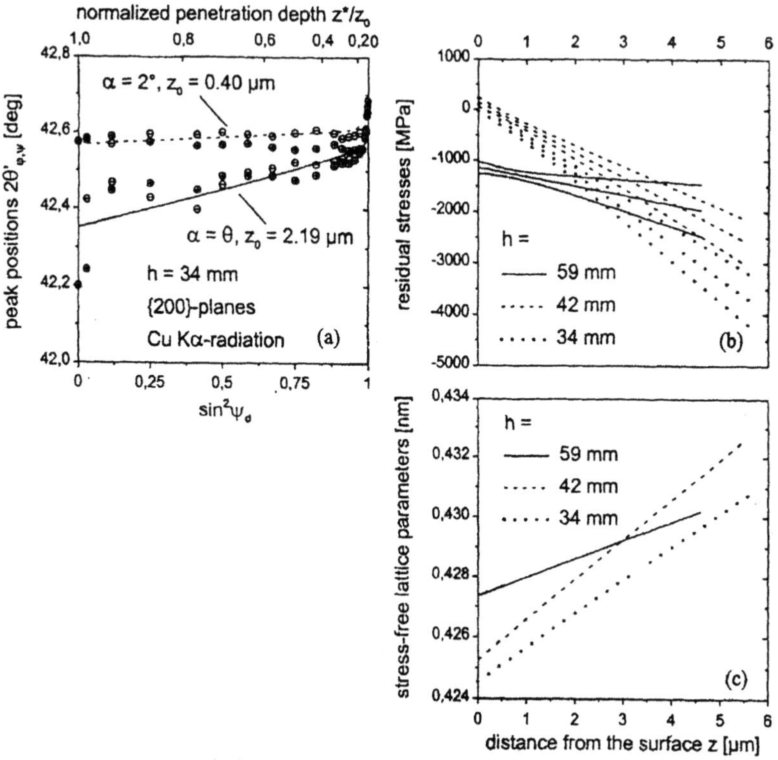

Fig. 18.16. The measured and recalculated peak positions of a sputtered PVD-Ti(C,N) film plotted versus $\sin^2 \psi_d$ (**a**), distributions of residual stresses with confidence intervals (**b**) and stress-free lattice parameters (**c**) for films sputtered at different heights between the substrate and the target (after [41])

Furthermore, the method is mainly restricted to untextured, or at least only weakly textured, materials.

Finally, as indicated in Table 18.2, it should be emphasised that the "mixed Ω–Ψ-mode technique" also permits a conventional $\sin^2\psi$ analysis at some predefined penetration depth, because one and the same penetration depth can be adjusted in relation to α by different effective inclination angles ψ given by (18.14). To the author's knowledge, such experiments have not yet been done, but they seem to be promising, especially with respect to the difficult problem of separating residual stress and compositional gradients.

Thin-Film Stress Analysis Using Strain Mapping

From X-ray stress analysis, it is well known that high residual stress fields lead to considerable peak shifts, which depend on the orientation (φ, ψ) of the diffraction vector. In thin films, these stress fields are often superimposed by a strong texture, the analysis of which is a necessary precondition for a correct analysis of the residual stress state, because texture leads to anisotropic elastic behaviour of the film on the macroscopic scale. In the presence of high residual strain fields, however, the texture cannot be evaluated by means of conventional pole figure measurements, since the peak position changes considerably with (φ, ψ)-tilting. In such cases, preferred use should be made of the strain mapping method suggested by [37,38]. It is based on the collection of 2θ–ψ-area maps, consisting of a set of θ–2θ scans performed at different inclination angles ψ. The example in Fig. 18.17 gives clear evidence for both the strong fibre texture with its main fraction [211] and the considerable

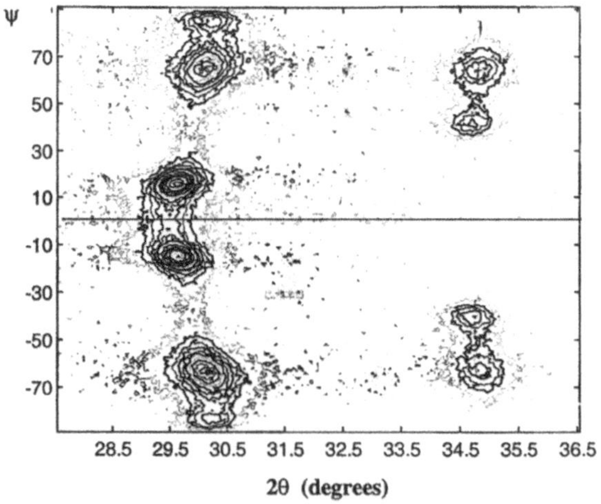

Fig. 18.17. A 2θ–ψ area map using Cu$K\alpha$ radiation for a r.f. magnetron sputtered CaO stabilised ZrO_2 film (CSZ) in the region of the (111) and (200) CSZ peaks (after [37])

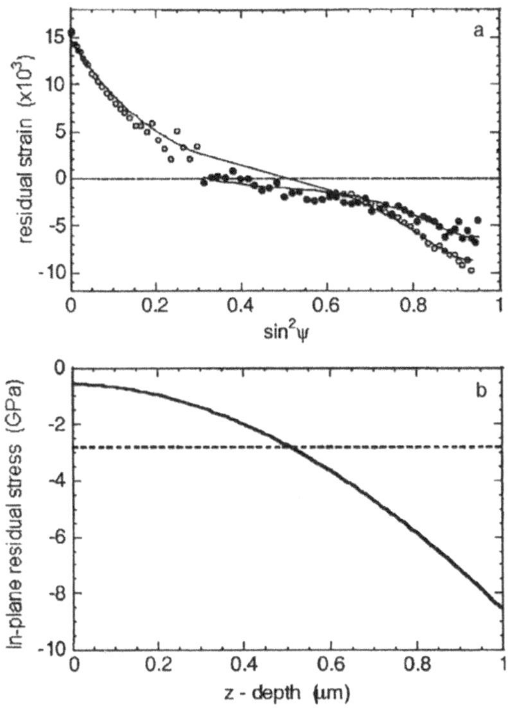

Fig. 18.18. (**a**) A $\sin^2\psi$ plot for (111) (open symbols) and (200) (closed symbols) with least-squares fitting results (lines) by means of the conventional $\sin^2\psi$ equation for a biaxial stress state of rotational symmetry. (**b**) The in-plane residual stress trend as a function of depth (position with respect to the surface of the CSZ film) (after [37])

peak shifts due to high residual stresses in a thin CaO stabilised ZrO_2 film deposited by r.f. sputtering on a $\langle h00\rangle$ Si wafer [37].

The analysis of residual stress distributions in textured samples involves some special problems, since the elastic behaviour of the materials becomes anisotropic on the macroscopic scale. Due to the preferred orientation of the crystallites, the DECs $s_1(hkl)$ and $1/2s_2(hkl)$ become dependent on the measurement direction as well, which leads to oscillating d_ψ versus $\sin^2\psi$ plots [56]. It can be demonstrated, however, that reflections of the type $\{h00\}$ and $\{hhh\}$ are not influenced by the texture within the Reuss hypothesis of crystallite interaction (homogeneous stress) [57, 58]. Therefore, if effects due to plastic deformation can be excluded, a non-linear curvature of the $d_\psi(h00)$- or $d_\psi(hhh)$ versus $\sin^2\psi$ plots as shown in Fig. 18.18a can be attributed to steep residual stress gradients within the penetration depth of the X-rays. Furthermore, it should be noted that the effect of elastic anisotropy leads to a significantly different slope of the data for (111) and (200) [37].

In order to gain an impression of the residual stress gradient, responsible for the curvature of the d_ψ versus $\sin^2\psi$ plots shown in Fig. 18.18a, it is convenient to fit an appropriate residual stress gradient model for $\sigma_\|(\tau)$ to the experimental data. Therefore, the conventional $\sin^2\psi$ equation is written as

$$\varepsilon_\psi(hkl,\tau) = \left[\frac{1}{2}s_2(hkl)\sin^2\psi + 2s_1(hkl)\right]\sigma_\|(\tau)\,, \tag{18.15}$$

where $\varepsilon_\psi(hkl,\tau)$ and $\sigma_\|(\tau)$ are the Laplace transforms of the actual z-depth profiles given by (18.11). In the present case, the authors used a third-order polynomial in z in order to perform a non-linear least-squares fit simultaneously on the data sets of both the (111) and the (200) reflections [37]. Figure 18.18b shows the calculated stress trend as well as the average stress value obtained by the conventional $\sin^2\psi$ method.

Hence the method of strain mapping is a powerful tool for thin-film residual stress analysis. It takes into account texture and the macroscopic anisotropy of the stress–strain state caused by the preferred crystallite orientation, as well as the depth dependence of the in-plane film stress. Limitations with respect to its applicability may arise for very marked textures, when the measurement directions (φ,ψ) are mainly restricted to the intensity poles. In this case the lack of experimental data and, therefore, of information on wide ranges of the d_ψ versus $\sin^2\psi$ curve, can lead to ill-conditioned systems of equations to be solved in the least-squares fit procedure described above.

18.4.3 Residual Stress Depth Profiling by Sample Rotation

The Scattering Vector Method

In previous sections it has already been mentioned that most of the X-ray methods for residual stress gradient analysis which are based on the measurement principle of the $\sin^2\psi$ method require untextured, or at least only weakly textured, materials. In order to overcome the difficulties related to this restriction, an alternative method has been proposed which uses the rotation of the sample around the scattering vector $\boldsymbol{g}_{\varphi\psi}$ at a fixed orientation (φ,ψ) with respect to the sample system in order to vary the X-ray penetration depth [59]. The τ-range covered by a rotation of $\pi/2$ starting from $\eta = 0°$ for a given value of $\psi \le \theta$ (cf. Fig. 18.6) corresponds to the difference $|\tau_\omega - \tau_\psi|$ between the penetration depths of the Ω- and Ψ-modes of conventional XSA (Fig. 18.19). For $\psi > \theta$ the lower limit for η is defined by the condition $\tau = 0$ in the expression given in Table 18.1, i.e. by $\eta_{\min} = \arcsin[\sqrt{\sin^2\psi - \sin^2\theta}/(\cos\theta\sin\psi)]$. Thus, taking diffraction profiles after stepwise rotation of the sample around $\boldsymbol{g}_{\varphi\psi}$ (the "η-mode"), τ is now varied by means of η, and $d_{\varphi,\psi}[hkl,\tau(\eta)]$-profiles are obtained even for the same orientation (φ,ψ) of the scattering vector with respect to the sample system $\boldsymbol{P}$. In the fundamental equation of XSA, the inclination angle ψ now

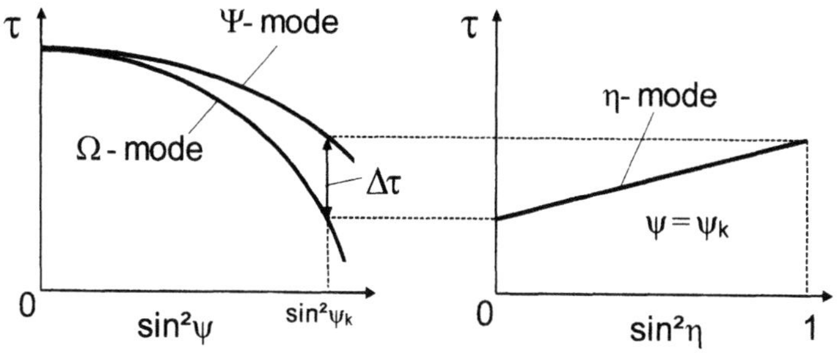

Fig. 18.19. Correlation of the penetration depths τ in Ω-, Ψ- and η- (scattering vector) geometry. For some fixed position ψ_k of the inclination angle, the rotation η around the scattering vector bridges the gap $\Delta\tau$ between the Ω- and Ψ-modes (after [26])

becomes a normal (fixed) parameter in the same sense as the azimuth angle φ and, therefore, individual components $\sigma_{ij}(\tau)$ in (18.10) and (18.12), respectively, can be separately evaluated, if η-mode scans performed at different values of φ and ψ are combined in an appropriate way [26–28, 44].

Assuming a biaxial stress state in the film (i.e. $\sigma_{i3} = 0$, $i = 1, 2, 3$), the in-plane stress distribution $\sigma_{\|}(\tau) = [\sigma_{11}(\tau) + \sigma_{22}(\tau)]/2$ can be calculated directly from one averaged lattice parameter depth profile $d_{\psi}^{+}(hkl, \tau) = 1/2[d_{0\psi}(hkl, \tau) + d_{90\psi}(hkl, \tau)]$ obtained in the η-mode at some inclination angle ψ by means of

$$\sigma_{\|}(\tau) = \frac{d_{\psi}^{+}(hkl, \tau) - d_0(hkl)}{d_0(hkl) F_{\|}(hkl, \psi)} = \frac{\varepsilon_{\psi}^{+}(hkl, \tau)}{F_{\|}(hkl, \psi)}, \qquad (18.16)$$

with $F_{\|}(hkl, \psi) = F_{11}(hkl, 0, \psi) + F_{11}(hkl, 90, \psi)$ [cf. (18.10), (18.12)]. Because the expression defined by (18.16) is very sensitive with respect to $d_0(hkl)$, it becomes obvious that two or more stress depth profiles $\sigma_{\|}^{(\psi_i)}(\tau)$ obtained by (18.16) at different positions ψ_i will only fit together in the sense of a "Universal Plot" [60] if the correct $d_0(hkl)$ is taken. This criterion, however, can be used in order to determine $d_0(hkl)$ itself (Fig. 18.20). A more detailed discussion of this "variation procedure", which even yields $\sigma_{33}(\tau)$ in the case of multiaxial residual stress states in the near-surface region, is given in [26, 61].

The example in Fig. 18.21 should demonstrate that the scattering vector method has special merits for residual stress gradient analysis in thin and strongly textured films. The investigations were performed on a new type of five-circle diffractometer (see Sect. 18.4.3 below) and refer to a PVD TiN film with a marked $\langle 111 \rangle$ fibre texture and a steep gradient of the in-plane stress component, $\sigma_{\|}$. Lattice spacing depth profiles were measured on both sides of the strain-free direction of the biaxial stress state defined by $\psi^* = \arcsin\sqrt{-2s_1(hkl)/1/2s_2(hkl)}$. For the large inclination of $\psi = 70.5°$ we observed the direct impact of the stress gradient in the form of smaller lattice

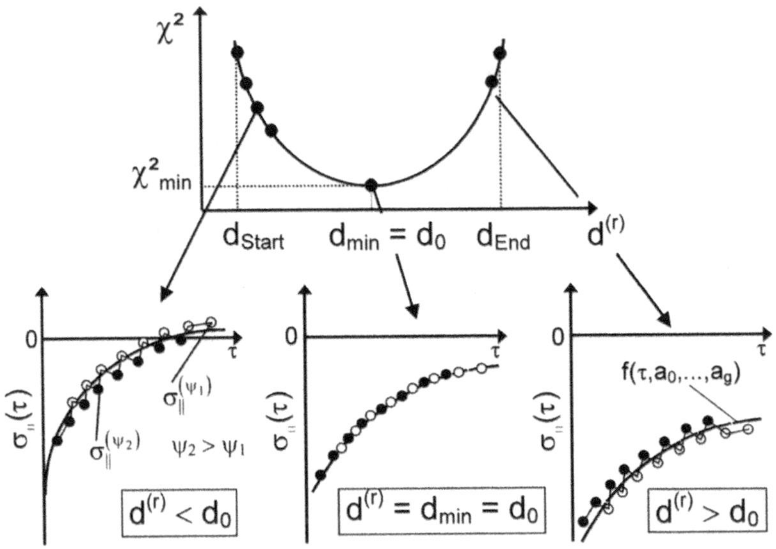

Fig. 18.20. The schematic view of the residual stress and $d_0(hkl)$-evaluation by means of the scattering vector method. The stress profiles $\sigma_{||}^{(\psi_i)}$ obtained by (18.16) for the individual inclination angles ψ_i only fit together without any offset if the correct $d_0(hkl)$ is used. Note that different values of $d^{(r)}(hkl)$ taken for $d_0(hkl)$ in the calculations lead additionally to a shift of the residual stress profiles on an absolute scale (after [26])

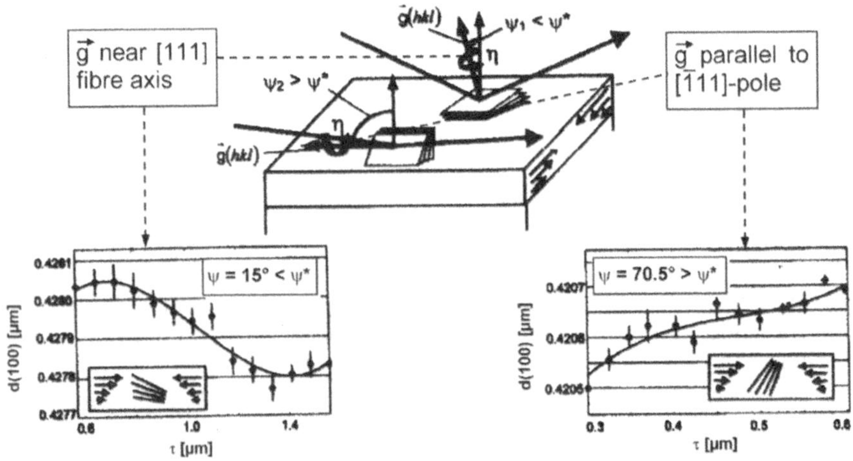

Fig. 18.21. Lattice spacing depth profiling by means of the scattering vector method at a 5 μm thick PVD TiN film deposited on a steel substrate. CoKa radiation, 111-reflection. See the text for details

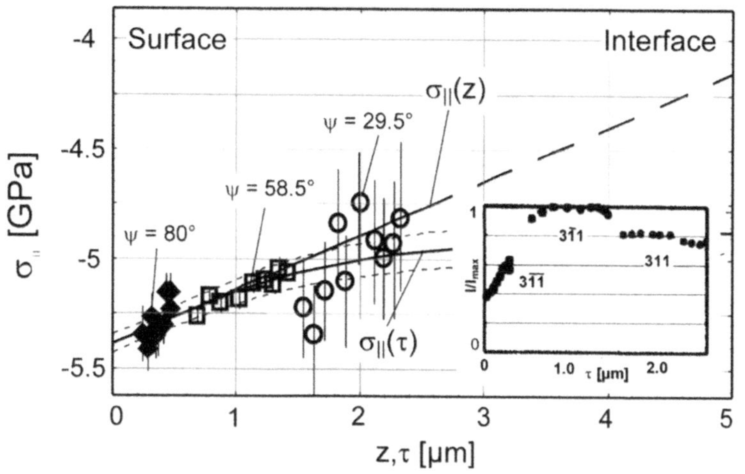

Fig. 18.22. Depth profiles of the in-plane stress in the Laplace- and the z-space, respectively, in a 5 μm thick TiN film with strong ⟨111⟩ fibre texture (cf. Fig. 18.21). The stress gradients were evaluated by means of the "variation procedure" (Fig. 18.20) from measurements performed in the scattering vector mode in the 311-texture poles. a_0^{opt} is the strain-free lattice parameter evaluated in the "variation procedure". (after [61])

spacings on the top of the film and larger values towards the interface between the film and the substrate. Consequently, the opposite effect due to transverse contraction was found for the small inclination angle $\psi = 15°$. The depth profiles of the film stress in both the τ- and the z-space evaluated by means of the "variation procedure" (cf. Fig. 18.20) from the 311-texture poles are shown in Fig. 18.22.

The scattering vector method has proved to be a suitable technique for the depth-resolved investigation of residual stresses in thin films with strong texture. In principle, the method is even applicable to single-crystalline materials. However, it should be noted that the method is very sensitive with respect to hkl-dependent microstresses. Therefore, it is recommended that at least two different reflections should be analysed. A geometrical restriction is given by the small accessible τ-range, which is the difference between the Ω- and Ψ-modes of XSA.

A Multipurpose Diffractometer for Surface Gradient and Texture Analysis

From the above considerations it becomes evident that the practical realisation of the methods developed for residual stress gradient analysis in thin films requires well-aligned diffractometers equipped with special optical tools in the primary and/or the diffracted beam for grazing or low incidence angle

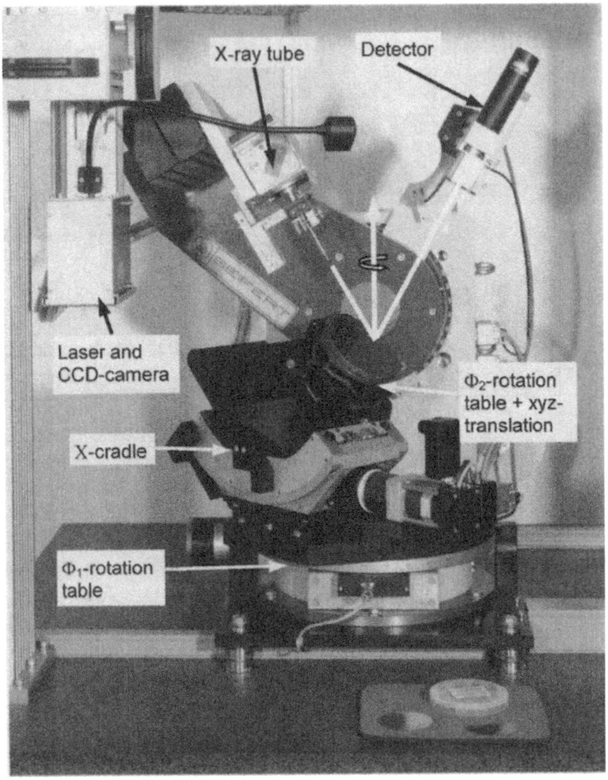

Fig. 18.23. The ETA diffractometer for X-ray stress and texture analysis. See the text for details

diffraction. Whereas the $\sin^2\psi$-based stress analysis can in principle be performed even on a two-circle diffractometer in the Ω-mode of diffraction [62], today the so-called Ψ-diffractometer [63] has become generally accepted, because it works in a symmetrical diffraction geometry, without any restrictions with respect to the available Ψ-range. Texture analysis, on the other hand, requires an additional rotation axis perpendicular to the Ψ-axis, which can be realised by means of a φ-rotation table or, in a more comprehensive way, using a conventional four-circle diffractometer (Eulerian cradle, Kappa diffractometer; see, for example, [64]).

Recently, a new type of X-ray diffractometer (the "ETA diffractometer") has been developed, which enables direct sample rotation around the scattering vector by means of one single physical axis (Fig. 18.23) [65,66]. It operates on the principle of the Seifert θ–θ diffractometer MZ VI in the vertical plane. All sample rotations are based on a second one-circle-goniometer which operates in the horizontal plane, i.e. its axis Φ_1 and the Θ/Θ'-axes (X-ray tube / detector) are perpendicular to each other. The angle ψ between the surface

Table 18.3. The realisation of different measurement techniques in X-ray stress and texture analysis in the ETA diffractometer

Method	Realisation in the ETA diffractometer
$\sin^2\psi$-based stress analysis	φ_1 fixed at 0° (Ψ-mode) or 90° (Ω-mode) sample tilt around the x-axis negative ψ-branch: $\varphi_1 = \varphi_1 + 180°$ and $\varphi_2 = \varphi_2 + 180°$
Scattering vector method	φ_2 and ψ fixed at the desired position of $\boldsymbol{g}(hkl)$ sample rotation around Φ_1
"Mixed Ω–Ψ-mode" stress analysis	φ_1 fixed between 0° and 90° sample tilt around the x-axis
Texture analysis	φ_1 fixed at 0° successive sample rotation around Φ_2 and ψ-tilt
Thin-film stress analysis	Use of grazing incidence diffraction tools (polycapillary semi-lens in the primary beam, soller + secondary monochromator in the diffracted beam)

normal and $\boldsymbol{g}(hkl)$ is adjusted by means of an x-cradle segment operating in an angular range between $0 \leq \psi \leq 90°$, which is placed on the Φ_1-goniometer in such a way that the x- and Φ_1-axes enclose an angle of 90°. Therefore, a rotation around Φ_1 corresponds to sample rotation around the scattering vector $\boldsymbol{g}(hkl)$. The azimuth φ of $\boldsymbol{g}(hkl)$ with respect to the sample system $\boldsymbol{P}$ (cf. Fig. 18.6) is adjusted by a rotation table Φ_2 inside the cradle segment.

From a generalised point of view, the ETA diffractometer is a five-circle diffractometer, which differs from a conventional diffractometer by the additional axis Φ_1 for rotation of the cradle (segment) perpendicular to the $\omega/2\theta$-axes. This Φ_1-axis is the necessary additional degree of freedom for the direct realisation of the η-rotation. For thin-film stress analysis at grazing incidence, a variety of optical elements such as polycapillary semi-lenses, soller slits and monochromators can be placed into both the incident and the diffracted beam, respectively. Sample positioning by means of a x-y-z translation table is controlled by a laser and a CCD camera system. Table 18.3 summarises some possible measurement techniques used in XSA (cf. also Table 18.1) and their adaptation in the ETA diffractometer.

18.5 Concluding Remarks

X-ray residual stress analysis of thin polycrystalline films involves some difficult problems, especially if the films are strongly textured and if the stress state is not uniform with respect to the film thickness. In most of these cases, the use of the conventional $\sin^2\psi$-method would fail, because the basic assumptions for its applicability are not fulfilled. Therefore, a series of promising methods have been developed in the last decade in order to meet the very special demands of thin-film XSA, and it was the main goal of this chapter to give a survey of some of the most interesting approaches in this connection. Special attention was directed to those methods that are especially suited for the detection of steep in-plane stress gradients within the films. From the theoretical point of view, such gradients can exist even in very thin films and their analysis is of great technical importance.

No method exists that solves all the sophisticated problems listed in Table 18.1 at once. For this reason, the individual experimental techniques were discussed with regard to their advantages as well their drawbacks for residual stress analysis in thin films. However, it can be concluded that those methods that are based on the measurement principle of the $\sin^2\psi$ method have some advantages in the case of untextured or weakly textured materials, whereas the techniques adopted mainly from single-crystal diffraction have some special merits for residual stress analysis in strongly textured films.

Therefore, it is up to the reader to decide which method is most suitable for the solution of his or her own (thin-film) problems. Finally, it should be emphasised that all methods require a very well-aligned diffractometer and, in the case of very thin films, additional optical elements for grazing incidence diffraction.

List of Principal Symbols

σ	macro-residual stress, homogeneous phase-specific stress
σ_{ij}	components of the stress tensor
$\sigma_{ij}(z)$	depth profile of the residual stress component σ_{ij} in the real space
$\sigma_{ij}(\tau)$	Laplace transform of the residual stress depth profile $\sigma_{ij}(z)$
$\varepsilon_{\varphi\psi}$	lattice strain in the measurement direction (φ, ψ)
$\varepsilon_{\varphi\psi}(\tau, hkl)$	Laplace depth profile of the lattice strain
$d_{\varphi,\psi}(\tau, hkl)$	lattice spacing depth profile in Laplace space
$d_0(hkl)$	strain-free lattice spacing
hkl	Miller indices
$s_1(hkl)$, $1/2s_2(hkl)$	diffraction elastic constants (DEC) of the quasi-isotropic polycrystal
$F_{ij}(hkl, \varphi, \psi)$	stress factors
φ, ψ	azimuth and inclination angle of the diffraction vector with respect to the sample system, respectively
η	rotation angle of the sample around the diffraction vector
$\boldsymbol{g}_{\varphi\psi}$	diffraction vector
τ	penetration depth of X-rays

τ_{eff}	effective penetration/information depth in the case of thin films
D	thickness of the investigated film
μ	linear absorption coefficient
θ	Bragg angle
α, β	angles of the primary and the diffracted beam, respectively, with the sample surface
α_c	critical incidence angle of total external reflection
λ	wavelength of the radiation

References

1. I.C. Noyan and J.B. Cohen: *Residual Stress Measurement by Diffraction and Interpretation.* (Springer, Berlin Heidelberg New York 1987)
2. B. Scholtes: *Eigenspannungen in mechanisch randschichtverformten Werkstoffen. Ursachen, Ermittlung und Bewertung.* (DGM Informationsgesellschaft, Oberursel 1991)
3. V. Hauk: *Structural and Residual Stress Analysis by Nondestructive Methods.* (Elsevier, Amsterdam 1997)
4. L.H. Schwartz and J.B. Cohen: *Diffraction from Material.* (Springer, Berlin Heidelberg New York Tokyo 1987)
5. E. Klockholm: IBM J. Res. Develop. **31**, 585 (1987)
6. D.S. Rickerby: J. Vac. Sci. Technol. A **4**, 2809 (1986)
7. J.-D. Kamminga, T.H. de Keijser, R. Delhez, E.J. Mittemeijer: J. Appl. Phys. **88**, 6332 (2000)
8. M.A. Herman: *Semiconductor Superlattices.* (Akademie, Berlin 1986)
9. G.H. Olsen: J. Cryst. Growth **31**, 223 (1975)
10. I.C. Noyan, T.C. Huang, B.R. York: Crit. Rev. in Solid State and Materials Sciences **20**, 125 (1995)
11. I.C. Noyan: Metall. Trans. A **14**, 1907 (1983)
12. J.H. van der Merwe: J. Microscopy **102**, 261 (1974)
13. D.S. Rickerby, A.M. Jones: Surface and Coatings Technol. **36**, 631 (1988)
14. A.J. Perry, M. Jagner, D. Sproul, P.J. Rudnik: Surface and Coatings Technol. **42**, 49 (1990)
15. S.N.G. Chu, A.T. Macrander, K.E. Strege, W.D. Johnston Jr.: J. Appl. Phys. **57**, 249 (1985)
16. E. Macherauch, H. Wohlfahrt, U. Wolfstieg: Härterei Tech. Mitt. **28**, 203 (1973)
17. T. Mura: *Micromechanics of Defects in Solids.* 2nd edn. (Nijhoff, Dordrecht Boston Lancaster 1987)
18. E. Kröner: *Kontinuumstheorie der Versetzungen und Eigenspannungen.* (Springer, Berlin Heidelberg 1958)
19. I.S. Sokolnikoff: *Mathematical Theory of Elasticity.* (McGraw-Hill, New York 1956)
20. S.G. Lekhnitskii: *Anisotropic Plates.* (Gordon and Breach, New York 1968)
21. B.K. Tanner: *X-Ray Diffraction Topography.* (Pergamon Press, Oxford 1976)
22. O. Brümmer, H. Stephanik (Eds.): *Dynamische Interferenztheorie.* (Akademische Verlagsgesellschaft Geest and Portig, Leipzig 1976)

23. C. Genzel: Thesis (Humboldt-Universität, Berlin 1986)
24. W.G. Sloof, B.J. Kooi, R. Delhez, T.H. de Keijser, E.J. Mittemeijer: J. Mater. Res. **11**, 1440 (1996)
25. C. Genzel: *Proc. of the 5th Int. Conf. on Residual Stress (ICRS-5).* (Linköping, Sweden, 16.–18.06.1997) Vol. 1, p. 514–521
26. C. Genzel: J. Appl. Cryst. **32**, 770 (1999)
27. C. Genzel, W. Reimers: Surface and Coatings Technol. **116-119**, 404 (1999)
28. C. Genzel: phys. stat. sol. (a) **159**, 283 (1997)
29. R. Delhez, T.H. de Keijser, E.J. Mittemeijer: Surf. Eng. **3**, 331 (1987)
30. V. Hauk: Härterei Tech. Mitt. **46**, 52 (1991)
31. H. Dölle, V. Hauk: Härterei Tech. Mitt. **34**, 272 (1979)
32. H. Behnken, V. Hauk: Z. Metallkde. **82**, 151 (1991)
33. M. van Leeuwen, J.-D. Kamminga, E.J. Mittemeijer: J. Appl. Phys. **86**, 1904 (1999)
34. U. Welzel, M. Leoni, and E.J. Mittemeijer: this volume, Chap. 14
35. H. Dölle, V. Hauk: Härterei Tech. Mitt. **31**, 165 (1976)
36. E. Macherauch, P. Müller: Z. angew. Physik **13**, 305 (1961)
37. M. Leoni, Y.H. Dong, P. Scardi: Mat. Sci. Forum **321–324**, 439 (2000)
38. P. Scardi, Y.H. Dong: J. Mater. Res. **16**, 233 (2001)
39. D.S. Rickerby, A.M. Jones, B.A. Bellamy: Surface and Coatings Technol., **36**, 661 (1988)
40. T. Dümmer, B. Eigenmann, M. Stüber, H. Leiste, D. Löhe, H. Müller, O. Vöhringer: Z. Metallkde **90**, 780 (1999)
41. T. Dümmer, B. Eigenmann, D. Löhe: Mat. Sci. Forum **321-324**, 81 (2000)
42. K. Van Acker, L. De Buyser, J.P. Celis, P. Van Houtte: J. Appl. Cryst. **27**, 56 (1994)
43. H. Mohrbacher, K. Van Acker, B. Blanpain, P. Van Houtte, J.-P. Celis: J. Mater. Res. **11**, 1776 (1996)
44. C. Genzel, M. Broda, D. Dantz, W. Reimers: J. Appl. Cryst. **32**, 779 (1999)
45. M.F. Doerner, S. Brennan: J. Appl. Phys. **63**, 126 (1988)
46. C.J. Shute, J.B. Cohen: J. Appl. Phys. **70**, 2104 (1991)
47. M. v. Laue: *Röntgenstrahlinterferenzen.* (Akademische Verlagsgesellschaft, Frankfurt am Main 1960)
48. L.V. Azaroff: *Elements of X-Ray Crystallography.* (McGraw-Hill, New York 1968)
49. A.C. Thompson, D. Vaughan (Eds.): *X-Ray Data Booklet.* (University of California, Berkeley 2001)
50. M.F. Toney, S. Brennan: Phys. Rev. B **39**, 7963 (1989)
51. P. Predecki, X. Zhu, B. Ballard: Adv. X-Ray Anal. **36**, 237 (1993)
52. A.J. Allen, M.T. Hutchings, C.G. Windsor, C. Andreani: Adv. Phys. **34**, 445 (1985)
53. K. Van Acker, H. Mohrbacher, B. Blanpain, P. Van Houtte, J.P. Celis: Mat. Res. Soc. Symp. Proceed. **308**, 677 (1993)
54. B. Eigenmann: Thesis (Karlsruhe, 1990)
55. T. Leverenz, B. Eigenmann, E. Macherauch: Z. Metallkde. **87**, 616 (1996)
56. V. Hauk: Adv. X-Ray Anal. **27**, 101 (1984)
57. H. Dölle, V. Hauk: Z. Metallkde **70**, 682 (1979)
58. H. Dölle, V. Hauk, H. Kockelmann, in: *Textures of Materials.* Ed. by G. Gottstein und K. Lücke, Vol. 2 (Springer Berlin Heidelberg New York 1978) p. 413

59. C. Genzel: phys. stat. sol. (a) **146**, 629 (1994)
60. H. Ruppersberg, I. Detemple, J. Krier: phys. stat. sol. (a) **116**, 681 (1989)
61. C. Genzel: Habilitationsschrift (Humboldt-Universität Berlin 2000)
62. G. Faninger: Härterei-Techn. Mitt. **31**, 16 (1976)
63. U. Wolfstieg: Härterei-Techn. Mitt. **31**, 19 (1976)
64. C. Giacovazzo (Ed.): *Fundamentals of Crystallography.* (Oxford University Press, Oxford 1998)
65. C. Genzel, W. Reimers, A. Haase: *Proc. of the 6th Int. Conf. on Residual Stress (ICRS-6).* (Oxford, UK, 10.–12.07.2000) Vol. 1, pp. 74–81
66. A. Haase, C. Genzel, D. Dantz, M. Löhmann, B. Wallis, C. Stock, W. Reimers: *Proc. of the 18th Int. Conf. of Appl. Crystallography.* Katowice, Poland, 04–07.09.2000 (World Scientific, 2001) pp. 97–100

19 Two-Dimensional XRD Profile Modelling in Imperfect Epitaxial Layers

A. Boulle, O. Masson, R. Guinebretière, and A. Dauger

A methodology is presented for the investigation of imperfect epitaxial layers, mainly oxides and ceramics. The data are recorded on a laboratory-built high resolution X-ray diffractometer that allows fast reciprocal space map acquisition even from poorly diffracting materials. The two-dimensional instrumental profile of the diffractometer is evaluated taking into account each optical element in the beam path in order to obtain reliable and useful information from the analysis of the peak profile shapes. Diffraction profiles from the scattering normal to the surface are modelled with a kinematical scattering model based on microstructural and instrumental parameters. The vertical grain size, size distribution and strain variation across the film thickness are recovered by profile simulation. In the direction parallel to the surface the effect of mosaic spread and finite domain size are considered. The mosaicity, as well as the in-plane crystallite size are deduced from profile modelling. Preliminary results are also given concerning the diffraction effects of a possible self-organized microstructure.

19.1 Introduction

X-ray diffraction (XRD) is a widely employed technique for characterizing epitaxial layers [1]. A vast amount of microstructural information such as the amount of strain and strain relaxation, the coherently diffracting domain shape, size and size distribution, the mosaic spread, defect densities, etc., can be recovered by the study of the shape of XRD line profiles. The so-called reciprocal space mapping technique (i.e. the representation of the scattered intensity in a two-dimensional (2-D) frame) [2] allows rigorous separation of line broadening effects acting along different directions of reciprocal space. Reciprocal space maps (RSMs) are usually recorded with double crystal diffractometers (DCD) equipped with a narrow detector slit [3], triple crystal diffractometers (TCD) [4] or multiple crystal diffractometers [2]. Such studies are now routinely performed on semiconductor layers and heterostructures and most of the diffractometer manufacturers provide software packages to simulate XRD line profiles from this type of materials.

In contrast with semiconductor and metallic layers which can be relatively easily grown as nearly-perfect single crystals, oxide and ceramic films are much less well-behaved in that they can rarely be obtained with a high crystalline-quality level. These materials are therefore referred to as imperfect materials. The poor crystalline quality of imperfect layers produces a smearing of the scattered intensity so that the methods of characterization used for semiconductor materials cannot be straightforwardly applied to imperfect materials. Since a series of θ-2θ scans (or ω scans) has to be undertaken to build up a complete RSM, the time required to record a single reciprocal lattice point (RLP) may become prohibitive, especially for poorly diffracting materials. The overall intensity can be increased, but this generally lowers the resolution of the diffractometer so that important reciprocal space features can be swamped by instrumental artifacts. A compromise has therefore to be found by keeping in mind that the resolution of the set-up has to match the problem of interest. Similarly, the simulation of the diffraction profiles has to take into account the characteristics of imperfect epitaxial layers, i.e. strain relaxation, islanding, island size distribution, etc.

In this chapter we present a methodology for the microstructural characterization of imperfect epitaxial layers. Data acquisition is performed with a laboratory-built high-resolution X-ray diffractometer suited to the study of imperfect materials in that it allows fast RSM acquisition even for very poorly diffracting materials [5]. Since 2-D RSMs are investigated, a deep understanding of the 2-D instrumental profile (IP) is of crucial importance if reliable and useful microstructural information is to be derived from XRD line shape analysis (Sect. 19.3). Diffraction profiles from the scattering normal to the surface are modelled with a kinematical scattering model based on microstructural and instrumental parameters (Sect. 19.4.1). In-plane diffraction profiles are also investigated and modelled (Sect. 19.4.2). The effect of mosaic spread and finite domain size are considered as well as the effect of a self-organized microstructure. Epitaxial yttria stabilized zirconia (YSZ) deposited on $(11\bar{2}0)$ sapphire substrates is chosen as a representative example of an imperfect epitaxial layer.

19.2 The Sample

Thin zirconia precursor films were formed on mirror polished $(11\bar{2}0)$ Al_2O_3 wafers by sol-gel dip-coating. The ZrO_2 precursor was prepared in the zirconium n-propoxide – acetylacetone – n-propanol – water system [6]. Yttrium nitrate ($Y(NO_3)_3(H_2O)_5$) dissolved in n-propanol was used as the Y_2O_3 precursor. The solution was prepared in such a way that the final ZrO_2-Y_2O_3 solid solution contained 10 mole % Y_2O_3, so that the cubic phase of YSZ is expected to form.

Coated samples were dried 1 h at 60°C and heat-treated at 600°C for 1 h to decompose the precursor and promote the crystallization of polycrystalline YSZ. As reported elsewhere [7, 8], further high temperature heat treatment of thin films causes the grains with the lower interfacial energy (i.e. epitaxial grains) to grow at the expense of unstable grains (i.e. randomly oriented grains). Parallel with grain growth, the polycristalline nature of the film, especially the presence of grain boundaries, is the root of capillary instability [9] forcing the film to split into more or less connected epitaxial islands. In the present study the samples were heat treated at 1550°C for 1 h. The YSZ islands are oriented on the substrate with the (200) planes of YSZ parallel to the $(11\bar{2}0)$ planes of the substrate [10]. The main in-plane orientation is $[001]_{YSZ}//[0001]_{Al_2O_3}$. The corresponding misfit strains are −40.8% (tensile) and −6.65% (tensile) in the $[001]_{YSZ}$ and $[010]_{YSZ}$ directions. Two other orientations have also been found: $[001]_{YSZ}//[1\bar{1}01]_{Al_2O_3}$ and $[001]_{YSZ}//[101]_{Al_2O_3}$.

19.3 XRD Equipment

19.3.1 Instrument Description

A dedicated high-resolution X-ray diffractometer was built to study imperfect materials [5]. The instrumental set-up, schematically represented in Fig. 19.1, is based on a 6 kW rotating anode generator with Cu target, a four bounce (220) Ge monochromator in the $CuK\alpha_1$ setting [11] and a curved position sensitive detector (PSD) [12]. A fine focus line source was used to generate an X-ray beam with dimensions 5 mm × 50 µm oriented in such a way that the line is perpendicular to the scattering plane. The high resolution monochromator provides a beam with a very low relative wavelength spread ($\Delta\lambda/\lambda = 1.4.10^{-4}$) and limited equatorial divergence ($\Delta\theta = 12$ arcsec) [13]. Samples are mounted on a five-axis sample holder. Each movement is controlled by stepper motors equipped with optical encoders, and the precision of each motor is 0.001° [14].

Line broadening effects acting along different directions of reciprocal space (i.e. along q_x and q_z in Fig. 19.2) cannot easily be separated by conventional θ-2θ scans or rocking curves because the investigated RLP is integrated over the Ewald sphere (i.e. along the arc AB in Fig. 19.2). To overcome this limitation, the range of integration of the RLP (i.e. the arc AB) is usually reduced by inserting a narrow slit in front of the detector [3] or preferably by inserting

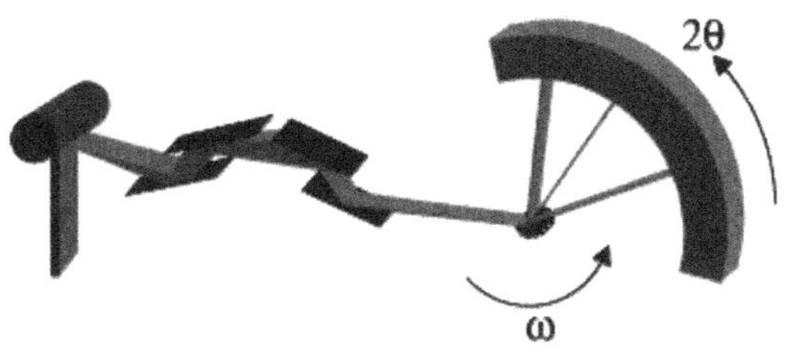

Fig. 19.1. Schematic drawing of the instrumental set-up

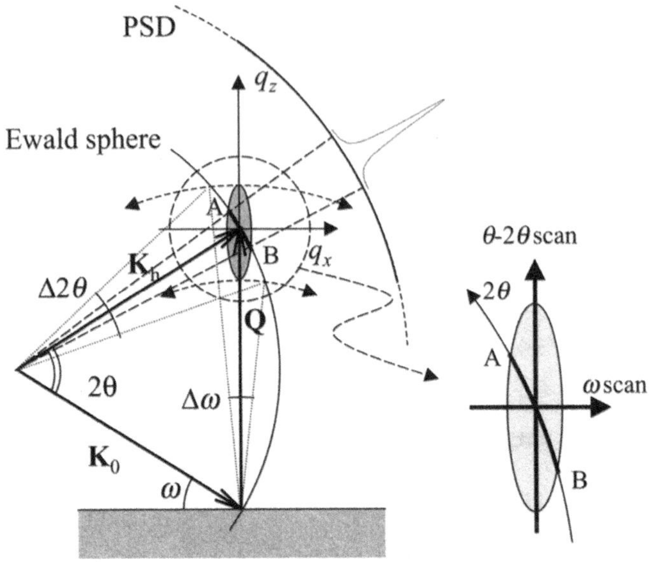

Fig. 19.2. Schematic representation in reciprocal space of data acquisition with a curved position sensitive detector. $\boldsymbol{K}_0$, $\boldsymbol{K}_h$ and $\boldsymbol{Q}$ are the incident wave vector, the scattered wave vector and the scattering vector, respectively. q_x and q_z are the x and z components of the reduced scattering vector $\boldsymbol{q} = \boldsymbol{Q} - \boldsymbol{h}$, with $\boldsymbol{h}$ the reciprocal lattice vector of the reflection. $\Delta\omega$ is the width of the classical rocking curve, and $\Delta 2\theta$ is the width of the θ–2θ scan

an analyzing crystal [4]. With the PSD, the angular range of integration is reduced to the width of one channel, which is close to 0.015°.

19.3.2 Reciprocal Space Mapping

As the sample is rocked through the Bragg position of a given reflection, the associated RLP passes through the Ewald sphere. At each scanning step the intensity distribution corresponding to the intersection of the RLP with Ewald sphere is recorded (Fig. 19.2). The superposition of the patterns collected at each incidence angle provides a RSM expressed in angular units, $I(\omega, 2\theta)$. A conventional RSM, $I(q_x, q_z)$, is then obtained by converting angular units $(\omega, 2\theta)$ into reciprocal lattice units (q_x, q_z) using the Ewald construction. q_x and q_z are the x and z components of the reduced scattering vector $\boldsymbol{q} = \boldsymbol{Q} - \boldsymbol{h}$. $\boldsymbol{Q}$ is the scattering vector ($Q = 4\pi \sin\theta/\lambda$) and $\boldsymbol{h}$ is the reciprocal lattice vector of the reflection ($h = 4\pi \sin\theta_B/\lambda$). It should be emphasised that owing to the presence of the PSD the complete RSM is recorded in one single sample scan, whereas a series of θ-2θ scans, or a series of ω scans, is required when a point detector is used [15]. Consequently the time needed to record a RSM is lowered by more than one order of magnitude as compared to conventional diffractometers. Typical RSM recording times

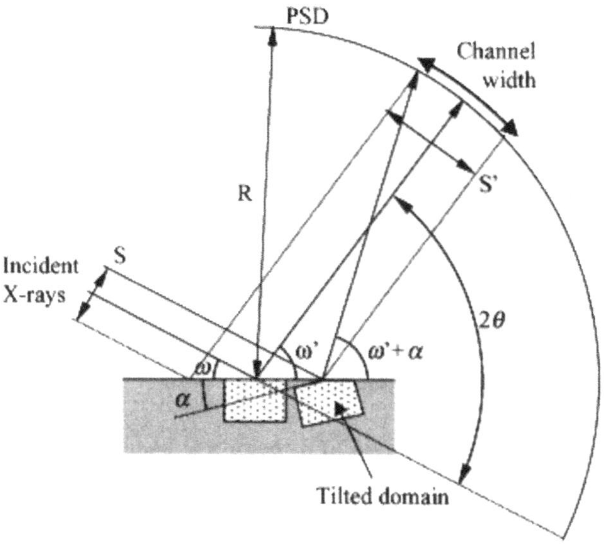

Fig. 19.3. Schematic representation of the effect of the scattering from a tilted domain. S is the size of the incident beam, and S' the size of the diffracted beam. ω, ω' and $\omega' + \alpha$ are the incidence angle, the exit angle for an oriented domain, and the exit angle for a domain tilted with an angle α, respectively. 2θ is the scattering angle, $2\theta = \omega + \omega'$. The dimensions are exaggerated

range between 20 min to 4 h for counting times per step ranging between 5 s and 1 min, respectively.

It should be emphasized that unlike an analysing crystal, a particular PSD channel is unable to discriminate diffracted X-rays from disoriented domains in a different part of the sample. This situation is schematically depicted in Fig. 19.3. Therefore the incident beam has to be kept as narrow as possible in order to limit the contribution of tilted domains. For a given beam size S, the maximum associated angular disorientation α is given by

$$\alpha = \frac{S}{R}\frac{\sin(\omega + \phi)}{\sin \omega},$$

where R is the radius of curvature of the detector and ϕ is the angle of asymmetry of the reflection, $\phi = 2\theta - 2\omega$. For a symmetrical reflection $\alpha = 0.011°$, which is close to the channel width, i.e. the contribution of disoriented domains is restricted to those domains tilted by less than $0.011°$. This value is one order of magnitude below the mosaicity generally observed in oxide thin films. This set-up is therefore very well suited to the study of imperfect materials. For asymmetrical reflections the angle of asymmetry ϕ has to be kept smaller than

$$\operatorname{asin}\left(c \times \sin \omega \times \frac{R}{S}\right) - \omega,$$

otherwise the diffracted beam size will exceed the channel angular width c, resulting in a larger contribution of disoriented grains according to the expression of α. However, in the case of very imperfect materials exhibiting a mosaicity of several tenths of degrees, the diffracted beam size may reasonably be increased to several channels without impairing the data.

There is another fundamental reason why the incident beam has to be as narrow as possible: the PSD directly measures the intensity distribution in the diffracted beam. Hence, increasing the incident beam size (or increasing the beam's footprint on the sample) will result in a broadening of the instrumental profile. This point will be discussed in Sect. 19.3.4.

19.3.3 Scans Across Reciprocal Space

The correspondence between real and reciprocal space is schematically represented in Fig. 19.4. For simplicity the discussion is restricted to the case of symmetrical reflections (i.e. the normal to diffracting planes is parallel to the sample surface normal), but it can be easily generalized to the case of asymmetrical reflections. The intensity distribution in reciprocal space is the Fourier transform of the self-correlation function of the sample's electron density [16]. Information concerning the defect structure of the sample can therefore be obtained by analysing the shape of the intensity distribution in reciprocal space or the shape of judiciously chosen sections of reciprocal space.

Information concerning the film thickness, thickness fluctuation, heterogeneous strain and strain gradients across the film thickness can be recovered by

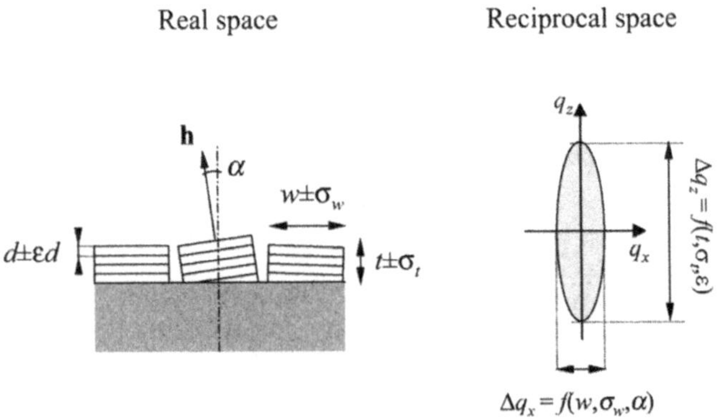

Fig. 19.4. Schematic representation of the effect of structural defects (**a**) in real space and (**b**) reciprocal space. d, ε, and $\boldsymbol{h}$ are the lattice spacing, the strain in the direction normal to the interface, and the normal to the atomic planes, respectively. α represents the angle between the normal to the atomic planes and the surface normal (i.e. mosaic spread). t and w are the layer thickness and lateral crystallite size, and σ_t and σ_w are the associated standard deviations

analysing sections along the q_z axis (Fig. 19.4). Such q_z-scans (or θ-2θ scans) are usually performed by moving the detector at twice the sample rotation rate. In the present case scans are carried out by selecting a line along the q_z direction of the RSM, i.e. $I(q_x = \text{constant}, q_z)$, which is strictly equivalent to selecting the channel corresponding to the angular position $2\theta = 2\omega$.

Information concerning the lateral extension of the crystallites, their size distribution, their angular distribution (mosaicity) as well as in-plane strain gradients can be recovered by analysing sections along the q_x axis. Such q_x-scans (or ω scans) are generally carried out by rocking the sample across the Bragg position while keeping the detector in a fixed position. In the present case, ω scans are performed by extracting a section along the q_x axis, i.e. $I(q_x, q_z = \text{constant})$, which is equivalent to record the intensity on a single channel.

In both cases if the number of channels is increased sufficiently to integrate the whole RLP (i.e. increasing the AB arc in Fig. 19.2) a classical rocking curve is obtained. Finally, in the case of highly imperfect materials, profile modelling is not straightforward and it is sometimes helpful to use line profile analysis procedures well-known in the field of powder diffraction such as the integral breadth method [17] or the Warren–Averbach method [18]. A powder is characterized by a random orientation of the grains so that in the scattering event a complete integration is carried out in the tangential (x, y) plane of reciprocal space [19]. With an epitaxial layer the integration is ensured in the y-direction by the large axial aperture of the detector associated with axial divergence. Integration in the q_x direction is easily carried out by integrating the recorded intensity over all the rocking range for each channel of the PSD. As a consequence of this integration counting statistics are significantly increased.

19.3.4 Instrumental Profile

As mentioned previously, the knowledge of the IP is essential in deriving microstructural information from the shape of the XRD line profiles. The 2-D IP (i.e. as a function of ω and 2θ, or q_x and q_z) can be calculated taking into account the transmittance function of each optical element in the beam path [20, 21]. The calculation is performed assuming a perfect, defect free sample. This approach avoids the use of "perfect" standard samples, which always contain a certain amount of defects. The 2-D IP is thus given by [5]

$$I(\Delta\omega, \Delta 2\theta) = \iint\limits_{\Delta\lambda\alpha} I(\Delta\lambda) R_m \left(\alpha - \frac{\Delta\lambda}{\lambda}\tan\theta_m\right)$$

$$\times R_s\left(\Delta\omega + \alpha - \frac{\Delta\lambda}{\lambda}\tan\theta_s\right) R_d(2\Delta\omega + \Delta 2\theta + \alpha) d\alpha d\Delta\lambda\,,$$

where $I(\Delta\lambda)$ is the source emission profile which in the present case is modelled by a Lorentzian function, R_m is the monochromator transmittance func-

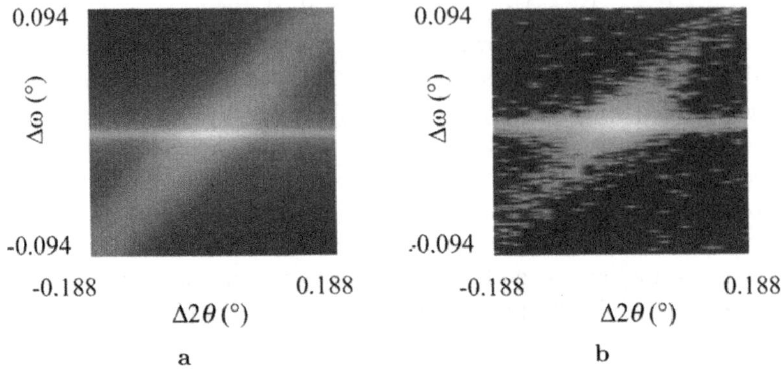

Fig. 19.5. (**a**) calculated 2-D instrumental profile corresponding to the (111) reflection of a perfect Si single crystal; (**b**) experimental 2-D instrumental profile. The detector streak is located along the 2θ-axis, the dynamical streak lies along the ω-2θ-axis (i.e. along the q_z direction of reciprocal space)

tion [21, 23] calculated using the dynamical reflection profile [24] of each Ge crystal, R_s is the dynamical reflection profile of the sample and Rd is the detector transmittance function, which can be very well modelled by a Voigt function [13]. The overall transmittance function of the PSD results as the convolution of its intrinsic resolution with the diffracted beam profile. The integral breadth of the latter is given by

$$\beta_{s'} = \beta_s \frac{\sin(\omega + \phi)}{\sin \omega} ,$$

where β_s is the breadth of the incident beam profile. θ_m, θ_s, λ, $\Delta\omega$ and $\Delta 2\theta$ are the monochromator and sample Bragg angle, the wavelength and the deviation from the peak maximum for the sample and the detector, respectively. Finally α and $\Delta\lambda$ are the equatorial divergence and wavelength spread. Further details concerning the set-up and the calculations elsewhere [5].

The result of the calculation for the symmetrical (111) reflection of a silicon wafer is plotted in a $(2\theta, \omega)$ frame in Fig. 19.5a. Two distinct features are observed. Firstly the transmittance function of the PSD results in a horizontal streak. This streak is equivalent to the so-called analyser streak when an analysing crystal is used [1], although it is about one order of magnitude wider in the present case. Secondly, the inclined streak is the so-called dynamical streak that originates from the long tails in the dynamical reflection profile of the sample. These tails are absent in the monochromator reflection profile because of the four successive reflections. A RSM plotted in a $(2\theta, \omega)$ frame of a real (111) Si sample is shown in Fig. 19.5b. The agreement between experimental and calculated data is good, and the detector streak is well reproduced. The actual dynamic range of the set-up does not allow recording the dynamical streak very far (up to 0.1° maximum), nevertheless this streak is clearly evidenced in Fig. 19.5b, in good agreement with the calculations.

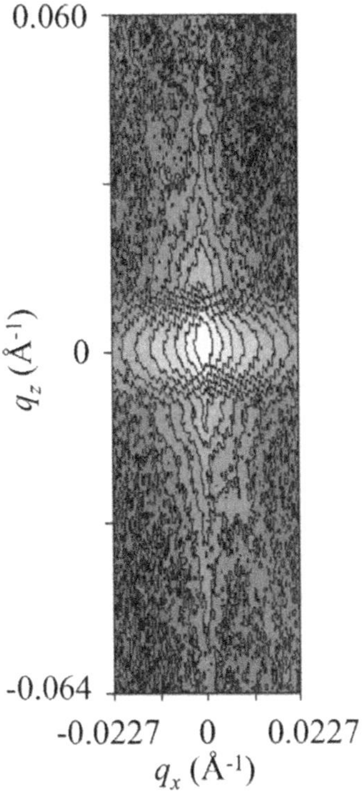

Fig. 19.6. Reciprocal space map recorded around the (200) RLP of the layer. q_x is parallel to the surface, q_z is parallel to the outwards surface normal

The IPs corresponding to specific scans across reciprocal space can be deduced from the 2-D IP [5]. In the ω scan mode, the IP is very close to a Gaussian function with Full Width at Half Maximum (FWHM) of 0.0029°. In the θ-2θ scan mode the profile exhibits a FWHM of 0.0058° on a 2θ scale. In the rocking-curve mode the resolution is 0.0029° and finally, in the powder scan mode, the IP is perfectly described by a Voigt function with $\beta_G = 0.0606°$ and $\beta_L = 0.0074°$. This loss of resolution is due to the detector transmittance function.

19.4 Profile Modelling

A RSM recorded close to the (200) reflection of the layer is shown in Fig. 19.6. Two noteworthy features should be stressed. Firstly, because of the limited film thickness, the RLP is noticeably broadened in the direction parallel to the outwards surface normal q_z. This point is discussed in the next section (Sect. 19.4.1). Secondly, the RLP is also broadened in the direction parallel to the surface, q_x. This point will be discussed in Sect. 19.4.2.

19.4.1 Out-of-Plane Profile Modelling

A projection of the RLP on the q_z axis is shown in Fig. 19.7. This profile exhibits thickness fringes attesting the flatness and the high structural quality of the layer. Such fringes are sometimes observed in oxide layers [25–27], but to our knowledge this is the first evidence of such a behaviour in sol-gel derived layers. The presence of thickness fringes is expected from both kinematical and dynamical theory of diffraction from flat crystals. However, a single Laüe function is here clearly unable to describe this profile accurately. In this respect we have to consider the effect of the microstructure of the

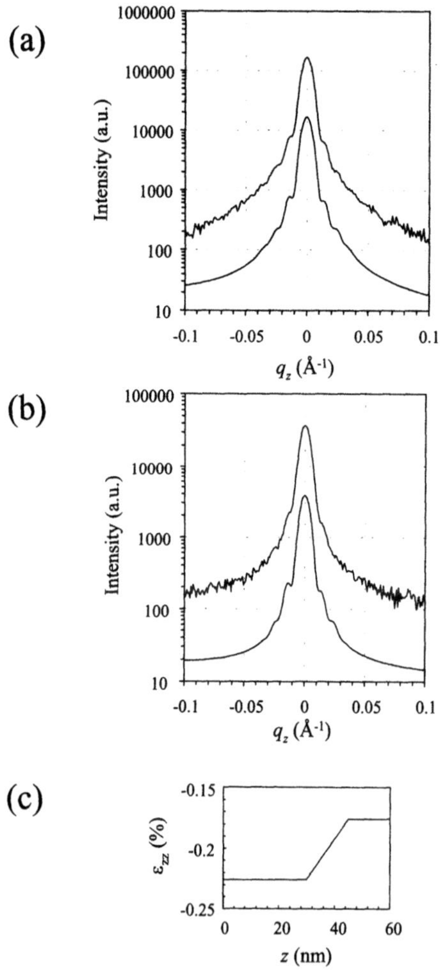

Fig. 19.7. Projection of the (200) (**a**) and (400) (**b**) RLPs on the q_z-axis; both experimental and calculated data are shown. For clarity the experimental curves have been multiplied by 10. The intensity is plotted logarithmically as a function of reciprocal lattice units q_z. (**c**) strain profile

layer and of the instrument described in the previous section. Because of the mosaicity of the layer and its nanometric dimensions, the kinematical theory of diffraction can be used. Within this theory, the basic equation of the scattered amplitude is

$$A(q_z, t) = A_0 \int \Omega(z, t) e^{i\boldsymbol{h}\boldsymbol{u}(z)} e^{iq_z z} dz\,. \tag{19.1}$$

Let us examine each term of this equation.

Size Effect

As thickness fringes are observed, and in agreement with TEM observations (not shown here), the YSZ islands are flat. The shape factor $\Omega(z, t)$ of a grain of thickness t, is just 1 if $z \leq t$, and 0 otherwise.

Strain Effect

The very existence of thickness fringes attests the low level of microstrain broadening. As mentioned in Sect. 19.2, the layer is subject to important tensile strain at the interface. The layer is therefore expected to be compressively strained in the direction perpendicular to the surface via Poisson contraction. Strain relaxation may thus occur across the film thickness resulting in the slight asymmetry observed on the (200) reflection profile [28]. The asymmetry is more pronounced in the (400) profile confirming that this effect is actually strain-related. The displacement vector $\boldsymbol{u}(z)$ is connected to the normal strain as $\varepsilon_{zz} = \partial u(z)/\partial z$ [29]. A step-like strain profile with a linear transition zone has been chosen to represent the actual strain distribution. Two parameters define the width of the transition zone, and two parameters define the height of the step.

The existence of this strain relaxation implies that (19.1) does not have a simple analytical solution and has to be numerically evaluated.

Size Distribution Effect

Oxide layers very often exhibit a significant grain size distribution, especially if they have been thermally treated. However, the thickness distribution function is very unlikely to exhibit long tails because of the limited film thickness. Moreover, in the case of thin films, a symmetrical distribution is expected [30]. We have therefore chosen a Gaussian thickness distribution function, $p(t)$, with the constraint: $p(t) = 0$ for $t \leq 0$. The scattered intensity is then

$$I(q_z) = \int p(t) I(q_z, t) dt\,. \tag{19.2}$$

Instrumental Broadening Effect

The instrumental broadening is taken into account by convoluting $I(q_z)$ with the calculated instrumental profile (see Sect. 19.3)

$$I_{\text{calculated}}(q_z) = \int I(\varepsilon)IP(q_z - \varepsilon)d\varepsilon\,. \qquad (19.3)$$

Equations (19.1)–(19.3) have been implemented in a computer program [31]. Integrations and convolutions are performed by means of a Gauss–Legendre quadrature formula, using at least 80 abscissa values. The model is fitted to the data by non-linear least squares. The result is shown in Fig. 19.7. Both profiles are plotted as a function of reciprocal lattice co-ordinates on a logarithmic scale to enhance the profile tails. Both profiles exhibit the same width, attesting the lack of strain broadening. The model perfectly describes the observed data even in the profile tails. Only two size parameters are sufficient to model the symmetric part of the profile:

- The average layer thickness is 62 nm, in agreement with the value expected from the elaboration parameters. Reflectivity measurements carried out before the 1550°C thermal treatment gave a 53 nm thickness. This increase of the average layer thickness shows that grain growth occurs upon thermal treatment.
- The standard deviation of thickness distribution function is only 6 nm. Although very narrow, this distribution is solely responsible for the fast damping of the oscillations.

The asymmetry of the profile can be accounted for with the strain profile shown in Fig. 19.7c:

- The average normal strain is −0.2%. This compressive strain is in agreement with the existence of a tensile interfacial strain.
- The strain profile is not constant. As the shape of the strain profile has been arbitrarily chosen, it is unlikely that the profile in Fig. 19.7c represents the *actual* state of strain within the layer. Nevertheless it clearly shows that strain relaxation occurs across the film thickness. Strain relaxation may be purely elastic, or it may also require the introduction of misfit dislocations depending on the elastic energy stored in the islands [32]. As the system is highly mismatched, misfit dislocations with the slip system $1/2\langle 110\rangle\{111\}$ are expected [33]. This particular point is still under investigation.

19.4.2 In-Plane Modelling

In the following the attention is focussed on diffraction features arising from the in-plane microstructure, i.e. mosaic spread and finite domain size. Results are preliminary.

Mosaic Spread

The effect of a mosaic structure on the ω scan profiles is well known [34,35]. The general expression of the scattered intensity is [18]

$$\int \Omega(\boldsymbol{r}-\boldsymbol{r}')\langle\exp[i\boldsymbol{h}\Delta\boldsymbol{u}(\boldsymbol{r}-\boldsymbol{r}')]\rangle \exp[i\boldsymbol{q}(\boldsymbol{r}-\boldsymbol{r}')]d(\boldsymbol{r}-\boldsymbol{r}')\,, \qquad (19.4)$$

where Ω is the block shape function, and $\Delta\boldsymbol{u}(\boldsymbol{r}-\boldsymbol{r}') = \boldsymbol{u}(\boldsymbol{r}) - \boldsymbol{u}(\boldsymbol{r}')$. For sufficiently small angular misorientations α, it is easily shown that $\Delta\boldsymbol{u}(\boldsymbol{r}-\boldsymbol{r}') = \alpha \times (\boldsymbol{r}-\boldsymbol{r}')$ [1]. Hence, assuming a given block shape function and orientation distribution function (ODF) of the mosaic blocks (i.e. the distribution of the α_s), the diffraction profile deriving from misoriented blocks of finite size can easily be computed. For the sake of simplicity, the spatial average $\langle\rangle$ in (19.4) is often performed considering a Gaussian ODF. However, the ω scan profile often exhibits non-Gaussian shapes [36,37]. Therefore, in order to be less restrictive on the actual profile shape the ω scan profile was modelled by a Voigt function, a convolution of a Lorentzian and a Gaussian function [38]. We implicitly assume that the shape of the spatial average $\langle\rangle$ is described by any function lying between a two-sided exponential and a Gaussian function. The integral breadth of the Voigt function expressed in reciprocal lattice units is a direct measure of the integral breadth of the ODF, $\Delta\omega$, such as

$$\beta_m = \Delta\omega \times Q\,, \qquad (19.5)$$

where Q is the length of the scattering vector. It is clear from (19.4) that the "size" and mosaicity profile are combined in a convolution relation. The previous Voigt function has therefore to be convoluted with the "size" profile to completely describe the ω scan profile. This implies the knowledge of the coherently diffracting domain shape. As it will be shown in the next section, this shape is to date not exactly established. We thus make the restrictive hypothesis that the "size" profile can be described by a Voigt function whose integral breadth in reciprocal lattice units is related to the coherently diffracting domain width w according to

$$\beta_w = \frac{2\pi}{w}\,. \qquad (19.6)$$

Unlike strain broadening in powder diffraction, the linear Q-dependence of β_m follows from a simple geometrical consideration. Similarly to the θ-2θ scan profiles, the ω scan profiles are broadened by the instrument following a convolution relationship [39]. Equation (19.4) may thus be rewritten in terms of the integral breadths β_G and β_L of a Voigt function

$$\beta_L = \frac{2\pi}{w_L} + \Delta\omega_L \times Q\,, \qquad (19.7a)$$

$$\beta_G^2 = \beta_{G,I}^2 + \left(\frac{2\pi}{w_G}\right)^2 + \Delta\omega_G^2 \times Q^2\,, \qquad (19.7b)$$

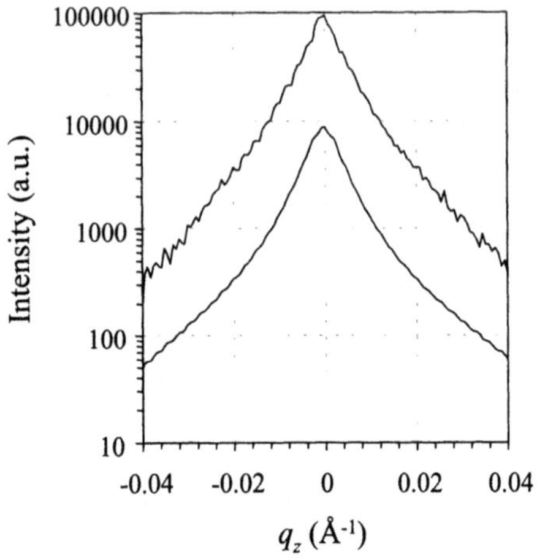

Fig. 19.8. ω scan through the centre of the (200) RLP. Both experimental and calculated data are shown. For clarity the experimental curves have been multiplied by 10. The intensity is plotted logarithmically as a function of reciprocal lattice units q_x

where the subscripts G and L refer to the Gaussian and Lorentzian components, and $\beta_{G,I}$ is the integral breadth of the instrumental profile (which is Gaussian in this scanning mode). The effect of size and mosaicity can hence be separated on the basis of their different Q-dependence by modelling the ω scan profiles of (at least) two orders of reflection of the same crystallographic plane family with a Voigt function, whose breadths are constrained by (19.7a+b). The lateral coherently diffracting domain size w, and the mosaicity $\Delta\omega$, can be obtained from their Gaussian and Lorentzian component using the relation given by Langford [38].

An ω scan performed through the maximum of the (200) RLP is shown in Fig. 19.8. A fit with Voigt function of the (200) and (400) ω scan profiles provides a 0.2° mosaicity and a 118 nm in-plane grain size. It should also be mentioned that the sample contribution to the line shape could solely be modelled by means of a Lorentzian function, demonstrating the non-Gaussian behaviour of the ODF.

The effect of random deformations taken into account in the spatial average $\langle\rangle$ in (19.3) is referred to as diffuse X-ray scattering (DXS) in the field of single crystal and epitaxial thin film XRD [34], whereas it is referred to as strain broadening in the field of powder diffraction [40]. The total scattered intensity from an imperfect sample is the sum of the diffuse and coherent scattering [41]. Its expression follows from (19.4):

$$I(\boldsymbol{q}) = \int \Omega(\boldsymbol{r}-\boldsymbol{r}')(\Gamma_C + \Gamma_D)\exp[i\boldsymbol{q}(\boldsymbol{r}-\boldsymbol{r}')]d(\boldsymbol{r}-\boldsymbol{r}'),$$

with

$$\Gamma_C = \langle \exp[i\boldsymbol{h}\Delta\boldsymbol{u}(\infty)] \rangle ,$$

$$\Gamma_D = \langle \exp[i\boldsymbol{h}\Delta\boldsymbol{u}(\boldsymbol{r}-\boldsymbol{r}')] \rangle - \langle \exp[i\boldsymbol{h}\Delta\boldsymbol{u}(\infty)] \rangle .$$

The term Γ_C is the contribution of coherent scattering for which the correlation length $|\boldsymbol{r}-\boldsymbol{r}'|$ reaches infinity (long range correlation). The term Γ_D is the contribution of diffuse scattering (short range correlation). The relative amount of each is directly dependent on the displacement difference correlation function $\Delta\boldsymbol{u}(\boldsymbol{r}-\boldsymbol{r}')$. In the case of strong disorder $\Delta\boldsymbol{u}(\infty)$ diverges and the so-called static Debye–Waller factor, Γ_C, reaches zero. However, because of the limited film thickness both contributions are frequently observed in thin films [36, 40]. The next section is devoted to the analysis of coherent scattering.

Coherent Scattering

As shown previously, a transverse scan performed through the centre of the RLP is dominated by diffuse scattering, although a discontinuous contribution can be distinguished. The effect of this contribution can be significantly enhanced by performing the scan slightly away from the peak maximum (e.g. at $q_z = -5.10^{-3}\,\text{Å}^{-1}$). A highly oscillating peak profile is obtained (Fig. 19.9).

The presence of satellites in the diffraction profile reflects the periodic nature of the scattering object, similarly to the diffraction by gratings [42]. The period of the grating can be estimated from the reciprocal satellite spacing. In the present case $\Lambda \approx 400\,\text{nm}$. This value is one order of magnitude greater than the film thickness and it also largely exceeds the in-plane coherently diffracting domain size. This observation confirms that this oscillating contribution effectively originates from a long-range correlation and can therefore

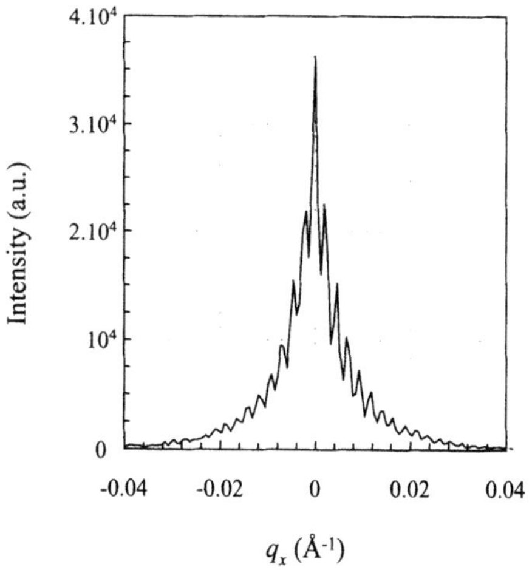

Fig. 19.9. ω scan through the (200) RLP performed at $q_z = -5.10^{-3}\,\text{Å}^{-1}$. The intensity is plotted linearly as a function of reciprocal lattice units q_x

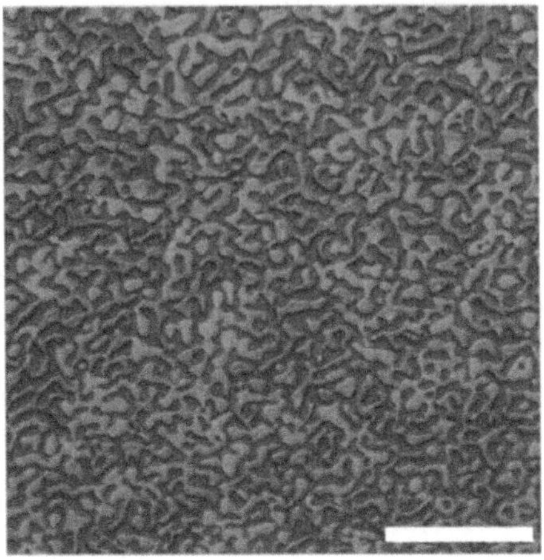

Fig. 19.10. Optical micrograph of a sample annealed 16 h at 1500°C showing the unusual microstructure of the samples (bar = 4 µm)

be attributed to coherent scattering. It also shows that the periodicity has to be maintained over a distance of several crystallites.

A very similar situation is encountered in periodically nanostructured materials such as quantum dots or quantum wires structures [1]. In such cases, grating diffraction occurs from the coherent scattering from islands periodically distributed over the surface of the substrate. The mechanisms of self-organization in semiconductor nanostructures have been thoroughly studied during the past few years [43]. In contrast, much less is known about such phenomena in oxide materials. To our knowledge, this is the first XRD evidence of self-organization in oxide layers. As mentioned in Sect. 19.2, capillary instabilities originating from the presence of grain boundaries [9] drive the film to split into individual islands. The excess of interfacial energy is the driving force of this dewetting mechanism [7]. However, the mechanism that yields an in-plane periodicity is to date not clearly identified. Because of the complicated interface, especially due to the existence of many in-plane variants, it is unlikely that this mechanism is strain-driven. Conversely, dewetting in thin films can take place by spinodal decomposition [44]. This hypothesis is supported by the fact that spinodal decomposition yields microstructures with preferential correlation lengths (see e.g. [45]). Such microstructures undergo characteristics evolutions as a function of time. We therefore annealed the sample at 1550°C for several time intervals.

In Fig. 19.10 an optical micrograph of a 16 h annealed sample is presented. This microstructure is drastically different from those known in self-organized semiconductor nanostructures, but very similar microstructures have been observed in organic thin films subject to spinodal decomposition [44]. The existence of an in-plane periodicity is clearly evidenced by the existence of a secondary maximum in the power-spectrum of the micrograph (Fig. 19.11)

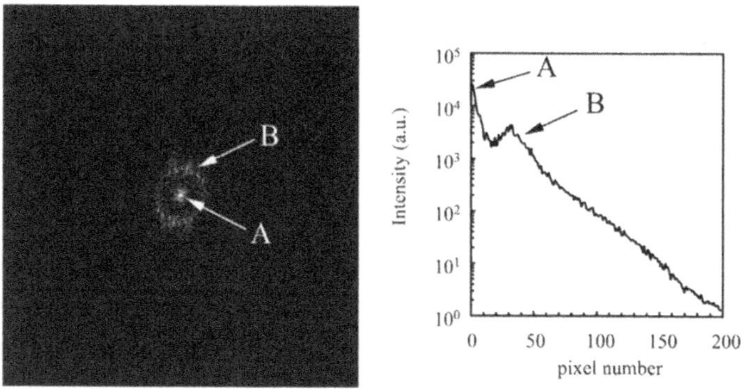

Fig. 19.11. (**a**) Power spectrum of the optical micrograph. The main peak is marked by A. The presence of a secondary maximum (marked B) indicates the existence of a preferential correlation length. (**b**) corresponding intensity profile obtained by circular integration

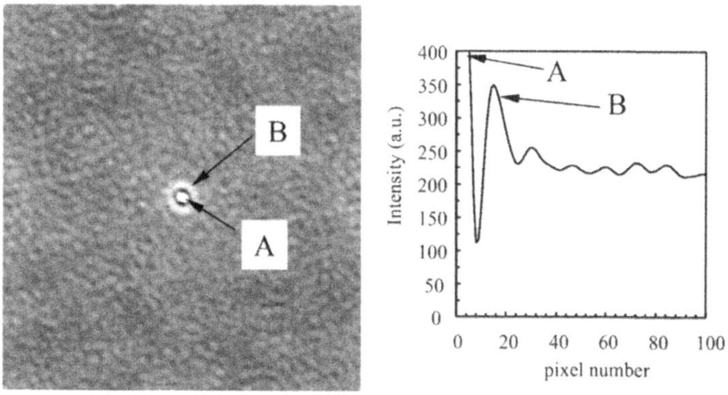

Fig. 19.12. (**a**) Self-correlation function of the optical micrograph. The main peak is marked by A. The position of the first order satellite (marked B) yields the in-plane periodicity (the main peak has been truncated). (**b**) corresponding intensity profile obtained by circular integration

corresponding to the first order grating peak. The lack of superior orders and the broadening of the first order peak attests the non-ideal ordering of the islands, i.e. ordering is limited to the first neighbours. The circular symmetry of the secondary maximum shows that the periodicity is isotropic over the sample surface, which provides further indication of the non-crystallographic origin of this phenomenon. The lateral periodicity in real space, Λ, is given by the reciprocal lateral position of the secondary maximum. In this case we have $\Lambda = 1\,\mu\text{m}$. Very similar conclusions can be drawn directly in real space from the analysis of the self-correlation function of the micrograph (see Fig. 19.12), the location of the first order peak giving the lateral periodicity (a microstructure without lateral ordering would give a continuously decreasing profile in both the power spectrum and in the self-correlation function). Whether this type of microstructure can give rise to the peak profile shown in Fig. 19.10 is discussed below.

The microstructure described above can be schematically viewed as a Dirac grating (responsible for the periodicity – in the case of short range order the coherence of the grating is limited to $N = 2\Lambda$) convoluted with island shape function. In a similar fashion to (19.1) the amplitude scattered by such a structure is given by the Fourier transform of its real space electron density distribution. The scattered amplitude is therefore simply given by the product of the Fourier transform of the Dirac grating and the Fourier transform of the island shape function. The intensity follows

$$\left\langle e^{i\boldsymbol{h}\Delta\boldsymbol{u}(\infty)}\right\rangle \left| \sum_{m=0}^{N/\Lambda} e^{iq_x m\Lambda} \sum_{n=-w/2a}^{w/2a} e^{iq_x na} \right|^2 .$$

Assuming square islands, the previous equation can simply be written (omitting the Debye–Waller factor) as

$$\frac{\sin^2 q_x N/2 \sin^2 q_x w/2}{\sin^2 q_x \Lambda/2 \sin^2 q_x a/2} .$$

This profile is depicted in Fig. 19.13. The first term is due to the grating and gives rise to satellites around the main reflection (Fig. 19.13a). The second term is due to the island shape and corresponds to the envelope of the profile (Fig. 19.13b). To account for the existence of an island size distribution and because the islands clearly do not exhibit square shapes, the latter term can be replaced by a Voigt function whose parameters are constrained to the values determined in the previous section. To describe the total scattered intensity, the coherent intensity has to be added to the diffuse scattering (modelled by a Voigt function as described in the previous section) and convoluted with the instrumental profile corresponding to the ω scan mode. Apart from the overall peak intensity and position, the only fitting parameters are the grating period Λ, and the static Debye–Waller factor. All others terms entering the model (in-plane coherently diffracting domain size and mosaic spread) are constrained to their values found in the previous section. The result of the fit on the (200) reflection is shown in Fig. 19.14. The model describes the data very well and the periodicity is found equal to $\Lambda = 410\,\text{nm}$.

This value is significantly different from the value deduced from the optical micrograph. This discrepancy is probably due to the fact that the optical micrograph has been taken on a 16 h-annealed sample, whereas the XRD diagram has been recorded on a 1 h-annealed sample. In order to quantitatively crosscheck the results obtained here we plan to use atomic force microscopy parallel with XRD. The physical origin of this in-plane periodicity also needs to be confirmed by further studies.

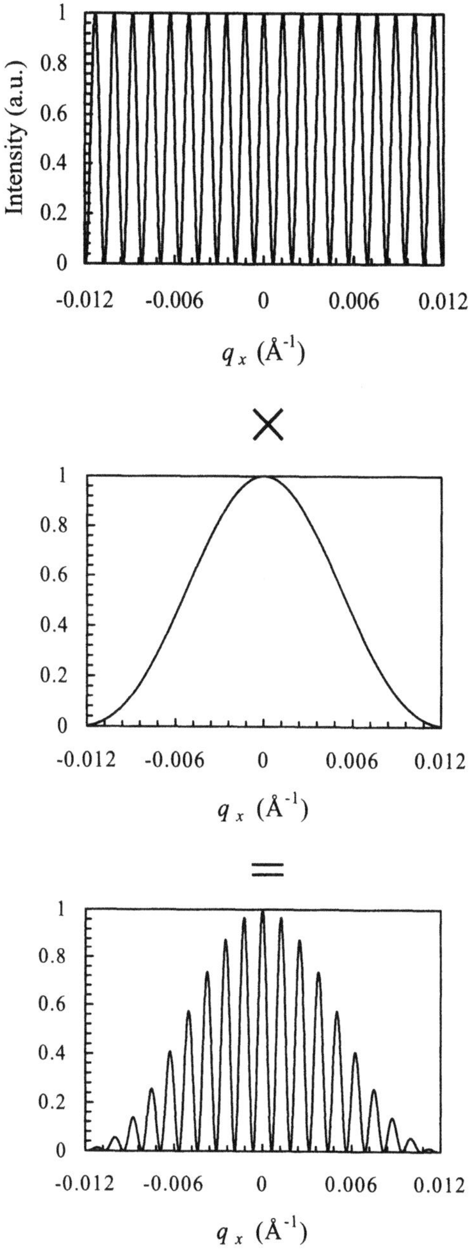

a

b

c

Fig. 19.13. Simulation of the experimental profile by using the product of the Fourier transform of a Dirac grating with the Fourier transform of the crystallite shape function

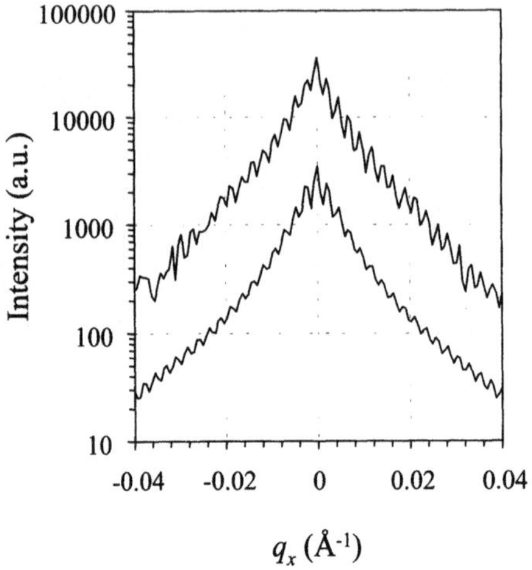

Fig. 19.14. ω scan through the (200) RLP performed at $q_z = -5.10^{-3}\text{Å}^{-1}$. Both experimental and calculated data are shown. For clarity the experimental curves have been multiplied by 10. The intensity is plotted logarithmically as a function of reciprocal lattice units q_x

19.5 Conclusions

We have presented a methodology for the investigation of imperfect epitaxial layers (mainly oxides and ceramics) using XRD. The experimental approach is based on a laboratory-built high-resolution X-ray diffractometer that allows fast RSM acquisition from defective samples. This technique allows the separation of defects acting along different directions of reciprocal space: normal size and strain in the direction normal to the surface and lateral size and mosaic spread in the direction parallel to the surface. All these parameters can be accessed by XRD line shape analysis. In the direction normal to the surface, the diffraction profiles are modelled taking into account both microstructural parameters and the instrumental profile of our set-up. The vertical grain size, size distribution and strain variation across the film thickness could be recovered by the simulation of the profile. In the direction parallel to the surface, the effect of mosaic spread has been modelled by a Voigt function. The assumption is thus made that the shape of the spatial average entering the scattered intensity lies between a two-sided exponential and a Gaussian function. Because of uncertainties concerning the in-plane grain shape, a Voigt function was used to simulate the effect of in-plane grain size. These two effects have been separated considering their dependence on the order of reflection. Finally a very unusual, highly oscillating additional intensity has been discovered in the RLP. The existence of many satellites around the main Bragg peak attests the existence of an in-plane periodicity, which has been modelled assuming a self-organized microstructure with short-range order.

References

1. V. Holý, U. Pietsch, T. Baumbach: *High Resolution X-Ray Scattering from Multilayers and Thin Films.* Springer tracts in Modern Physics Vol. 149 (Springer Berlin Heidelberg New York 1999) pp. 1
2. P.F. Fewster: Crit. Rev. Solid State Mater. Sci. **22**, 69 (1997)
3. N. Itoh, K. Okamoto: J. Appl. Phys. **63**, 1486 (1988)
4. A. Iida, K. Kohra: Phys. Stat. Sol. A **51**, 533 (1979)
5. A. Boulle, O. Masson, R. Guinebretière, A. Lecomte, A. Dauger: J. Appl. Cryst. **35**, 606 (2002)
6. R. Guinebretière, A. Dauger, A. Lecomte, H. Vesteghem: J. Non-cryst. Solids **147**, 542 (1992)
7. K.T. Miller, F.F. Lange, D.B. Marshall: Mater. Res. Soc. Symp. Proc. **121**, 823 (1988)
8. M.G. Cain, F.F. Lange: J. Mater. Res. **9**, 674 (1994)
9. D.J. Srolovitz, S.A. Safran: J. Appl. Phys. **60**, 247 and 255 (1986)
10. A. Boulle, L. Pradier, O. Masson, R. Guinebretière, A. Dauger: Appl. Surf. Sci. **188**, 80 (2002)
11. W.J. Bartels: J. Vac. Sci. Tech. B **1**, 338 (1983)
12. J. Ballon, V. Comparat, J. Pouxe: Nucl. Instr. Meth. **217**, 213 (1983)
13. O. Masson, R. Guinebretière, A. Dauger: Mater. Sci. Forum **278–281**, 115 (1998)
14. A. Boulle, O. Masson, R. Guinebretière, A. Dauger: Appl. Surf. Sci. **180**, 322 (2001)
15. M.A. Pick, K. Bickmann, E. Pofahl, K. Zwoll, H. Wenzl: J. Appl. Cryst. **10**, 450 (1977)
16. J.M. Cowley: *Diffraction physics.* (North-Holland, Amsterdam 1975) pp. 95
17. H.P. Klug, L.E. Alexander: *X-ray Diffraction Procedures for Polycrystalline and Amorphous Materials.* (John Wiley, New York 1974) pp. 618
18. B.E. Warren: *X-Ray Diffraction.* (Addison-Wesley, Reading 1969) pp. 251
19. R.W. James: *The optical principles of the diffraction of X-rays.* (Bell, London 1967) pp: ??
20. P. Zaumseil, U. Winter: Phys. Stat. Sol. A **70**, 497 (1982)
21. P.F. Fewster: J. Appl. Cryst. **22**, 64 (1989)
22. M.O. Möller: J. Appl. Cryst. **27**, 369 (1994)
23. S.E.G. Slusky, A.T. Macrander: J. Appl. Cryst. **20**, 522 (1987)
24. A. Authier: *Dynamical theory of X-ray diffraction.* IUCr monograph on crystallography 11 (Oxford University Press, Oxford 2001) pp. 173
25. A.G. Zaitsev, G. Ockenfuss, D. Guggi, R. Wördenweber: J. Appl. Phys. **81**, 3069 (1997)
26. X. Castel, M. Guilloux-Viry, A. Perrin, J. Lesueur, F. Lalu: J. Cryst. Growth **187**, 211 (1998)
27. M.H. Cho, D.H. Ko, Y.K. Choi, I.W. Lyo, K. Jeong, T.G. Kim, J.H. Song, C.N. Whang: J. Appl. Phys. **89**, 1647 (2001)
28. I. Vartanyants, C. Ern, W. Donner, H. Dosch, W. Caliebe: Appl. Phys. Lett. **77**, 3929 (2000)
29. D.J. Dunstan: J. Mater. Sci. **8**, 337 (1997)
30. S. Rao, C.R. Houska: Acta Cryst. A **46**, 6 (1986)

31. O. Masson: *Peakoc, Profile fitting computer program.* (SPCTS, Limoges 1998) (olivier.masson@unilim.fr)
32. S.C. Jain, A.H. Harker, R.A. Cowley: Philos. Mag. A **75**, 1461 (1997)
33. C.H. Chen, N. Wakiya, A. Saiki, T. Kiguchi, K. Shinozaki, N. Mizutani: Physica B **308–310**, 1050 (2001)
34. V. Holý, K. Wolf, M. Kastner, H. Stanzl, W. Gebhardt: J. Appl. Cryst. **27**, 551 (1994)
35. Y.I. Nesterets, V.I. Punegov: J. Appl. Cryst. **56**, 540 (2000)
36. P.F. Miceli, C.J. Palmstrom: Phys. Rev. B **51**, 5506 (1995)
37. A. Gibaud, R.A. Cowley, D.F. McMorrow, R.C.C. Ward, M.R. Wells: Phys. Rev. B. **48**, 14463 (1993)
38. J.I. Langford: J. Appl. Cryst. **11**, 10 (1978)
39. P.D. Healey, J.E. Ayers: Acta Cryst. A **52**, 245 (1996)
40. R.I. Barabash, W. Donner, H. Dosch: Appl. Phys. Lett. **78**, 443 (2001)
41. M.A. Krivoglaz: *Theory of X-ray and thermal neutron scattering by real crystals.* (Masson, Paris 1969) French translation, pp. 95
42. M. Born, E. Wolf: *Principles of Optics.* (Cambridge University Press, Cambridge 1999) pp. 446
43. C. Teichert: Phys. Rep. **365**, 335 (2002)
44. S. Joly, A. Raquois, F. Paris, B. Hamdoun, L. Auvray, D. Ausserre, Y. Gallot: Phys. Rev. Lett. **77**, 4394 (1996)
45. J.W. Cahn: Acta Metall. **9**, 195 (1961)
46. J.I. Langford: Natl. Inst. Stand. Technol. Spec. Pub. **846**, 110 (1992)
47. J.G.M. van Berkum, R. Delhez, T.H. de Keijser, E.J. Mittemeijer: Acta Cryst. A **52**, 730 (1996)
48. G.K. Williamson, W.H. Hall: Acta Mettal. **1**, 22 (1953)

20 Three-Dimensional Reciprocal Space Mapping: Application to Polycrystalline CVD Diamond

M. Golshan, D. Laundy, P.F. Fewster, and M. Moore

The principles and practice of three-dimensional reciprocal-space mapping (3D-RSM) for studying polycrystalline materials are outlined and compared with powder diffraction methods. Results obtained from applying 3D-RSM to polycrystalline chemical vapour-deposited (CVD) diamond films are presented and discussed. The experiments were performed using synchrotron radiation at SRS and ESRF.

20.1 Introduction

Studying polycrystalline materials requires a well-defined probe that can give detailed information at the microstructural level. In reciprocal space, Bragg scattering from a single crystallite gives rise to a three-dimensional intensity distribution around the reciprocal lattice point. If the crystal is strained, there will be a variation in interplanar spacing causing broadening of the feature in the longitudinal direction. If the atomic planes are bent, the broadening will be in the lateral direction. Conventional diffractometry results in the averaging of the scattering along the longitudinal direction and a loss of the resolution along the perpendicular direction in reciprocal space. This is due to the large number of crystallites, each one having a different orientation that contribute to the scattering. Reciprocal space mapping in contrast uses a probe that has high resolution both parallel and perpendicular to the scattering vector and in principle allows the scattering from a single crystal to be isolated in reciprocal space. CVD diamond films are of great interest to the diamond, aerospace, semiconductor and microprocessor industries and many novel applications are based on these materials. Surface coatings with CVD diamond films are also finding wide application in a number of areas where surfaces possessing the characteristics of extreme hardness, low wear rate and low coefficient of friction are required. Being radiation hard, CVD diamond films are also good candidates for high-energy particle detectors (e.g. CVD diamond film is being used for the Large Hadron Collider (LHC) detector at CERN). Having a high thermal conductivity, they are also widely used for thermal management devices; and as they transmit IR and UV radiation, they are used for windows in hostile environments.

The ability to produce diamond coatings and free-standing layers by CVD methods is one of the most exciting advances in material science. The quality of the material continues to improve, but the strength properties are still not fully optimised. Therefore, quantifying strain within the individual crystallites in CVD diamond films is of crucial importance in the technological advancement of these materials. Here, we have applied three-dimensional diffraction-space mapping to study individual grains in as-grown polycrystalline CVD diamond films grown using a microwave plasma technique [1].

20.1.1 X-Ray Methods Used for Strain Determination

Powder diffraction is one of the most favoured X-ray diffraction techniques for analysing polycrystalline material [2]. It is a powerful method for what might be referred to as "good powder material". In this method, a standard θ–2θ scan over a diffraction peak is used for determining the specimen strain. Such a scan can be thought of as an average of longitudinal scans in reciprocal space over many crystallites, each scan being offset in sample rotation angle (ω). The difficulty arises when analysing textured polycrystalline material or specimens consisting of few crystallites. In such a case let us consider the reflections from a series of crystallites. Each reflection will have similar 2θ positions but will have positions offset in ω and χ (tilt angle out of the diffraction plane) caused by the variations in crystallite orientations. When the reflections from individual crystallites are sharp, or the number of crystallites is small, then the probability of any crystallite being in the correct ω position to reflect becomes low. In this case, the scan only passes through the tails of the scattering in reciprocal space where the pattern is dominated by diffuse scattering [3].

Another limitation of the powder method, arising from the averaging inherent in the method, is that it is not possible to measure separately the strain variation within each crystallite and the strain variations between crystallites.

Microbeam diffraction is a method used to study individual grains in polycrystalline materials by using a small beam to isolate crystals in real space [4,5]. If the beam is too large compared to the crystallite size, there is a chance that more than one crystallite will be illuminated. If the beam is too small, the scattering may not include the whole crystallite.

Diffracted fluorescence, or studying Kossel lines, is another high-resolution technique used for determining strain [6,7]. This method can produce accurate results, but its use is limited to the K edges that can be reached in the hard X-ray regime. The diffracted intensity also is strongly affected by the quality of the material. Therefore, the technique is difficult to use in the case of most polycrystalline materials which diffract weakly.

To isolate the scattering from individual crystallites in polycrystalline films of CVD diamond we have used high-resolution three-dimensional reciprocal-space mapping (3-D RSM) [8].

20.2 Principle of Three-Dimensional-Reciprocal Space Mapping (3-D RSM)

Reciprocal-space mapping [8–12] can be described briefly as creating a small probe in reciprocal space by reducing the angular spread of the incident beam and the angular acceptance of the detector and by reducing the wavelength spread of the incident beam. It is a general term and can be used in both low- and high-angular resolution modes. The low-resolution technique is suitable for surveying very large areas or volumes of reciprocal space and for highly imperfect crystals, whereas the high-resolution technique is used for mapping the reciprocal space in great detail around a sharp reflection from a good crystal by restricting the diffraction-space probe [10].

Figure 20.1 shows a two-dimensional array of diffraction spots in the diffraction plane with the incident beam vector $\boldsymbol{k}_0$ pointing towards the origin in reciprocal space and the scattered beam vector $\boldsymbol{k}_h$ drawn from the origin at an angle of 2θ to the incident beam vector. The angle between $\boldsymbol{k}_0$ and the sample surface is ω. The surface normal is given by the vector $\boldsymbol{q}_z$ and a vector parallel to the sample surface is $\boldsymbol{q}_x$. By adjusting the detector angle to pick up radiation with wave vector $\boldsymbol{k}_h$ any feature at position $\boldsymbol{Q} = \boldsymbol{k}_h - \boldsymbol{k}_0$ may be investigated. $\boldsymbol{Q}$ must lie within the limiting sphere in reciprocal space and for reflection geometry must lie outside the two smaller spheres which are the loci of points generated when either $\boldsymbol{k}_0$ or $\boldsymbol{k}_h$ lie along the sample surface.

The presence of lattice parameter variations and lattice plane tilting results in the X-ray scattering from a single crystallite forming a broadened three-dimensional feature around the reciprocal lattice point. This reciprocal space feature therefore contains information about the strain fields present within the crystallite and can be mapped by a small three-dimensional probe scanned through the feature. In two-dimensional mapping, the probe is small in two dimensions but large in the direction perpendicular to the scattering plane. In three-dimensional mapping, the probe is also reduced in the perpendicular direction and this prevents unrelated features in reciprocal space from overlapping on the map. A small probe can only be achieved at the

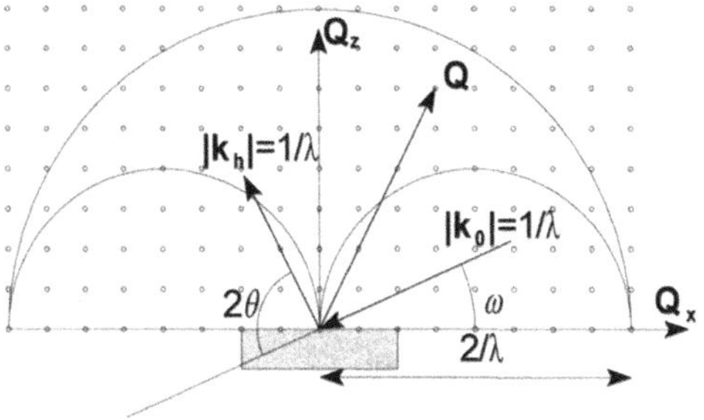

Fig. 20.1. The limiting sphere showing the region of reciprocal space accessible to X-rays [11]. $\boldsymbol{k}_0$ and $\boldsymbol{k}_h$ are the incident and the scattered beam vectors respectively

expense of X-ray flux. Synchrotron radiation is therefore advantageous for these measurements.

For the measurements presented in this article, the conventional synchrotron diffraction geometry with a vertical scattering plane was used. The diffractometer had a three-bounce analyser crystal and a detector mounted on a 2θ-arm and a sample goniometer consisting of an Eulerian cradle with an ω-circle rotating about an axis perpendicular to the scattering plane on which an orthogonal χ-circle was mounted. Finally, the sample stage was mounted on a j-circle. The probe position $\boldsymbol{Q}$ may be expressed in terms of the three Euler angles [13].

The size of the probe depends on many factors. The treatment of resolution can be simplified by giving the size of the probe volume in three orthogonal directions – (a) along the diffraction vector $\boldsymbol{Q}$, (b) perpendicular to $\boldsymbol{Q}$ in the diffraction plane and (c) perpendicular to $\boldsymbol{Q}$ and the diffraction plane. These can be referred to as the longitudinal size, the lateral in-plane size and the lateral out-of-plane size, respectively.

20.3 Instrumental Requirements for Performing High-Resolution Reciprocal-Space Mapping

At a synchrotron source, the X-rays are well collimated but have a large wavelength spread. In high resolution diffraction experiments, perfect crystal optics are used to give a beam with low wavelength spread, while preserving the high collimation. As an example, Fig. 20.2 shows a schematic diagram of the high-resolution diffraction Station 16.3 at the SRS, Daresbury Laboratory [14].

For simplicity, the various contributions to the size of the reciprocal-space probe are considered separately. For two-dimensional mapping, the size of the lateral out-of-plane probe is large and therefore need not be considered. In the three-dimensional case, we also need to take into account the lateral out-of-plane resolution but, to begin with, we concentrate on the resolution in the scattering plane (as in the 2-dimensional case). The contributing elements are:

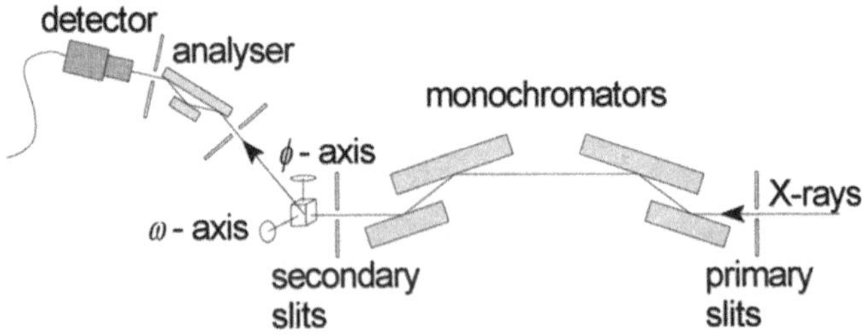

Fig. 20.2. A schematic diagram of the six-circle high-resolution diffractometer at Station 16.3, Daresbury Laboratory, used for reciprocal-space mapping

1. *Divergence of the incident beam in the vertical plane.* This is caused by the vertical size of the source and by the vertical size of the diffracting region of the sample. It should be noted however that the vertical divergence can be reduced by an energy dispersive monochromator arrangement. For a source to sample distance L_1, vertical source size v_s and vertical diffracting size v_c we can write the angular divergence in the vertical plane as approximately $\delta\omega_1 = \sqrt{v_s^2 + v_c^2}/L_1$. The diffracting size is given by the "footprint" of the beam on the crystallite under investigation. The divergence of the incident beam in the vertical plane gives an angular spread to $\boldsymbol{k}_0$ in the vertical plane which therefore affects $\boldsymbol{Q} = \boldsymbol{k}_h - \boldsymbol{k}_0$. This results in a broadening of the probe in the diffraction plane in the direction perpendicular to $\boldsymbol{k}_0$ by an amount $\delta\omega/\lambda$.
2. *Wavelength spread in the incident beam.* For perfect crystal optics the wavelength spread is determined by the bandwidth of the crystal reflection. For the silicon 111 reflection, the bandwidth is $\Delta\lambda/\lambda = 1.7 \times 10^{-4}$. There is also a contribution to the wavelength spread from the vertical divergence. This contribution can be calculated using Bragg's Law to give $\Delta\lambda/\lambda = \delta\omega_1/\tan\theta_{\text{mono}}$, where θ_{mono} is the monochromator angle. The wavelength spread of the beam causes a spread in the magnitude of $\boldsymbol{k}_0$ and $\boldsymbol{k}_h$ and a change in the direction of $\boldsymbol{k}_h$ when an analyser crystal is used. The broadening of the probe in the longitudinal direction is approximately $\Delta\lambda/\lambda^2$.
3. *Detector angular acceptance in the vertical plane.* This is given by the rocking curve width of the analyser reflection. In our case, with a silicon (111) triple-bounce analyser this is $\delta\omega \approx 20$ µradians at a wavelength of 0.1 nm. This gives an angular spread to $\boldsymbol{k}_h$ and results in a broadening of the diffraction probe in the direction perpendicular to $\boldsymbol{k}_h$ in the diffraction plane by an amount $\delta\omega_1/\lambda$.

Finally, we have to consider the out-of-plane lateral resolution. This can be calculated simply using Fig. 20.3. We assume that the source has a Gaussian distribution of standard deviation σ_S in the horizontal direction. The hori-

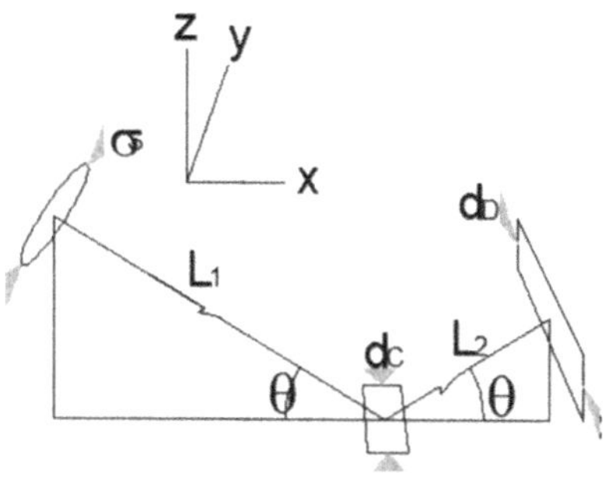

Fig. 20.3. Reciprocal-space resolution in the direction perpendicular to scattering plane

zontal size of the diffracting region is d_C and the detector has horizontal slits of width d_D. The distance between the source and sample is L_1 and from sample and detector is L_2. A ray displaced horizontally from the centre of the source by yS diffracts from the crystallite at a position displaced horizontally from the centre by y_C and enters the detector at a position displaced horizontally from the centre y_D. This defines a position in reciprocal space which is offset horizontally from the scattering plane by

$$|\Delta \boldsymbol{Q}| = \frac{y_D}{L_2} + \frac{y_S}{L_1} - y_C\left(\frac{1}{L_1} - \frac{1}{L_2}\right) . \qquad (20.1)$$

We then calculate the standard deviation of $|\Delta \boldsymbol{Q}|$ as y_S is integrated over the Gaussian horizontal source profile, y_C is integrated over the horizontal diffracting region on the crystallite and y_D is integrated between the detector horizontal slit limits. Dividing this by $|\boldsymbol{Q}|$ gives $\sigma(\omega_2)$, the standard deviation in the out-of-plane angle ω_2

$$\sigma(\omega_2)^2 = \frac{\left\{\frac{d_D^2}{12L_2^2} + \frac{\sigma_S^2}{L_1^2} + \frac{d_C^2}{12}\left(\frac{1}{L_1} + \frac{1}{L_2}\right)^2\right\}}{4\sin^2\theta} . \qquad (20.2)$$

20.4 Studies of Polycrystalline CVD Diamond

Reciprocal-space mapping has been used to study three different grades of free-standing CVD diamond films. The measurements were performed at Station 16.3, Daresbury Laboratory and ID10B, ESRF. Nitrogen incorporated in one diamond during growth appeared to relieve the strain and give a sharper texture to the film.

20.4.1 Scanning Electron Microscopy

Three specimens are discussed in this section. The samples are labelled CVD1, CVD2 and CVD3. These specimens were grown using a microwave plasma technique in a reactor based on the ASTeX model 5400 5.0 kW source and chamber. All these specimens are free standing layers. CVD1 was doped with nitrogen. The dominant surface orientation of the crystallites was determined by powder diffraction and showed a strong preference for alignment along a [100] direction, but random azimuthal angles about this direction ("strong {100} texture"), which is seen also in the SEM image (see Fig. 20.4). Surface steps can be seen on all the crystallites. Occurrence of these large crystallites is the result of higher temperature at those points which caused faster growth. This is because the protrusion from the surface enters the growth environment (plasma) and therefore, has a longer path back to the heat-sink (substrate). The growth surface of the protrusion can be hotter than the surrounding (thinner) material, since there is a net heat flux (ca.

Fig. 20.4. An SEM image of CVD1 with nitrogen doping

100 Wcm^{-2}) at the growth surface from the chemical reactions and the heat of the plasma. This phenomenon, coupled with a regime where the rate limiting growth process has a low activation energy, causes a spatial instability in the growth and "thermal runaway". A cross sectional view of the film (not shown here) showed the columnar growth of the crystallites. This specimen was grown on a W substrate, which had subsequently been detached from the layer.

CVD2 was grown on a W substrate and nitrogen was unintentionally present during the growth. Predominant octahedron and occasional large crystallites could be seen on the surface. This specimen is also "{100}-textured".The third specimen, CVD3, was an intrinsic sample with a "strong {110} texture". Star twinning was a frequent occurrence on the surface. This is where the diamond twins on four non-parallel {111} planes, to give five members radially disposed about a common [1–10] axis.

20.4.2 Powder Diffraction Studies

A series of powder diffraction data was collected with the standard θ–2θ geometry to measure the strain across the sample using 0.078 nm X-rays at Station 9.1 [15], Daresbury Laboratory. Figure 20.5a shows a powder diffraction profile across the 004 reflection obtained from CVD2. The FWHM

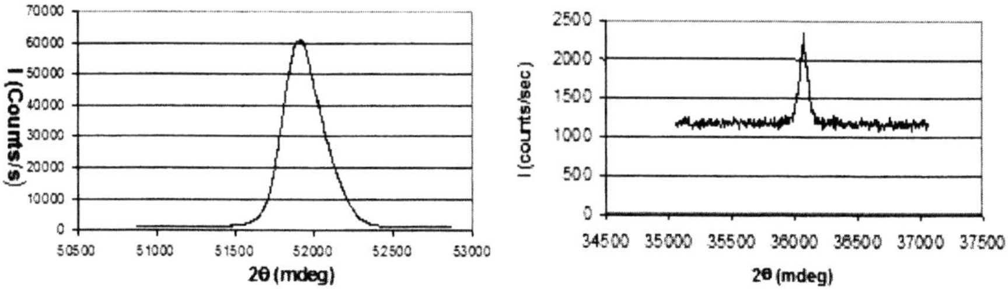

Fig. 20.5. A powder diffraction profile surrounding (**a**) the 004 reflection and (**b**) the 220 reflection. The 0.078 nm X-rays were selected for these measurements

along the 2θ-scan direction is $0.251° \pm 0.002°$, which gives an apparent strain value, $\Delta d/d$, of $(4.50 \pm 0.03) \times 10^{-3}$ derived simply from Bragg's law. (This high value may be compared with the instrumental resolution of $\approx 10^{-3}$). Although the sample was spinning, this rotation was about the surface normal (φ angle) and therefore had only a small effect on bringing the crystallites into the diffraction position, since 004 was the reflection used and the dominant orientation of the crystallites on the surface was [001].

The diffraction data presented in Fig. 20.5b show the 220 reflection profile obtained from the same sample. The calculated strain gave the large value of $(1.20 \pm 0.03) \times 10^{-3}$ for $\Delta d/d$. In this geometry the peak intensity of the 220 reflection was lower than that of the 400 reflection by an order of magnitude, whereas for randomly oriented diamond powder, the intensity of the 220 reflection is greater than that of the 004 reflection by a factor approximately 3.9. This is because the crystallites are predominantly oriented with the [001] direction close to the normal to the sample surface and therefore more crystallites are in the correct 2θ position for the 004 reflection.

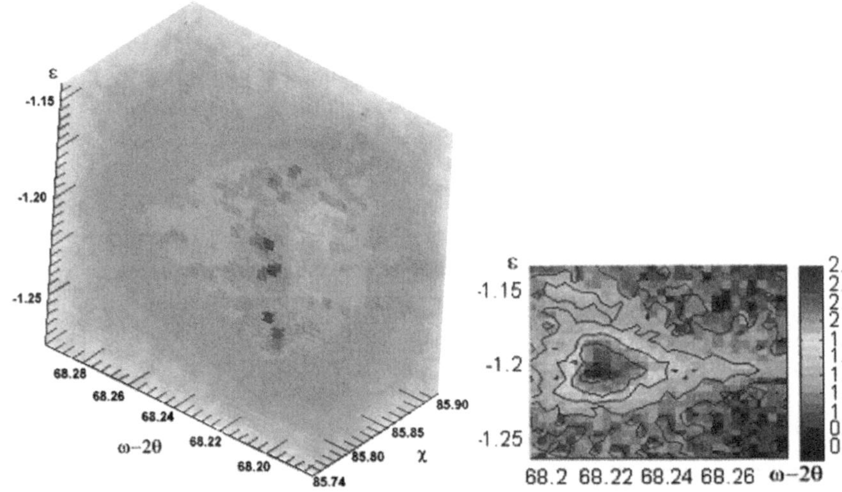

Fig. 20.6. (**a**) A three-dimensional isointensity presentation of a series of reciprocal-space maps obtained from CVD2 surrounding the 004 reflection using 0.1 nm X-rays. The highest intensity is shown with dark blue. The ω–2θ scan shows the direction of the diffraction vector and $varepsilon = (2\theta/2) - \omega$, shows the off-set in the ω-scan direction. The tilt angle, χ was increased by steps of 0.01° between each map. (**b**) The map with the highest intensity from this series, showing only one crystallite in diffraction position. The colour scale is logarithmic to base 10

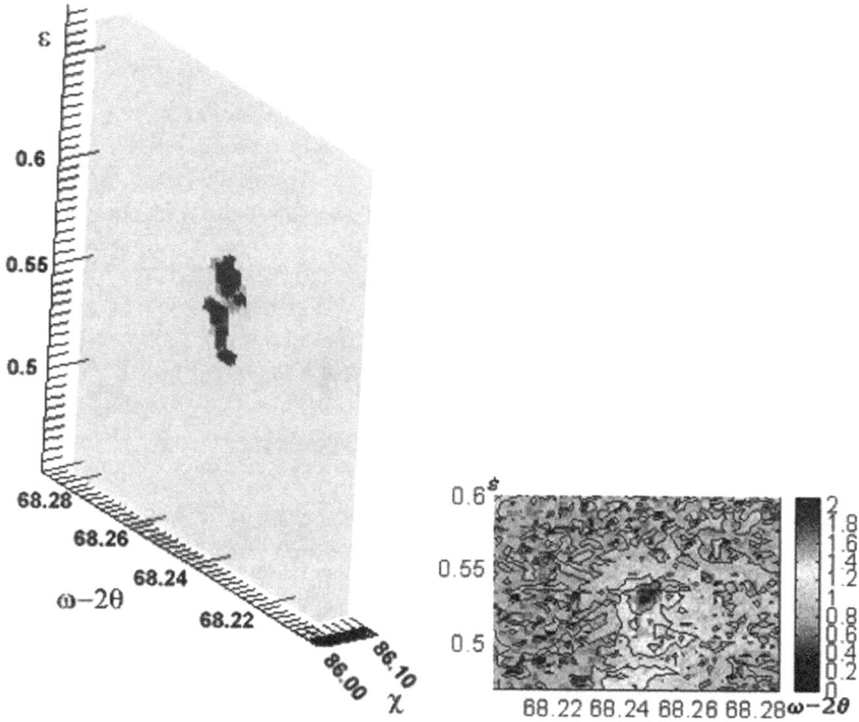

Fig. 20.7. (**a**) A 3-D isointensity presentation of a series of reciprocal-space maps at 68% of the maximum intensity around the 004 reflection obtained from CVD1 using 0.1 nm X-rays. χ was increased by steps of 0.02° between each map. (**b**) The map with the highest intensity showing only one diffraction spot. The colour scale is logarithmic to base 10

20.4.3 Three-Dimensional Reciprocal-Space Mapping

A series of reciprocal space maps was taken surrounding the 004 reflection (see Fig. 20.6a) from CVD2 using 0.1 nm X-rays [16]. The experimental set-up is shown in Fig. 20.2. The maps clearly display Bragg reflections from several crystallites appearing at very similar 2θ and w positions but separated in χ. The map with the highest intensity from this series, shows one peak (400 counts per second) associated with only one crystallite (see Fig. 20.6b). A horizontal scan extracted from the centre of the peak gives the value of 0.01° (1.7×10^{-4} radians) for the FWHM in 2θ-scan direction: (the theoretical width of the diffraction curve from a perfect crystal for the 004 reflection being 1.16×10^{-6} radians). If we assume the crystallite size effects are minimal, the strain along the diffraction direction, $\Delta d/d$, is therefore ($9.7 \pm 0.3 \times 10^{-5}$). The scattering planes have a broadening caused by lattice tilting (as shown by Dw) of $(5.2 \pm 0.2) \times 10^{-5}$ radians, provided the crystallite size parallel to the surface is greater than 1 µm. The difference between 2θ angles indi-

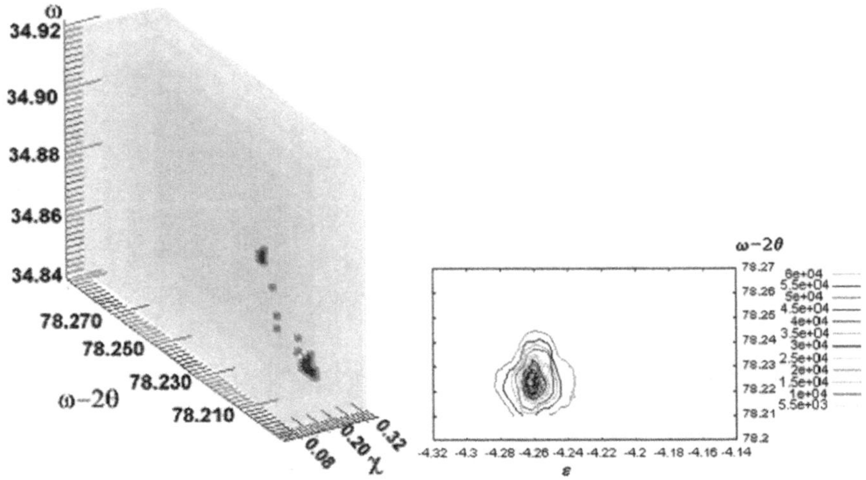

Fig. 20.8. (**a**) A 3-D isointensity presentation of a series of RS-maps surrounding the 220 reflection from CVD3 at 68% of the maximum intensity. The wavelength used was 0.16 nm. χ was increased by steps of 0.02 between each map. (**b**) The map with the highest intensity showing one diffraction spot. The colour scale is linear

cates a different lattice parameter for each crystallite, whereas the difference in w shows a difference in orientation of these grains with respect to each other. Similar measurements were performed (see Fig. 20.7a) for a nitrogen-doped sample (CVD1) surrounding the symmetric 004 reflection giving a value of $(7.7 \pm 0.4) \times 10^{-5}$ for strain along the diffraction direction (calculated from Fig. 20.7b) [15]. The lattice tilting calculated, gave the value of $(3.5 \pm 0.3) \times 10^{-5}$ radians. Reciprocal space maps collected from other regions of the samples gave different values for lattice parameter variations and lattice bending for this sample. This means that the strain variation could differ from one crystallite to another.

A series of reciprocal-space maps (see Fig. 20.8a) was collected from the intrinsic specimen, CVD3, around the 220 reflection using 0.16 nm synchrotron X-rays. These experiments were performed at ID10 B, ESRF, which uses a pair of diamond (111) monochromators in non-dispersive geometry. A number of diffraction spots can be observed in these maps. The horizontal axis shows the direction of the ω–2θ scan and the vertical axis shows the direction of the ω-scans. The tilt angle was changed by 0.02° between each map. The lattice parameter variation, $\Delta d/d$, was calculated from the map shown in Fig. 20.8b (the map with the highest intensity showing one diffraction spot) giving the value of $(1.70 \pm 0.03) \times 10^{-4}$. The lattice bending calculated from this map was $(1.00 \pm 0.03) \times 10^{-4}$ radians. These results show a higher amount of lattice bending and lattice parameter variation in this specimen compared to the values obtained from the specimens containing nitrogen impurities.

20.5 Discussion and Conclusion

Strain in CVD diamond is difficult to probe as the crystal quality is high and the individual crystallites are relatively large. The high crystal quality gives sharp features in ω–2θ space which are well spaced because of the large crystallite size. Conventional powder diffraction methods are inappropriate because the probability of a significant number of crystallites meeting the Bragg condition is low and therefore θ–2θ powder scans show only diffuse scattering with artificially large peak widths. RS-mapping using high-resolution optics and synchrotron radiation gives sufficient resolution to allow the sharp scattering around a reciprocal lattice point to be probed. Reflections must be first located by searching in reciprocal space. Three-dimensional reciprocal space mapping allows overlapping maps to be separated.

In the case of CVD diamond, extinction may be important (the extinction depth for the 004 reflection in diamond being approximately 30 µm using 0.1 nm X-rays), in which case the technique will become surface sensitive as in Bragg geometry, the X-rays do not penetrate far into the crystallites. This means that the technique would probe the growth surface of the crystal.

The values of strain obtained from reciprocal-space mapping are approximately 60 times smaller than those obtained from powder diffraction as they give quantitative estimates of strain within individual grains, rather than across a whole population of grains.

The broadening seen in the ω–2θ scan direction may correspond to dislocations, since point defects would distribute more or less evenly leading to a shift in the peak position rather than a distribution of strain. The broadening seen in the w-scan direction could be caused by lattice curvature or finite size of the coherently diffracting regions. However, the size of the coherent regions would need to be smaller than 1 micron to have any measurable effect on the profile broadening. In order to obtain a rough estimate of the broadening caused by crystal size, the Scherrer equation could be used

$$B(\theta) = \frac{\lambda}{L\cos(\theta)}\,, \tag{20.3}$$

where λ is the wavelength, L is the lateral size of the crystal and θ is the Bragg angle. The average size of the crystallites obtained from SEM pictures is approximately 50 µm and the broadening around the 004 reflection obtained from (20.3) is approximately 4.7×10^{-6} using 0.1 nm X-rays surrounding the 004 reflection. This figure is at least 20 times smaller than those given by the RS maps, suggesting that the crystallite size contribution to the profile broadening is negligible. The crystallites should be significantly smaller than about 1 micron to contribute to the broadening.

The experiments performed at Station 16.3, SRS and ID10B, ESRF showed that the sample with nitrogen doping has lower $\Delta d/d$ (microstrain) compared with the other specimensSEM observations and powder diffraction results showed that such nitrogen doping results in well oriented ⟨100⟩ crystallites on the surface. Introducing a small amount of nitrogen (1–4%) during the growth results in reducing the strain along the growth direction: a conclusion which, as it relates to individual grains, is probably only obtainable by reciprocal-space mapping.

Acknowledgements

The authors would like to acknowledge Dr. S.P. Thompson (Daresbury Laboratory) for three-dimensional presentaion of the RS-maps, Dr. O. Konovalov (ESRF) and Dr. M.A. Roberts (Daresbury Laboratory) for their help during the data collection, Dr. J.E. Butler (Naval Research Laboratory) and Dr. A. Whitehead (Element Six Ltd.) for the provision of the CVD diamond specimens.

References

1. P.K. Bachmann, in *Microwave Plasma Chemical Vapor Deposition of Diamond.* Handbook of Industrial Diamonds and Diamond Films, 1st Edn. Ed. by M.A. Prelas, G. Popovici and L.K. Bigelow (Marcel Dekker, New York 1997) pp. 821–850
2. B.E. Warren: *X-ray Diffraction.* (Dover Publications Inc., New York 1990) pp. 51–74
3. P.F. Fewster: Commission on Powder Diffraction (IUCr) Newsletter **24**, 17 (2000)
4. T. Ungar, L.S. Toth, J. Illy and I. Kovac: Acta Metall **34**, 1257 (1986)
5. H.F. Poulsen. S.F. Nielsen, E.M. Lauridsen, U. Lienert, R.M. Suter and D.J. Jensen: J. Appl. Cryst. **34**, 751 (2001)
6. H.J. Ullrich, M. Schlaubiz, F. Friedel, T. Spann, J. Bauch, T. Wroblewski, S. Garbe, G. Gaul, A. Knöchel, F. Lechtenberg, E. Rossmanith, G. Kumpat, G. Ulrich: Nuclear Instruments and Methods in Physics Research A **349**, 269 (1994)
7. J. Bauch, J. Brechbühl, H.J. Ullrich, G. Meinl, H. Lin, W. Kebede: Cryst. Res. Technol. **34**, 71 (1999)
8. P.F. Fewster and N.L. Andrew, in *Defects and Microstructure Analysis by Diffraction.* Ed. by R.L. Snyder, J. Fiala and H.J. Bunge (Oxford University Press, London 1999) pp. 346–364
9. P.F. Fewster: J. Phys. D: Appl. Phys. **28**, A97 (1995)
10. P.F. Fewster: Critical reviews in solid state and material science **22**, 69 (1997)
11. P.F. Fewster: *X-ray scattering from semiconductors.* (Imperial College Press, London 2000) pp. 215–218
12. P.F. Fewster, in *X-ray and Neutron dynamical Diffraction; Theory and Application.* Ed. by A. Authier, S. Lagomarsino, B.K. Tanner (Plenum Press in co-operation with NATO Scientific Affairs Division, 1996) pp. 269–289
13. W.R. Busing and H.A. Levy: Acta Cryst. **22**, 457 (1967)
14. S.P. Collins, R.J. Cernik, B. Fell, C.C. Tang, N.W. Harris, M.C. Miller and G. Oszlanyi: J. Synchrotron Rad. **5**, 1263 (1998)

15. G. Bushnell-Wye and R.J. Cernik: Review of Scientific Instruments **63**, 999 (1992)
16. M: Golshan: PhD thesis *High-resolution X-ray diffraction studies of synthetic diamond.* (University of London, 2001)

Index

Springer Series in

MATERIALS SCIENCE

Editors: R. Hull R. M. Osgood, Jr. J. Parisi H. Warlimont

10 **Computer Simulation of Ion-Solid Interactions**
By W. Eckstein

11 **Mechanisms of High Temperature Superconductivity**
Editors: H. Kamimura and A. Oshiyama

12 **Dislocation Dynamics and Plasticity**
By T. Suzuki, S. Takeuchi, and H. Yoshinaga

13 **Semiconductor Silicon**
Materials Science and Technology
Editors: G. Harbeke and M. J. Schulz

14 **Graphite Intercalation Compounds I**
Structure and Dynamics
Editors: H. Zabel and S. A. Solin

15 **Crystal Chemistry of High-T_c Superconducting Copper Oxides**
By B. Raveau, C. Michel, M. Hervieu, and D. Groult

16 **Hydrogen in Semiconductors**
By S. J. Pearton, M. Stavola, and J. W. Corbett

17 **Ordering at Surfaces and Interfaces**
Editors: A. Yoshimori, T. Shinjo, and H. Watanabe

18 **Graphite Intercalation Compounds II**
Editors: S. A. Solin and H. Zabel

19 **Laser-Assisted Microtechnology**
By S. M. Metev and V. P. Veiko
2nd Edition

20 **Microcluster Physics**
By S. Sugano and H. Koizumi
2nd Edition

21 **The Metal-Hydrogen System**
By Y. Fukai

22 **Ion Implantation in Diamond, Graphite and Related Materials**
By M. S. Dresselhaus and R. Kalish

23 **The Real Structure of High-T_c Superconductors**
Editor: V. Sh. Shekhtman

24 **Metal Impurities in Silicon-Device Fabrication**
By K. Graff 2nd Edition

25 **Optical Properties of Metal Clusters**
By U. Kreibig and M. Vollmer

26 **Gas Source Molecular Beam Epitaxy**
Growth and Properties of Phosphorus Containing III–V Heterostructures
By M. B. Panish and H. Temkin

27 **Physics of New Materials**
Editor: F. E. Fujita 2nd Edition

28 **Laser Ablation**
Principles and Applications
Editor: J. C. Miller

29 **Elements of Rapid Solidification**
Fundamentals and Applications
Editor: M. A. Otooni

30 **Process Technology for Semiconductor Lasers**
Crystal Growth and Microprocesses
By K. Iga and S. Kinoshita

31 **Nanostructures and Quantum Effects**
By H. Sakaki and H. Noge

32 **Nitride Semiconductors and Devices**
By H. Morkoç

33 **Supercarbon**
Synthesis, Properties and Applications
Editors: S. Yoshimura and R. P. H. Chang

34 **Computational Materials Design**
Editor: T. Saito

35 **Macromolecular Science and Engineering**
New Aspects
Editor: Y. Tanabe

36 **Ceramics**
Mechanical Properties, Failure Behaviour, Materials Selection
By D. Munz and T. Fett

37 **Technology and Applications of Amorphous Silicon**
Editor: R. A. Street

38 **Fullerene Polymers and Fullerene Polymer Composites**
Editors: P. C. Eklund and A. M. Rao

Springer Series in
MATERIALS SCIENCE

Editors: R. Hull R. M. Osgood, Jr. J. Parisi H. Warlimont

39 **Semiconducting Silicides**
Editor: V. E. Borisenko

40 **Reference Materials in Analytical Chemistry**
A Guide for Selection and Use
Editor: A. Zschunke

41 **Organic Electronic Materials**
Conjugated Polymers and Low Molecular Weight Organic Solids
Editors: R. Farchioni and G. Grosso

42 **Raman Scattering in Materials Science**
Editors: W. H. Weber and R. Merlin

43 **The Atomistic Nature of Crystal Growth**
By B. Mutaftschiev

44 **Thermodynamic Basis of Crystal Growth**
P–T–X Phase Equilibrium and Non-Stoichiometry
By J. Greenberg

45 **Thermoelectrics**
Basic Principles and New Materials Developments
By G. S. Nolas, J. Sharp, and H. J. Goldsmid

46 **Fundamental Aspects of Silicon Oxidation**
Editor: Y. J. Chabal

47 **Disorder and Order in Strongly Nonstoichiometric Compounds**
Transition Metal Carbides, Nitrides and Oxides
By A. I. Gusev, A. A. Rempel, and A. J. Magerl

48 **The Glass Transition**
Relaxation Dynamics in Liquids and Disordered Materials
By E. Donth

49 **Alkali Halides**
A Handbook of Physical Properties
By D. B. Sirdeshmukh, L. Sirdeshmukh, and K. G. Subhadra

50 **High-Resolution Imaging and Spectrometry of Materials**
Editors: F. Ernst and M. Rühle

51 **Point Defects in Semiconductors and Insulators**
Determination of Atomic and Electronic Structure from Paramagnetic Hyperfine Interactions
By J.-M. Spaeth and H. Overhof

52 **Polymer Films with Embedded Metal Nanoparticles**
By A. Heilmann

53 **Nanocrystalline Ceramics**
Synthesis and Structure
By M. Winterer

54 **Electronic Structure and Magnetism of Complex Materials**
Editors: D.J. Singh and D. A. Papaconstantopoulos

55 **Quasicrystals**
An Introduction to Structure, Physical Properties and Applications
Editors: J.-B. Suck, M. Schreiber, and P. Häussler

56 **SiO_2 in Si Microdevices**
By M. Itsumi

57 **Radiation Effects in Advanced Semiconductor Materials and Devices**
By C. Claeys and E. Simoen

58 **Functional Thin Films and Functional Materials**
New Concepts and Technologies
Editor: D. Shi

59 **Dielectric Properties of Porous Media**
By S.O. Gladkov

60 **Organic Photovoltaics**
Concepts and Realization
Editors: C. Brabec, V. Dyakonov, J. Parisi and N. Sariciftci

Druck: betz-druck GmbH, D-64291 Darmstadt
Verarbeitung: Buchbinderei Schäffer, D-67269 Grünstadt